Salmonid Ecosystems of the North Pacific

Salmonid Ecosystems of the North Pacific

Edited by

William J. McNeil
Daniel C. Himsworth

Published by

OREGON STATE UNIVERSITY PRESS
AND
OREGON STATE UNIVERSITY SEA GRANT COLLEGE PROGRAM
Corvallis, Oregon

Library of Congress Cataloging in Publication Data
Main entry under title:

Salmonid ecosystems of the North Pacific.
1. Pacific salmon—Ecology. 2. Fishes—Ecology.
3. Fish-culture—Northwest, Pacific. I. McNeil,
William J. II. Himsworth, Daniel C., 1953-
QL638.S2S24 597'.55 80-14800
ISBN 0-87071-335-3

The Oregon State University Sea Grant College Program is supported cooperatively by the National Oceanic and Atmospheric Administration, U.S. Department of Commerce, by the State of Oregon, and by participating local governments and private industry.

This publication was supported in part by the National Marine Fisheries Service.

Contents

Appendix: Abstracts of Soviet Salmonid Studies

Contributing Authors

Nicholas J. Bax
Fisheries Research Institute
University of Washington
Seattle, Washington

Robert L. Burgner
Fisheries Research Institute
University of Washington
Seattle, Washington

Donald Campton
Washington Department of Game
Olympia, Washington

D. M. Eggers
Fisheries Research Institute
University of Washington
Seattle, Washington

Clifford H. Fiscus
National Marine Mammal Laboratory
National Marine Fisheries Service
Seattle, Washington

R. A. Fredin
Northwest and Alaska Fisheries Center
National Marine Fisheries Service
Seattle, Washington
presently at
Natural Resources Consultants
Seattle, Washington

Stewart Grant
Northwest and Alaska Fisheries Center
National Marine Fisheries Center
Seattle, Washington

Allan C. Hartt
Fisheries Research Institute
University of Washington
Seattle, Washington
presently at
Makah Indian Tribe
Neah Bay, Washington

M. C. Healey
Pacific Biological Station
Nanaimo, British Columbia

L. A. Hobson
University of Victoria
Victoria, British Columbia

Herbert Jaenicke
Auke Bay Fisheries Laboratory
National Marine Fisheries Service
Auke Bay, Alaska

William J. Kinney
Fisheries Research Institute
University of Washington
Seattle, Washington

James E. Lannan
Marine Science Center
Oregon State University
Newport, Oregon

Stephen B. Mathews
Fisheries Research Institute
University of Washington
Seattle, Washington

John D. McIntyre
National Fisheries Research Center
U.S. Fish and Wildlife Service
Seattle, Washington

William J. McNeil
Oregon Aqua-Foods, Inc.
Springfield, Oregon

George Milner
Northwest and Alaska Fisheries Center
National Marine Fisheries Service
Seattle, Washington

Randall M. Peterman
Institute of Resource Ecology
University of British Columbia
Vancouver, British Columbia
presently at
Simon Fraser University
Burnaby, British Columbia

Donald E. Rogers
Fisheries Research Institute
University of Washington
Seattle, Washington

Ernest O. Salo
Fisheries Research Institute
University of Washington
Seattle, Washington

James Seeb
Washington Department of Fisheries
Olympia, Washington

Charles A. Simenstad
Fisheries Research Institute
University of Washington
Seattle, Washington

John R. Smith
Marine Science Center
Oregon State University
Newport, Oregon

Bruce P. Snyder
Fisheries Research Institute
University of Washington
Seattle, Washington

Richard R. Straty
Auke Bay Fisheries Laboratory
National Marine Fisheries Service
Auke Bay, Alaska

Fred M. Utter
Northwest and Alaska Fisheries Center
National Marine Fisheries Service
Seattle, Washington

Lavern J. Weber
Marine Science Center
Oregon State University
Newport, Oregon

Lisa Wishard
Pacific Fisheries Research
Olympia, Washington

Preface

The North Pacific Ocean and contiguous seas together delineate a vast pasture for anadromous salmonids. Juvenile salmon from streams and lakes in Asia and North America feed at sea a few months to several years. Their feeding migrations carry them thousands of kilometers from their home waters. Stocks from Asia and North America often intermingle at sea before the adult fish return to their streams and lakes of origin to reproduce.

Some stocks of salmonids have been fished heavily since before the turn of the century, and most have been heavily fished since the 1930s. Freshwater spawning and nursery habitats have also been degraded by land and water use activities. Most natural populations have declined from the combined effects of overfishing and environmental damage.

Natural processes affecting salmon survival have been studied more in freshwater than in estuarine and marine habitats. We possess a better understanding, therefore, of processes regulating production from natural spawning and freshwater nursery areas than from estuarine and marine nursery areas.

Although most salmonids are still produced from natural spawning, artificial propagation is becoming widespread due to a better understanding of processes affecting freshwater survival. Major fisheries are sustained by artificial propagation - sometimes referred to as "salmon ranching." Salmon ranching is presently practiced most intensively in Japan, in the Pacific Northwest states and on Sakhalin Island. Large-scale salmon ranching programs are also emerging in British Columbia and Alaska.

Exponential growth of salmon ranching in the last two decades has led to a quadrupling in the number of juvenile salmonids entering the North Pacific and contiguous seas from hatcheries and spawning channels. Approximately three billion juveniles are now released annually by salmon ranchers. Although this number is considerably smaller than the number of juveniles from natural propagation, another doubling of juveniles from salmon ranching appears to be a possibility in the 1980s. It is timely for the scientific community to devote more attention to questions related to the capacity of marine waters to grow salmon.

A symposium entitled "Salmonid Ecosystems of the North Pacific Ocean" was held at Otter Crest, Oregon, in May 1978. The meeting was organized at the suggestion of Dr. Robert L. Edwards, Director of the National Marine Fisheries Service Northeast Fisheries Center, as an activity of the Joint U.S.-U.S.S.R. Working Group on Biological Productivity and Biochemistry of the World Ocean. The symposium was organized to accommodate subjects relevant to freshwater, estuarine and marine ecosystems; population and community dynamics; and artificial propagation.

These proceedings of the symposium include papers presented by American and Canadian scientists and abstracts of papers submitted by Soviet scientists. Several of the papers summarize many years of productive research by the authors.

The National Marine Fisheries Service of the National Oceanic and Atmospheric Administration provided financial support for the editing of these proceedings. Oregon Aqua-Foods, Inc. (a subsidiary of the

Weyerhaeuser Company) provided financial support for the symposium.

Format and content of the symposium was suggested largely by faculty and staff at the University of Washington, Seattle, and Oregon State University, Corvallis. Editorial assistance was provided by the Oregon State University Sea Grant Program and the Oregon State University Press.

William J. McNeil
May 5, 1980

Testing for Density-Dependent Marine Survival in Pacific Salmonids

Randall M. Peterman

INTRODUCTION

The catch of salmon *(Oncorhynchus* sp.) in both British Columbia and all Pacific rim countries has dropped 50 percent in the last forty years (Larkin 1975, MacLeod 1977). In addition, several studies have shown that major mortality factors on salmon populations operate in the freshwater environment, and that if these bottlenecks are removed, salmon production can be greatly increased (e.g. Foerster 1968, Johnson 1965). These two observations have provided the stimulus to embark on major salmonid enhancement projects. British Columbia already has considerable experience with projects designed to increase freshwater survival rates of salmon. Many of these programs have concentrated on controlling stream flows, improving the amount and quality of spawning beds and enhancing food quality for juvenile salmon. Most of these projects have successfully increased the average fry production (e.g. Cooper 1977, Ginetz 1977, Larkin 1974, Paine 1974), as well as adult returns (Manzer 1976, Cooper 1977).

Future enhancement projects are planned on a much larger scale. For example, British Columbia aims to double present salmon production back to historic levels (MacLeod 1977) and Alaska plans to quadruple catches to the 100 million fish annually caught in the 1930s (Groot 1976). Washington, Oregon, Japan and the U.S.S.R. are also embarking on large scale enhancement programs.

Originally published in the Journal of the Fisheries Research Board of Canada, Vol. 35, No. 11. Reprinted with permission.

Underlying Assumptions

This large scale perturbation in salmonid abundances emphasizes the need to examine some of the underlying assumptions of enhancement. The first major assumption is that the abundance of juvenile salmon does not significantly affect their marine survival rate; that is, a doubling of smolt numbers should result in the same proportion returning as adults. This assumption could be wrong if marine food supplies were unexpectedly limiting, if predators aggregate to consume a larger portion of the higher abundances of salmon, or if parasites and diseases spread faster in denser populations.

The second major assumption of enhancement is that conditions in the marine environment (physical characteristics, timing and abundance of food or predators, abundance of competitors) have not changed enough to become limiting since the time when salmon catches were much larger than present levels. Although this assumption will not be explicitly tested here, there is substantial literature which demonstrates that many characteristics of the marine environment have changed within the last 50 years not only in the Pacific, but also the Atlantic and North Sea (Cushing 1975, Ketchen 1956, Sutcliffe et al. 1977). For example, the North Atlantic zooplankton growing season has shortened considerably, and phytoplankton blooms are occurring later in the season (Glover et al. 1974). Many of these changes have been related to abundance, survival rate or recruitment of various fish species including salmon (Glover et al. 1974, Godfrey 1958, Iles 1973). Relatively few papers deal directly with conditions in the North Pacific, but these do show that physical as well as biological conditions there have changed, often showing long term trends (Ketchen 1956, Parsons and LeBrasseur 1968, Wickett 1973) and that marine survival rates of fish species such as sockeye *(O. nerka)*, pink salmon *(O. gorbuscha)* and Pacific herring *(Clupea pallasi)* are often positively correlated (Godfrey 1958, Iles 1973). Present evidence does not address the question whether these marine changes have recently become limiting for salmon, but their dynamic nature should caution us that such limitation could occur in the future.

There have been few direct attempts to relate marine survival of a cohort of fry or smolts to characteristics of that cohort. Most studies merely report estimates of marine survival rates and compare the magnitudes to freshwater survival rates or marine survival rates of other stocks. The few exceptions, to be discussed later, are Foerster (1954), Peterman (1975), Ricker (1962), Staley (1975), and Ward and Larkin (1964) where ocean survivals are related to smolt body size and/or abundance.

The present paper attempts to test for the existence of density-dependent marine survival both within and between salmon populations using data on nine different stocks. In order to test the first major assumption, that there is no density-dependence in marine survival of salmon, it is simpler to break it down into four separate null hypotheses.

Null Hypothesis 1 (Within-cohort interaction). Within a cohort of smolts, there is no relation between smolt abundance or other smolt characteristics and their marine survival rate; i.e. there is a linear relation between smolt abundance and adult returns.

Null Hypothesis 2 (Between-cohort interaction). There is no relation between marine survival rate of a cohort of smolts and the abundance of all smolt cohorts of the same stock that go to sea in the same year.

Null Hypothesis 3 (Between-stock interaction on one or among nearby rivers). For a given river or nearby rivers, there is no relation between a cohort's marine survival rate and the smolt abundance of other stocks or species.

Null Hypothesis 4 (Interactions between stocks or species of distant river systems). There is no relation between a cohort's marine survival and the abundance or marine survival of other species or stocks that go to sea in the same year as the cohort of interest.

METHODS

Data

Several independent sets of data on production of seaward migrating juvenile salmon and adult returns were utilized in the present analysis. "Marine survival rates" as discussed here actually include survival during the downstream migration period, since virtually all smolt or migrant fry counts are made in freshwater. But these data provide the best estimates we have of true marine survival rates.

Babine Lake, B.C. Sockeye Salmon: Estimates of smolt counts for these sockeye are made at a counting fence at the outlet of the lake. Total adult returns by brood year are estimated by summing, for the appropriate years, the age 4_2 and 5_2 salmon in the escapement, native Indian catch, and commercial catch of statistical areas 4, 3X, and 3Y. "Ages 4_2 and 5_2" indicate that fish spent one year in the rearing lake. Data were obtained from R. Ginetz (1977), and F. E. A. Wood and E. Zyblut (personal communication, Canadian Department of Fisheries and Environment, Fisheries and Marine Service). Because age 5 returns from the 1972 brood year have not yet been completely verified, statistical tests were done including and excluding this one point.

Chilko Lake, B. C., Sockeye Salmon: Smolt estimates are for the one-year-in-freshwater smolts, which make up the bulk of this stock's smolt production. Brood year adult returns are catch plus escapement. Data are from Ward and Larkin (1964), annual reports of the International Pacific Salmon Fisheries Commission (IPSFC), and a 1975 workshop (Peterman 1975).

Fraser River Pink Salmon: All Fraser River pink stocks are combined into one group, and data are from the 1975 IPSFC annual report.

Hokkaido, Japan Hatchery Chum Salmon *(O. keta)*: Fry counts are the total number of fry released from all Hokkaido chum hatcheries in a given year. Brood year adult returns are escapement plus coastal fishery catches (Atkinson 1976, Moberly and Lium 1977).

Lakelse Lake, B.C. Pink Salmon: Fry counts were made by downstream sampling gear and pre-emergent fry sampling. Brood year adult returns are escapement plus Area 4 catch, the latter quantity being estimated by taking into account the weekly exploitation pattern and run compositions in the Skeena River commercial fishing area (personal communication, D. R. Harding and E. Zyblut, Canadian Department of Fisheries and Environment, Fisheries and Marine Service).

Kvichak River, Alaska Sockeye Salmon: Smolt counts are broken down by age group. Smolts that have spent one year in freshwater are called Age I smolts and two-year freshwater fish are Age II. Brood year adult returns are catch plus escapement sorted by Age I and Age II smolts (personal communication, K. P. Parker, Alaska Fish and Game Department).

Naknek River, Alaska Sockeye Salmon: The same type of data are used as for Kvichak River sockeye. For this, and possibly the other two Alaskan sockeye systems, there are obvious problems with estimating smolt abundance because a few marine survival estimates are unrealistically high (personal communication, K. Parker). Smolt samples have been taken in the past with fyke nets. In years with large smolt migrations, these nets sample a smaller proportion of the stock, which leads to too low an estimate of smolts. This in turn gives too high an index of marine survival (adults/smolts). Thus observed survival rates for the Alaskan sockeye systems may be too large, especially at high smolt densities. Because of this bias, any density-dependent marine survival shown for these stocks will underestimate the actual severity of density dependence.

Ugashik River, Alaska Sockeye Salmon: The same type of data are used as for Kvichak River sockeye.

Washington State Steelhead Trout *(Salmo gairdneri)*: Good estimates of smolt releases are available; however, the only index of brood year adult production is catch by sport fishermen (Royal 1972). Numbers of fishing licenses are also available, which permit calculation of catch per fisherman. This is perhaps a better index of adult returns.

Fraser River, B.C. Enhanced Salmon Stocks: Cooper (1977) provides data for juvenile salmon output and brood year adult returns for three sockeye and two pink stocks. However, pink adult returns are calculated in a biased and perhaps misleading way. Returns are prorated on the basis of the relative contribution of fry of each stock to the total Fraser River pink fry production in that year. In other words, all Fraser River pink stocks are assumed to have the same marine survival rates. There is a similar bias for sockeye because, for a given stock, the relative contributions of channel and natural fry to adult returns are also calculated on a prorated basis.

Statistical Methods

My purpose was to test whether density-dependent marine survival exists in a given set of data. "Density dependent" here simply means that as fish density increases, survival rate changes, presumably decreases. Two different tests for density dependence have been widely used, especially in entomology (Morris 1959, 1963*a*,*b*; Varley and Gradwell 1960, 1963). Both techniques have been tangled in a controversy for which there is no generally accepted solution. Morris' method is too liberal in rejecting the null hypothesis of no density dependence (Luck 1971, Maelzer 1970, St. Amant 1970) while the other technique is too conservative (Kuno 1971). Because there is no single accepted method at present, and because we do not know how measurement errors are biased with density for these salmon data, both of these techniques are applied in order to bracket the possible number of cases of density-dependent marine survival.

The present study examines the survival relation from the seaward migrating lifestage (fry or smolts) to adult returns. Such a relation can be expressed as:

$$A = f(S) \cdot \exp(v) \tag{1}$$

where A = abundance of adults returning
S = abundance of the seaward migrating lifestage henceforth called smolts for simplicity
v = normally distributed random variant with mean zero and variance σ^2.

exp(v) = multiplicative, log-normal noise term as argued for by Allen (1973) and Walters and Hilborn (1976) and as shown by the Central Limit Theorem (see Appendix).

The null hypothesis of no density-dependent survival assumes that $f(S) = c \cdot S^k$ where c and k are two constants and k has the value 1.0. Substituting into Equation 1 gives:

$$A = c \cdot S^k \cdot \exp(v) \tag{2}$$

This null hypothesis corresponds graphically to the solid line in Figure 1. Taking natural logarithms (ln) of both sides gives:

$$\ln A = \ln c + k \ln S + v \quad (3)$$

which is in the form of a linear regression equation with a normalized error term:

$$Y = a + bX + \varepsilon \quad (4)$$

The slope b_{YX} of this regression provides an estimate of the curvature k of the smolt-to-adult relation. We cannot interpret with the usual standard error statistics a regression of an index of marine survival, say lnA/lnS on lnS since the X and Y variates are not independent.

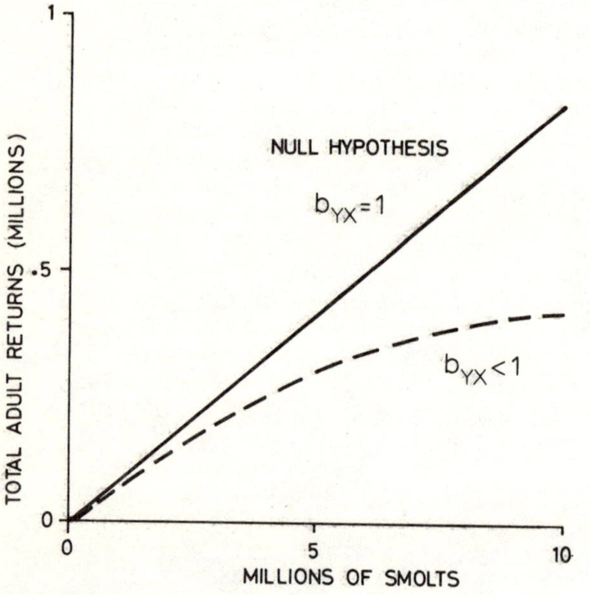

Figure 1. Two possible types of relations between smolt abundance and total adult returns. The solid line represents the linear, null hypothesis of no density-dependent marine survival and the dashed line indicates density dependence. The value b is an estimate of k in equation (2).

Method 1: In the special case of Morris' method that deals with within-generation survivals, data are tested to see if the estimated slope b_{YX} of Equation 4 is significantly less than 1.0 by using a one-tailed *t*-test. If it is, then the relation has decreasing slope as shown by the dashed line in Figure 1. Numerous authors have criticized the validity of this method on the ground that errors in measurement of the independent variate will bias the predicted slope below 1.0, thereby causing users to incorrectly categorize some relations as density dependent when they compare their calculated slope with b = 1.0 (St. Amant 1970, Benson 1973, Eberhardt 1970, Ito 1972, Luck 1971). This bias is especially true when sample size is small ($n < 20$, Maelzer 1970). However, Kuno (1971) shows that as long as the sampling variance is less than about 10 percent of the range of values observed, the effect of this bias is negligible. The sampling errors for many of the sets of salmon data used here are not known so the "Morris plot" method is used only as one means to identify cases of density-dependent marine survival. Results from applying this method therefore represent an upper limit to the number of cases that do, in reality, have density dependence. If sampling errors are relatively large, then the null hypothesis of density-independent survival is rejected more often with the Morris method than is justified.

Quantitatively, these measurement errors ε_1 and ε_2 come into Equation 2 in the following way, if we assume that they are proportional rather than additive errors:

$$A \cdot \varepsilon_1 = c \cdot (S \cdot \varepsilon_2)^k \cdot \exp(v) \quad (5)$$

$$\text{or} \quad \ln A + \ln \varepsilon_1 = \ln c + k \ln S + k \ln \varepsilon_2 + v \quad (6)$$

$$\text{or} \quad \ln A = \ln c + k \ln S + k \ln \varepsilon_2 - \ln \varepsilon_1 + v$$

$$Y = a + bX + \varepsilon$$

If $\ln\varepsilon_1$ and $k \ln\varepsilon_2$ are small, then the regression statistics still apply.

Method 2: The second method of testing for density dependence, attributed to Varley and Gradwell (1963, 1968) is similar to Morris' technique except that one additional step is designed to circumvent the difficulties which arise from the observation errors referred to above (Ito 1972, Varley and Gradwell 1968). Their method is equivalent to determining the functional or geometric mean regression, a procedure sometimes used in fisheries (Ricker 1973). $\log_e N_{t+1}$ is regressed on $\log_e N_t$ and the reverse is also performed, $\log_e N_t$ is regressed on $\log_e N_{t+1}$, where N_t and N_{t+1} are population sizes at successive lifestages. If the slopes of both lines b_{YX} and b_{XY} are significantly different from 1 and lie on the same side of b = 1 when plotted on the same graph of points, then the true relation (which lies somewhere between the two regression lines) is assumed to be density dependent. That is, if $b_{YX} < 1$ and $1/b_{XY} < 1$, then the "true" b is assumed less than

1 and the smolt-to-adult relation is curvilinear like the dashed line in Figure 1.

Kuno (1971) and Luck (1971) discuss and give examples that this so-called "proof of density dependence" of Varley and Gradwell's is too conservative, especially if the process noise or random variation is large. In other words, this test may classify a survival relation as density independent when in actual fact there was a density-dependent process operating, but its effects were masked by the random process. However, Kuno argues, the Varley and Gradwell method only errs on the conservative side; density-independent processes are never classed as density dependent, no matter how large the noise. This has been confirmed by my own "Monte Carlo" simulations where the Varley-Gradwell test was applied to output generated by a simple stochastic model with known density-dependent or density-independent survival functions.

Because we do not know the error structure of the fish population estimates discussed here, we cannot interpret the results of these two methods without ambiguity. Nevertheless, results for both approaches will be presented for comparison.

RESULTS AND DISCUSSION

Within-Cohort Interaction

Table 1 shows the results of the regressions for the Morris, and Varley and Gradwell methods. One-tailed *t*-tests were applied to determine if slopes were significantly less than 1. Using the "liberal" Morris criterion for density dependence ($b_{YX} < 1.0$), seven systems show significant ($P < 0.05$) density dependence: Babine Lake sockeye, Lakelse Lake pinks, Kvichak River sockeye (both age I and age II smolts), Naknek River sockeye (age I and age II smolts), and age II smolts of Ugashik River. When the more stringent Varley and Gradwell test is applied, only one of the 12 systems shows significant density dependence, even-year Lakelse Lake pinks. Figure 2 graphically illustrates results of the Varley-Gradwell method for the Babine sockeye and even-year Lakelse Lake pink salmon. Note that while both Babine sockeye regressions are significantly different from 1, they are on opposite sides of the b = 1 line. Therefore, the noise in the data does not permit us to conclude density-dependent survival. However, for the Lakelse pinks, the true relation is significantly different from b = 1. Raw data for all stocks are plotted by brood year in Figures 3-8 in order to permit readers to interpret them in other ways.

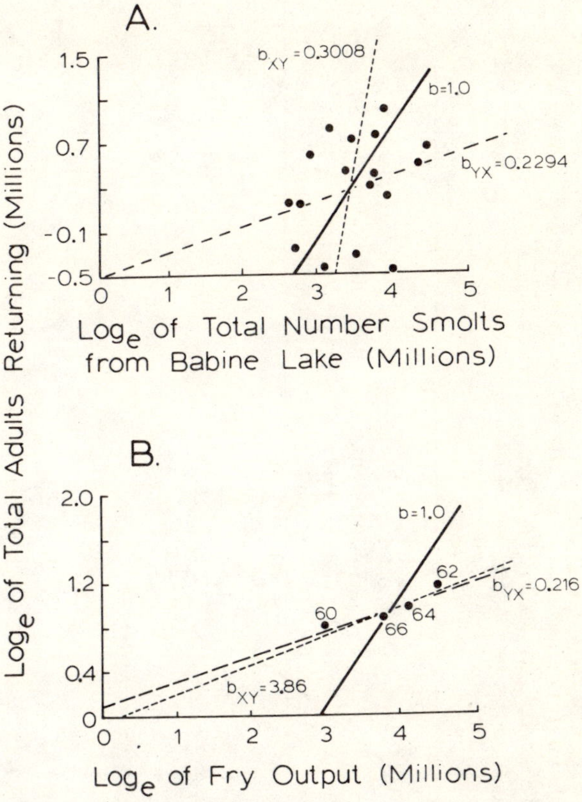

Figure 2. $\log_e$ - $\log_e$ plots for smolt-to-adult relations, graphically illustrating the Varley-Gradwell method of testing for density dependence. A line with slope b_{XY} is the line for regressing $\log_e$ of smolts on $\log_e$ of adults, but this line is plotted back on the present graph in order to illustrate whether it is on the same side of b = 1 as b_{YX}. (A) For Babine Lake sockeye, these lines are opposite sides, therefore the true relation cannot be distinguished from b = 1. (B) For even-year Lakelse Lake pinks the true relation, which is between the two dashed lines, is significantly different from b = 1. Brood years are indicated beside data points.

In addition, data for Washington State steelhead and Fraser River enhanced stocks are plotted in Figures 9-12. Statistical tests were not applied to these sets of data because of the relatively incomplete way in which adult returns were estimated.

Table 1. Results of Statistical Tests for Within-Cohort Interactions that Determine Marine Survival Rates. Regressions were Done Using Natural Logarithms of Raw Data (See Text). Probabilities were were Determined by T-Tests. Last Column Refers to the Relation Between Marine Survival Rate and Smolt Body Size. "N.A." Indicates Body Size Data were Not Available.

Stock	Sample size (years)	b_{YX} and SE	$P_{b=1}$ (one-tailed test)	b_{XY} and SE	$P_{b=1}$ (one-tailed test)	Density dependent with Morris' method?	Density dependent with Varley-Gradwell method?	$P_{b=0}$ for body size relation (two-tailed test)
Babine L. Sockeye								
All data	17	0.23±0.21	<0.002	0.35±0.31	<0.03	Yes	No	>0.1
omitting 1972 brood year	16	0.23±0.23	<0.005	0.30±0.29	<0.02	Yes	No	>0.1
Chilko L. Sockeye	24	0.80±0.15	=0.093	0.73±0.13	=0.025	No	No	N.A.
Fraser R. Pinks	7	-0.38±1.1	>0.1	0.07±0.19	<0.005	No	No	N.A.
Hokkaido Chums	20	0.55±0.28	=0.067	0.31±0.16	<0.001	No	No	N.A.
Lakelse L. Pinks								
Even-year stock	4	0.22±0.004	<0.001	3.86±0.69	<0.03	Yes	Yes	N.A.
Odd-year stock	4	-0.22±0.66	<0.025	-0.12±0.17	<0.025	No(neg. slope)	No(neg. slope)	N.A.
Kvichak R. Sockeye								
Age I smolts	19	0.71±0.13	<0.02	0.89±0.17	>0.2	Yes	No	>0.25
Age II smolts	19	0.71±0.10	<0.005	1.03±0.16	>0.2	Yes	No	<0.02
Naknek R. Sockeye								
Age I smolts	17	0.29±0.20	<0.005	0.43±0.30	=0.042	Yes	No	>0.1
Age II smolts	14	0.51±0.18	<0.01	0.82±0.08	>0.2	Yes	No	>0.2
Ugashik R. Sockeye								
Age I smolts	14	1.04±0.13	>0.2	0.82±0.08	=0.029	No	No	>0.2
Age II smolts	14	0.70±0.12	=0.013	1.0±0.19	>0.2	Yes	No	>0.2

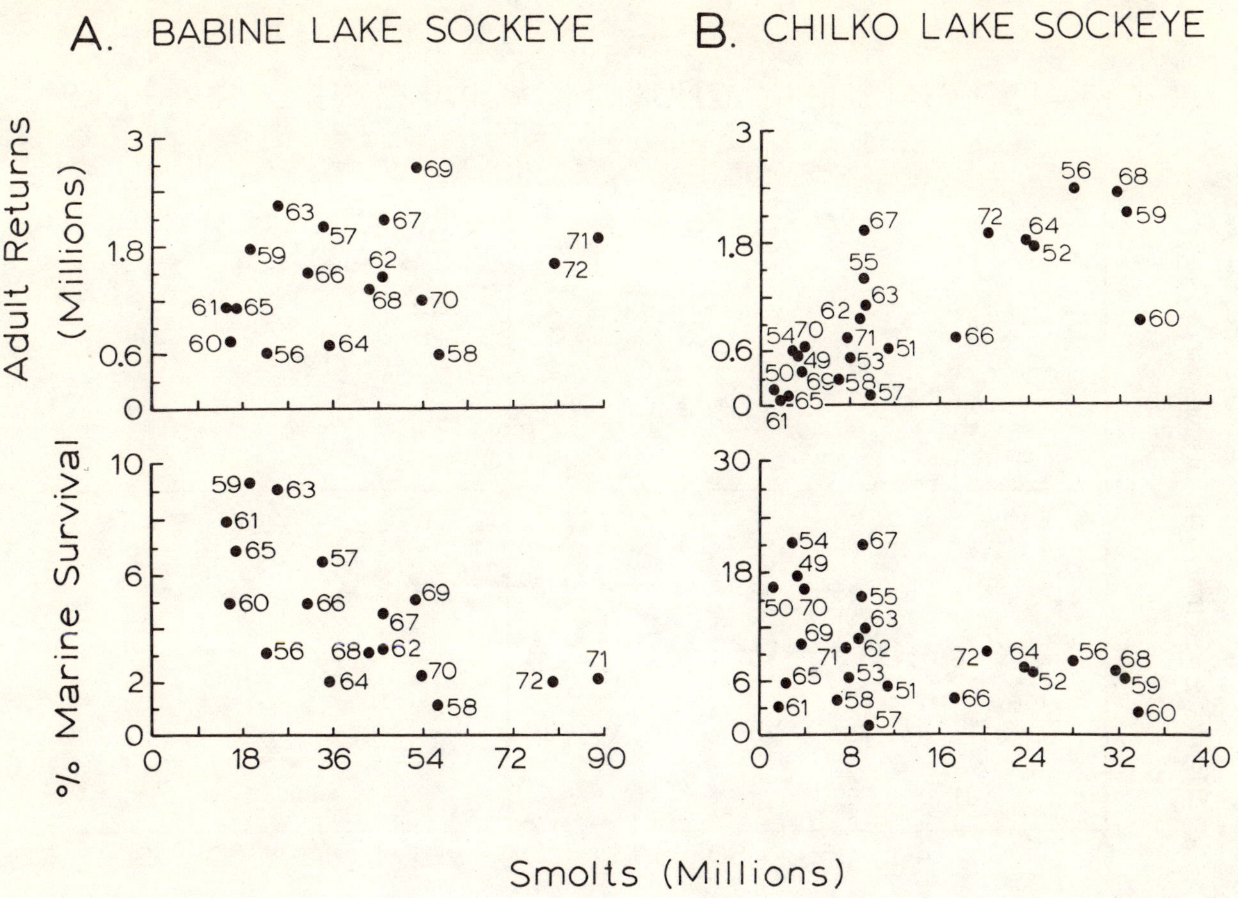

Figure 3. Adult returns and percent marine survival as a function of smolt abundance, by brood year. (A) Babine Lake sockeye, (B) Chilko Lake sockeye.

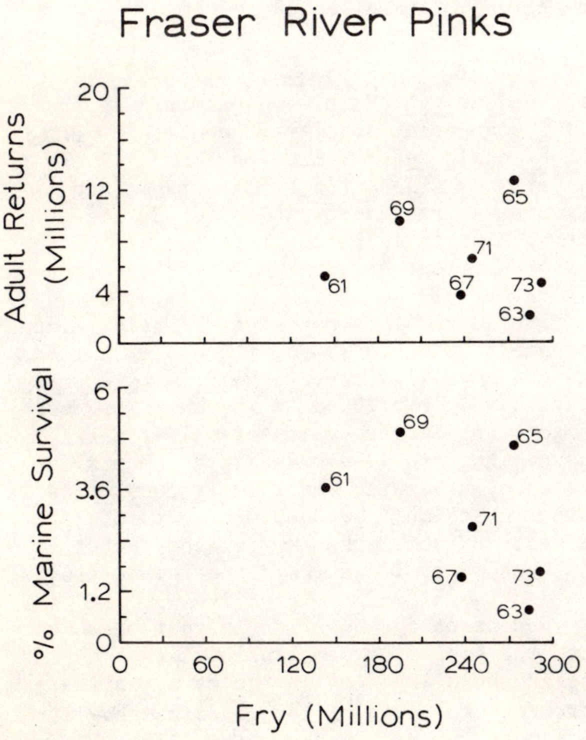

Figure 4. Fraser River pink salmon adult returns and percent marine survival as a function of fry abundance, by brood year.

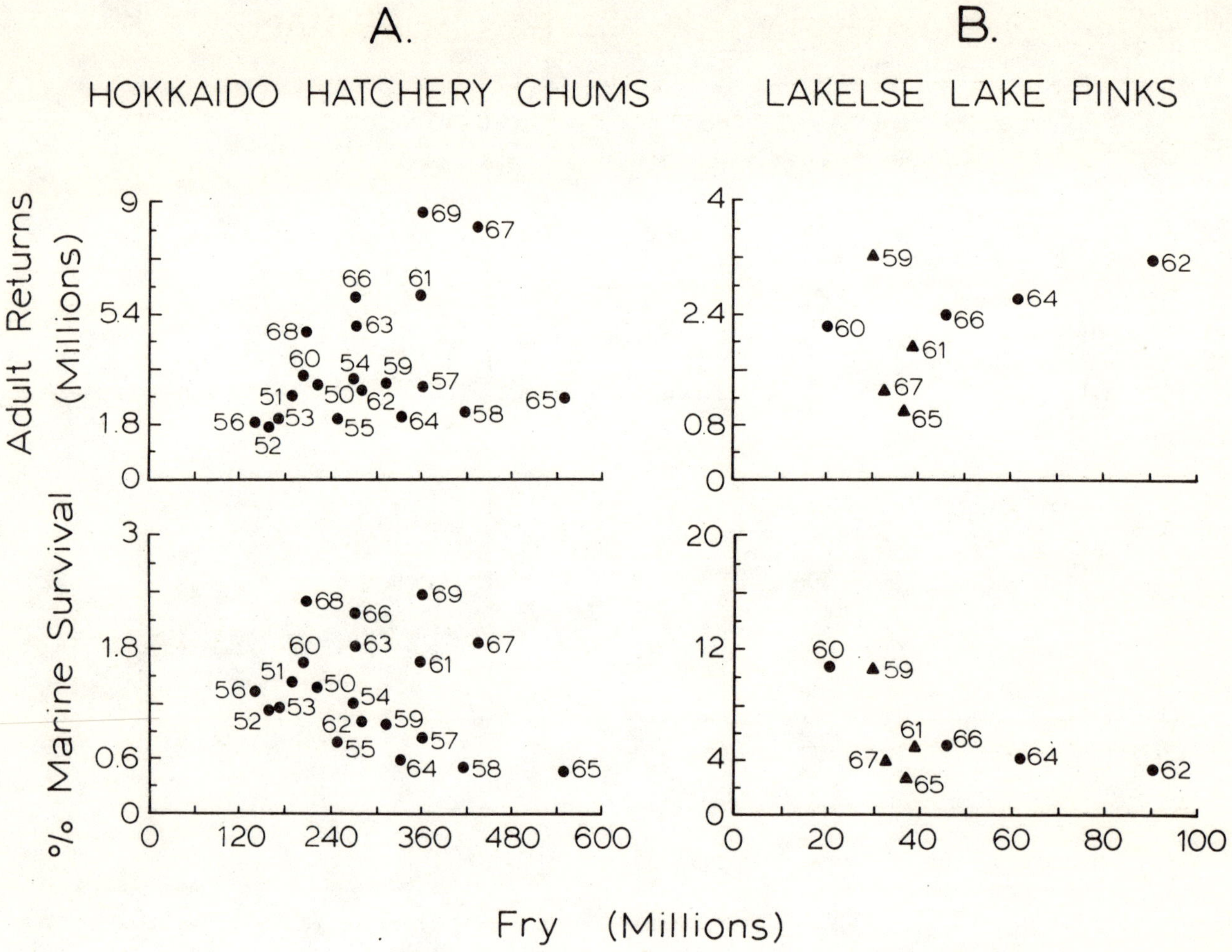

Figure 5. Adult returns and percent marine survival as a function of fry abundance, by brood year. A. Hokkaido Hatchery chum salmon, B. Lakelse Lake pinks.

Additional information is obtained by inspecting Figures 3-12, and noting the large proportion of percent marine survival graphs in which there is an apparent hyperbolic relation and in which there are no data points in the upper right-hand corner. Thus, most large smolt or fry runs are associated with below-average survivals. The so-called "leader-follower" effect, and Watt's (1964, 1968 pp. 33-34) criticism that one expects a hyperbolic relation in such graphs, are relevant only to between-generation changes in populations. These arguments do not apply here, since within-generation changes are being discussed.

Foerster (1954) showed that the apparent within-stock density dependence in Cultus Lake sockeye marine survival could be explained on the basis of the inverse relation between smolt abundance and smolt body size. The smaller the smolt population, the larger the average body size, and therefore the larger the marine survival (Ricker 1962). Therefore, the present marine survival data were examined to see if the apparent density-dependent situations could be explained on the basis of smolt body size. $\log_e$ of the ratio adults/smolts was regressed on $\log_e$ (body size). Results in Table 1 show that for all of the systems except one that are density dependent by either of the statistical criteria, marine survival is not significantly related to smolt body size. For the one exception, Kvichak age II sockeye, the density dependence can be explained on the basis of body size since larger smolts of this stock have higher survivals and large body size is associated with small smolt runs. The other cases must be due to effects of density or some other correlated factor such as water flows (Ward and Larkin 1964).

Depending on the criteria that one uses and the amount of sampling bias that one believes is inherent in the data, anywhere from one to seven of the 12 cases show within-

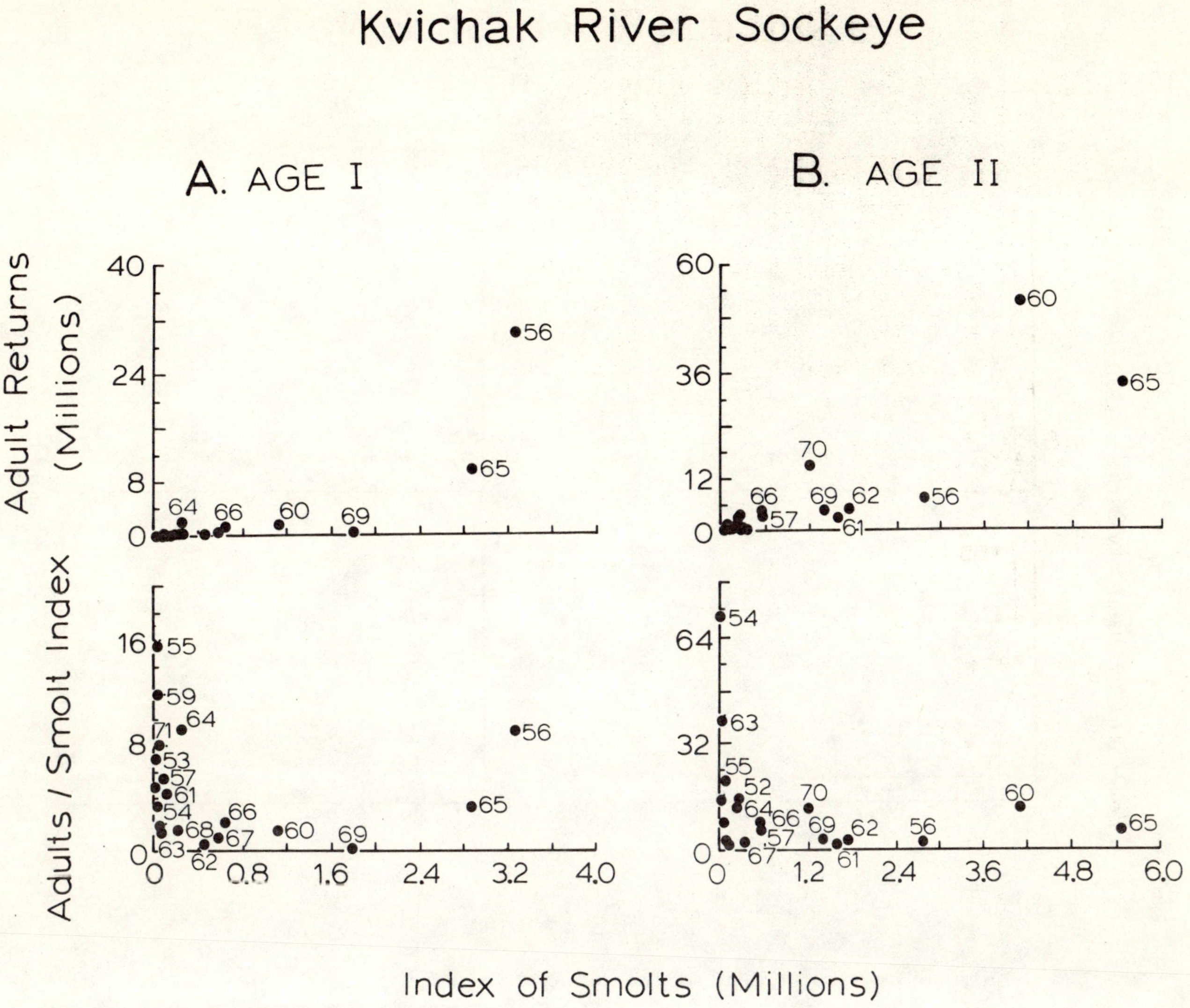

Figure 6. Adult returns and an index of marine survival as a function of an index of smolt abundance for Kvichak River sockeye by brood year. A. Age I smolts, B. Age II smolts.

cohort density-dependent marine survival. All of these cases show a negative influence of increased smolt abundance such that a smaller proportion of smolts survive as numbers increase. The Washington steelhead data (Fig. 9) certainly show the same trend. These findings argue against the implicit assumption made by proponents of salmonid enhancement that there is a linear relation between smolt output and adult returns. Staley's (1975) analysis, which inspired the present study, also resulted in predictions of decreased marine survival as smolt abundance increased, based only on pre-enhancement Babine sockeye data. As seen in Figure 3A, this general prediction was supported by subsequent events. The possible mechanisms for such within-cohort density dependence will be discussed later.

The relation between marine survival rate and freshwater survival rate (egg-to-fry, or fry-to-smolt) was also examined to determine if there was some characteristic of fish that might affect survival at both stages. Freshwater data were available for only four systems: Fraser and Lakelse pinks, and Babine and Chilko sockeye. Two of the four regressions showed a significant relation; Lakelse pinks b = 0.55 $\pm$ 0.18, r = 0.798, $P_{b=0} < 0.03$ and Babine sockeye b = -0.09 $\pm$ 0.04, r = -0.565, $P_{b=0}$ = 0.052 (Fig. 13). Because these slopes have different signs, no biological explanation is consistent. However, biologists should examine this relationship for other systems in order to help managers predict adult return rates.

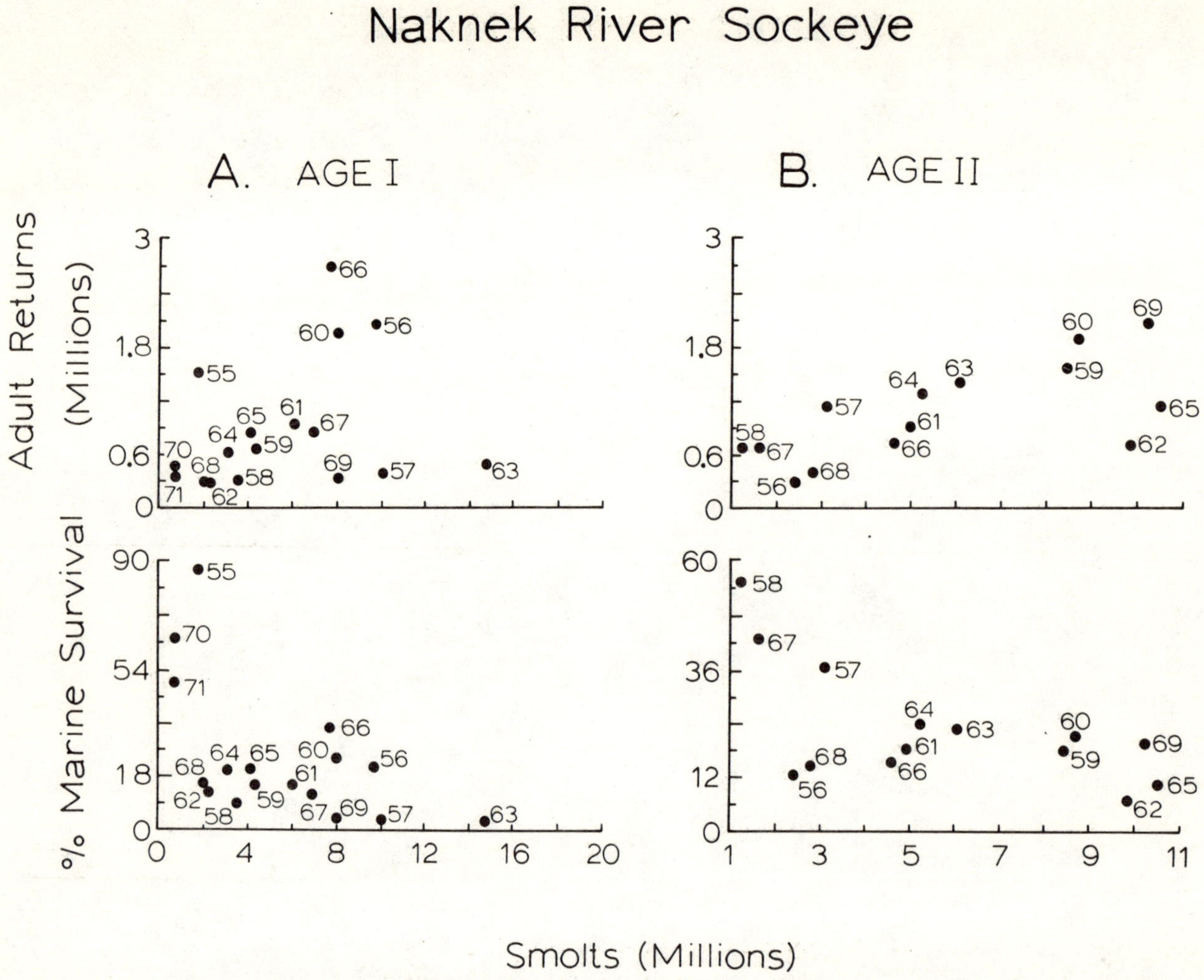

Figure 7. Adult returns and percent marine survival as a function of smolt abundance for Naknek River sockeye, by brood year. A. Age I smolts, B. Age II smolts.

Between-Cohort Interactions

For the Alaskan sockeye discussed here, there is the potential for interaction between different aged smolts of the same stock. In any year when, say, age I Kvichak smolts are migrating, their marine survivals could be affected by the total Kvichak smolts present, both age I and age II's. To test the significance of such an interaction, I applied Morris, and Varley-Gradwell tests to the data. Strict regressions of marine survival of stock X on total abundance of that river's smolts going to sea at the same time are not interpretable with the usual standard error statistics because the independent variate again forms part of the dependent variate (Varley and Gradwell 1968). That is, smolts of a given age class are well correlated with total smolts of that river system ($P_{b=0} < 0.01$ except for Ugashik smolts) so the marine survival index (adults/smolts) is not a measure independent from the total smolts. Table 2 shows results of applying these tests to the Alaskan sockeye data. Four of six cases are density dependent according to Morris' criterion and none by the Varley-Gradwell method (because in the latter cases, the two slopes b_{YX} and b_{XY} end up on opposite sides of the $b = 1$ line when plotted on the same graph, as happened in Fig. 2A). Figure 14 shows the between-cohort relation for three cases that are significant by Morris' criterion. Note that for the Naknek age II smolts (Fig. 14C), a greater proportion of the variance in marine survival is accounted for by this between-cohort relation than by the within-cohort one ($r_{within} = -0.632$, $r_{between} = -0.861$; cf. Fig. 7B).

As for the within-cohort analysis, the predominant trend for the effect of total stock smolt abundance on cohort marine survival is negative. This may provide an

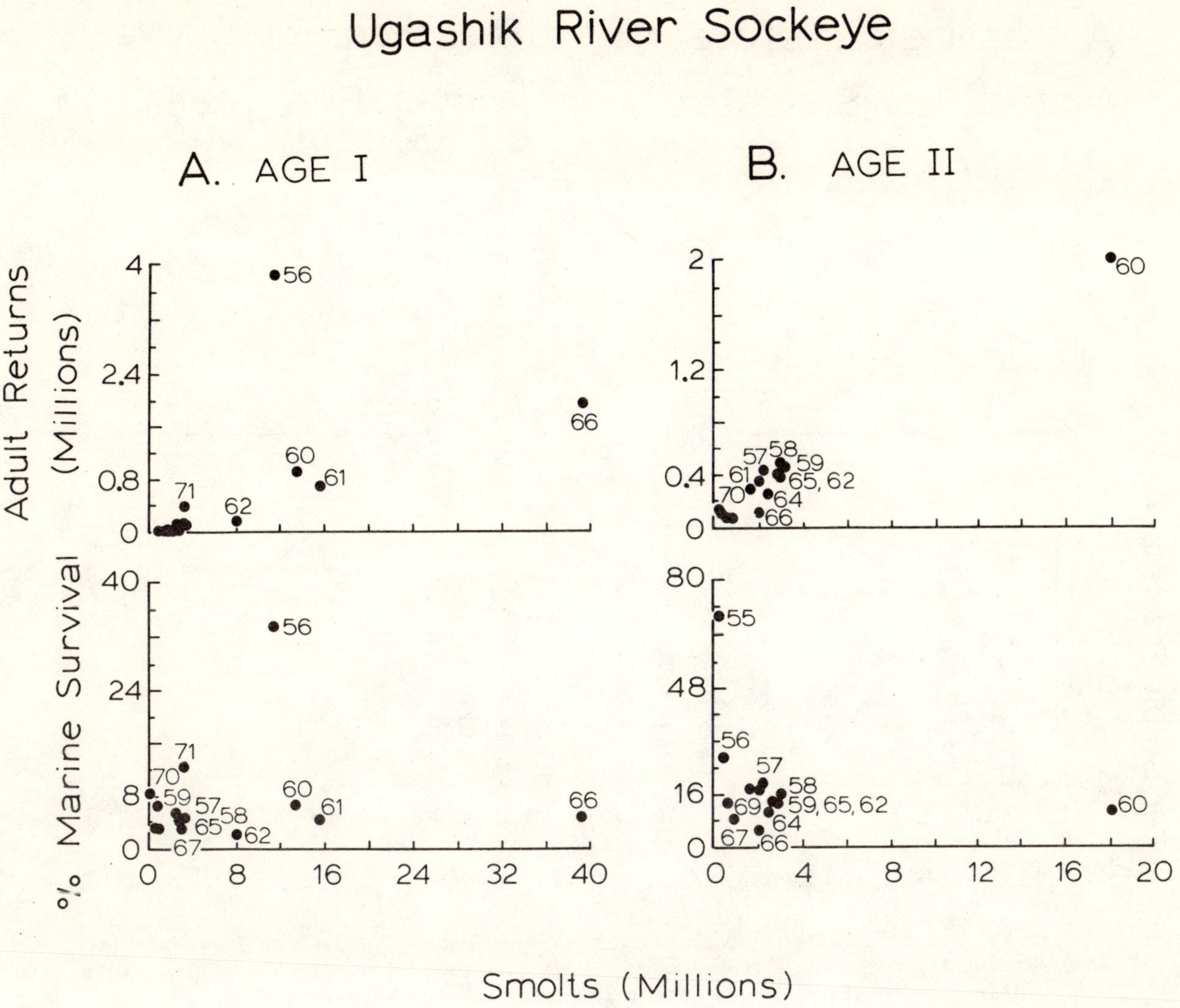

Figure 8. Adult returns and percent marine survival as a function of smolt abundance for Ugashik River sockeye by brood year. A. Age I smolts, B. Age II smolts.

additional predictive tool for managers who must make pre-season run forecasts.

Interactions Between Stocks on Same or Nearby Rivers

The dependent variate here is $\log_e$ (adults/smolts) of stock Y and the independent variate is $\log_e$ of smolt abundance of a different stock, X. Marine survival of each Alaskan sockeye stock was regressed on each of the other two Alaskan systems' smolt abundance, and Chilko sockeye survival was regressed on Fraser River pink fry since the latter are about ten times as abundant as Chilko smolts. In addition, Lakelse pink survival was regressed on Babine smolt abundance and vice versa. Out of the 15 regression slopes, only one was significantly different from zero, Naknek age II smolt survival as a function of total Kvichak smolts going to sea in the same year ($n = 14$, $b = -0.29 \pm 0.09$, $P_{b=0} < 0.01$, $r = -0.71$). Biologically, such a relation is plausible because the Naknek and Kvichak rivers empty into Bristol Bay only 20 km apart. Although Kvichak smolts reach the ocean two weeks earlier than Naknek smolts, both stocks are found together a few weeks later because of the larger size and faster swimming speed of Naknek fish (Straty 1974).

Another between-stock relation similar to the one above is adult production of one stock as a function of the total smolt abundance of that stock plus other nearby stocks. Both Morris, and Varley and Gradwell tests were applied to data of this type. The Fraser and Skeena river systems were the only ones examined; Kvichak smolts are estimated as an index of production, not as smolt abundance, and hence could not be

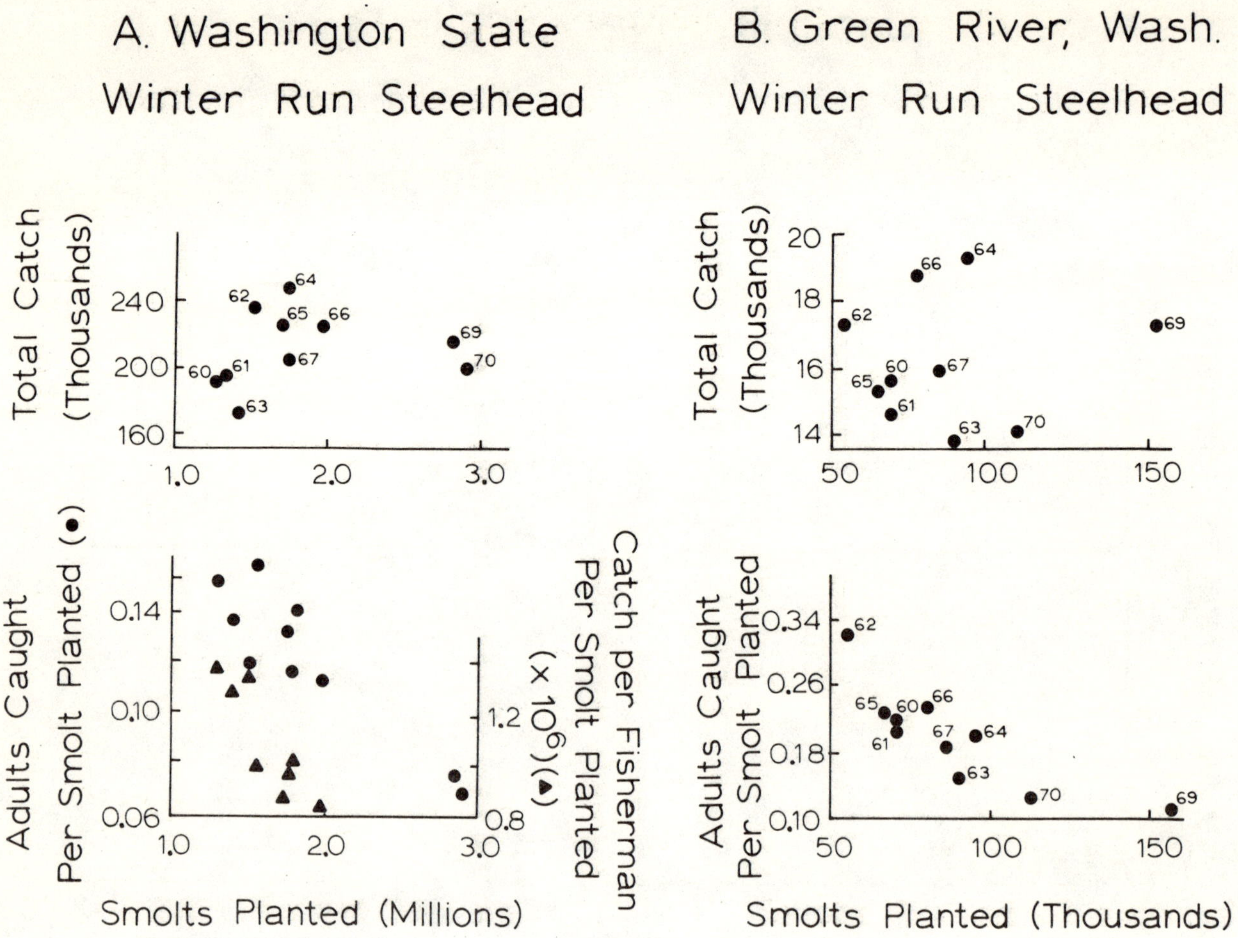

Figure 9. Total sports fishermen catch, catch per smolt planted and catch per fishermen per smolt planted as a function of smolts planted for A. all Washington State winter run steelhead, and B. for Green River only, by year of smolt planting.

combined with the other Alaskan stocks. Fraser pink and Chilko sockeye adults were related to total Fraser pink and Chilko smolt output in each year, and Lakelse pink and Babine Lake sockeye adults were related to total Lakelse fry and Babine sockeye smolt output. Only one of these five tests showed significant density dependence, and that was for even-year Lakelse Lake pinks (n = 4, $b_{YX} = 0.32 \pm 0.06$, $P_{b=1} < 0.01$, $b_{XY} = 2.62 \pm 0.11$, $P_{b=1} < 0.01$).

The third class of between-stock interactions on the same or nearby rivers is shown by close inspection of the Babine Lake sockeye data in Figure 3A. A colleague, F. E. A. Wood, noted that almost without exception, the odd-brood year data points fall above a curvilinear regression line and the even-year points fall below it. This immediately suggests an interaction with pink salmon stocks. At least three different types of interactions can be imagined: Returning adult pinks may predate heavily on Babine sockeye smolts. Large numbers of seaward migrating pink fry may depress sockeye smolt survival through attraction of predators, competition for food, or spreading of diseases or parasites. Finally, seaward migrating pink fry may increase sockeye smolt survival by swamping predators.

The life history of Babine Lake sockeye is such that most odd-brood year smolts, say of 1957, went to sea in 1959, and even-brood year fish migrate seaward in even years. On the Skeena River, into which Babine Lake drains, odd-year pink escapements and total run sizes have averaged almost the same as the even-year stock (Fisheries and Marine Service data). So interaction with pinks would have to be due to some characteristic other than abundance of even and odd year pink stocks. Low sockeye survival in even brood years is consistent with the hypothesis that even-year pink adults, which start to

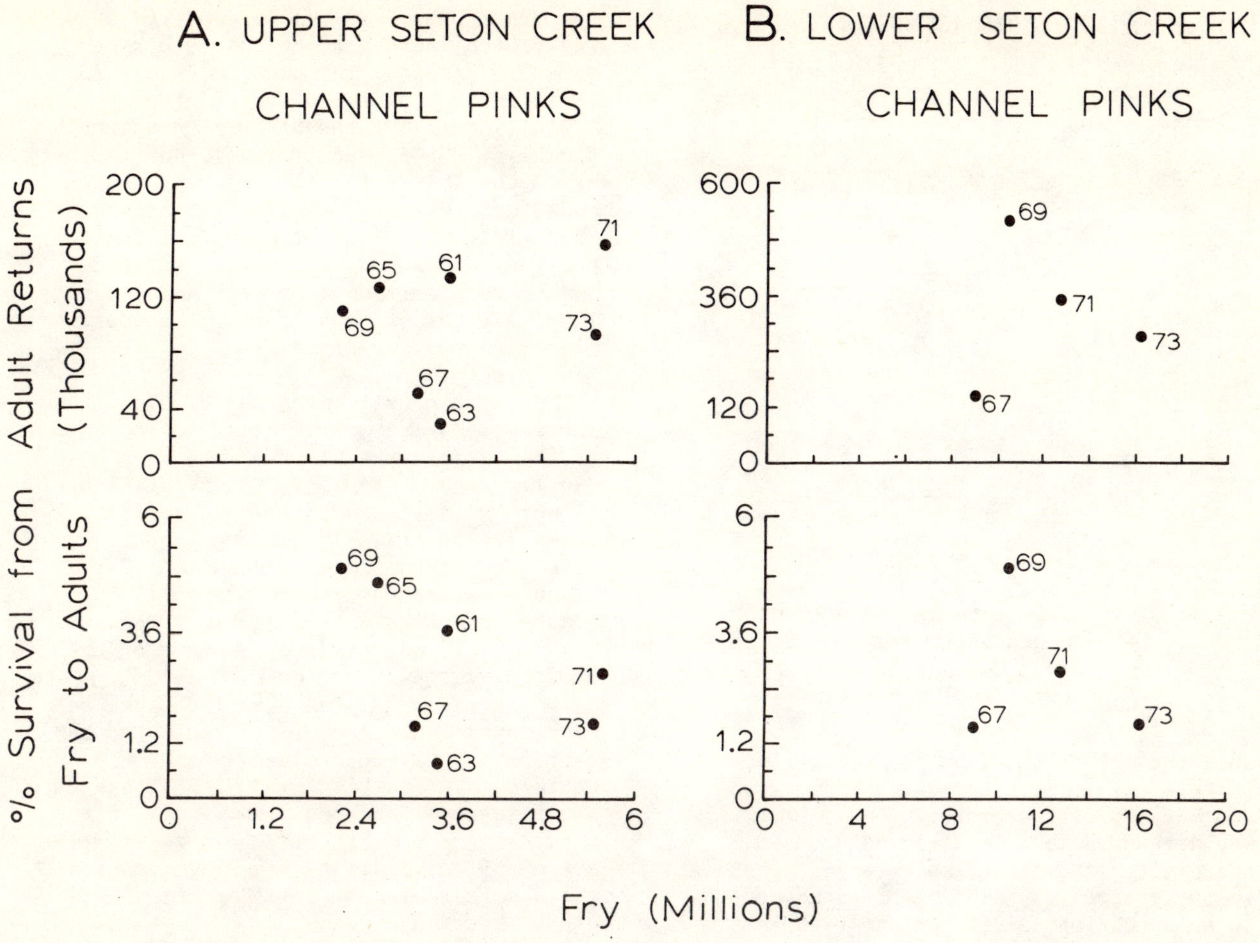

Figure 10. Adult returns and percent survival from fry to adults for two Fraser River enhanced stocks. Note that this is fry-to-adult survival, not smolt-to-adult. Data, by brood year, from Cooper (1977).

return in mid-June, are feeding more heavily on the sockeye smolts in coastal waters than odd-year adults, which return a few weeks later in the season. If the observed difference between even-and odd-year sockeye smolt survival is due to swamping of predators, we would expect that the downstream migrating fry from the even-year pink stock would go to sea at the same time of year as Babine sockeye smolts. If the difference is due to attraction of predators, then odd-year pink fry should be the ones that go to sea with the sockeye smolts. Last, if the observed differences in survival are due to competition for food, odd-year pink fry should be migrating sooner than or with Babine sockeye, while even-year pinks should migrate later than the sockeye. Unfortunately, not enough is presently known to distinguish between these hypotheses, and there may be others not yet considered. In any case, the observed difference in even and odd year Babine sockeye survival is strongly suggestive of between-stock interactions.

Note that the enhancement process itself could be used as an experiment in order to sort out the possible alternative hypotheses outlined above. Run timings could be altered with certain predicted effects, and future operations of those facilities could be modified to take the findings into account. This point will be reemphasized later.

Interactions Between Stocks of Distant Rivers

This hypothesis is intended to deal with both direct and indirect interactions between stocks. Direct interactions are where stocks are at the same location in the ocean at some point in time. Competition for food could occur to one stock's benefit or both could be affected similarly by attraction of predators to that location or swamping of them. Indirect interactions occur when, for example, one stock depletes the local food supply and moves away while shortly afterward, another stock moves into that

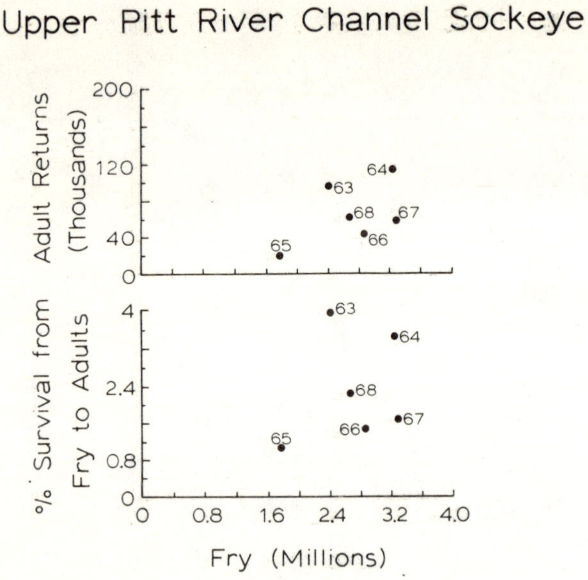

Figure 11. Adult returns and percent fry-to-adult survival for Upper Pitt River Channel sockeye, by brood year, from Cooper (1977).

location before the food has recovered. In this case, the effect of one stock on another would be negative. If predators are attracted to a location and remain there after the first stock leaves, then an incoming stock may also be affected severely by predation, unless predators were satiated. If parasites and diseases can be spread between fish in the ocean then an incidence of such agents would result in a positive correlation in marine survivals between those stocks present at that time. The effect of indirect space-time interactions is unknown for parasites or diseases. The following table summarized this framework for interpreting correlations between survival rates:

Direction of expected correlation between marine survival rates of two stocks for different types of interactions.

	Competition for Food	Predation	Parasites and Disease
Direct interaction	-	+	+
Indirect interaction	-	+	?

Correlations were made between the time series of survival rates of different stocks, after normalizing the data with arcsin transforms. Values were paired according to the year in which each stock went to sea. Results are shown in Figure 15. There is no significant correlation between marine survival rates of British Columbia stocks at the 5 percent level, but two out of 24 possible relations between Alaskan and B.C. stocks are significant and positive (Fig. 16). Five out of 15 correlations between Alaskan sockeye marine survival rates are significant and all are positive. Thus, a total of seven out of 55 cases show significant correlation.

Due to the smolt sampling difficulties mentioned earlier for Alaskan sockeye, some of the marine survival estimates are unusually high. The correlations of Figure 15 were reestimated after all survival rates greater than 30 percent were eliminated for these stocks. Results are slightly different; three of the five Alaskan vs. Alaskan correlations remain significant and positive, and one of the two B.C. vs. Alaska correlations remains the same (Naknek age II vs. Lakelse pinks becomes non-significant). In addition, however, a negative correlation between Hokkaido chum survival and Ugashik age II smolt survival becomes significant at the 5 percent level (Fig. 17). In total, five out of 55 correlations are significant at the 5 percent level, still more than expected by chance alone.

The predominance of positive correlations in Figure 15 suggests that the interactions between Alaskan stocks are more likely mediated by predation, parasitism or disease than by competition for food, according to the criteria discussed above. For the remaining negative and positive correlations, there may be no "most common" type of interaction.

Godfrey (1958), Larkin (1975), and Iles (1973) all investigated salmon marine survival, each using different indices. Larkin found no evidence for negative interactions between stocks, and Godfrey showed strong positive correlations in survivals. Iles demonstrated a positive relation between salmon survival and British Columbia herring survival. These findings are corroborated by most results presented in Figure 15. The common positive correlations may also be due to the overriding influence of physical oceanographic conditions. The one negative correlation may be fortuitous or an indication of biological situations that deserve further study. Too little is known at present to interpret this particular case.

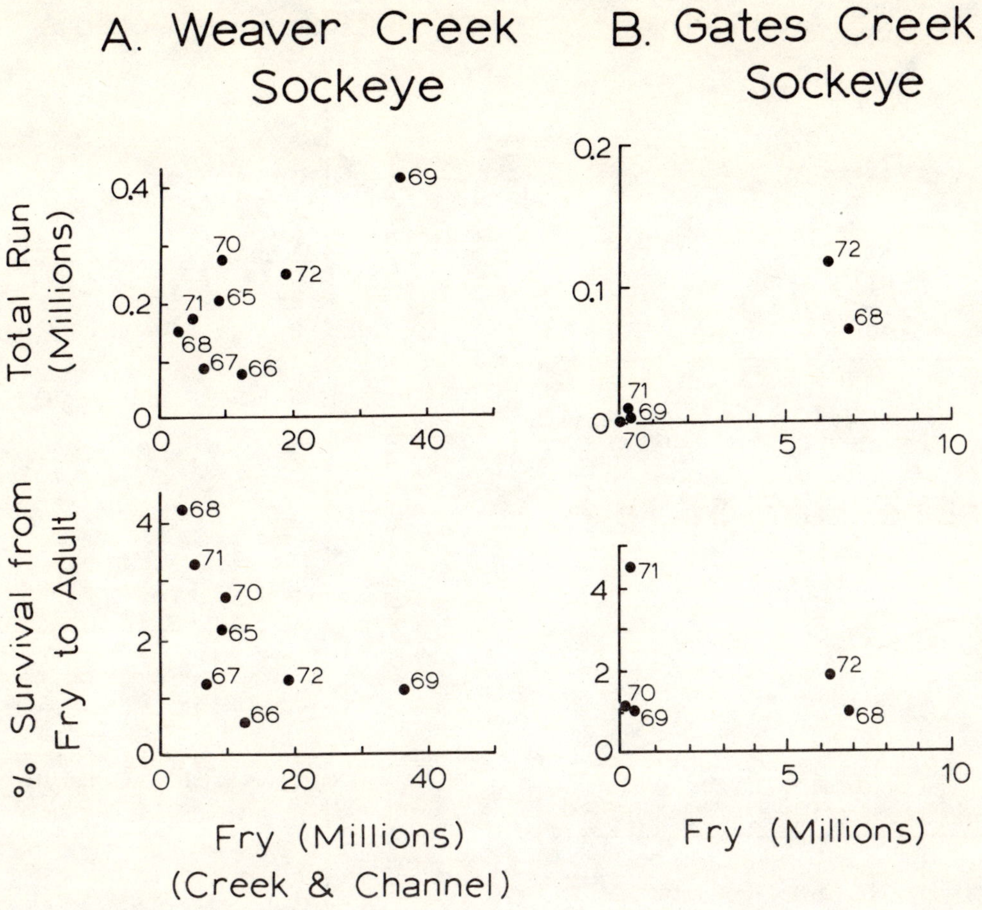

Figure 12. Adult returns and fry-to-adult survival for Weaver Creek and Gates Creek sockeye, by brood year. Redrawn from Cooper (1977).

GENERAL DISCUSSION

The foregoing analyses suggest that there is reason to doubt the prevailing belief that marine survival of salmon is not density dependent. At the least, the belief does not apply to all stocks examined here. Between one and seven of 12 cases show within-cohort density dependence in a negative direction, between zero and four of 6 cases show the same for between-cohort interactions, and two of 19 show negative density dependence between stocks on the same or nearby river systems. Between five and seven of 55 between-distant stock interactions are significant and most of these are positively correlated. These relatively few cases of density dependence may imply that present stock abundances are too low to result in significant interaction, but this says nothing about the magnitude of the interactions after abundances are increased through enhancement. Also, if density-dependent survival is more common in the stocks for which we presently have no data, then results from enhancement may be significantly smaller than expected.

Relatively few authors have dealt explicitly with the three possible mechanisms for density dependence mentioned here: Competition for food, predation, and parasitism and disease. Bailey et al. (1975) examined a rearing estuary for Alaskan pink and chum fry to determine if the salmon were near carrying capacity based on food supply. They found that the estuary could in fact support about 10 times existing fry numbers. Walters et al. (1978) constructed a simulation model of some major British Columbia salmon stocks to investigate the effect of smolt feeding on the food supply in coastal waters. The model included downstream migration timing, movement rates of fish along the coast, zooplankton dynamics, and feeding, growth and mortality of fish. Results show that observed changes in mortality rates cannot be explained solely on the basis of competition for food. Finally, LeBrasseur (1972)

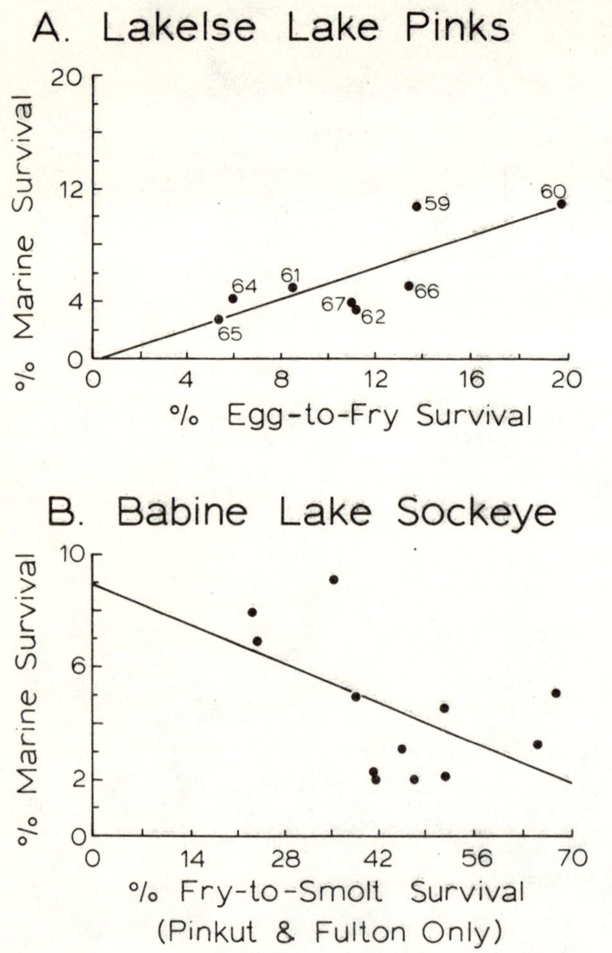

Figure 13. Relation between percent marine survival and freshwater survival rate for A. Lakelse Lake pink salmon and B. Babine Lake sockeye, by brood year.

estimated on the basis of food chain efficiency calculations that in the northeast Pacific Ocean, salmon remove less than 20 percent of the food production available to them.

Predation on juvenile salmon in the marine environment, has, understandably enough, received little attention. Ricker (1968) shows that Pribilof Islands fur seals, which have numbered up to 3 million, can consume 800,000 salmon per year. Participants at a workshop in 1975 calculated on the basis of different assumptions that two to 60 million salmon could be eaten (Peterman 1975). Laevastu and Favorite (1977) calculate that pinnipeds consume 45 million kg of returning adult salmon annually in the eastern Bering Sea alone (or about 10 million fish of 4.5 kg each). Such depredations can have enormous significance, especially if they are concen-

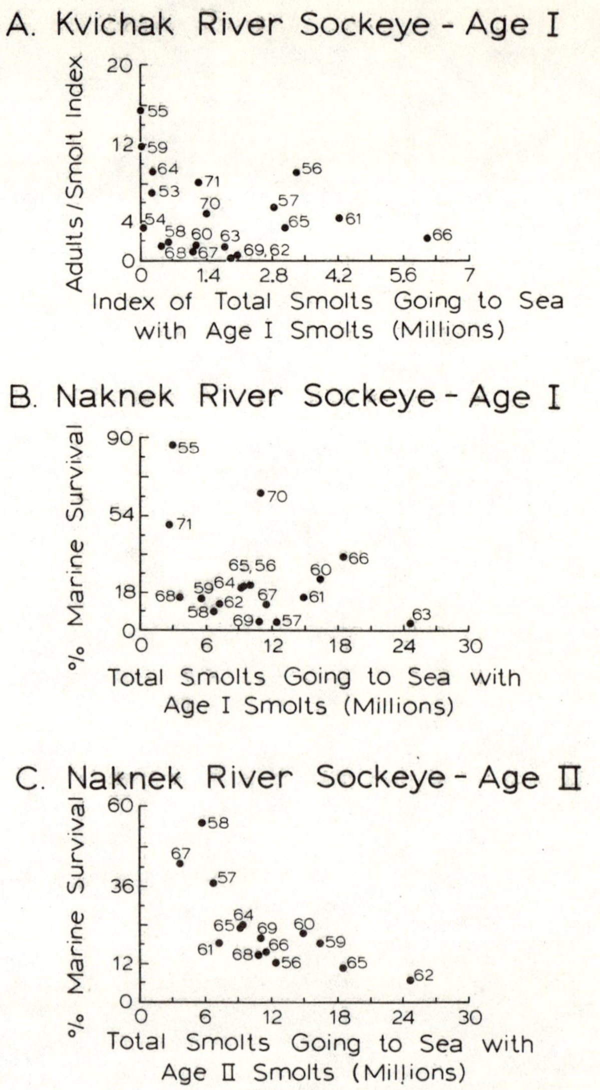

Figure 14. Marine survival rate of a specific cohort of Alaskan sockeye smolts as a function of the total abundance of all sockeye smolts on that river going to sea in the same year. A. Age I Kvichak sockeye, B. Age I Naknek sockey, C. Age II Naknek sockeye.

trated on particular stocks or at certain times. The essential question to ask is how the percent predation mortality varies as salmon abundance varies. Parker's studies (1968, 1971) show that between 55 and 77 percent of pink salmon fry of one stock are eaten in the estuarine lifestage by juvenile coho salmon *(O. kisutch)*. While he does estimate fry abundance for three years, there is no clear relation between abundance and mortality rate due to this early marine predation. As shown in a review of existing data on freshwater predation on salmon, there are surprisingly few

Table 2. Results of Statistical Tests for Density Dependence Between Smolt Cohorts of Alaskan Sockeye.

Stock	Sample size (years)		$P_{b=1}$ (one-tailed)		$P_{b=1}$ (one-tailed)	Density dependent by Morris	Density dependent by Varley-Gradwell
Kvichak R.							
Age I smolts	19	0.37±0.23	<0.01	0.35±0.23	=0.005	Yes	No
Age II smolts	19	0.65±0.20	=0.052	0.59±0.19	<0.025	No	No
Naknek R.							
Age I smolts	18	0.32±0.29	<0.02	0.22±0.21	<0.005	Yes	No
Age II smolts	14	0.15±0.32	=0.012	0.13±0.26	<0.005	Yes	No
Ugashik R.							
Age I smolts	15	0.78±0.44	>0.25	0.27±0.15	<0.001	No	No
Age II smolts	15	0.51±0.24	<0.04	0.53±0.24	<0.04	Yes	No

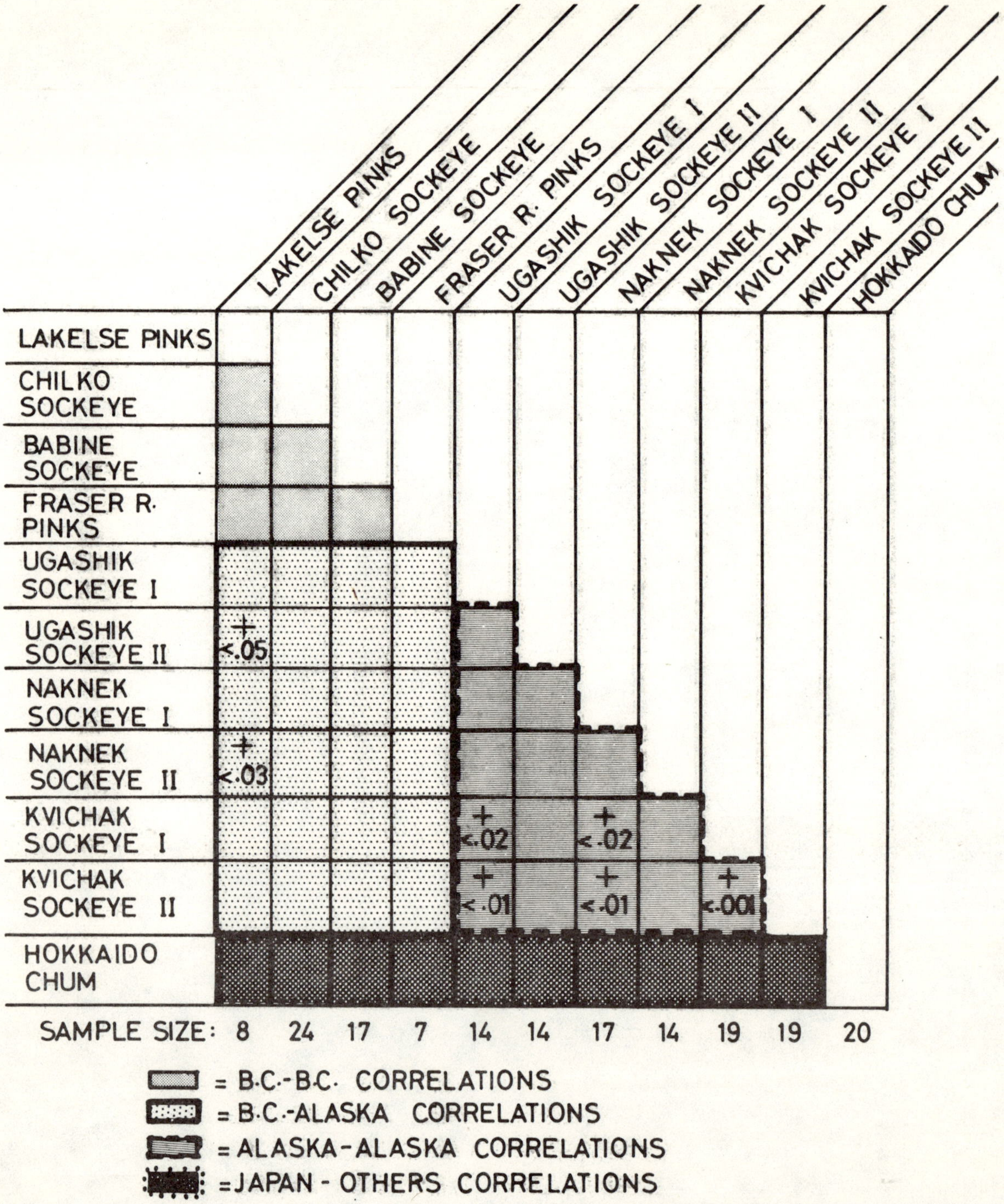

Figure 15. Correlations of marine survival rates of different salmon socks. Data are paired by the year when smolts went to sea. The sign of the correlation coefficient is shown as well as the probability that r = 0.

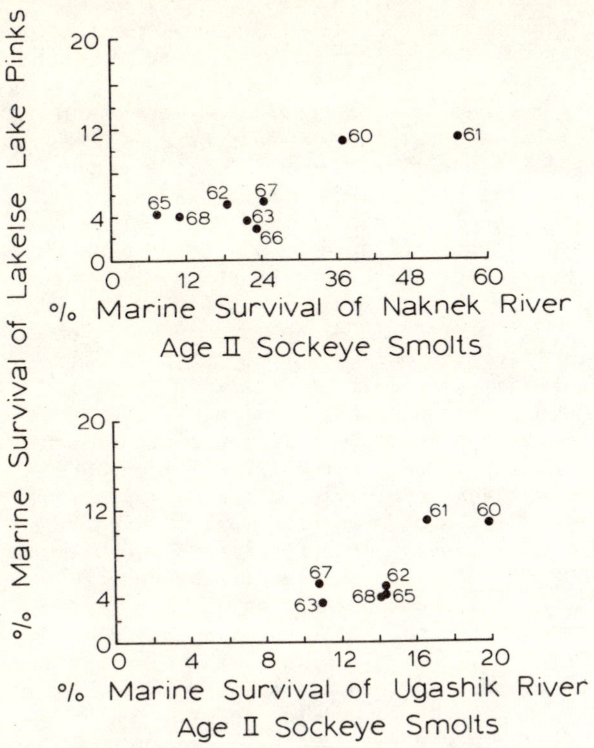

Figure 16. Two statistically significant relations between marine survival rates of British Columbian and Alaskan salmon stocks, by year of seaward migration.

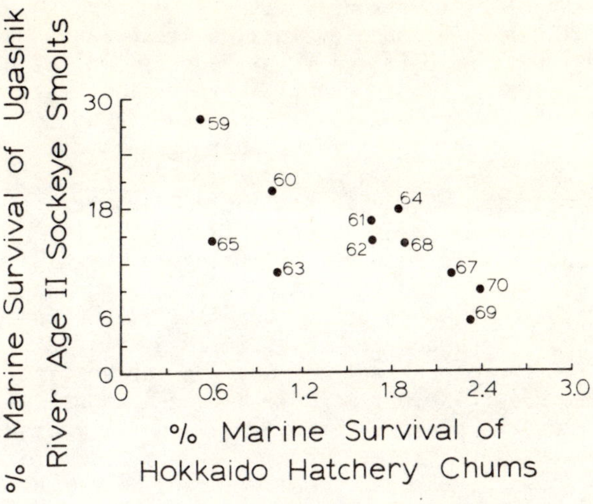

Figure 17. Relation between marine survival rates of Hokkaido Hatchery chum and Ugashik River age II sockeye, by year of seaward migration.

cases where predators have been swamped by high prey densities (Peterman and Gatto 1978). An increase in prey numbers merely increases percent mortality. The same may be true of marine predators, which may have even larger ratios of body size to prey body size.

Parasitism and fish disease form an even larger black box than predation. Fisheries biologists presume that higher smolt and adult densities result in greater rates of transmission of these mortality agents, but no one knows if such agents can have as large an effect on the older, marine lifestages as on the younger.

A number of authors argue that the highest ocean mortality rates of salmon occur early in the marine life (Godfrey 1958, Foerster 1968, Parker 1968, Ricker 1976). This is also what one would expect from prevailing ideas of fish growth and behavior. As fish grow larger, they become less severely affected, in terms of mortality, by a given length of food deprivation. Such fish merely have a slower growth rate. Also, larger fish in the marine lifestage become susceptible to predation by a smaller number of predator types (Ricker 1976).

In sum, there is presently insufficient information to sort out the causes of observed density-dependent marine survival. We cannot even tell where in the marine life history such factors are operating. However, the relatively few cases of density-dependent marine survival demonstrated in this paper should not lead one to the conclusion that limitation in the ocean is unimportant for salmonids. At some large abundance, there has to be density-dependent survival, just as there is in freshwater. The question is, where is that point for the marine environment of each stock, and has that point changed since the period when salmon catches were much larger?

Enhancement and Management Implications

The statistical tests applied earlier were designed merely to demonstrate that there are indeed cases of density-dependent marine survival of salmon and that we should be careful in managing or enhancing salmon stocks. But the threshold criteria established by those tests are almost meaningless; in the real world, there are degrees of density dependence, not just presence or absence. Perhaps a better criterion that we can use to judge the marine survival data is whether the change in adult returns when smolt numbers are increased is worth the amount of money that is spent to provide that change in smolt abundance. This is a difficult yet critically important question which needs to be quantified, since it makes several assumptions concerning short versus long-term benefits, discount rates, amortization periods, etc. Economists and sociologists strongly disagree on the structure of such evaluation criteria and I am in no position to provide my own. The following simple example merely illustrates the need to reexamine the economics of enhancement after returns are generated.

Approximately 10 million dollars have been spent on enhancement facilities for Babine Lake sockeye, beginning in the mid-1960s, and adults from the first large increase in smolt production (the 1971 brood year) returned beginning in 1975 (Fig. 3A). The increased smolt production has resulted in about 150,000 more fish caught annually than the average for the previous 15 years. Using a net value for each caught fish of $8.74 (J. Barclay, Department of Fisheries and Environment, Vancouver), ($4.37 1966-dollars) gives an increase in net annual benefit of $655,500 in 1966-dollars for this enhancement project. If the expected life span of the facilities is 20 years, and if the present increase in catch is maintained, then the total net benefits from the project will be $13.11 million, only slightly larger than the $10 million capital cost. But once high variability in return rate, operating costs, and the effect of discounting are included in the analysis, the benefits would become marginal.

No cost figures were available for the Washington steelhead smolt planting program, but an inspection of Figure 9 clearly demonstrates that an increase in smolts above a certain point does not result in greater catch. While this and the Babine sockeye enhancement project may have been sound investments given the data available at initiation of construction, reevaluation of benefits should be made after return data accumulate. Highly variable marine survival rates will cloud the evaluation process, but the final reassessment could provide the opportunity for modifying other projects planned for the future, or channelling operating costs of current non-beneficial projects into more useful areas.

One final point on these evaluation criteria. Ideally, adult returns from all stocks should be considered, not just the returns from enhanced stocks because of the possible interaction between stocks. The enhanced stock may just be replacing other natural populations.

If the stocks analyzed in this paper are a representative sample of all stocks, then we will clearly have to expect that some enhancement projects will produce high net returns while others will not. We cannot yet determine with certainty before enhancement what the results will be, although Staley's (1975) general predictions for the Babine sockeye were borne out.

Because of this uncertainty about outcomes, we must manage and enhance salmon stocks in a flexible and adaptive manner (Holling et al. 1978, Larkin 1974). Enhancement should be viewed as a series of experiments designed to distinguish between alternative hypotheses about underlying biological relationships (Loftus 1976, Peterman 1975, Walters 1977). For example, the noise in some existing data makes it difficult to determine if there is a linear or non-linear relation between smolts and adults. A large scale perturbation experiment in the form of increased smolt abundance would, after many years, help distinguish which underlying relation is most likely correct (Walters and Hilborn 1976). In fact, the larger the perturbation, the shorter the span of time required to distinguish (Peterman et al. 1978, Walters 1977). Thus, one of the main values of enhancement, in addition to increasing catch, can be to provide information that will help the future mangement of stocks (Peterman 1977, Walters 1977). The value of this information should be explicitly considered when weighing management alternatives (Silvert 1978).

Finally, enhancement experiments could be designed in such a way that it would be possible to distinguish between various mechanisms that alter marine survival rates. For example, lake fertilization provides an excellent means to sort out the effects of smolt body size from smolt abundance. Size of smolts can be increased without necessarily increasing smolt abundance. Fertilization of slow-mixing estuaries might also provide clues about the relative importance of predation or competition for food by smolts.

To separate the relative importance of early versus later marine life, it might be possible to enhance a stock which is known to mix with other stocks early in ocean life but which matures in an area separate from other stocks. Comparison of marine survivals of this stock, a stock with the reverse characteristics, and unenhanced stocks of those same types could yield valuable information on the significance of early and late ocean mortality.

Another important conclusion emerges from this study. A high priority area for research is to determine the extent of measurement errors (Ricker 1973) for a wide range of smolt abundances on each system. Then regression statistics from the tests for density dependence can be adjusted (Kuno 1971) and exact tests of the null hypothesis can be performed.

ACKNOWLEDGMENTS

I am extremely grateful to K. P. Parker, R. Ginetz, E. Zyblut, F. E. A. Wood, and D. R. Harding for providing me with some of their unpublished data. I also thank Carl Walters, John Schnute, Mike Staley, Jay Barclay, Neil Gilbert, Ray Hilborn, and especially Ralf Yorque for numerous discussions on this topic. C. J. Walters, M. Staley, P. A. Larkin, R. R. Parker, and an anonymous reviewer provided many useful suggestions on the manuscript.

LITERATURE CITED

Allen, K. R. 1973. The influence of random fluctuations in the stock-recruitment relationships on the economic return from salmon fisheries. Pages 350-359 *in* Parrish, B. B. (ed.), Fish stocks and recruitment. Rapp. P.-V. Reun. Cons. Int. Explor. Mer. 164.

Atkinson, C. E. 1976. Salmon aquaculture in Japan, the Koreas and the U.S.S.R. Pages 79-154 *in* Rosenberg, D. H. (ed.), Proceedings of the conference on salmon aquaculture and the Alaskan fishing community, Cordova, Alaska. Alaskan Sea Grant Rep. 76-2.

Bailey, J. E., B. L. Wing, and C. R. Mattson. 1975. Zooplankton abundance and feeding habits of fry of pink salmon, *Oncorhynchus gorbuschus*, and chum salmon, *Oncorhynchus keta*, in Traitor's Cove, Alaska, with speculations on the carrying capacity of the area. Fish. Bull. 73:846-861.

Benson, J. F. 1973. Some problems of testing for density-dependence in animal populations. Oecologia (Berl.) 13:183-190.

Cooper, A. C. 1977. Evaluation of the production of sockeye and pink salmon at spawning and incubation channels in the Fraser River system. Int. Pacific Salmon Fish. Comm. Prog. Rep. No. 36. 80 pp.

Cushing, D. H. 1975. Marine ecology and fisheries. Cambridge University Press, Cambridge. 278 pp.

Eberhardt, L. L. 1970. Correlation, regression, and density dependence. Ecol. 51:306-310.

Foerster, R. E. 1954. On the relation of adult sockeye salmon *(Oncorhynchus nerka)* returns to known smolt seaward migrations. J. Fish. Res. Bd. Can. 11:339-350.

Foerster, R. E. 1968. The sockeye salmon, *Oncorhynchus nerka*. Fish. Res. Bd. Can. Bull. 162. 422 pp.

Ginetz, R. M. J. 1977. A review of the Babine Lake development project 1961-1976. Fish. and Marine Serv., Envir. Can. Tech. Rep. Series No. PAC T-77-6. 192 pp.

Glover, R. S., G. A. Robinson, and J. M. Colebrook. 1974. Marine biological surveillance. Envir. and Change 2(6): 395-402.

Godfrey, H. 1958. Comparisons of the index of return for several stocks of British Columbia salmon to study variations in survival. J. Fish. Res. Bd. Can. 15: 891-908.

Groot, C. 1976. An overview of salmonid enhancement programs in countries bordering the north Pacific Ocean. Appendix II 19 to Internal Fish. and Marine Serv. Document, Envir. Can.

Holling, C. S. (ed.), A. Bazykin, P. Bunnell, W. C. Clark, G. C. Gallopin, J. Gross, R. Hilborn, D. D. Jones, R. M. Peterman, J. E. Rabinovich, J. H. Steele, and C. J. C. J. Walters. *In press*. Adaptive environmental assessment and management. John Wiley and Sons, London.

Iles, T. D. 1973. The interaction of environment and parent stock size in determining recruitment in the Pacific sardine, as revealed by analysis of density-dependent O-group growth. Pages 228-240 *in* Parrish, B. B. (ed.), Rapp. P.-V. Reun. Cons. Int. Explor. Mer. 164.

International Pacific Salmon Fisheries Commission. 1959-1976. Annual reports. New Westminster, B.C.

Ito, Y. 1972. On the methods for determining density-dependence by means of regression. Oecologia (Berl.) 10:347-372.

Johnson, W. E. 1965. On mechanisms of self-regulation of population abundance in

Oncorhynchus nerka. Mitt. Internat. Verein. Limnol. 13:66-87.

Ketchen, K. S. 1956. Climatic trends and fluctuations in yield of marine fisheries of the Northeast Pacific. J. Fish. Res. Bd. Can. 13:357-374.

Kuno, E. 1971. Sampling error as a misleading artifact in "key factor analysis." Res. Popul. Ecol. XIII:28-45.

Laevastu, T. and F. Favorite. 1977. Preliminary report on dynamical numerical marine ecosystem model (DYNUMES II) for eastern Bering Sea. Northwest and Alaska Fish. Center Processed report, August 1977. Seattle, Wash. 81 pp.

Larkin, P. A. 1974. Play it again Sam - an essay on salmon enhancement. J. Fish. Res. Bd. Can. 31:1433-1459.

Larkin, P. A. 1975. Some major problems for further study on Pacific salmon. Int. North Pacific Fish. Comm. Bull. 32: 3-9.

LeBrasseur, R. J. 1972. Utilization of herbivore zooplankton by maturing salmon. Pages 580-588 *in* Takenouti, A. Y. (ed.), Biological oceanography of the northern North Pacific Ocean. Idemitsu Soten, Tokyo, Japan.

Loftus, K. H. 1976. Science for Canada's fisheries rehabilitation needs. J. Fish. Res. Be. Can. 33:1822-1857.

Luck, R. F. 1971. An appraisal of two methods of analyzing insect life tables. Can. Ent. 103:1261-2171.

MacLeod, J. R. 1977. Enhancement technology: A positive statement. Pages 137-147 *in* Ellis, D. V. (ed.), Pacific salmon, management for people. Univ. Victoria Press, Victoria, B.C.

Maelzer, D. A. 1970. The regression of log N_{n+1} on log N_n as a test of density dependence: An exercise with computer-constructed density-independent populations. Ecol. 51:810-822.

Manzer, J. I. 1976. Preliminary results of studies on the effects of fertilization of an oligotrophic lake on adult sockeye salmon *(Oncorhynchus nerka)* production. Fish. Marine Serv. Res. Dev. Tech. Rep. 678. 25 pp.

Moberly, S. A., and R. Lium. 1977. Japan salmon hatchery review. Alaska Dept. Fish and Game, Div. Fisheries Rehabilitation, Enchancement and Development. 124 pp.

Morris, R. F. 1959. Single-factor analysis in population dynamics. Ecol. 40(4): 580-588.

Morris, R. F. (ed.). 1962*a*. The dynamics of epidemic spruce budworm populations. Mem. Entom. Soc. Can. No. 31. 332 pp.

Morris, R. F. 1963*b*. Predictive population equations based on key factors. Pages 16-21 *in* Leroux, E. J., et al. (eds.), Population dynamics of agriculatural and forest insect pests. Mem. Entom. Soc. Can. No. 32.

Paine, J. 1974. The Big Qualicum River artificial spawning channel for chum salmon. Pages 72-78 *in* Harding, D. R. (ed.), Proceedings of the 1974 northeast Pacific pink and chum salmon workshop. Dep. Envir., Fish. Mar. Serv., Vancouver, B.C.

Parker, R. R. 1968. Marine mortality schedules of pink salmon of the Bella Coola River, Central British Columbia. J. Fish. Res. Bd. Can. 25:757-794.

Parker, R. R. 1971. Size selective predation among juvenile salmonid fishes in a British Columbia inlet. J. Fish. Res. Bd. Can. 28:1503-1510.

Parsons, T. R., and R. J. LeBrasseur. 1968. A discussion of some critical indices of primary and secondary production for large-scale ocean surveys. Calif. Mar. Res. Comm. Cal. COFI Rep. 12:54-63.

Peterman, R. M. 1975. "Ocean effects" in salmon. PR-3, Inst. Res. Ecol., Univ. British Columbia, Vancouver, B.C. 37 pp.

Peterman, R. M. 1977. A simple mechanism that causes collapsing stability regions in exploited salmonid populations. J. Fish. Res. Bd. Can. 34:1130-1142.

Peterman, R. M., and M. Gatto. 1978. Estimation of functional responses of predators on juvenile salmon. J. Fish. Res. Bd. Can. 35:797-808.

Peterman, R. M., C. J. Walters, and R. Hilborn. *In press*. Pacific salmon management. *In* Holling, C. S. (ed.) et al., Adaptive environmental assessment and management. John Wiley and Sons, London.

Ricker, W. E. 1962. Comparison of ocean growth and mortality of sockeye salmon during their last two years. J. Fish. Res. Bd. Can. 19:531-560.

Ricker, W. E. 1968. Background information and theory related to management of the Skeena sockeye salmon. Fish. Res. Bd. Can. Manuscript Rep. Series No. 961. 38 pp.

Ricker, W. E. 1973. Linear regressions in fishery research. J. Fish. Res. Bd. Can. 30:409-434.

Ricker, W. E. 1976. Review of the rate of growth and mortality of Pacific salmon in saltwater, and non-catch mortality caused by fishing. J. Fish. Res. Bd. Can. 33:1483-1524.

Royal, L. A. 1972. An examination of the anadromous trout program of the Washington State Game Department. Wash. State Dep. Game. Internal report. 176 pp.

Silvert, W. 1978. The price of knowledge: Fisheries management as a research tool. J. Fish. Res. Bd. Can. 35:208-212.

St. Amant, J. L. S. 1970. The detection of regulation in animal populations. Ecol. 51:823-828.

Staley, M. 1975. Pre-season prediction of sockeye salmon runs on the Skeena River. Int. Inst. Appl. Systems Analysis, Laxenburg, Austria. WP-75-16. 10 pp.

Straty, R. R. 1974. Ecology and behavior of juvenile sockeye salmon *(Oncorhynchus nerka)* in Bristol Bay and the eastern Bering Sea. Pages 285-320 *in* Wood, D. W., and E. J. Kelly (eds.), Oceanography of the Bering Sea. Inst. Mar. Sci., Univ. Alaska, Fairbanks.

Sutcliffe, W. H., R. H. Loucks, and K. R. Drinkwater. 1977. Correlations of fish catch and environmental factors in the Gulf of Maine. J. Fish. Res. Bd. Can. 34:19-30.

Varley, G. C., and G. R. Gradwell. 1960. Key factors in population studies. J. Anim. Ecol. 29:399-401.

Varley, G. C., and G. R. Gradwell. 1963. Predatory insects as density-dependent mortality factors. Page 240 *in* Moore, J. A. (ed.), Proceedings of XVI International Congress Zool. Vol. 1.

Varley, G. C., and G. R. Gradwell. 1968. Population models for the winter moth. Pages 132-142 *in* Southwood, T. R. E. (ed.), Insect abundance. Blackwell Sci. Publ., London.

Walters, C. J. 1977. Management under uncertainty. Pages 261-297 *in* Ellis, D. V. (ed.), Pacific salmon, management for people. Univ. Victoria Press, Victoria, B.C.

Walters, C. J. and R. Hilborn. 1976. Adaptive control of fishing systems. J. Fish. Res. Bd. Can. 33:145-159.

Walters, C. J., R. Hilborn, R. M. Peterman, and M. Staley. *In press*. A model for examining early ocean limitations of Pacific salmon production. J. Fish. Res. Bd. Can.

Ward, F. J., and P. A. Larkin. 1964. Cyclic dominance in Adams River sockeye salmon. Int. Pac. Salmon Fish. Comm., Prog. Rep. No. 11. 114 pp.

Watt, K. E. F. 1964. Density dependence in population fluctuations. Can. Ent. 96:1147-1148.

Watt, K. E. F. 1968. Ecology and resource management. McGraw-Hill Book Co., N.Y. 450 pp.

Wickett, W. P. 1973. Our changing ocean environment. Fish. Res. Bd. Can. Pac. Biol. Station, Nanaimo, B. C., Circular No. 94. 23 pp.

APPENDIX

The Central Limit Theorem applies to Equation 1 in the following way:

$$A = S \cdot s_1 \cdot s_2 \cdot s_3 \ldots s_n$$

where A and S are adult and smolt abundances, respectively, and where $s_1 \ldots s_n$ are survival rates through successive periods of the salmon's life, e.g. lake-to-estuary, early estuary, early coastal period etc.

Taking logarithms, we obtain

$$\ln A = \ln S + \ln s_1 + \ln s_2 \ldots \ln s_n$$

If $s_1 \ldots s_n$ are independent and random variates, then by the Central Limit Theorem, the sum of these random variates is a normally distributed random variate, n.

$$\ln A = \ln S + n$$

Taking antilogs

$$A = S \cdot \exp(n)$$

Thus, we conclude that there should be a multiplicative, log-normal noise term.

Juvenile Salmonids in the Oceanic Ecosystem—The Critical First Summer

Allan C. Hartt

INTRODUCTION

The life history of salmon in the ocean has been studied much less than that in freshwater and "inside" marine ecosystems, such as Puget Sound, primarily because of priorities and cost. And among the oceanic studies of salmon that have been done, research on juvenile salmon during their first summer at sea has probably received the least attention. Yet, the early juvenile stage is most critical for the majority of organisms. This report summarizes what we presently know about the distribution, migrations and growth of juvenile salmon in the eastern North Pacific during their first summer in the open sea.

METHODS AND RESULTS

Data Base

Material summarized in this paper primarily comes from the published bulletins of the International North Pacific Fisheries Commission (INPFC) and a research summary that Hartt and Dell *(in press)* of the Fisheries Research Institute (FRI) prepared for the INPFC. A synopsis of the primary data indicates that sampling was conducted throughout the eastern north Pacific and Bering sea from 1956-70. A total of 3,073 sets were made with a large, fine-meshed purse seine. Figure 1 shows geographic features referred to in the text. Figure 2 shows the distribution of purse seine fishing effort by season and location.

Contribution No. 489, College of Fisheries, University of Washington, Seattle, Washington.

About 110,000 juvenile salmon and steelhead trout were caught. Of these, 35,259 were tagged (10,411 sockeye, *O. nerka;* 4,412 chum, *O. keta;* 13,060 pink, *O. gorbuscha;* 7,015 coho, *O. kisutch;* 276 chinook, *O. tshawytscha;* and 85 steelhead trout, *Salmo gairdneri*). A majority of the catch was sampled for length. Subsamples were taken for freshwater age and stomach contents. All tagged fish were measured.

The number of tag recoveries by species were: sockeye, 41 or 0.4 percent; chum, 6 or 0.1 percent; pink, 56 or 0.4 percent; coho, 245 or 3.5 percent; chinook, 12 or 4.3 percent; steelhead, 1 or 1.2 percent.

From these data the general distribution, migrations, growth, density and abundance of some major juvenile salmon stocks during their first summer at sea are described. The principal items of analysis are:

1. Catch per unit effort (CPUE) by location, date and species.
2. Catch according to direction of the seine set.
3. Lengths of fish sampled by location, date and species.
4. Tag returns including length at tagging.

Distribution

The general summer distribution of juvenile salmon as shown by the CPUE of the purse seine sampling is illustrated in Figures 3-8.

Gulf of Alaska: In the Gulf of Alaska, the distributions and timing of juvenile sockeye, chum and pink salmon were fairly similar (Figs. 3-5). These species typically occurred together in the catches.

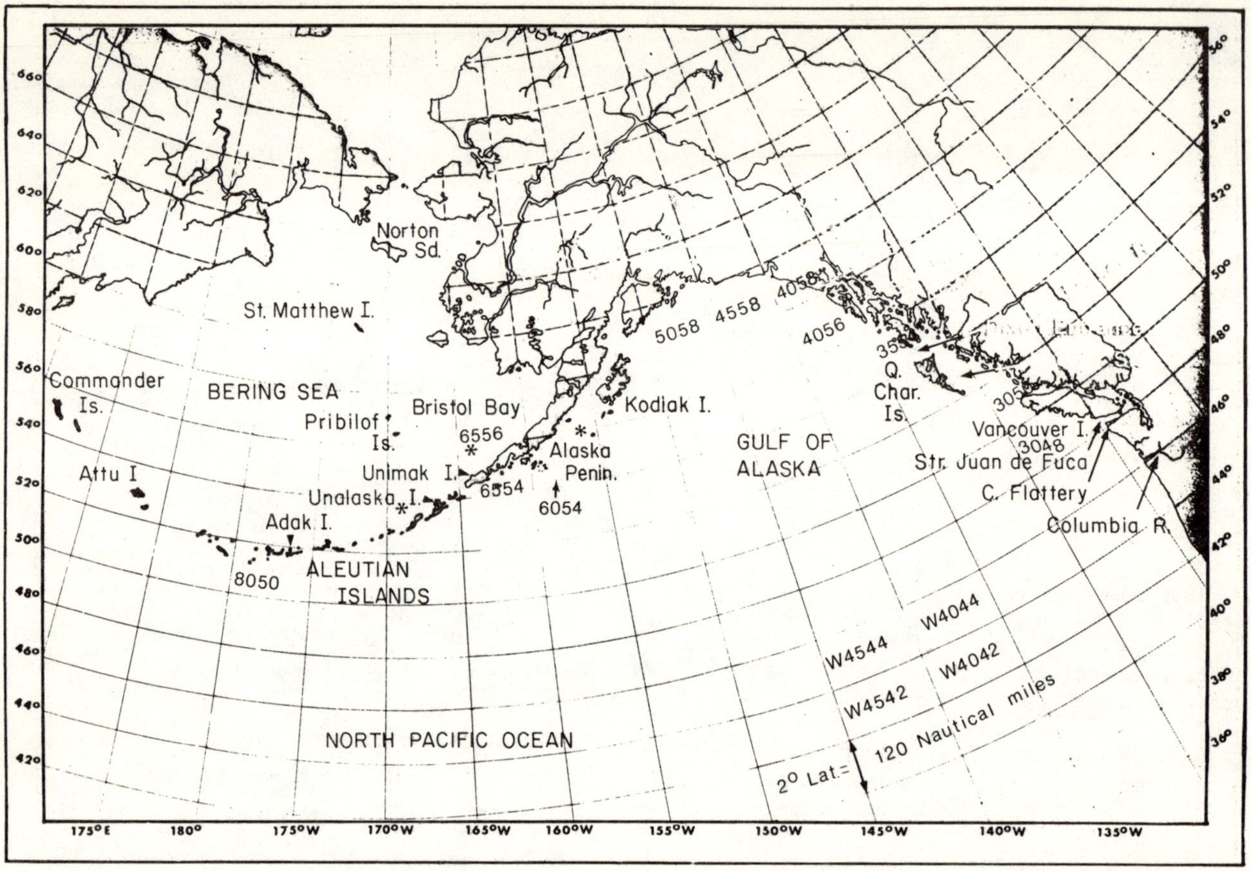

Figure 1. Base map showing study area and geographic locations referred to in the text. The INPFC code for $2^{o} \times 5^{o}$ statistical rectangles is illustrated for several key areas. Code numbers are determined by the geographic coordinates of the S.W. corner of a rectangle. Longitude, prefixed by E or W, is given first, omitting the hundreds, and then latitude, omitting N.

*The borders of areas W7054, W6556, and W6054 were extended north or south to include part of an adjoining areas as shown by asterisks. This was to avoid including both Bering Sea and Pacific waters in one statistical area.

Entry into the sea was earliest in the southern part of the range as indicated by catches in the southwest portion of the gulf off Vancouver Island during the April-June period. From July-September, the numbers of these three species generally increased in all areas. The increase progressed from south to north, and to the west along the coastal belt of the Alaska Peninsula. Few were caught in offshore areas.

Pink salmon catches by Canadian longline gear in November and December 1966 illustrate that juveniles are distributed widely in late fall far offshore in substantial numbers (Fig. 5). We do not know the stocks involved, their density compared to the coastal purse seine catches or the routes that they followed. No juvenile sockeye or chum salmon were caught along with the pinks in these offshore catches. This suggests that these two species may not move offshore as early as pink salmon do.

Catch distributions of juvenile coho and chinook salmon, and steelhead trout are illustrated in Figures 6-8. Their distributions differ from those of sockeye, chum and pink salmon as they occurred both in coastal as well as offshore areas. The offshore catches of coho, chinook and steelhead become more significant because they are much less numerous than sockeye, chum and pink salmon. By the same token, the scarcity of sockeye, chum and pink salmon in offshore waters during their first summer at sea reinforces evidence indicating that they primarily adhere to the coastal belt. We do not know what portion of the coho, chinook and steelhead migrate offshore in the summer. Typically, substantial numbers of juvenile coho and chinook remain in coastal waters and even in inside waters for their entire first summer or longer. The scarcity of steelhead in coastal waters, and their general occurrence over wide areas at sea, suggests that a majority of this species migrates directly to sea rather than remaining along the coastal belt.

The band of juvenile sockeye, chum and pink salmon along the coastal belt was apparently very narrow, at least off southeastern Alaska where a transect of stations indicated the width to be less than 20 nautical miles (nm) (Table 1). Within 4-9 nm of shore, the mean catch was 434 juvenile salmon. Between 11-16 nm, the mean catch was 267. Between 23-27 nm, only five juvenile salmon were caught in four sets, or 1.3 per set. At this sampling point, the continental shelf narrows to less than 20 nm. Perhaps this affects the distribution of juvenile salmon. The width of the band was considerably wider in the northern area of the gulf where the continental shelf widens (Royce et al. 1968).

Bering Sea: The sampling distribution in the Bering Sea was more limited than in the Gulf of Alaska. It was sufficient, however, to show that juvenile salmon were present in substantial numbers throughout the summer in an area north of the Alaska Peninsula between 160° W and 164° W (Figs. 3-7). The distribution extended to about 50 nm offshore. Straty (1974) showed that the distribution extended close to shore and even into major bays such as Port Moller. Sockeye dominated, as would be expected, based on salmon production in Bristol Bay rivers. Chum, pink, coho and chinook were also present in small numbers. No juvenile steelhead were caught, as was expected, because of the absence of seagoing stocks in rivers that empty into the Bering Sea (Sutherland 1973) (Fig. 8).

Migrations

During their first summer at sea, juvenile salmon migrated rapidly and extensively. Evidence came from three sources: (1) the catch distribution of juvenile salmon by date and location; (2) catches with respect to the direction that the seine was opened; and (3) tag returns. In addition, the length of the fish sampled, by date and location, and the length at release of recovered fish, provided indirect information on migrations.

Migrations Inferred from Catch Data: By making appropriate assumptions, as Hartt and Dell *(in press)* discussed, the seasonal catch distribution may be interpreted to indicate the seasonal migrations of salmonids during the early phase of their life history in the ocean (Figs. 3-8).

Gulf of Alaska: The data for salmon in the Gulf of Alaska only show the general trend and the approximate timing of migrations between June and October. The migrations of individual stocks and stock mixing can only be inferred from tag returns and length data, which will be discussed later.

Data were more consistent for the more abundant sockeye, chum and pink salmon (Figs. 3-5). These species tended to remain near shore. Stocks from southern locations apparently entered the ocean first, followed by stocks in more northern and western locations as the season progressed. Later in the season, catches in southern areas diminished at the same time that catches in northern and western areas continued large, or increased. Thus, the seasonal catches of these three species are consistent with a northerly, westerly and southwesterly migration along the coastal belt of the gulf.

The catch data for coho and chinook salmon indicate that at least some of these species migrate as do the sockeye, chum and pink salmon (Figs. 6-7). However, their presence in offshore waters early in the season, and their continued presence in southern inshore water late in the season, suggested a more diverse migratory pattern.

The distribution of juvenile steelhead differed vastly from that of any salmon species (Fig. 8). It follows that their migration must also have been different. Migratory data for juvenile steelhead from other lines of evidence are scarce or lacking. Their migration must be based primarily on comparisons between their time-space distribution at sea and the locations of their main production areas. Based on catch distribution, steelhead apparently migrate directly offshore early in the season, and not northward along the coastal belt.

Bering Sea: In the eastern Bering Sea, juvenile sockeye migrated in substantial numbers (presumably from rivers that empty into Bristol Bay) to an area between 160° W and 165° W by July. Within this area, they remained abundant until at least mid-September. Sampling late in the season was inadequate to determine where and when the sockeye migrated farther seaward from this area of concentration.

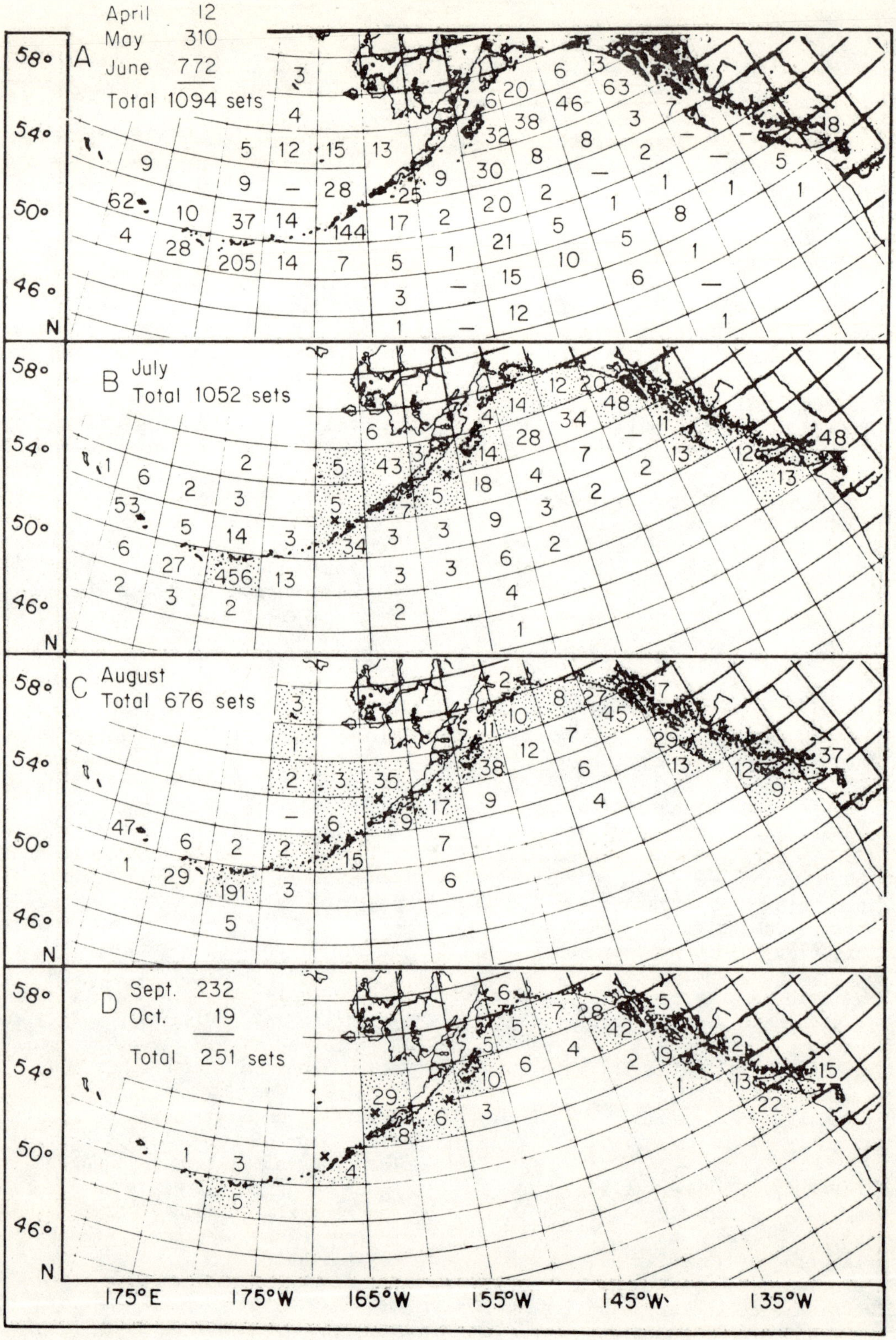

Figure 2. Distribution of purse seine fishing effort by time and areas, 1956-1970 (3073 sets). Shaded areas were fished specifically for juvenile salmon, 1964-1968.

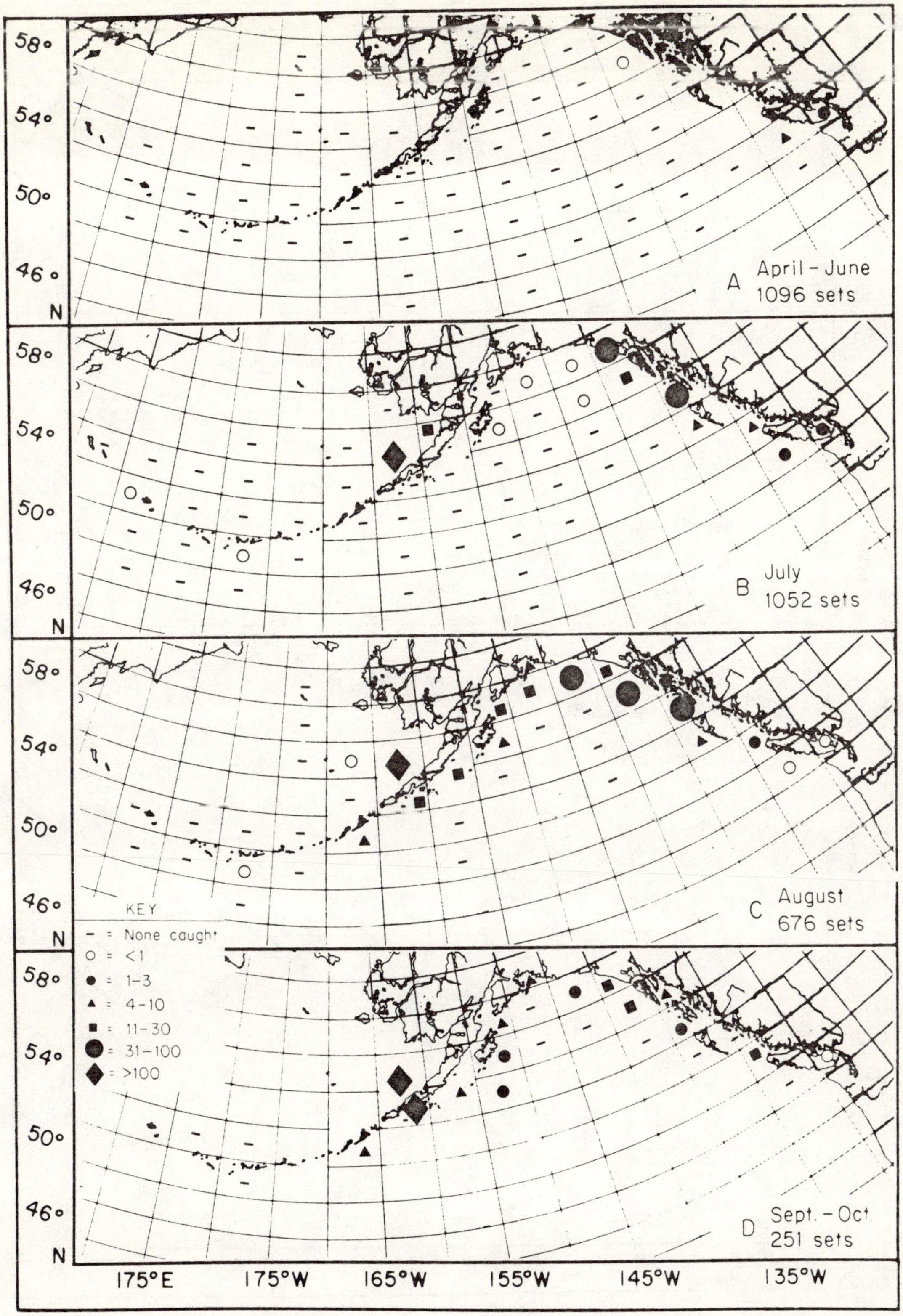

Figure 3. Catch per seine set of age .0 sockeye salmon by area and by time period; 3,073 sets, 1956-1970 (number of sets per area shown in Figure 2).

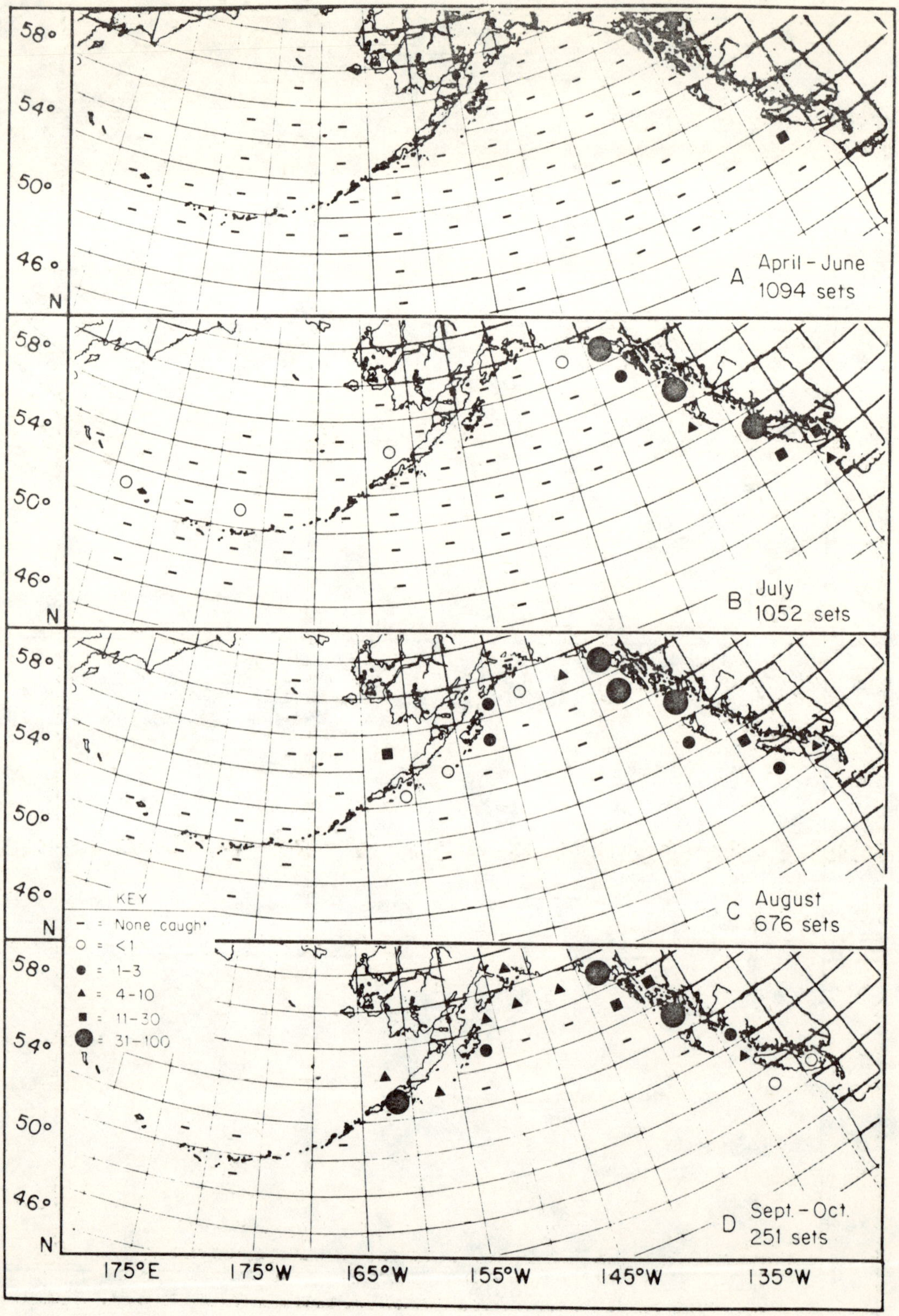

Figure 4. Catch per seine set of age .0 chum salmon by area and by time period; 3,073 sets, 1956-1970 (number of sets per area shown in Figure 2).

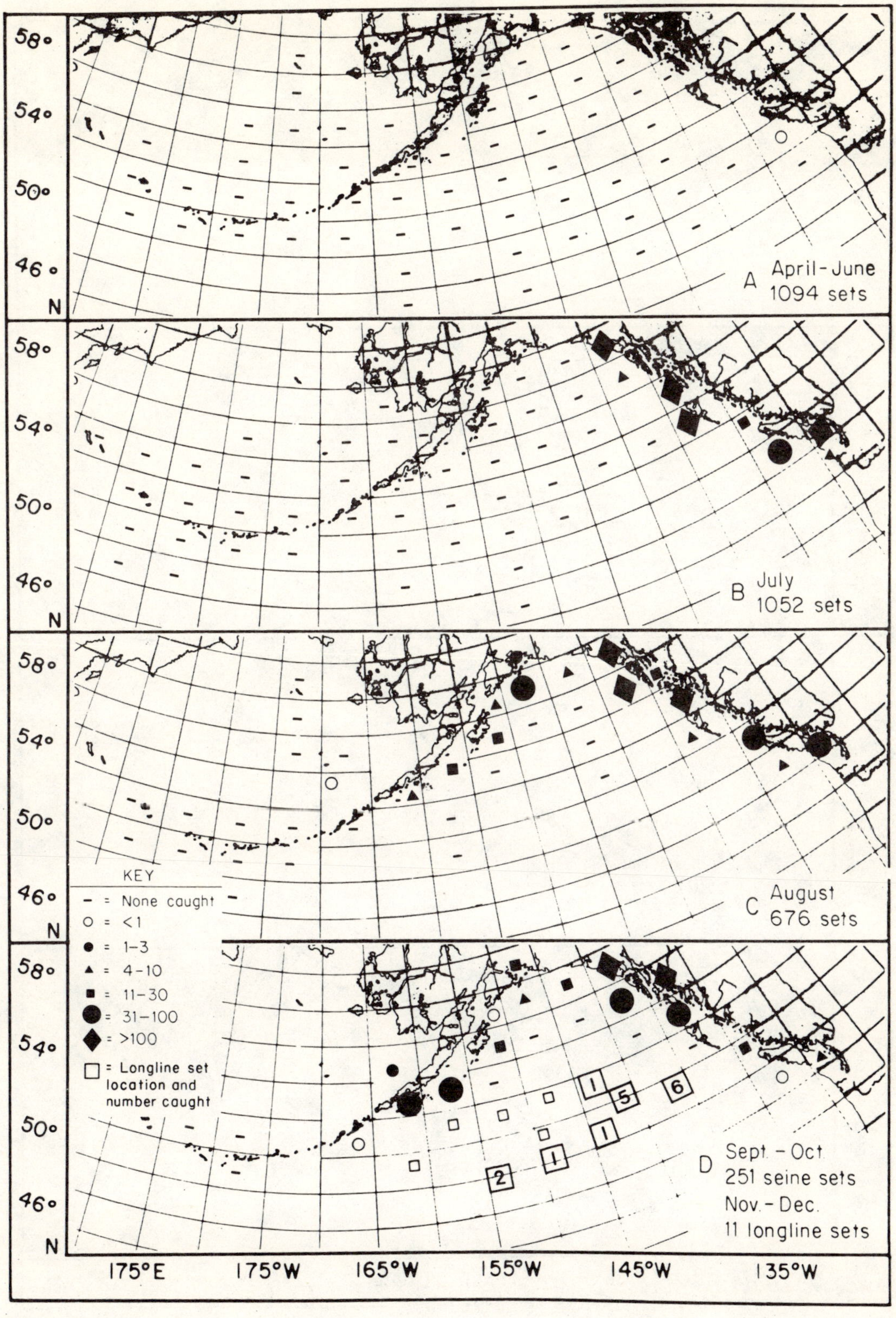

Figure 5. Catch per seine set of age .0 pink salmon be area and by time period; 3,073 sets, 1956-1970. Also catches during Canadian offshore longline operations at 11 stations 21 November-7 December 1966. Source: Fish. Res. Bd. Can. (1969).

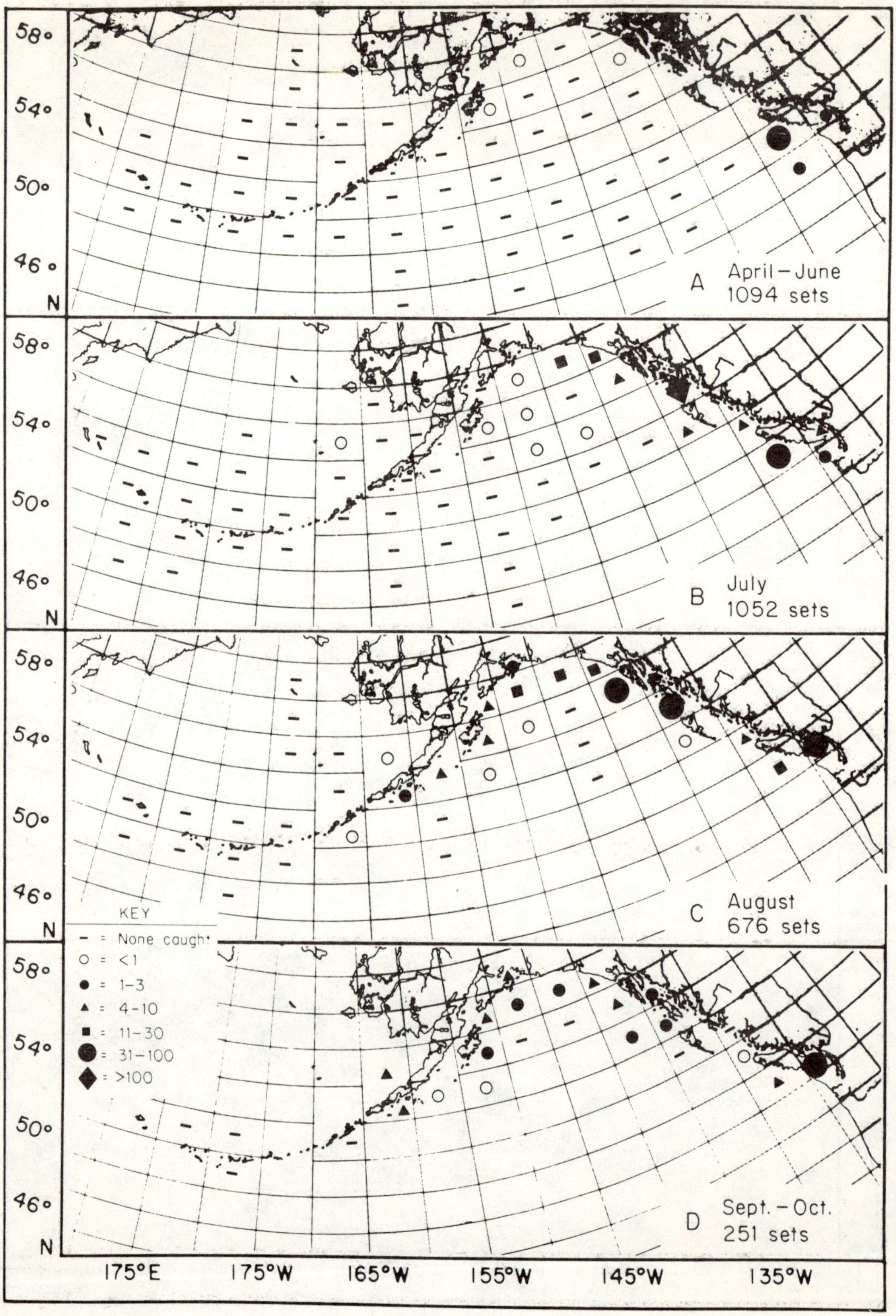

Figure 6. Catch per seine set of age .0 coho salmon by area and by time period; 3,073 sets, 1956-1970 (number of sets per area shown in Figure 2).

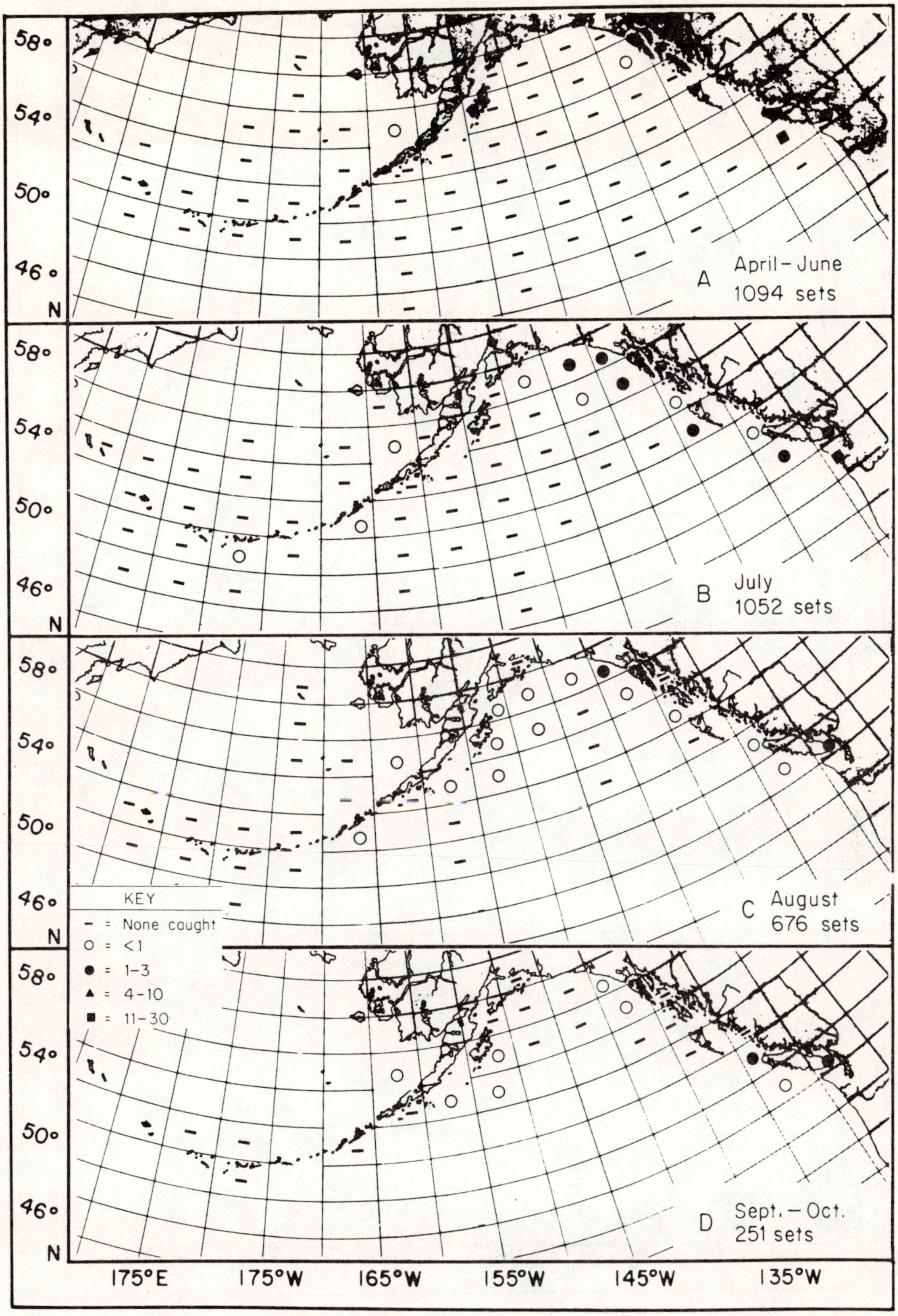

Figure 7. Catch per seine set of age .0 chinook salmon by area and by time period; 3,073 sets, 1956-1970 (number of sets per area shown in Figure 2).

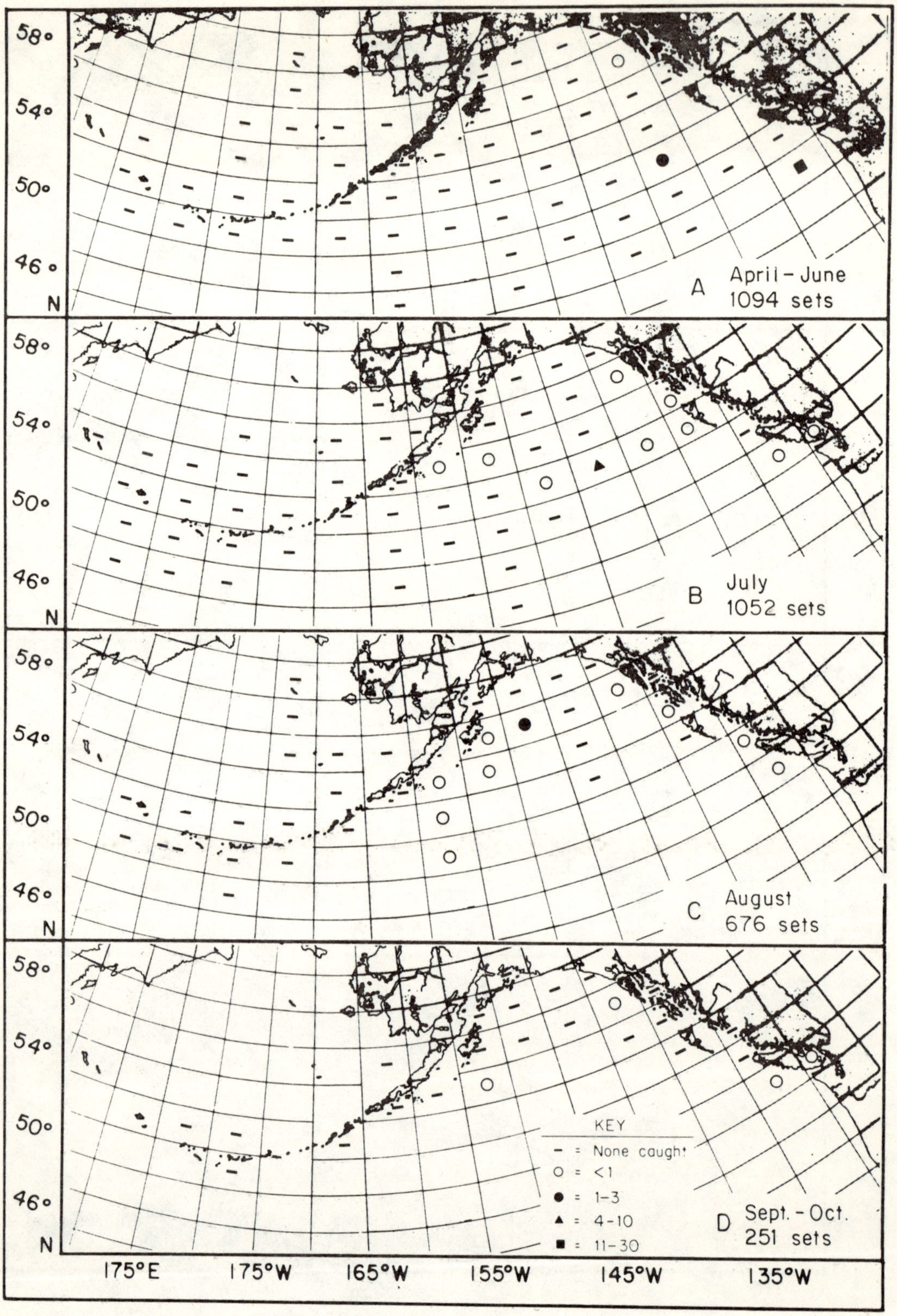

Figure 8. Catch per seine set of age .0 steelhead trout by area and by time period; 3,073 sets, 1956-1970 (number of sets per area shown in Figure 2).

Table 1. Catch Per Set of Age .0 Salmonids, Distance Offshore, and Surface Temperatures in Area W4056 off Baranof and Chichagof Islands, Southeastern Alaska, During August and September 1964 and 1965. (Includes All 29 Sets Open South or Southeast and Excludes Four Sets Open Northwest.)

Set No.	Date	Location N. Lat.	Location W. Long.	Distance offshore (naut. miles)	Catch of age .0 salmonids	Surface Temp. °C
X-103	8/28/65	57°16'	135°59'	4	975	13.3
C-96	8/20/64	56°30'	135°13'	5	49	12.2
C-102	8/30/64	56°47'	135°37'	5	978	12.7
X-68	8/5/65	56°57'	136°00'	5	369	13.8
X-106	9/3/65	56°31'	135°12'	5	262	13.5
C-87	8/10/64	56°17'	135°03'	6	74	11.3
X-100	8/26/65	57°49'	136°38'	6	1,244	13.0
X-118	9/14/65	57°50'	136°41'	6	204	13.0
X-102	8/27/65	57°29'	136°14'	7	452	12.6
X-104	9/1/65	56°49'	135°47'	7	107	13.1
X-111	9/10/65	57°28'	136°12'	7	34	11.5
C-103	8/31/64	56°39'	135°30'	8	1,237	12.6
C-107	9/4/64	56°39'	135°30'	8	102	11.9
C-108	9/5/64	57°24'	136°09'	8	666	12.5
X-101	8/27/65	57°43'	136°37'	8	96	13.4
X-66	8/2/65	56°16'	135°08'	9	91	10.6
X-107	9/3/65	56°29'	135°22'	11	54	12.4
C-104	9/2/64	56°36'	135°37'	12	591	12.4
C-105	9/2/64	56°35'	135°36'	12	453	12.7
C-94	8/17/64	57°20'	136°15'	13	280	11.9
C-97	8/21/64	56°30'	135°27'	13	424	12.7
X-67	8/2/65	56°16'	135°22'	15	154	10.6
C-106	9/3/64	56°34'	135°42'	16	299	12.6
X-69	8/5/65	56°57'	136°19'	16	80	16.3
X-108	9/4/65	56°26'	135°30'	16	66	11.4
X-70	8/5/65	56°57'	136°37'	23	3	16.7
C-93	8/17/64	57°23'	136°46'	25	2	12.6
C-100	8/22/64	56°22'	135°46'	25	0	13.2
X-109	9/4/65	56°19'	135°43'	27	0	12.7

Summary-- 4-9 miles 16 sets mean catch = 433.8
11-16 miles 9 sets mean catch = 266.8
23-27 miles 4 sets mean catch = 1.3

Chum salmon in the Bering Sea apparently migrated as did the sockeye, but were much scarcer in our catches than sockeye. This supports the presumption that both sockeye and chum in the Bering Sea were of Bristol Bay origin where such a density relationship could be expected based upon relative production of the two species.

Catches of juvenile pink salmon in the Bering Sea were too small to interpret migrations, except that their small numbers suggested that they were of Bristol Bay origin. Relatively small runs occur there in even-numbered years.

Strait of Juan de Fuca: Our sampling in the Strait of Juan de Fuca during 1968 provided important information on the timing and migration of juvenile salmon entering the sea through the strait. In 1968, we sampled at three stations on a transect across the strait (Fig. 9). Our objectives were to locate areas where juvenile salmon were concentrated and to test for any north-south stratification of juveniles as they migrated seaward. As shown in Table 2, we made 37 sets from 3-17 July and 13 sets from 11-23 August. The distribution of fishing effort at the three stations during the two time periods was inadequate

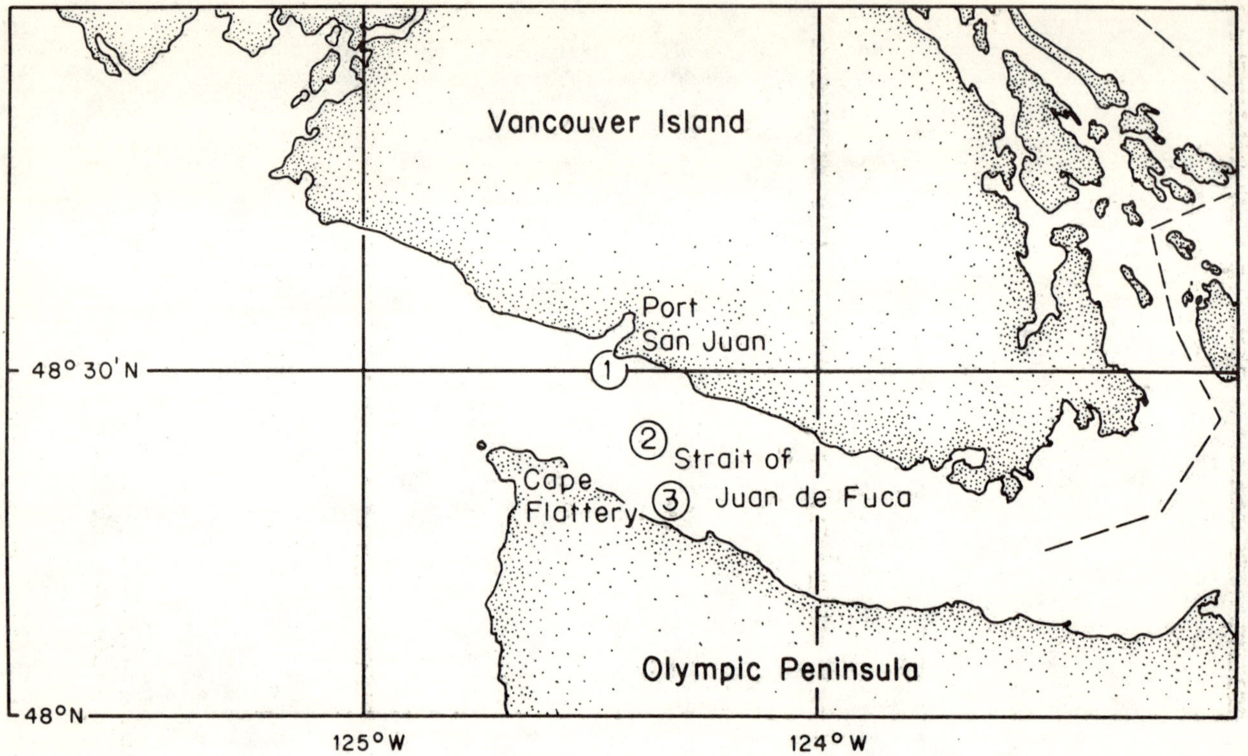

Figure 9. Locations of the three fishing stations in the Strait of Juan de Fuca referred to in Table 2.

for drawing firm conclusions, but the catches consistently indicated that juvenile chum, pink and coho were most abundant at Station 3 on the south shore of the strait. Sockeye catches were too small to be indicative, but the limited catches suggested that juvenile sockeye, in contrast with the other three species, were most abundant at Station 1 near the north shore. Thus, the data suggested that juvenile chum, pink and coho apparently tend to favor the south shore of the strait, and sockeye the north shore. More extensive sampling earlier in the season and at other transects is needed to test the validity and the general applicability of these conclusions.

The catch per set of juvenile sockeye, chum and pink salmon decreased between the July and August sampling periods. The catch per set of coho increased. This indicated that the migration of juvenile sockeye, chum and pink salmon peaked prior to August. A comparison of CPUE values suggested that the juvenile sockeye migrated sometime prior to the 3-17 July sampling period. Sockeye apparently preceded chum and pink salmon in their seaward migration from this location.

The substantial increase in the coho catch during August indicated that this species migrates into the outer strait later than the other species. However, as will be shown in the discussion of tag returns, not all coho salmon migrate seaward as do sockeye, chum and pinks. Coho migrated both eastward and westward after tagging in the strait, with some migrating eastward as far as Puget Sound (Hartt and Dell *in press*).

Migrations Inferred from Direction of Seine Sets: The direction that salmon were moving at the point of sampling could be determined by comparing the numbers caught according to the direction of the seine opening. Hartt and Dell *(in press)* and Hartt (1975) discuss the assumptions and details in doing this.

The information inferred from this technique proved to be a valuable supplement to migration data derived from other lines of evidence. In previous research on mature and immature salmon, it was found that migration direction in some areas was strongly undirectional. In other areas, however, migrations were random (Hartt 1962*a*, 1966, 1975; Royce et al. 1968). The most consistent evidence of directional movement was seen in our sampling along the coastal belt from Vancouver Island to Yakutat where paired sets were frequently

Table 2. Catch Per Purse Seine Set of Age .0 Salmon at Three Fishing Stations on a Transect Across the Outer Strait of Juan de Fuca During July and August 1968 (See Fig. 9)

Period	Station	Number of sets	Catch/set by species Sockeye	Chum	Pink	Coho	Total
July 3-17	1	6	3.2	2.5	8.8	2.8	17.3
	2	3	0.6	7.0	32.7	3.0	43.3
	3	28	1.0	16.1	162.6	6.2	185.9
August 11-23	1	4	0.5	2.0	33.3	13.8	49.6
	2	0	-	-	-	-	-
	3	9	0.0	3.0	54.4	80.1	137.5

Table 3. Numbers of Juvenile and Maturing Salmonids Caught in 19 Pairs of Purse Seine Sets in Which the Net Was Held Open in Opposite Directions. (Consecutive Sets Made on the Same Dates and at the Same Locations Along the Outer Coastal Belt from Cape Flattery to Yakutat 1964-1968).

Area	Date	Set numbers	Juvenile salmonids: Net open SE	Juvenile salmonids: Net open NW or N[1]	Maturing salmonids: Net open SE	Maturing salmonids: Net open NW or N[1]
W3048	7/14/66	H-46, 47	57	25	43	33
W3048	7/26/66	H-57, 58	1,213	0	0	5
W3050	7/26/65	X-54[1], 55	75	0[1]	9	20[1]
W3050	7/27/66	H-59, 60	241	0	7	74
W3050	7/6/67	X-8, 9	387	4	22	9
W4056	8/20/64	C-95, 96	49	3	4	0
W4056	8/21/64	C-97, 98	424	0	19	129
W4056	9/3/65	X-105, 106	262	5	4	2
W4056	8/5/66	H-69, 70	110	0	3	17
W4056	8/11/66	H-72, 73	426	1	33	34
W4056	8/18/66	H-77, 78	116	1	6	56
W4056	8/21/66	H-81, 82	258	14	22	27
W4056	8/29/66	H-92, 93	119	167	1	8
W4056	9/9/66	H-94, 95	31	0	0	0
W4056	7/26/67	X-27, 28	231	92	0	57
W4056	9/1/67	X-61, 62	23	95	0	14
W4058	8/15/64	C-89, 90	14	4	7	67
W4058	8/16/64	C-91, 92	1,216	9	14	5
W4058	9/10/66	H-97, 98	36	688	0	24
Total catch 19 sets			5,288	1,108	194	581
Average catch/set			278	58	10	31

[1]Set X 54 (area W3050) open N; all others open NW.

made with purse seines opened in opposite directions.

Table 3 summarizes the catches in 19 pairs of purse seine sets. The sets were made to determine the directional movements of fish and each pair was made consecutively on the same day and at the same location. The table shows the catch data separately for juvenile and maturing salmonids according to the direction of the net opening. Species were combined since the general trend prevailed for all.

In 19 sets where the net was open to the southwest, the catch was 5,288 juvenile salmon (mean catch 278). In the 19 paired sets with the net open to the north or northwest, the catch was 1,108 (mean catch 58). The northwestward trend of migration was dominant and probably even greater than the combined data indicated. This is because the few substantial catches in sets open to the northwest likely resulted from a temporary reverse direction of migration due to tidal or other influences. Hartt (1966, 1975) discusses factors that can influence catches. If in fact there had been a strong and continuous southeast migration, it should have been noted in areas farther south, or in tag returns. However, this was not the case. No zero catches were made with the nets open to the southeast. Several zero catches were made with the nets open to the northwest, and these zero catches frequently corresponded to large catches when the net was open to the southeast (Table 3). This evidence supports the positive nature of the northwest migration from Cape Flattery to Yakutat.

The main migration of maturing salmon was opposite to that of juvenile salmon. The catch per set of maturing fish was 31 when the net was opened to the northwest and only 10 when the net was opened to the southeast. These results alleviate any concern that the seine's efficiency was affected by the direction it was opened.

In the eastern Bering Sea, the catches of juvenile salmon with respect to the direction the seine was opened indicated a variable direction of migration (Table 4). This sharply contrasted the situation in the Gulf of Alaska. In eight pairs of sets, substantial catches were made with the net open either to the northeast or southwest. No zero catches occurred. The average catch was 204 in sets open to the northeast, and 128 in sets to the southwest. Fish from Bristol Bay estuaries undoubtedly exhibit general movement to the southwest (Straty 1974), but the migration pattern is apparently not as directed or positive as in the Gulf of Alaska. Differences in oceanographic conditions could account for different behavior of salmon in the gulf and Bering Sea. The broad, flat continental shelf in the eastern Bering Sea, with depths less than 30 fathoms, contrasts with the steep and narrow shelf of the eastern Gulf of Alaska and south of the Alaska Peninsula. In the Bering Sea, the shelf extends 50-100 miles offshore. Strong and variable tidal currents result (Hebard 1959). The dominant oceanographic feature in this part of the sea is the counterclockwise gyre that flows northeastward along the north side of the Alaska Peninsula. The gyre extends to the north and northwest along the northern coast of Bristol Bay. Juvenile salmon must migrate seaward against currents rather than with currents as in the Gulf of Alaska.

Feeding conditions may also affect the pattern and speed of seaward migration. Dense schools of euphausiids and larval fish were seen much more frequently in the eastern Bering Sea than in other areas (Hartt et al. 1970). These rich feeding conditions may explain the lingering behavior of juvenile salmon there.

Migrations Determined from Tag Returns: Tag returns provided the most definitive information on juvenile salmon migrations.

Table 4. Catches of Juvenile Salmonids According to Direction of Set of the Purse Seine Using Paired Sets on the Same Dates and at the Same Locations in the Eastern Bering Sea (Area W6556) 1966, 1968.

Date	Set Nos.	Net open NE	Net open SW
7/7/66	C-53, 54	835	62
7/28/66	C-91, 92	154	550
8/9/66	C-104, 105	91	286
8/11/66	C-108, 109	6	27
8/18/68	C-69, 70	496	17
8/25/68	C-77, 78	7	20
9/8/68	C-104, 105	9	42
9/12/68	C-107, 108	31	10
Total catch 8 sets		1,629	1,014
Average catch/set		204	127

The returns also corroborated the migrations inferred from seasonal catch data and from catches with respect to the direction that the seine was opened. Although the number of tag returns was limited, it was sufficient to identify the major stocks present in areas where juvenile salmon were abundant, and show that migration can be rapid and extensive during the first summer in the ocean. Tags were returned in the same year fish were tagged, and up to four years later, depending on the species and age at maturity.

Sockeye Salmon: Figure 10 diagrams release and recovery locations for the 41 returns of 10,411 tagged sockeye. Release location is symbolized with a circle. Arrows are drawn to the recovery location. The actual migrations during the two to three years between release and recovery are not known, although considerable information is known about the annual migrations of major sockeye salmon stocks during their total oceanic sojourn (French et al. 1976, Royce et al. 1968). The key information the diagram conveys is that the juvenile salmon must have migrated from their home estuary to the point of tagging soon after entering the sea. Thus, for those fish recovered in terminal fishing areas, the migration arrows should be viewed in reverse (i.e. the juvenile fish had migrated from the arrowhead to the circle during the interval between leaving their home estuaries and their capture at sea for tagging).

The tag returns illustrated in Figure 10 show that juvenile sockeye salmon tagged in the eastern Bering Sea originated from Bristol Bay, as would be expected. The recovery from the Japanese high seas fleet far to the west demonstrates the extensive ocean migrations of Bristol Bay stocks. The one tagged fish that was recovered south of the Alaksa Peninsula showed the typical east-to-west migration route of some maturing Bristol Bay sockeye as they return from the eastern North Pacific (Hartt 1966).

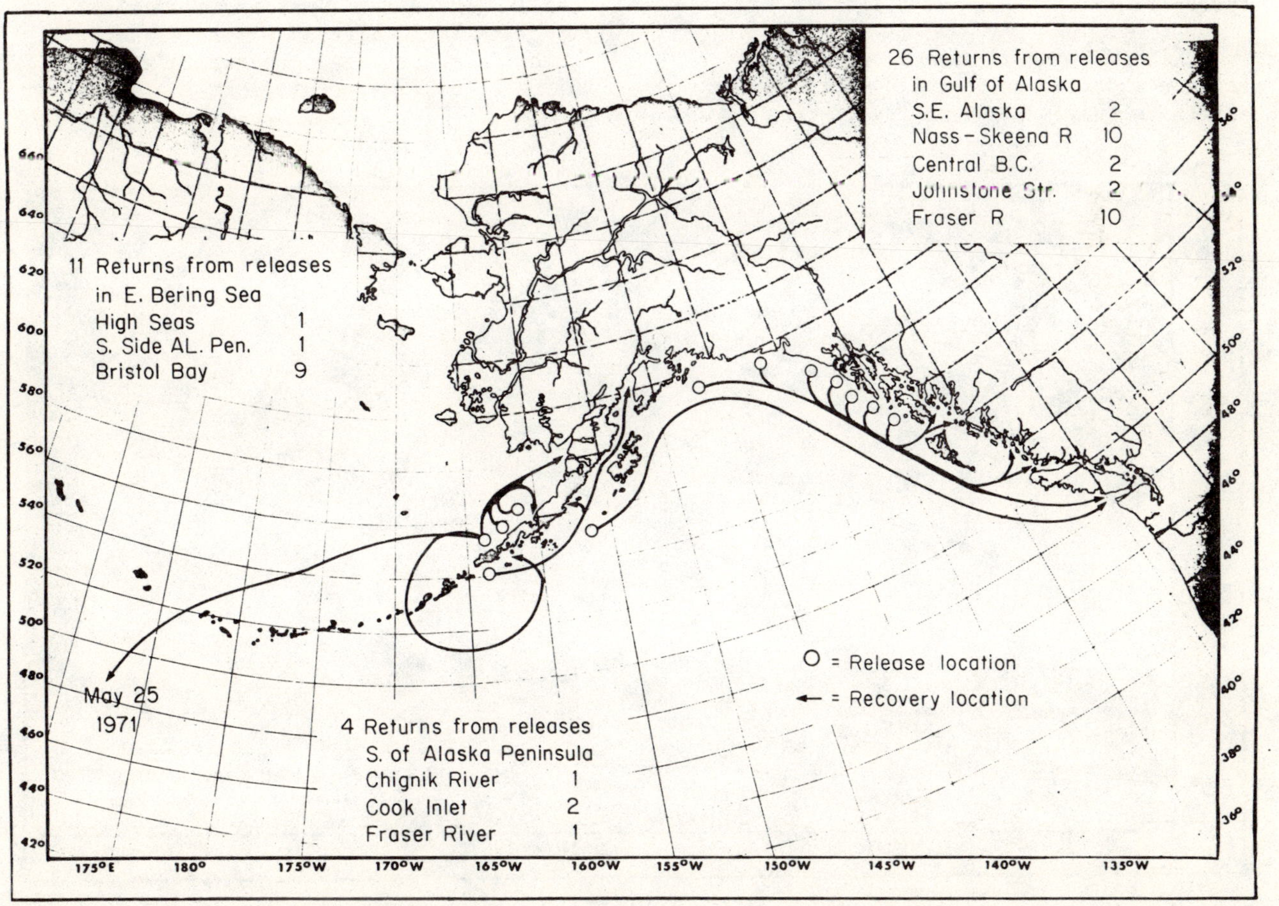

Figure 10. Release and recovery locations for 41 sockeye salmon tagged as juvenile fish and recovered 2 or 3 years later.

The tag returns from fish released in the Gulf of Alaska showed that juvenile sockeye during their first summer migrated along the coast to the north, west and southwest. One Fraser River fish, tagged southwest of Kodiak Island, had travelled about 1,500 nm by 30 August of its first summer. Another Fraser River sockeye, tagged near the coast south of Prince William Sound on 9 August, had migrated about 1,100 nm by that date.

Chum Salmon: Figure 11 diagrams the release and recovery locations for the six chum salmon returns from among the 4,412 that were tagged and released. Dates of release and recovery are included in the figure.

Chum salmon migrate in a pattern that resembles that of sockeye. The recovery locations were generally south of the release locations. Juvenile chum migrated north along the coast prior to being captured and tagged. The one fish tagged in Shelikov Strait had apparently moved west or southwest of its point of origin in a manner similar to that of the sockeye tagged in this area (Fig. 10).

Pink Salmon: Figure 12 shows the release and recovery locations for 55 pink salmon recovered a year after tagging from 13,060 that were released. The pattern in both odd and even-numbered years showed that juvenile pink salmon had migrated in a manner that resembles the sockeye and chum salmon migrations. Two pinks from the Fraser River and Puget Sound areas had migrated northward to about 58° N (960 nm) by 29 July and 12 August, respectively.

Coho Salmon: Tagging coho salmon gave more information than tagging other species because there were more returns. A total of 245 tagged fish were returned from the 7,015 released in all years. The overall return rate was 3.5 percent, and the annual rate ranged from 1.3 to 5.8 percent. Sixty-four were recovered in the year of release (0.9 percent) and 181 in the year after (2.6 percent).

Some obvious reasons explain the relatively high return rate of tagged coho. First, juvenile coho are larger than juvenile sockeye, chum or pinks. Presumably the coho retain their tags better, and suffer lower tagging and natural mortalities. Second, tagged coho can be recovered in the year of tagging as well as a year later. Finally, intensive sport and commercial fisheries in both fresh and saltwater commonly recover tagged coho. A substantial number of coho were tagged in the Strait of Juan de Fuca, a popular fishing area.

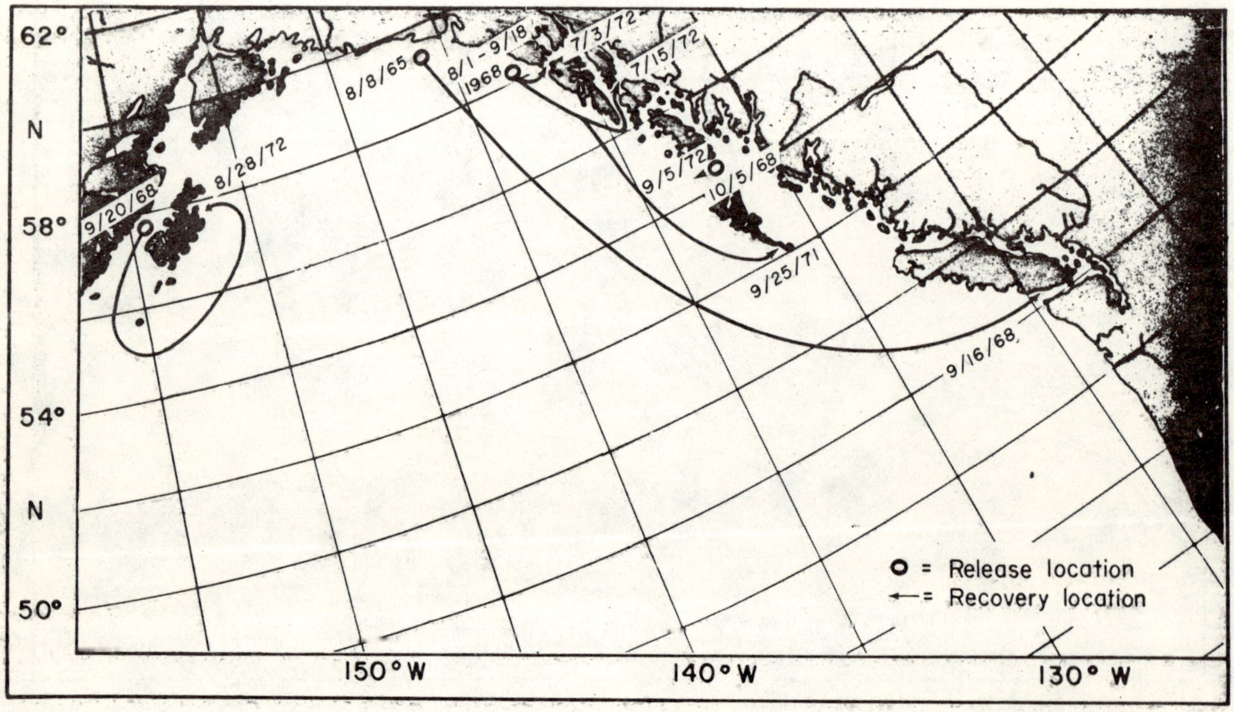

Figure 11. Release and recovery locations for 6 chum salmon tagged as juvenile fish and recovered 3 or 4 years later.

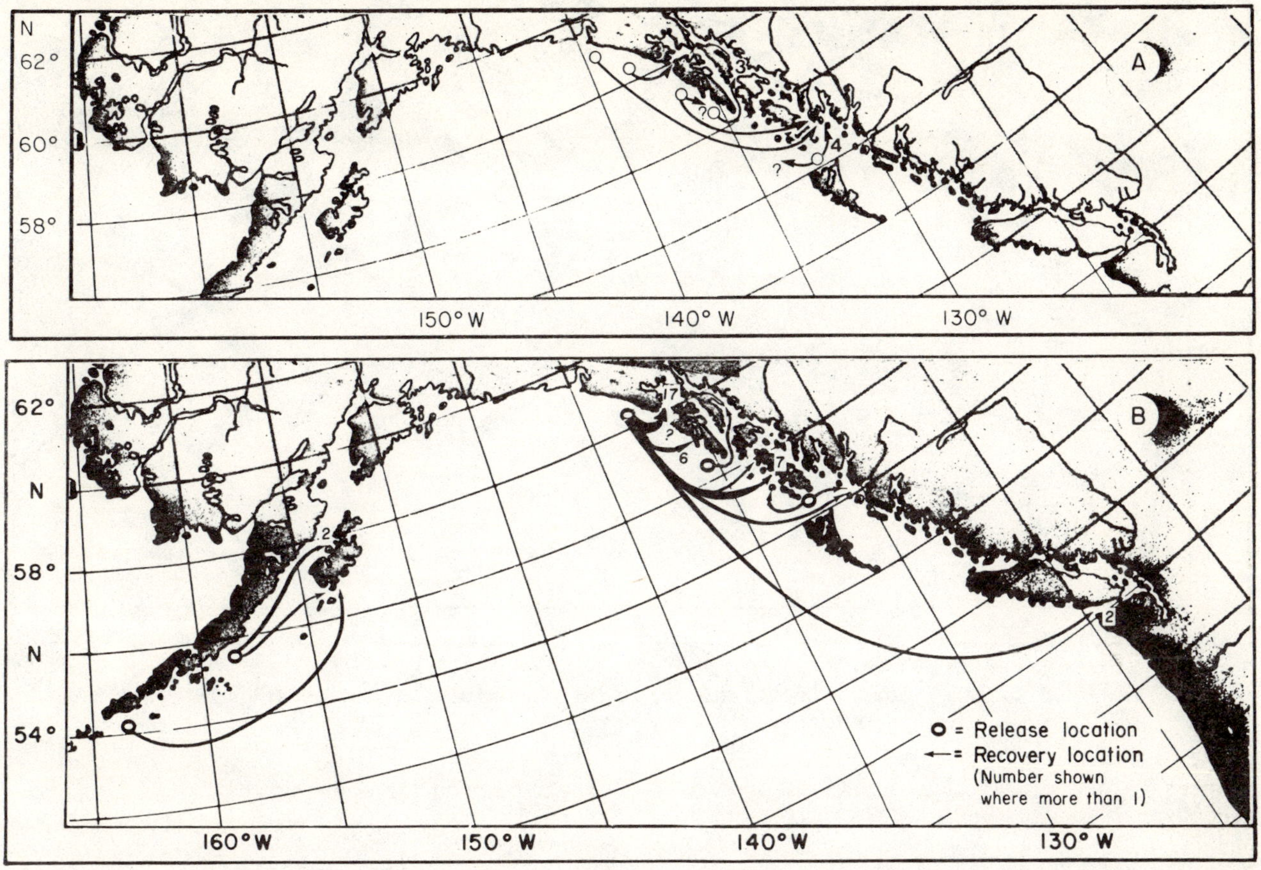

Figure 12. Release and recovery locations for 55 pink salmon tagged as juvenile fish and recovered one year later. A - 14 fish released in odd-numbered years; B - 41 fish released in even-numbered years.

The tag return efficiency there was probably high.

Figure 13 shows the release and recovery locations for 120 coho that were tagged along the coast from Cape Flattery to the Shumagin Islands. Coho tagged in the Strait of Juan de Fuca are not shown. As in the case of sockeye, chum and pink salmon, coho move rapidly and extensively along the coast to the north, west and southwest during their first summer.

Table 5 summarizes returns of coho that were tagged along the coastal belt from Cape Flattery to Kodiak Island according to major release and recovery areas. This provided a means to estimate the proportions of the major stocks tagged. Subtotals are provided for release areas between Kodiak Island and 56° N, and for areas between 56° N and Cape Flattery. More tag returns are shown in Table 5 (130) than in Figure 14 (12) because returns are grouped only by general recovery area in the table. In Figure 14, only those returns were shown for which the recovery location was specific.

For the present gross comparison, the possible biases in tag return rates between areas will be ignored. Bias could be caused by time and location of tagging, by non-uniform return efficiency between recovery areas, and by higher tag return rates for larger fish. The latter factor would favor returns from southern production areas because salmon there were larger at tagging.

Considering first the fish released in the northernmost part of the Gulf of Alaska (Kodiak Island to 56° N), there were returns one year later from all five production areas--Alaska, British Columbia, Washington, the Columbia River and Oregon. Percentages of returns ranged from 31.4 percent (British Columbia) to 10.0 percent (Washington). The most distant release areas (W4558 and westward, Fig. 1) yielded no returns from Washington or the Columbia River. This may be significant at least in the case of Washington as overall returns from there were low. This agrees with the earlier observation that Puget Sound coho were practically absent among

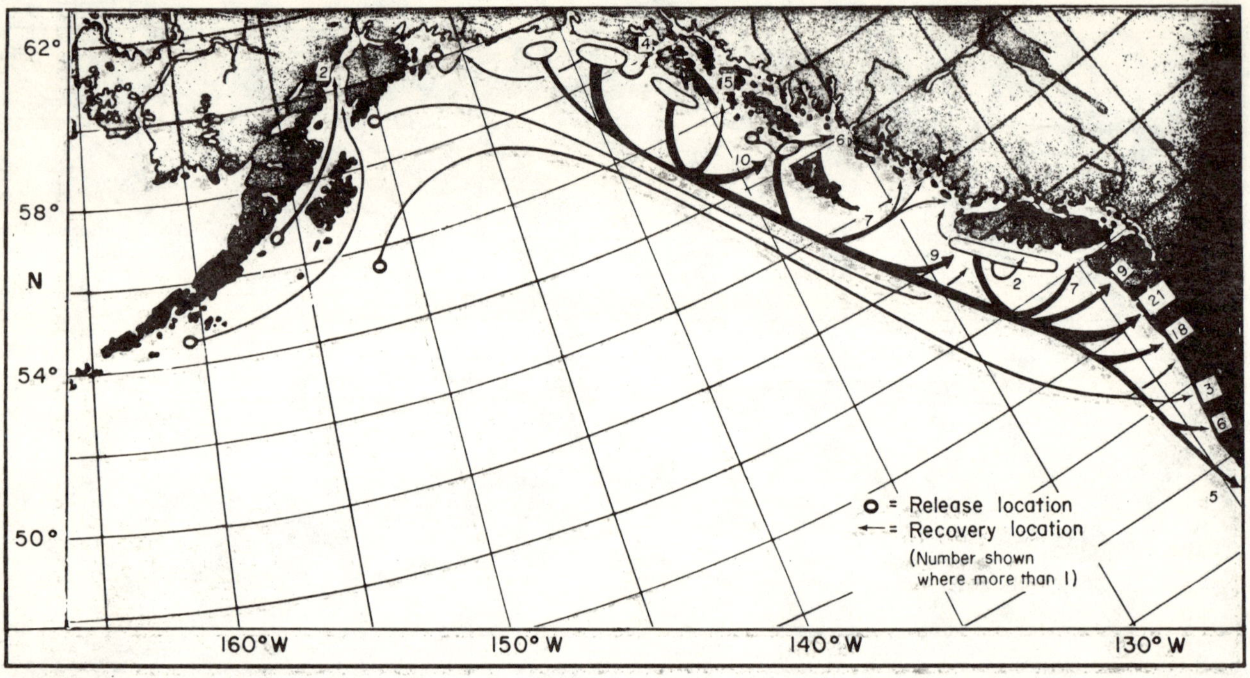

Figure 13. Release and recovery locations for 120 coho salmon tagged as juvenile fish and recovered one year later.

samples tagged along the coast from Vancouver Island to Kodiak Island (Hartt and Dell *in press*). The main point to be made is that Oregon and California coho returned in relatively high proportions (20 percent) from these distant release areas. At face value, the proportion of fish from the Columbia River and Oregon-California would be 37 percent. Some of these fish may have been destined for production areas other than those in which they were recovered, but it is probably safe to assume that they were destined for areas between the outer Washington coast and California. Essentially none were destined for Puget Sound.

Juvenile coho tagged in coastal areas between 56° N and Cape Flattery (Table 5) yielded a similar array of returns, except for a lack of Alaska returns. This lack would be expected based on the tagging locations and the fact that the fish move north in their early oceanic migration. As in the more northerly releases, the largest return was from British Columbia (35 percent), followed by a 31.7 percent return from Oregon-California. The total recovery rate from the Columbia River and south was 46.7 percent.

The data in Table 5 indicated that juvenile coho from the Columbia River, and Oregon-California form a large portion of the coho stocks that migrate north along the coastal belt each summer. Although the data did not indicate what portion of these southern stocks make the northerly migration, a large proportion probably do. Juvenile coho were consistently caught in substantial numbers in most seine sets throughout the study area (Hartt and Dell *in press*).

Chinook Salmon and Steelhead Trout: Figure 14 illustrates the release and recovery locations for 12 chinook salmon and one steelhead trout of the 276 juvenile chinook and 85 juvenile steelhead that were tagged.

Three of the 29 juvenile chinook salmon that were released in area W4058 in the northeastern Gulf of Alaska (Fig. 1) were recovered in the Columbia River. The fish were recovered one, two and three years after release (Hartt and Dell *in press*). The fish were released on 6 August 1965 and 13 August 1966. Thus, juvenile chinook salmon from the Columbia River can migrate 1,000 nm by August of their first summer at sea. All three returns must have been from spring-run stocks since recovery dates in the river ranged from 17 March to 25 May. In addition, a juvenile chinook tagged in the northern part of Hecate Strait on 20 July 1968 was recovered in the Columbia River on 26 April 1970. This specimen was also a spring-run fish.

Not enough chinook were returned to draw firm conclusions, but the four tag returns

*Table 5. Return of Juvenile Coho Salmon Tagged in Northeastern Gulf of Alaska According to Major Release and Recovery Areas.**

Areas of release	Number of returns by area of recovery: Alaska	British Columbia	Wash. State	Columbia River	Oregon-Calif.	Total
Kodiak Island** to 56°N.						
W4558***	3	3	0	0	3	9
W4058	8	4	2	4	3	21
W4056	4	15	5	8	8	40
Subtotal	15	22	7	12	14	70
Percent	21.4	31.4	10.0	17.1	20.0	100.0
56°N. to Cape Flattery						
W3554	0	15	7	6	11	39
W3050	0	4	1	0	0	5
W3048	0	2	3	3	8	16
Subtotal	0	21	11	9	19	60
Percent	0	35.0	18.3	15.0	31.7	100.0
Totals	15	43	18	21	33	130
Percent	11.5	33.1	13.8	16.2	25.4	100.0

*Excluding one fish for which recovery location was unknown. Strait of Juan de Fuca recoveries arbitrarily placed in Washington.

**Refer to Figure 1.

***Also including one return each from areas W5060, 5554 and 5558.

suggested that spring-run chinook stocks from the Columbia River migrate more extensively than fall-run stocks.

Another eight chinook salmon were recovered that had been tagged in the outer Strait of Juan de Fuca in 1968 (Fig. 14). Three were recovered in the year of release, two in 1969, two in 1970 and one in 1971. One of the three 1968 returns was recovered near the point of release after about a month. The other two were captured after migrating extensively eastward to the San Juan Islands and the Skokomish River in Hood Canal, respectively. The latter fish must have been a maturing jack. The two recoveries in 1969 were captured in the Strait of Juan de Fuca and off west-central Vancouver Island. One of the two fish captured in 1970 was caught in Puget Sound. The other was recovered in the Columbia River on 23 September, indicating that it was a fall-run fish. Finally, the specimen recovered three years after tagging was caught in the Strait of Juan de Fuca near where it was released.

Although we do not know the actual migrations of these eight juvenile chinook, results suggest that they do not migrate extensively. Some of them had obviously migrated east toward inside waters during their first summer or fall rather than migrating seaward.

Based on these few chinook tag returns, it appears that this species, as with coho from Puget Sound, do not migrate extensively to the north. However, Columbia River chinook stocks, as do some of the Columbia River and coastal Washington, Oregon and California coho stocks, apparently migrate far to the north during their first summer at sea.

The single steelhead trout return was tagged southeast of Kodiak Island (Fig. 14) on 5 September 1958 and recovered on 5 February 1960 in the Alsea River, Oregon. In addition to the tag, it bore a fin clip that identified it as being released from the Alsea River Hatchery in April 1958. In its first four or five months at sea, it had travelled at least 1,200 nm at an

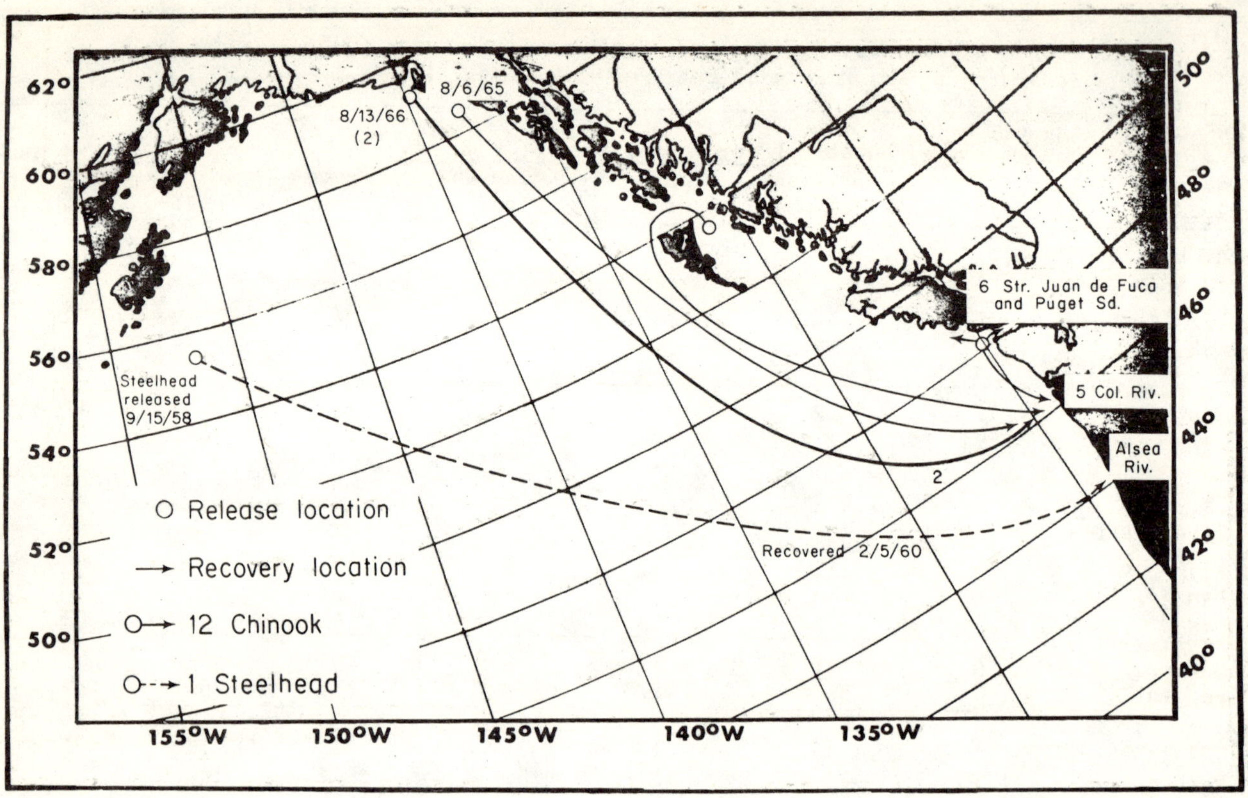

Figure 14. Release and recovery locations for 12 chinook salmon and one steelhead trout tagged as juvenile fish and recovered up to 3 years later.

average rate of 8-10 nm a day. Little more can be concluded from a single tag return except that steelhead trout can migrate as extensively as salmon during their first summer at sea.

Migration Models

The data from all three lines of evidence--seasonal catch distribution, direction that the seine was set and tag returns--provided a fairly coherent picture of juvenile salmon migrations during their first summer at sea. Migrations were more obvious for some species than others because of more sharply defined distribution patterns or their abundance. Comparing results between species helped interpret single species migrations.

Sockeye, Chum and Pink Salmon: Since basic migration patterns of juvenile sockeye, chum and pink salmon were similar, their migrations are diagrammed together in Figure 15. These three species, by far the most abundant of the Pacific salmonids, generally were mixed in the catches and usually similar in size.

Stippled arrows in Figure 15 illustrate the migrations of Bristol Bay and Gulf of Alaska stocks during their first summer at sea. Bristol Bay stocks migrated slowly in their first summer. By September, substantial numbers of juveniles were only 250-300 nm from their estuaries of origin. They tended to favor the north side of the Alaska Peninsula, remaining mainly within 50-60 nm of shore. The direction of migration varied. Tidal currents strongly influenced this direction, but the net movement was to the southwest. The full extent of migration in September remains unknown because not enough samples were taken farther west or northwest after August. Extended sampling in September and October is needed to determine the western limit of migration in late summer or early fall.

Dashed lines in Figure 15 indicate the presumed fall and winter migrations. The ends of the arrows depict the general distribution of age .1 sockeye of Bristol Bay origin during their second spring and summer. The locations where arrows end is based on seasonal distribution patterns and migration as determined by purse seining, longline and gillnet sampling, and tag returns (French et al. 1976). The actual timing and migration routes from the Bering Sea to the North Pacific remain unknown. We also do not know if some fish

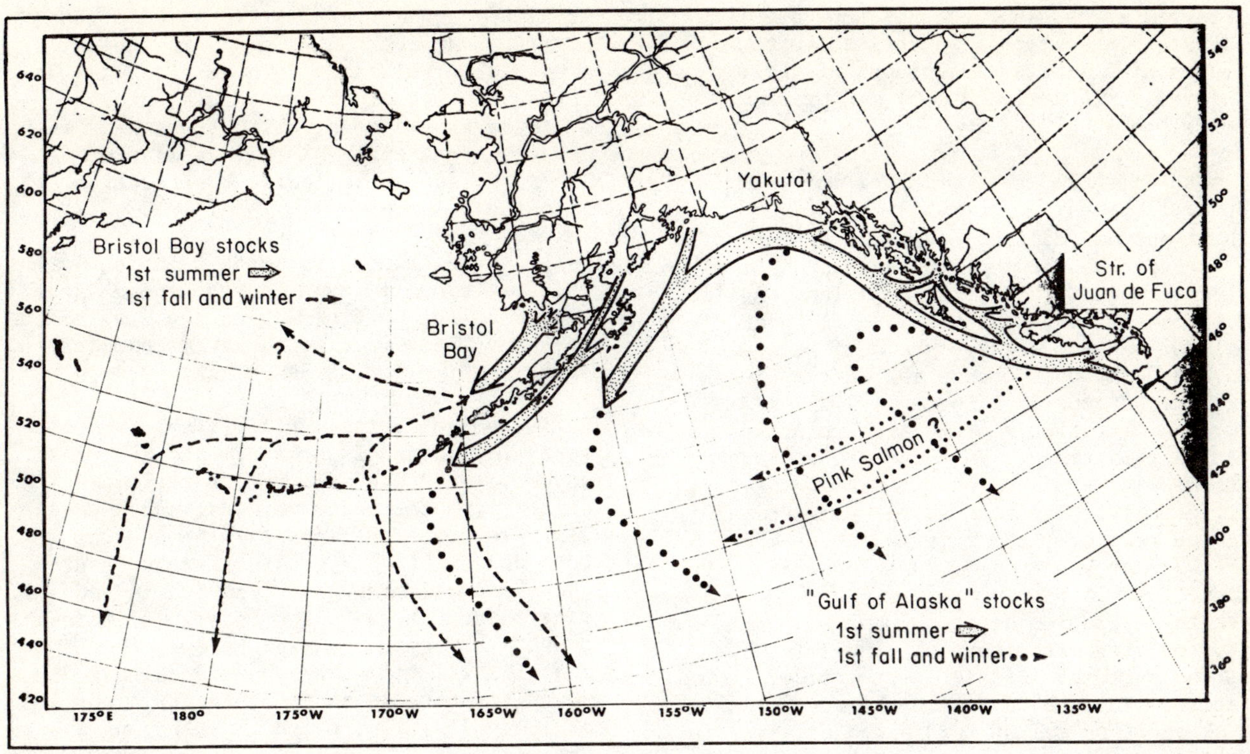

Figure 15. Diagram of oceanic migration patterns of some major stocks of North American sockeye, chum and pink salmon during their first summer at sea, plus probable migrations during their first fall and winter.

remain in the Bering Sea during winter. Gillnet sampling showed the presence of age .1 sockeye in the south-central Bering Sea in January, February and March (French and Mason 1964). The fish probably migrated from the Bering Sea to the North Pacific in an elongated band, or bands, similar to the long band observed along the Gulf of Alaska coast. The fish in the vanguard could be 1,000 nm or more ahead of fish in the rear.

Gulf of Alaska stocks apparently migrate much more rapidly and extensively than those from Bristol Bay. By August, substantial numbers of all three species from as far south as the Strait of Juan de Fuca migrated at least as far as Yakutat (900 nm). Some sockeye reached waters near Kodiak Island (1,500 nm). They also tend to remain much closer to shore than Bristol Bay stocks do, particularly off southeast Alaska where the continental shelf narrows, and the band of migrating juvenile salmon is less than 20 miles wide. In the northern Gulf of Alaska, the band of salmon tends to widen, presumably because of the wider continental shelf in the area. The extended migration past Kodiak is based on only one sockeye return and may not apply to chum and pinks.

A similar band of migrating sockeye, chum and pink salmon, moves southwestward along the south side of the Alaska Peninsula and the eastern Aleutian Islands. Tag returns showed that the band includes stocks from Cook Inlet, the Alaska Peninsula and Kodiak Island, and possibly stocks from farther east and southeast in the Gulf of Alaska. The width of the band along the south side of the peninsula and eastern Aleutian Islands has not been measured during late summer when abundance is high, and offshore distribution may be expected. Since the continental shelf is wide here, the band of salmon may be correspondingly wide. The migration's westward limit remains unknown; however, purse seine catch data indicated that Unalaska Island (167° W) was about the western limit through September. Furthermore, extensive purse seine fishing effort showed that juvenile salmon were extremely rare in the central Aleutian Island area through September (Figs. 3-5).

In view of the above, it seems that juvenile sockeye, chum and pink salmon depart from the coastal belt in the Gulf of Alaska at some point, or points, and disperse offshore. We do not know the locations and times of offshore migration, but the dotted lines extending from the stippled arrows in Figure 15 show possible routes. These routes would lead to the distributions of age .1 groups of these species that Hartt and Dell *(in press)* described.

The offshore dispersions of these three species may differ somewhat in their locations and timing because of different maturity schedules. As discussed earlier with respect to Figure 5, Canadian longline catch data indicated that juvenile pink salmon, but not sockeye or chum, were distributed far offshore in the Gulf of Alaska by late November. Accordingly, Figure 15 shows a migration route unique to pinks, at least in timing. Moreover, different stocks within a species may migrate differently. The fish in the vanguard and in the rear of the migration may proceed offshore at vastly different locations. In fact, fish in the vanguard may have made a lengthy migration along the coast and started offshore before the late migrants of the same species have left inside waters. The substantial number of juvenile pink and chum salmon, and a few sockeye, in Hecate Strait (LeBrasseur and Barner 1964) and Puget Sound (Hartt and Dell *in press*) in November showed that many fish enter the sea too late to become part of the distribution and migration summarized in Figure 15. The late migrants may not make the extensive migrations to the north and west that Figure 15 depicts. They may either proceed directly offshore from their point of embarkation, or migrate along the coast some abbreviated distance before proceeding seaward. Fall and winter sampling and tagging at several key points along the coast and in major channels leading to the sea would resolve these important aspects of juvenile salmon migrations in the ocean.

The data in Figure 15 also indicate the probable area in which juvenile salmon stocks from Bristol Bay and the Gulf of Alaska begin to overlap. This occurred near the eastern Aleutian Islands (165° W) probably in the late fall or early winter. Later in their life, stock distributions from these two areas overlap much more widely. Bristol Bay salmon, particularly sockeye, distribute eastward in the Gulf of Alaska to about 140° W and westward in the North Pacific and Bering Sea to about 165° E. Gulf of Alaska stocks, particularly sockeye, distribute west to about 175° E (French et al. 1976). Chum and pink salmon do not overlap as extensively. Chum range approximately from the central Aleutian Islands to the central Gulf of Alaska, and pink salmon from the eastern Aleutian Islands to Kodiak Island (Neave et al. 1976, Takagi et al. *in press*). Thus, the overlap that Figure 15 depicts only begins to indicate a much wider mixing. This results in widespread sharing of common feeding grounds by these two major stocks. The mixing is, however, apparently minimal until the winter or spring following their entrance into the ocean.

<u>Coho and Chinook Salmon</u>: Oceanic migration patterns of juvenile coho and chinook salmon in the Gulf of Alaska are diagrammed in Figure 16. Data were unavailable for depicting the Bering Sea stocks.

In the gulf, the basic migration pattern resembles those of sockeye, chum and pink salmon. Juvenile salmon migrate in a narrow band to the north, west and southwest along the gulf coast. The band of coho and chinook originates farther south than bands of other species, and includes stocks from Oregon and probably California coastal streams. Unlike sockeye, chum and pink salmon, some coho and chinook are found offshore in the north-central portion of the gulf (Figs. 6 and 7). Appropriate arrows in Figure 16 show this. Since the origins of these fish are unknown, question marks accompany these arrows. The extended dotted lines indicate probable southward and eastward migrations toward areas in the southern Gulf of Alaska where numerous age .1 coho and a few chinook salmon were found in spring and early summer (Hartt and Dell *in press*). I must reiterate that the oceanic migrations of juvenile salmon shows in Figure 16 apply to only a portion of the coho and chinook stocks. Many of these species remain in inside waters or in nearshore oceanic waters for the entire marine phase of their life history.

The migrations shown in Figure 16 probably include the maximum extent that Gulf of Alaska coho migrate. Tag returns from age .1 coho show a maximum westward and southward migration of about 155° W and 42° N (Godfrey et al. 1975) except for one coho tagged south of the central Aleutian Islands near 176° W. This fish was recovered near Kodiak Island.

Most juvenile chinook salmon originating in the Gulf of Alaska apparently remain within the area illustrated in Figure 16 for their entire life. However, a few migrate west of 165° W. Two chinook (age .2) tagged south of the central Aleutian Islands near 177° W were recovered in areas tributary to the Gulf of Alaska. One was recovered in southeastern Alaska and the other in the Salmon River, a Columbia River

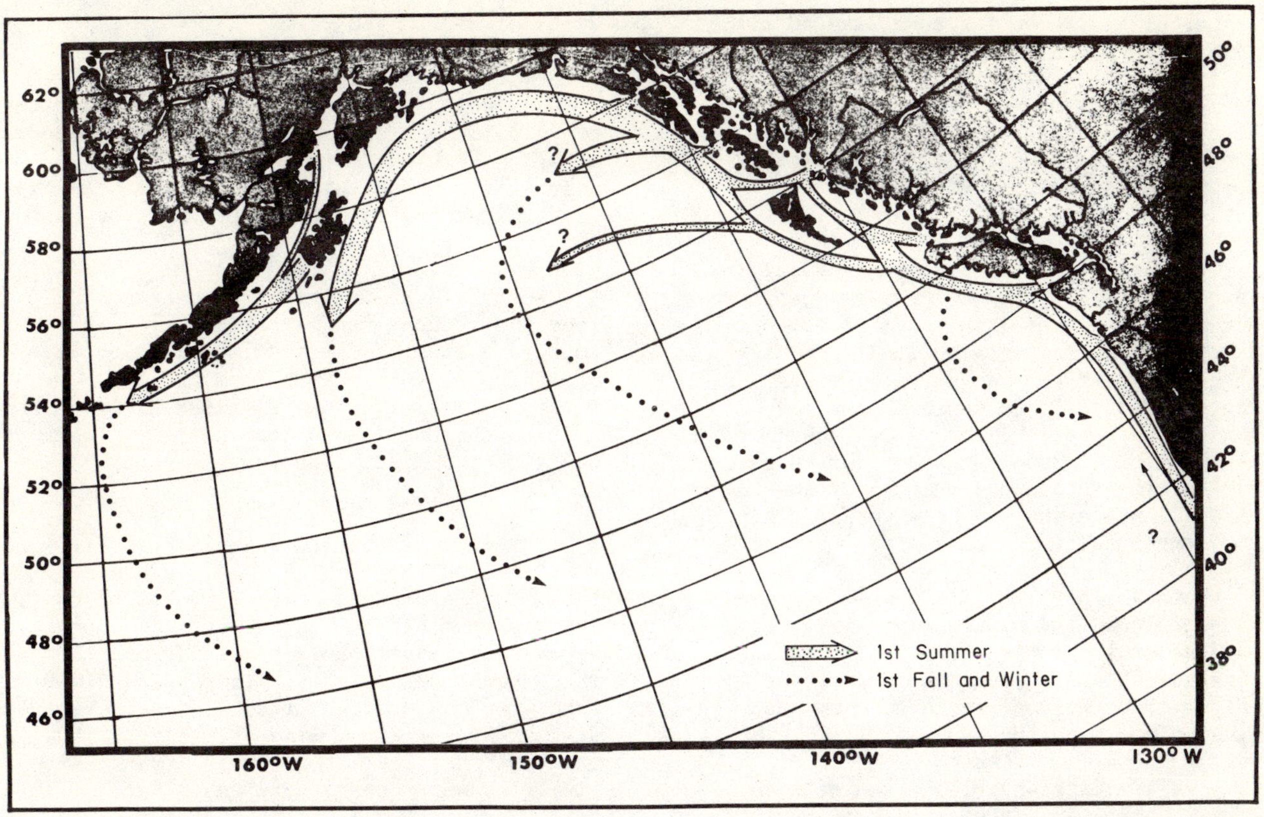

Figure 16. Diagram of oceanic migration patterns of some major stocks of coho and chinook salmon during their first summer at sea, plus probable migration during their first fall and winter.

tributary in Idaho (Hartt 1962*a*). We still do not know what portion of the long westward migration these fish made during their first summer and what portion was made during subsequent summers as age .1 or age .2 fish. The specimen recovered in the Salmon River was a spring-run fish. It was recovered on 7 July and adds to the evidence discussed with regard to Figure 14 that juvenile spring-run chinook from the Columbia River tend to migrate more extensively at sea than fall-run fish.

The oceanic distribution of coho salmon from the Gulf of Alaska overlaps considerably with Bering Sea stocks from Bristol Bay and northwest Alaska. Stocks from northwest Alaska extend east to about 140° W in the Gulf of Alaska and overlap with stocks from the gulf between 140° W and 155° W (Godfrey et al. 1975). We do not know if the overlap occurs among juvenile coho during their first summer or fall at sea, or if the overlap begins later in the winter or spring when the fish are maturing at age .1.

Overlap between chinook stocks from the Bering Sea and Gulf of Alaska at the juvenile stage, or any stage, is apparently minimal. The only two chinook salmon tagged in the North Pacific and recovered in northwestern Alaska were tagged close to the eastern and central Aleutian Islands. As mentioned earlier, tag returns indicated that a few chinook salmon of Gulf of Alaska origin migrated as far as the central Aleutian Islands where overlap could occur. No evidence exists that they overlap at the juvenile stage.

Steelhead Trout: As Hartt and Dell *(in press)* discussed, any model of the early oceanic migrations of steelhead trout would have to be based primarily on catch distribution of the age .0 and .1 groups, respectively. Few data are available from tag returns or from the direction the seine was set. Accordingly, any model at this stage of our knowledge would merely indicate an offshore migration.

Length and Growth

The lengths of salmon and steelhead caught in our purse seine provided substantial data on the sizes and growth of juvenile salmonids during their first summer at sea. The length of tagged fish when they were released, and their length at recovery, provided additional information on relative sizes and growth of a given stock according to the year of maturity and the distance migrated. The main observations pertaining to length are summarized below.

The minimum length at which any salmon species enters the ocean proper is approximately 10 cm. The mean sizes of sockeye, chum and pink salmon were usually similar at given locations and dates. Mean sizes ranged from 12-20 cm depending on the season and location. Apparently chum and pink salmon grow more rapidly during early marine life than sockeye and overcome the size advantage that sockeye possess when they enter saltwater as smolts.

The mean fork lengths of coho, chinook and steelhead were typically similar to each other and ranged from 15-27 cm. Fork lengths of these species were substantially larger than those of sockeye, chum and pink salmon from the same samples.

In a series of samples taken between Cape Flattery and Yakutat in a 10-day period, the fish were clearly larger in the northern areas. However, these samples included a much broader range of sizes than in southern areas, and were frequently bimodal.

Tag returns showed that the larger fish tagged in northern areas were usually from southern production areas. The smaller fish were from local production areas. Thus, the samples in the northern Gulf of Alaska included larger fish that had been migrating at sea for some time, plus smaller fish that had only recently entered the open sea from production areas adjacent to the sampling location.

Juvenile salmon of all species caught in the North Pacific on the south side of the Alaska Peninsula were substantially larger than those caught in the eastern Bering Sea just north of the peninsula.

Finally, samples taken at a given location over a long period of time showed that the mean length generally increased with time. However, there were some cases where sizes decreased periodically, dominant size groups shifted, and bimodality occurred, presumably as various stocks passed through the sampling area.

Rates of Growth and Travel

Selected tag release and recovery data were used to estimate growth and travel rates of juvenile salmon during their first summer at sea. Such estimates, of course, only approximate these rates because the data and length at entry into saltwater must be based on average values.

Growth and travel rates were estimated from tag returns for 11 Fraser River and 10 Skeena River sockeye salmon (Table 6). Growth rates for sockeye from the two rivers were similar. The Fraser River sockeye growth rate ranged from 0.11-0.16 cm/day. Skeena River fish ranged from 0.10-0.18 cm/day.

Travel rates were substantially faster for Fraser River fish (7.6-14.4 nm/day) than for Skeena River fish (3.5-7.5 nm/day). The apparent slower travel rate of Skeena River fish was probably partially due to their fewer days at sea. The Skeena River fish spent an average of 69.7 days in the sea and Fraser River fish 85.8 days. Thus, any time that the Skeena River fish spent lingering near the estuary would have formed a larger proportion of their time at sea. This would reduce their apparent travel rate relative to Fraser River fish. Also, young salmon probably travel more rapidly as they grow larger. Thus, the earlier migrating, and larger, Fraser River fish could have migrated more rapidly than the Skeena River fish even after the two stocks intermingled, presumably in the vicinity of Dixon Entrance (Fig. 10).

Six Bristol Bay sockeye returns were used to estimate growth and travel. The fish were caught and tagged in the Bering Sea between Port Moller (161° W) and Unimak Island (164° W), and within 50 nm north of the Alaska Peninsula (Fig. 10).

Growth rates of Bristol Bay sockeye ranged from 0.05-0.10 cm/day. This was much less than the rates of the Fraser-Skeena river fish (0.10-0.18 cm/day). The slower growth rate may be due to poor feeding conditions in the extreme eastern portion of Bristol Bay (Straty 1974). Migrants probably grew slowly until they reached the richer feeding grounds west of 160° W. There, however, they presumably grew rapidly because food organisms were abundant (Hartt et al. 1969, 1970).

Travel rates of Bristol Bay sockeye ranged from 2.1-3.6 nm/day as compared to 3.5-14.4 nm/day for Fraser-Skeena fish. The slower travel rate of Bristol Bay

Table 6. Estimated Growth and Migration Rates for Sockeye Salmon Tagged at Sea as Juveniles and Subsequently Recovered in the Fraser and Skeena Rivers. Lengths in CM and Distances in Nautical Miles (See Fig. 10 for Release Areas). Source: Hartt and Dell (MS) Figures 52 and 53.

	Fraser River Returns	Skeena River Returns
Number of fish	11	10
Dates of release	7/20 - 9/15	7/24 - 8/28
Days enroute	66-123*	54-89**
Lengths at release	18.5 - 23.0	16.6 - 19.0
Growth (from 8 cm)	10.5 - 15.0	8.6 - 11.0
Growth per day	0.11 - 0.16	0.10 - 0.18
Distance travelled	640 - 1,540	240 - 520
Nautical miles/day	7.6 - 14.4	3.5 - 7.5

*Fraser River estimated saltwater entry May 15.
**Skeena River estimated saltwater entry June 1.

sockeye is in agreement with observations that Bristol Bay fish move more randomly than Gulf of Alaska fish.

One chum salmon, tagged in the northern Gulf of Alaska and recovered in the Strait of Juan de Fuca, grew at an estimated rate of 0.16 cm/day (Fig. 11). It travelled an estimated 9.6 nm/day. These rates fell within the ranges estimated for Fraser River sockeye, the nearest stock available for comparison.

Two tagged pink salmon were recovered, one in the San Juan Island area and the other in the Fraser River. They grew at estimated rates of 0.19 and 0.18 cm/day and travelled at rates of 9.2 and 10.7 nm/day. Pinks grew at a greater estimated rate than Fraser River sockeye (Table 6). However, a greater growth rate would be expected because, as mentioned earlier, pinks typically grow more rapidly than sockeye or chum. Pink salmon travelled at a rate that fell within the range of travel rates estimated for Fraser River sockeye, the nearest stock available for comparison.

Density and Abundance

In this section, purse seine catch data will be examined to estimate the density and abundance of juvenile salmon in tagging and sampling areas.

Density: The method that the purse seine was operated, and the seine's efficiency, makes it possible to roughly estimate the minimum number of juvenile salmon per unit of surface area from purse seine catch data. The purse seines used were about 385 fathoms long on the corkline and about 25 fathoms deep. The seine was set in an approximate semicircle for 30 minutes, then closed and pursed. The mesh sizes of the seine retained essentially all sizes of juvenile salmon present in the areas where density was examined. Thus, the density estimate was based on the area the seine enclosed plus any additional area that salmon migrated from during the 30-minute set (Fig. 17).

An average CPUE of 350 juvenile salmon per set was used as a typical catch from August to September in areas W4056 and W4058 (Fig. 1) in the northern part of southeast Alaska during several of our sampling years (Hartt and Dell *in press*). We chose these areas to estimate density because tag returns and catches in opposed seine sets showed that juvenile salmon typically migrate to the north (Table 3).

For this average catch of 350 juveniles, the density was calculated at 0.0015 salmon /m^2, or 681 m^2/salmon. To estimate density, several assumptions were made:

1. The seine was 100 percent efficient in catching juvenile salmon that crossed the imaginary line at the entrance of the seine's semicircle.
2. All salmon migrated in a direction perpendicular to the diameter line of the seine.
3. Juvenile salmon travel at a uniform rate of 10 nm/day.

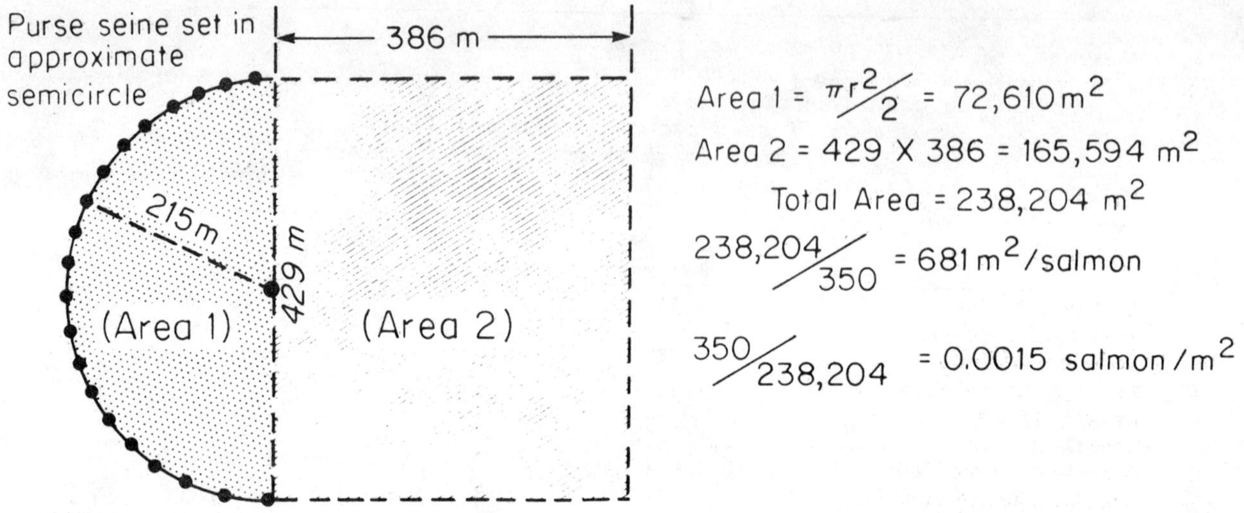

Figure 17. Diagram of area sampled by purse seine during a 30-minute set and assuming a rate of travel of 10 nm/day.* Also resultant estimates of fish density where mean catch was 350/set.
*Int'l Naut. Mi. = 1,852 m.

Because the first and second assumptions are obviously not met, the calculated density figures must be considered minimal. The migration direction varies substantially (Table 3) so that the seine was probably seldom set for the optimum catch in a given area. Even when the seine was set in an optimum direction, there were probably juvenile salmon migrating temporarily in directions that would make them invulnerable to capture. The seine certainly does not operate with 100 percent efficiency even for fish within the net. Salmon could possibly escape either under the seine, or around its ends during both the towing and pursing phases of the set. Under ideal visual conditions, salmon can sometimes be seen escaping around the net ends.

Error in the estimated travel rate (assumption 3) could cause either a positive or negative error in the density estimate. If the travel rate were actually greater than 10 nm/day, then the density of fish caught in a set would be overestimated. If the travel rate was less than 10 nm/day, density would be underestimated. The travel rate probably varies substantially according to various physical and biological factors. The estimate of 10 nm/day was approximately that observed for Fraser River sockeye tagged in the northern Gulf of Alaska as discussed with respect to Table 6.

Because the accuracy of assumption 3 cannot be evaluated, and error in assumptions 1 and 2 yield reduced density estimates, we conclude that the density of all juvenile salmon species in W4056 and W4058 in August and September is at least 0.0015 salmon/m^2, or 1 salmon/681 m^2. Due to the shortcomings in assumptions 1 and 2, the actual density could be substantially greater than the figures calculated.

The dispersion pattern of the fish is probably in small patches or schools uniformly distributed rather than in large dense schools with vacant or sparsely populated areas in between. As mentioned while discussing Tables 1 and 3, seine sets with zero catches very seldom occurred except offshore, or when the seine was set in a non-optimum direction.

The density of salmon in the eastern Bering Sea was probably similar to the density calculated for the northeastern Gulf of Alaska. Assumption 2 becomes even less valid in the Bering Sea because the direction juvenile salmon migrate varies more here than in the northeastern portion of the gulf as was discussed in explaining Tables 3 and 4. In order to derive a more equitable average CPUE for the eastern Bering Sea, the catches in sets open to the northeast and southeast should be summed before applying the density formula. This gives a mean catch of 331/set (Table 4), which yielded a density estimate similar to the one calculated for the northeastern Gulf of Alaska. Because juvenile salmon in the eastern Bering Sea travelled at a rate of 2-3 nm/day, the actual density could have been greater than the estimate for the northeastern portion of the gulf.

The maximum density of salmon in areas that we sampled can also be calculated from catches in individual seine sets. The largest catch (4,551) was in area W3554 (Dixon Entrance) on 14 July 1967. Other individual seine catches ranged from 1,000-2,000 occasionally in both the northeastern Gulf of Alaska and eastern Bering Sea.

The record catch of 4,551 juvenile salmon yielded a density estimate of 0.0193 salmon/m^2, or 1 salmon/53 m^2. This catch was probably the result of a fortuitous set in which a large school, or several smaller schools, were momentarily concentrated by a tide rip or feeding conditions. This set was made within six miles of Cape Muzon, Dall Island in northern Dixon Entrance. Juvenile salmon from nearby production areas tend to concentrate in this area.

In offshore waters of the eastern Bering Sea and northeastern Gulf of Alaska, catches occasionally reached 1,000-2,000 per set. This indicated that minimum densities of 0.0042-0.0084 salmon/m^2, or 1 salmon/238 m^2 to 1 salmon/119 m^2, do occur in offshore areas.

At this stage of our knowledge, we cannot relate these density estimates to food availability and feeding competition. Juvenile salmon are small. The food requirements per fish are much less than for larger fish at later life history stages. Nevertheless, juvenile salmon must graze food organisms at a substantial rate in epipelagic waters where juveniles concentrate. This would be particularly so in areas off southeastern Alaska where catch data indicated that masses of juveniles migrate continuously to the north within a narrow coastal belt for two or three months.

Populations of juvenile sockeye salmon concentrated in the eastern Bering Sea for two to three months must also have high food requirements. Substantial numbers of food organisms were seen frequently from the vessel during the purse seine operations (Hartt et al. 1969). In the Bering Sea, euphausiids and larval fish were particularly abundant. They were observed both from the surface and underwater. Throughout large areas of the eastern Bering Sea, juvenile salmon swim through a "soup" of food organisms.

It is perhaps significant that the largest catch of older-aged salmon corresponded closely to the largest catch of juveniles. We caught 4,819 older salmon from a set on 13 July 1969, seven miles south of Adak Island in the central Aleutian Islands. Our largest catch of juveniles, discussed above, was 4,551.

Most of the older fish were sockeye and chum. A few were maturing, but the majority were age .1 and .2 immature fish.

Occasionally we caught 1,000-2,000 older fish in sets south of the central Aleutian Islands. Catches of juveniles in the northeastern Gulf of Alaska and eastern Bering Sea were occasionally this large, too. These results suggest that there may be an upper limit on the density of salmon from all age-groups during their oceanic migrations, especially when considering the numerous seine sets made over vast areas (3,073 sets, Fig. 2).

It should be remembered that the density of older-aged salmon for a given CPUE may be substantially less than for juveniles because older fish travel faster. Average travel rates of immature age .1 and .2 salmon in the central Aleutian Island ranged from 10 to 30 nm/day as indicated by precise tagging data (Hartt 1966). Average travel rates of maturing salmon during their last 30 days at sea ranged from 25-30 nm/day. Thus, the density that a given CPUE indicates would be considerably less for older salmon than juveniles because a greater area was sampled when older salmon were purse seined.

Abundance: By extrapolating the appropriate density estimates over time and space, the purse seine CPUE can also be used to estimate the number of salmon in a sampling area. Ideally, this method requires adequate sampling to establish density isopleths over the area in which a population is being estimated. Although our sampling was inadequate for this precise method, gross population estimates could be made for limited areas where repeated sampling was done. In these areas approximate width and length of the fish distribution was known. From these dimensions, a population estimate was made for a coastal strip in the northern Gulf of Alaska extending from the center of W4058 to the southern boundary of area W4056 (Fig. 1). A conservative estimate of the area based on our seining yielded a strip 150 nm long by 15 nm wide, or 2,250 nm^2. At a density of 0.0015 juvenile salmon m^2, we calculated that there were 5,145 salmon nm^2 or 11,576,150 juvenile salmon in the 2,250 nm^2 coastal strip. This figure would be larger if corrected for purse seine inefficiency and directional movement that was discussed earlier. The travel rate of juveniles could either increase or decrease the figure.

The estimate would, of course, be for an instantaneous population, and constantly fluctuate because of the population's dynamic nature during the two or three month migration through the area. A population of this magnitude would seem a minimum because stock from all production areas extending southward at least to Washington and Oregon were present. Furthermore, at a travel rate of 10 nm/day, the population would renew itself every 12 days as the migration proceeded through the coastal strip.

There is another method of estimating population size in the northern Gulf of Alaska. Due to the pronounced and relatively consistent migratory direction, the number of salmon migrating daily past a given line perpendicular to the coast can be calculated and multiplied by time. The assumption must be made that the seine's CPUE estimates the number of salmon passing the line of the seine's set during a 30-minute period. Thus, for an estimate in area W4056, we might use these parameters:

C: Catch per set = 350 juvenile salmon;
w: width of seine opening = 0.25 nm;
W: width of band of fish = 15 nm;
T: daily period of passage = 24 hours;
t: length of set - 0.5 hours.
Total passing/60 days - 60,480,000.

On either an area-related or time basis, the volume of salmon in the sampling areas off the northern part of southeastern Alaska is substantial. This volume must form an important portion of juvenile salmon stocks that originate in production areas from southeastern Alaska to at least Oregon.

DISCUSSION

In general, despite gaps in the data, sampling and tagging juvenile salmon has provided a working model of the distribution, migrations and growth of some major North American salmonid stocks during their first summer at sea. This model, combined with distribution and migration information of older age-groups, makes it possible to develop a preliminary model of the ocean distribution and migrations of these stocks throughout their entire life history in marine waters. Royce et al. (1968) proposed a combined generalized model based on earlier sampling. Models for sockeye, chum and pink salmon, respectively, were proposed in the recent series of INPFC joint comprehensive reports (French et al. 1976, Neave et al. 1976, Takagi et al. *in press*).

The summer distribution and migration features of juvenile salmon imply some important biological principles pertaining to their life history, growth and survival during their critical first summer in the sea. Some of the most obvious principles are:

1. The high concentration of juvenile salmon in a limited area would seem to make them more vulnerable to predation and disease than if they dispersed widely. Although not quantified, the main predator scars noted in our sampling were from lampreys, seals, sea lions, sharks and other predaceous pelagic fishes such as daggertooth *(Anotopterus pharao)*. The incidence of external injuries on juvenile salmon was not notably higher or lower than on older age-groups sampled along with the juveniles, or sampled by themselves elsewhere. Thus, as measured by the incidence of injuries on fish that survived to be caught, the restricted distribution of juvenile salmon did not result in a high predation rate. Probably, however, juvenile salmon are less able to survive any injuries, including lamprey attacks, due to their small size and fragility.

Although we conducted no formal parasite and disease study, we noted any external evidence of infections. A uniquely high incidence of dermal lesions was recorded among juvenile sockeye sampled in the eastern Bering Sea in 1968 (Hartt et al. 1970). This may have been associated with the high abundance of fish in that year class. The infection was characterized by open, necrotic lesions on various body parts, and frequently involved the operculum or fins. In some of the more advanced cases, the fish were emaciated. The infection rate was approximately 2-3 percent as compared to negligible rates in all other areas and sampling years. Although the disease was probably contracted earlier in life, any additional stress associated with increasing density might increase mortality. In future sampling of juvenile salmon at sea, a predation and disease study would be worthwhile.

2. The nearshore distribution of juvenile salmon during their first summer in the Gulf of Alaska results in minimal overlap with age .1 and older immature salmon. This behavior may minimize feeding competition and possibly cannibalism, although by the time juvenile salmon reach the open sea they are large enough that cannibalism would be unlikely. The nearshore distribution also minimizes overlap with large predator species such as

albacore *(Thunnus alalunga)*, pomfret *(Brama japonica)* and jack mackerel *(Trachurus symmetricus)*. These predators migrate annually to the north in offshore waters of the gulf during August and September.

3. The distribution of juvenile salmon occupies waters directly in the path of maturing salmon en route to their spawning streams. Furthermore, maturing salmon in the Gulf of Alaska typically move to the southeast along the coast on their homing migration (Neave 1964), and traverse many miles of the belt that juvenile salmon concentrate in. Although some cannibalism may occur at this stage, the size of juveniles by the time they reach the open sea probably minimizes this as previously mentioned. In our operations, we frequently examined stomachs from mature salmon. We found that predation on juvenile salmon was extremely rare, even among the larger coho and chinook. Although speculative, perhaps behavioral patterns such as swimming depth or feeding habits minimize predation at this critical crossroads of salmon life history.

4. Large masses of juvenile salmon migrate continuously through a relatively restricted coastal belt. This raises interesting questions about stock and species interactions, and their impact on the food supply. These concerns are particularly important because the band of fish is over 1,000 nm long, and the migration continues for at least three months. It could be asked if the fish in the vanguard enjoy better feeding conditions than those bringing up the rear, or if the fish from southern production areas have a significant feeding advantage. These southern fish are larger than fish just entering the sea in northern areas.

The coastal belt that juvenile salmon occupy is a turbulent ecosystem characterized by upwelling, eddies from promontories and coastal entrances, and a northerly flow in a counterclockwise gyre (Favorite et al. 1976). Food organisms likely occur in patches because of these factors. Therefore, the availability of food organisms to epipelagic fish throughout the coastal belt probably changes continually in species composition, size, life stage, depth and horizontal distribution. If so, the salmon's position in the migration may not be critical in regard to the food supply.

It might be conjectured that where juvenile salmon concentrate most heavily, they would limit the food supply. In our qualitative examination of stomach contents, the types and quantities of food varied greatly. In no sampling area did we find evidence of widespread paucity of food in stomachs, nor fish in obviously emaciated or starved condition. If size offers an advantage in food competition, then coho and chinook should fare better than sockeye, chum and pink salmon. However, it must be remembered that the smaller species outnumber coho and chinook. In addition, many coho and chinook remain in inside waters such as Puget Sound and Johnstone Strait during much of their first summer and even longer. These factors tend to limit competition between small and large juvenile salmon in the coastal belt.

5. It remains unknown the extent that juvenile salmon use inside passages during their coastal migration (e.g. Johnstone Strait, Hecate Strait, Chatham Strait, Prince William Sound and Shelikof Strait). The young fish could possibly re-enter some of these passages after having migrated extensively on the outside. The lack of juvenile salmon in our sampling west of Queen Charlotte Islands suggested that the fish may use Hecate Strait rather than the outer coast in that area. If juvenile salmon from southern production areas do re-enter inside passages, they would compete with much smaller salmon from local production areas. These smaller fish would be less than the 10 cm minimum size we found in the open sea. Sampling throughout the season in some major passages, and tagging both the large and small juvenile salmon, could answer this question of re-entry.

6. It has been well established that many coho and chinook remain in inside waters for a year or longer. There has been little study on the degree that sockeye, chum and pink salmon do the same.

In their report, Hartt and Dell *(in press)* cited examples where juvenile sockeye, chum and pink salmon were still present in northern Hecate Strait and in central Puget Sound during November. The species composition was primarily chum and pink salmon in Hecate Strait, and primarily chum in Puget Sound. Puget Sound samples were clearly local fish that were still residual in inside waters. The Hecate Strait samples were in rather open water and could have been within a few days of entering the open sea. Nevertheless, these fish were probably of local origin based upon their small size.

Present evidence indicates that the majority of juvenile sockeye, chum and pink salmon go to sea during summer, but some, particularly chum and pinks, remain in inside waters until late fall at least. As Hartt and Dell *(in press)* discussed, the pink salmon in Puget Sound were probably part of a residual group that typically remains in the sound to maturity. Sockeye and chum, however, probably go to the sea at some stage of their life history.

Studying the relationship between the abundant summer migrants and the less abundant residuals could yield useful information on the basic biology of the species. For example, it would be useful to know whether or not the residual chum salmon in Puget Sound were a particular stock which typically lingers and feeds in inside waters. The length of the residuals (23 cm in November) indicated that they had experienced good feeding conditions in inside waters, and that growth, and therefore food availability, must have been comparable to that in the open sea. This would imply that food availability was not necessarily an impetus for the seaward migration. However, population pressure and food availability at some earlier stage may have been factors in the main body of fish migrating to sea earlier.

It would also be useful to compare the relative survival rates of residuals to migratory groups, and to determine when the residual groups migrate. As discussed earlier, the fish that linger apparently migrate at some time prior to age .2.

LITERATURE CITED

Birman, I. B. 1969. Distribution and growth of young Pacific salmon of the genus *Oncorhynchus* in the sea. Voprosy Ikhtiologii t. 9, Vyp. 5/58. (Soviet literature on North Pac. Fish. Collec. 65.) English translation *in* Problems of ichthyology. Amer. Fish. Soc. 9(5): 651-666.

Carlin, B. 1955. Tagging of salmon smolts in the River Lagan. Rep. Inst. Freshwater Res. Drottningholm 36:57-74.

Dell, M. B. 1968. A new fish tag and rapid, cartridge-fed applicator. Trans. Amer. Fish. Soc. 97(1):57-59.

Dell, M. B. 1972. A comparison of five types of tags used for tagging juvenile Pacific salmon at sea. Univ. Wash. Fish. Res. Inst. Unpubl. manuscr.

Dell, M. B. 1974. Tag returns and movements of rainbow trout *(Salmo gairdneri)* and rainbow-steelhead trout released in the Lake Washington system. Trans. Amer. Fish. Soc. 103(2):250-254.

Dodimead, A. J., F. Favorite, and T. Hirano. 1963. Salmon of the North Pacific Ocean. Part II. Review of oceanography of the subarctic Pacific region. Int. North Pac. Fish. Comm. Bull. 33. 195 pp.

Favorite, F., A. J. Dodimead, and K. Nasu. 1976. Oceanography of the subarctic Pacific region. Int. North Pac. Fish. Comm. Bull. 33. 187 pp.

Fisheries Research Board of Canada. 1964*a*. Progress in 1962 in Canadian research on problems raised by the protocol. Pages 30-53 *in* Int. North Pac. Fish. Comm. Annu. Rep. 1962.

Fisheries Research Board of Canada. 1964*b*. Progress in 1963 in Canadian research on problems raised by the protocol. Pages 46-59 *in* Int. North Pac. Fish. Comm. Annu. Rep. 1963.

Fisheries Research Board of Canada. 1966. Progress in 1964 in Canadian research on problems raised by the protocol. Pages 29-47 *in* Int. North Pac. Fish. Comm. Annu. Rep. 1964.

Fisheries Research Board of Canada. 1969. Progress in 1967 in Canadian research on problems raised by the protocol. I. Results of high seas fishing and tagging. Pages 29-53 *in* Int. North Pac. Fish. Comm. Annu. Rep. 1967.

French, R. R., and J. E. Mason. 1964. Salmon distribution and abundance on the high seas - winter season 1962 and 1963. Pages 131-141 *in* Report on the investigations by the United States for the International North Pacific Fisheries Commission, 1963. Int. North Pac. Fish. Comm. Annu. Rep. 1963.

French, R. R., D. R. Craddock, R. G. Bakkala, and J. R. Dunn. 1969. Ocean distribution, abundance, and migration of salmon. Pages 85-96 *in* Investigations by the United States for the International North Pacific Fisheries Commission, 1967. Int. North Pac. Fish. Comm. Annu. Rep. 1967.

French, R. R., and W. B. McAlister. 1970. Winter distribution of salmon in relation to currents and water masses in the northeastern Pacific Ocean and migrations of sockeye salmon. Trans. Amer. Fish. Soc. 99(4):649-663.

French, R. R., T. Bilton, M. Osako, and A. C. Hartt. 1976. Distribution and origin of sockeye salmon in offshore waters of the North Pacific Ocean. Int. North Pac. Fish. Comm. Bull. 34. 113 pp.

Fukuhara, R. M., S. Murai, J. J. LaLanne, and A. Sribhibhadh. 1962. Continental origin of red salmon as determined from morphological characters. Int. North Pac. Fish. Comm. Bull. 8:15-109.

Godfrey, H., J. E. Mason, and S. Tanaka. 1965. Salmon of the North Pacific Ocean. Part IX. Coho, chinook and masu salmon in offshore waters. Int. North Pac. Fish. Comm. Bull. 16. 135 pp.

Godfrey, H., K. A. Henry, and S. Machidori. 1975. Distribution and abundance of coho salmon in offshore waters of the North Pacific Ocean. Int. North Pac. Fish. Comm. Bull. 31. 80 pp.

Hartt, A. C. 1962*a*. Movements of salmon in the North Pacific Ocean and Bering Sea as determined by tagging--1956-1958. Int. North Pac. Fish Comm. Bull. 6. 157 pp.

Hartt, A. C. 1962*b*. Observations on pink salmon in the Aleutian Islands area 1956-1960. Pages 123-133 *in* Symposium on pink salmon, H.R. MacMillan lectures in fisheries, 1960. Univ. British Columbia, Vancouver, B.C.

Hartt, A. C. 1966. Migrations of salmon in the North Pacific Ocean and Bering Sea as determined by seining and tagging, 1959-1960. Int. North Pac. Fish. Comm. Bull. 19. 141 pp.

Hartt, A. C. 1975. Problems in sampling Pacific salmon at sea. Paper presented at Symposium on evaluation of methods of estimating abundance and biological attributes of salmon on the high seas, Vancouver, B.C., 24-25 October 1972. Int. North Pac. Fish. Comm. Bull. 32: 165-231.

Hartt, A. C., M. B. Dell, and S. B. Mathews. 1966. Tagging studies. Pages 81-91 *in* Report on the investigations by the United States for the International North Pacific Fisheries Commission, 1964. Int. North Pac. Fish. Comm. Annu. Rep. 1964.

Hartt, A. C., M. B. Dell, and L. S. Smith. 1969. Tagging and sampling. Pages 78-85 *in* Investigations by the United States for the International North Pacific Fisheries Commission, 1967. Int. North Pac. Fish. Comm. Annu. Rep. 1967.

Hartt, A.C., M. B. Dell, and L. S. Smith. 1970. Tagging and sampling. Pages 68-79 *in* Investigations by the United States for the International North Pacific Fisheries Commission, 1968. Int. North Pac. Fish. Comm. Annu. Rep. 1968.

Hartt, A. C. and M. B. Dell. *In press*. Migrations and growth of juvenile salmon and steelhead trout during their first summer in the open sea. Int. North Pac. Fish. Comm.

Haw, F., H. O. Wendler, and G. Deschamps. 1967. Development of Washington State salmon sport fishery through 1964. Wash. Dep. Fish, Res. Bull. 7. 191 pp.

Hebard, J. R. 1959. Currents in south-eastern Bering Sea and possible effects upon king crab larvae. U.S. Fish Wildl. Serv., Spec. Sci. Rep., Fish. 293. 11 pp.

Jensen, H. M. 1953. Migrations of silver salmon of Puget Sound. Wash. Dep. Fish., Fish. Res. Pap. 1(1):13-21.

Jensen, H. M. 1956. Migratory habits of pink salmon found in the Tacoma Narrows area of Puget Sound. Wash. Dep. Fish., Fish. Res. Pap. 1(4):21-24.

Koo, T. S. Y. 1962. Age designation in salmon. Pages 37-48 *in* Koo, T. S. Y. (ed.), Studies of Alaska red salmon. Univ. Wash. Publ. Fish., New Ser. Vol. 1. Univ. Wash. Press, Seattle.

Lander, R. H. 1970. Distribution in marine fisheries of marked chinook salmon from the Columbia River Hatchery Program, 1963-1966. Res. Reps. Fish. Comm. Oregon 2(1):28-55.

Lebrasseur, R.J., and L.W. Barner. 1964. Salmon catches in northern Hecate Strait, November 1963. Fish. Res. Bd. Can. Manuscript Rept. Series (Oceanog. and Limnol.) No. 176. 13 pp.

Machidori, S. 1970. On summer distribution of immature sockeye salmon in the northwestern North Pacific and its adjacent waters. Int. North Pac. Fish. Comm. Doc. 1311. 59 pp. (In Japanese with preliminary translation of 15 pp.)

Major, R., J. Ito, S. Ito, and H. Godfrey. *In press*. Chinook salmon in offshore waters. Int. North Pac. Fish. Comm. Bull.

Manzer, J. I. 1956. Distribution and movement of young Pacific salmon during early ocean residence. Fish. Res. Bd. Can., Prog. Rep., Pac. Coast Biol. Sta. 106:24-28.

Margolis, L., F. C. Cleaver, Y. Fukuda, and H. Godfrey. 1956. Salmon of the North Pacific Ocean. Part VI. Sockeye salmon in offshore waters. Int. North Pac. Fish. Comm. Bull. 20. 70 pp.

McCurdy, M. L., and T. R. Schroder. 1972. Bristol Bay sockeye salmon *(Oncorhynchus nerka)* 1971. Alaska Dep. Fish Game Tech. Data Rep. 7. 102 pp. (Processed)

Milne, D. J. 1950. The difference in the growth of coho salmon on the east and west coasts of Vancouver Island in 1950. Fish. Res. Bd. Can., Prog. Rep., Pac. Coast Biol. Sta. 85:80-82.

Milne, D. J. 1957. Recent British Columbia spring and coho salmon tagging experiments and a comparison with those conducted from 1925 to 1930. Fish. Res. Bd. Can. Bull. 113. 56 pp.

Milne, D. J., E. A. R. Ball, H. M. Jensen, and E. Jewell. 1958. Progress report on the 1957 coho salmon studies in the Strait of Juan de Fuca by Canada and the United States. 68 pp. Unpubl. mimeo rep.

Neave, F. 1964. Ocean migrations of Pacific salmon. J. Fish. Res. Bd. Can. 21(5):1227-1244.

Neave, F. 1966*a*. Salmon of the North Pacific Ocean. Part III. A review of the life history of North Pacific salmon. 5. Pink salmon in British Columbia. Int. North Pac. Fish. Comm. Bull. 18: 71-79.

Neave, F. 1966*b*. Salmon of the North Pacific Ocean. Part III. A review of the life history of North Pacific salmon. 6. Chum salmon in British Columbia. Int. North Pac. Fish. Comm. Bull. 18: 81-85.

Neave, F., T. Ishida, and S. Murai. 1967. Salmon of the North Pacific Ocean. Part VII. Pink salmon in offshore waters. Int. North Pac. Fish. Comm. Bull. 22:1-39.

Neave, F., T. Yonemori, and R. G. Bakkala. 1976. Distribution and origin of chum salmon in offshore waters of the North Pacific Ocean. Int. North Pac. Fish. Comm. Bull. 35. 79 pp.

Ogi, H. 1973. Ecological studies on the juvenile salmon, *Oncorhynchus nerka* (Walbaum), in Bristol Bay, with special reference to its distribution and population. Bul. Pac. Fish., Hokkaido Univ. 4(1):14-41.

Paulus, R., and M. Nelson. 1972. Bristol Bay sockeye salmon *(Oncorhynchus nerka)* 1970. Alaska Dep. Fish Game Tech. Data Rep. 6. 112 pp. (Processed)

Rogers, D. E. 1976. Forecast of the sockeye salmon run to Bristol Bay in 1976. Univ. Wash. Fish. Res. Inst. Circ. 76-1. 48 pp. (Processed)

Royce, W. F., L. S. Smith, and A. C. Hartt. 1968. Models of oceanic migrations of Pacific salmon and comments on guidance mechanisms. Natl. Ocean. Atmos. Admin. Fish. Bull. 66(3):441-462.

Sakagwa, G. T. 1972. The dynamics of juvenile salmon with particular emphasis on pink salmon *(Oncorhynchus gorbuscha)* during their early marine life. Ph.D. Dissertation. Univ. of Wash., Seattle. 352 pp.

Shepard, M. P., A. C. Hartt, and T. Yonemori. 1968. Salmon of the North Pacific Ocean. Part VIII. Chum salmon in offshore waters. Int. North Pac. Fish. Comm. Bull. 25:1-69.

Straty, R. R. 1974. Ecology and behavior of juvenile sockeye salmon *(Oncorhynchus nerka)* in Bristol Bay and the eastern Bering Sea. Pages 285-320 *in* Oceanography of the Bering Sea. Int. Sympos. Bering Sea Study, Hakodate, Japan, 3 January-4 February 1972.

Straty, R. R., and H. W. Jaenicke. 1971. Studies of the estuarine and early marine life of sockeye salmon in Bristol Bay, 1965-1967. Manuscript Rep. 83, Natl. Ocean. Atmos. Admin., Natl. Mar. Fish. Serv. Biol. Lab., Auke Bay, Alaska. 137 pp. (Processed)

Sutherland, D. F. 1973. Distribution, seasonal abundance, and some biological features of steelhead trout, *Salmo gairdneri*, in the North Pacific Ocean. Natl. Ocean. Atmos. Admin., Natl. Mar. Fish. Serv., Fish. Bull. 73(1):787-826.

Takagi, K., V. Aro, A. C. Hartt, and M. B. Dell. *In press*. Distribution and origin of pink salmon in offshore waters of the North Pacific Ocean. Int. North Pac. Fish. Comm. Bull.

Thorson, K. N. 1967. A new high-speed tagging device. Calif. Fish Game 53(4):289-292.

Verhoeven, L. A. 1952. A report to the salmon fishing industry of Alaska on the results of the 1947 tagging experiments. Univ. Wash., Fish. Res. Inst. Circ. 28. 21 pp. (Processed)

Washington State Department of Fisheries. 1959. Contributions of western states, Alaska, and British Columbia to salmon fisheries of the North American Pacific Ocean. Fish. Vol. 2. Wash. Dep. Fish., Olympia. 81 pp.

Wright, S., and J. Bernhardt. 1969. A review of fin-marking experiments with chinook and coho salmon on the Pacific coast. Rep. 2: experiments on 1964 brood year coho salmon. Wash. Dep. Fish., Res. Div. Suppl. Prog. Rep. Mar. Fish. Invest. Ocean Troll Fishery. 32 pp.

Trends in North Pacific Salmon Fisheries

R. A. Fredin

INTRODUCTION

The annual world harvest of Pacific salmon *(Oncorhynchus* sp.) has declined from about 1.7 billion pounds in the 1930s to about 0.9 billion pounds in the 1970s. Countries currently harvesting salmon are Japan (about 34 percent of the total catch), United States (about 29 percent), U.S.S.R. (about 21 percent) and Canada (about 16 percent). The principal species harvested are pink *(O. gorbuscha)* (about 40 percent of the total catch); chum *(O. keta)* (about 30 percent); sockeye *(O. nerka)* (about 14 percent); coho *(O. kisutch)* (about 9 percent); and chinook *(O. tshawytscha)* (about 7 percent).

This report describes catch trends of Pacific salmon by the four major salmon-producing nations from 1920 through 1977. Tables 1-5 in the appendix give catch statistics, in numbers of fish, for sockeye, pink, chum, coho and chinook salmon, respectively. Table 6 gives catch statistics for coho and chinook in North America.

Figure 1 shows coastal and high seas fishing areas. High seas catches by Japan have been allocated to the continent (North America or Asia) where the fish originated.

Sockeye Salmon

North America produces about three times more sockeye salmon than Asia. Catches of this species have declined. The recent decline appears to be more severe for Asian stocks (Fig. 2).

Japanese fisheries operating on the high seas harvest the great majority of Asian sockeye (Fig. 3). They also intercept some sockeye originating in western Alaska (Fig. 4).

North American sockeye primarily originate from streams and lakes in Alaska and British Columbia. Canadian fisheries harvest fish originating primarily from British Columbia, but a significant portion of the Washington State catch (Fig. 5) originates from the Fraser River in southern British Columbia.

Pink Salmon

Asia produces about twice as many pink salmon as North America does (Fig. 6). The harvest on both North American and Asian stocks has severely declined.

Japanese fisheries on the high seas take a significant portion of Asian pink salmon (Fig. 7). The U.S. harvest has declined precipitously, but the Canadian harvest has not (Fig. 8). These declines have been most severe in scuthwestern Alaska and Washington State (Fig. 9).

Chum Salmon

Almost three times more chum salmon are caught in Asia than in North America (Fig. 10). As with sockeye and pink salmon, the Japanese high seas fisheries harvest a high percentage of Asian chum salmon (Fig. 11). Inspection of Figures 11 and 12 reveals that chum salmon catches are trending downward in the U.S.S.R., U.S. and Canada, but upward in Japan. In Japan, a large hatchery program contributes to the increased Japanese harvest. North American chum salmon catches have been declining from southeast Alaska to Oregon, but are increasing in western Alaska where fisheries north of Bristol Bay have expanded (Fig. 13).

Coho Salmon

Commercial fishermen in North America harvest about twice as many coho salmon than Asian fishermen catch (Fig. 14). The high seas fishery of Japan takes the largest portion of the Asian catch of coho (Fig. 15). Figures 15 and 16 suggest that the harvest is trending downward for the U.S.S.R. fisheries, relatively stable for Japanese fisheries, and upward for U.S. and Canadian fisheries. British Columbia, Washington, Oregon and California catches of coho salmon have all increased in recent years (Fig. 17), probably as a result of expanding hatchery programs. Sport fisheries for coho salmon have become more important in British Columbia, Washington, Oregon and California (Fig. 18). Although smaller than the commercial catch, the sport harvest contributes significantly to the total catch (Fig. 19).

Chinook Salmon

Commercial fishermen from North America catch about 15 times more chinook salmon than their Asian counterparts (Fig. 20). The Japanese high seas fisheries harvest more Asian chinook than U.S.S.R. fisheries (Fig. 21). Japanese high seas fisheries also catch North American chinook salmon (Fig. 22).

Catch trends have varied among geographic regions of North America (Fig. 23). The upward trend in western Alaska represents increased pressure on wild stocks. Increased catches in British Columbia are probably due to interception of fish from hatcheries in the Pacific Northwest. Sport fisheries for chinook salmon are becoming increasingly important in British Columbia, Washington, Oregon and California (Fig. 24), and the sport catch now contributes significantly to the total catch (Fig. 25).

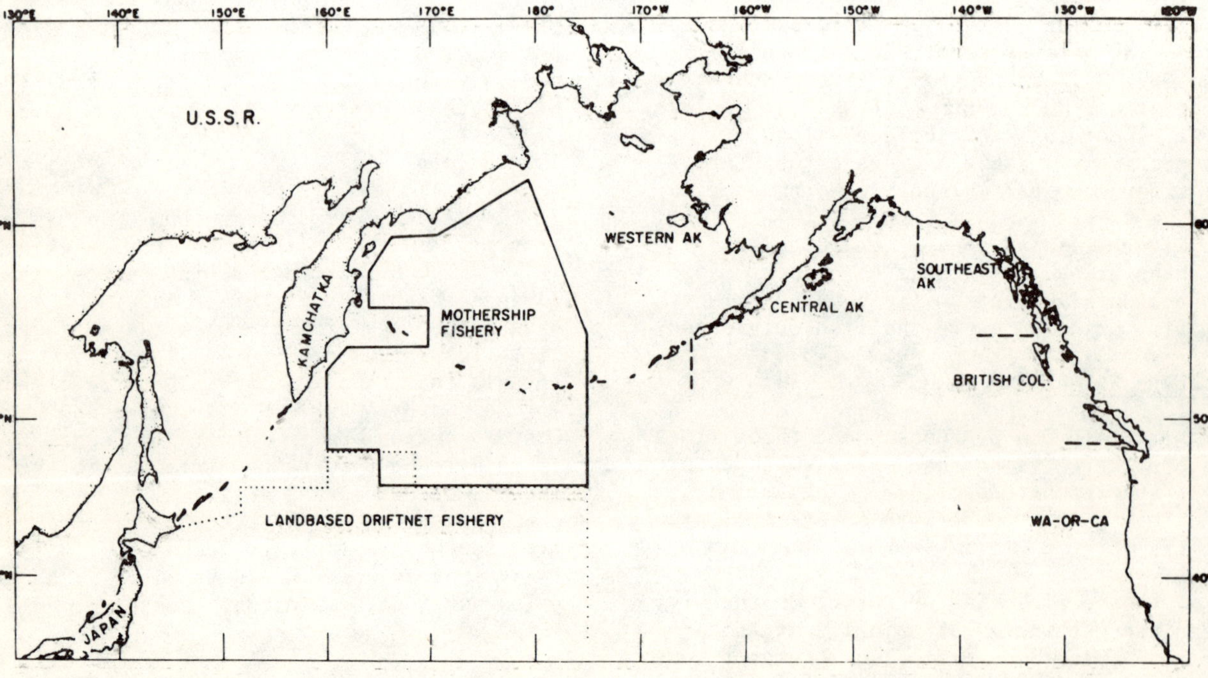

Figure 1. Major fishing areas for Pacific salmon.

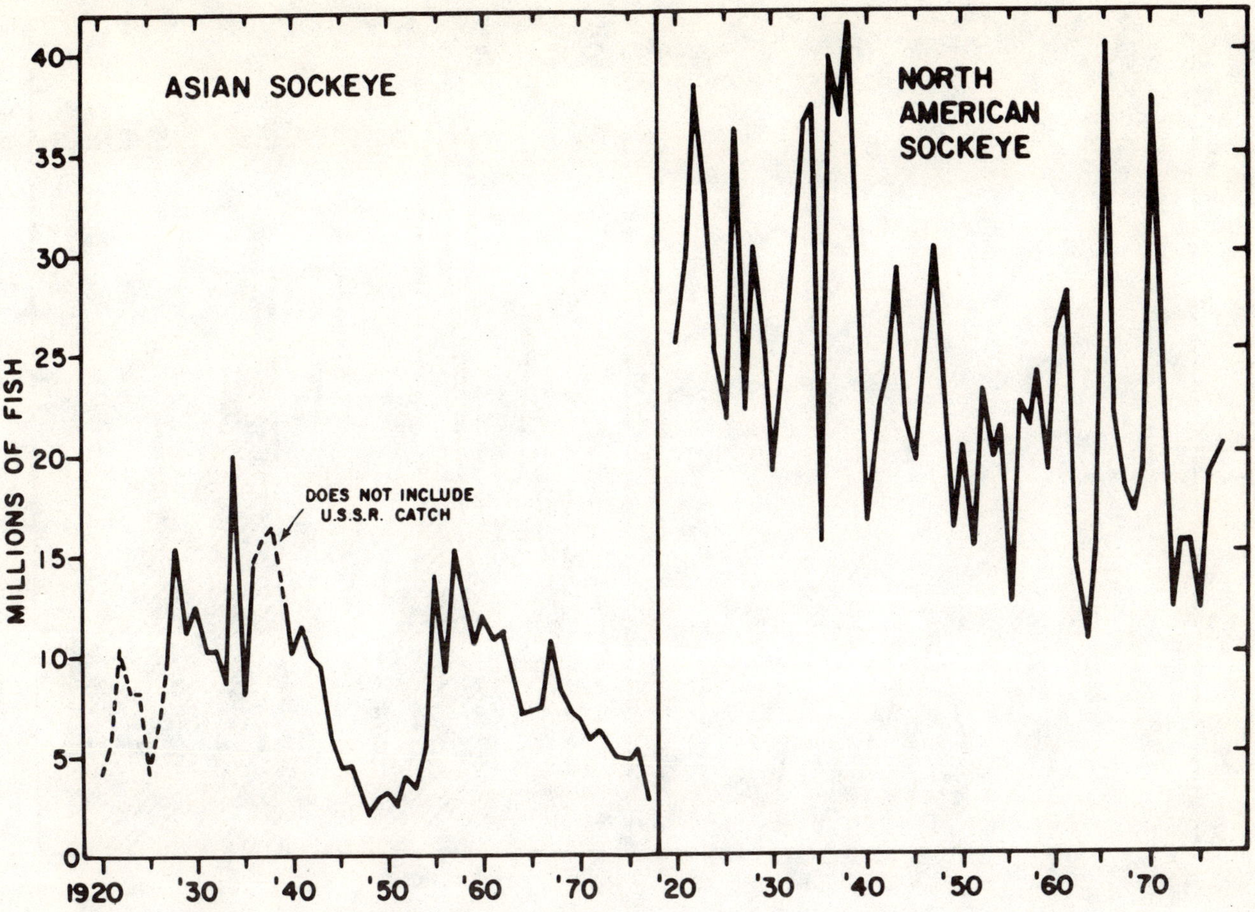

Figure 2. Catch of sockeye salmon of Asian and North American origins by commercial fisheries of the U.S.S.R., Japan, U.S. and Canada, 1920-77. (Data are from Table 1.)

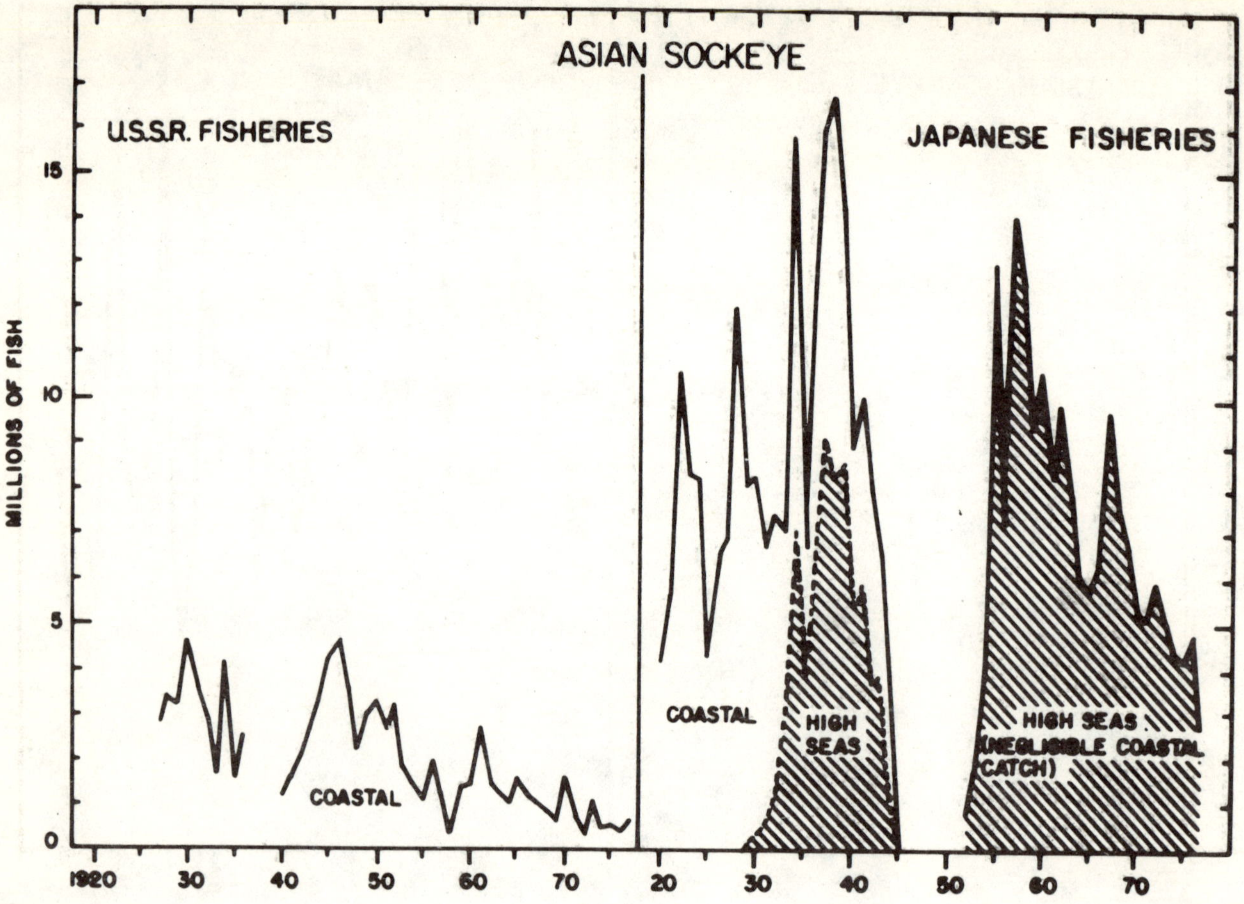

Figure 3. Catch of Asian sockeye salmon by commercial fisheries of the U.S.S.R. and Japan, 1920-77. (Data are from Table 1.)

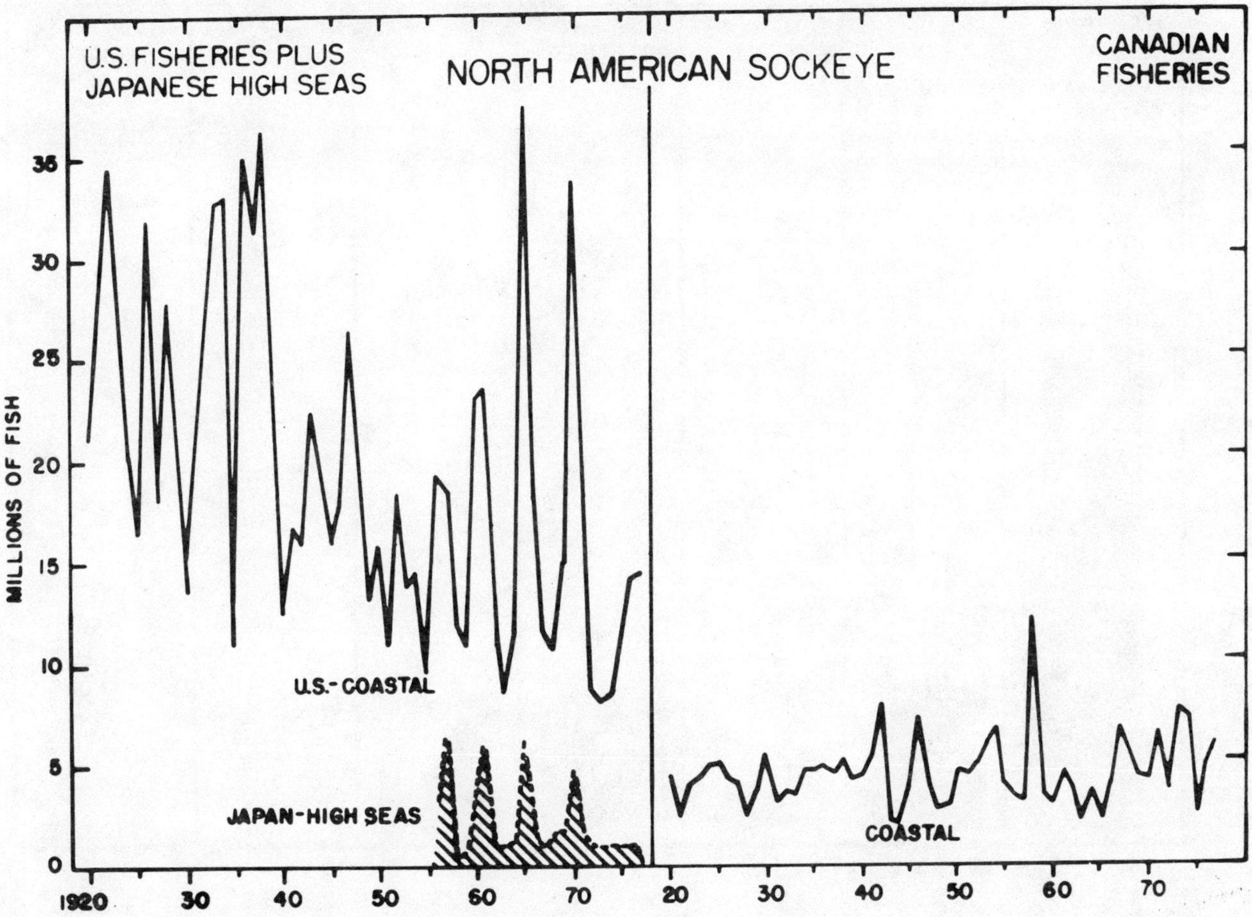

Figure 4. Catch of North American sockeye salmon by commercial fisheries of Japan, U.S. and Canada, 1920-77. (Data are from Table 1.)

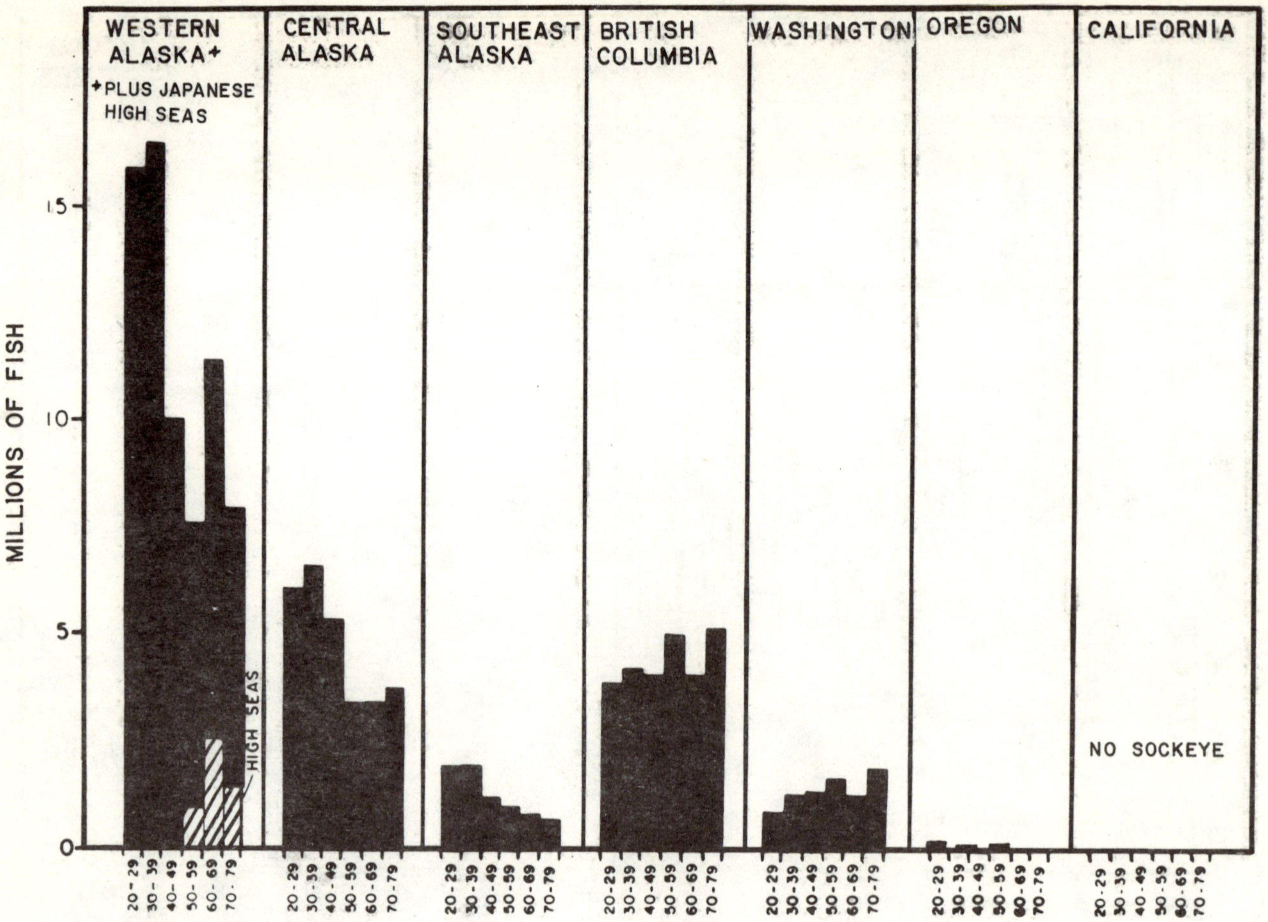

Figure 5. Average annual commercial catch of North American sockeye salmon, by region and decade, 1920-29 to 1970-77. (Data are from Table 1.)

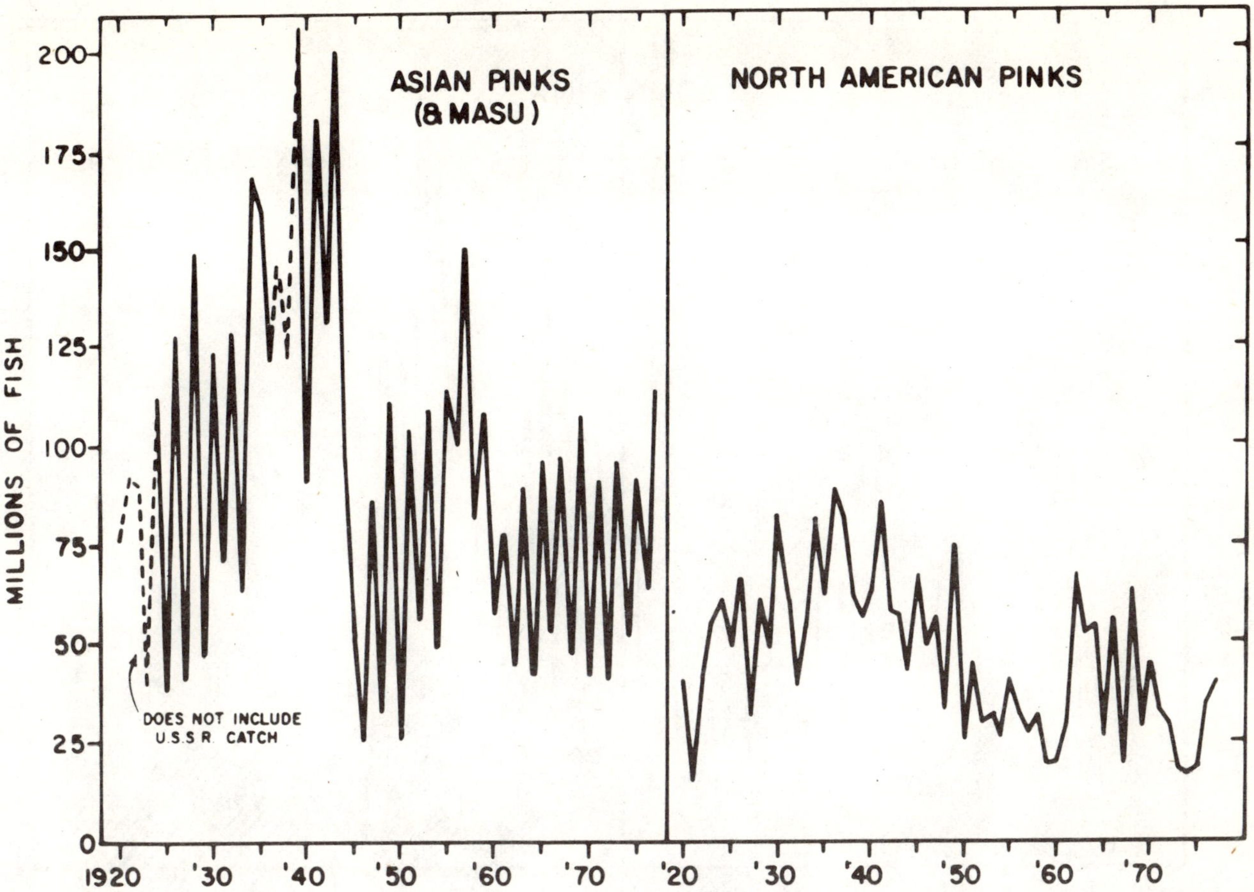

Figure 6. Catch of pink salmon of Asian and North American origins by commercial fisheries of the U.S.S.R., Japan, U.S. and Canada, 1920-77. (Includes masu caught by the Japanese coastal fishery. Data are from Table 2.)

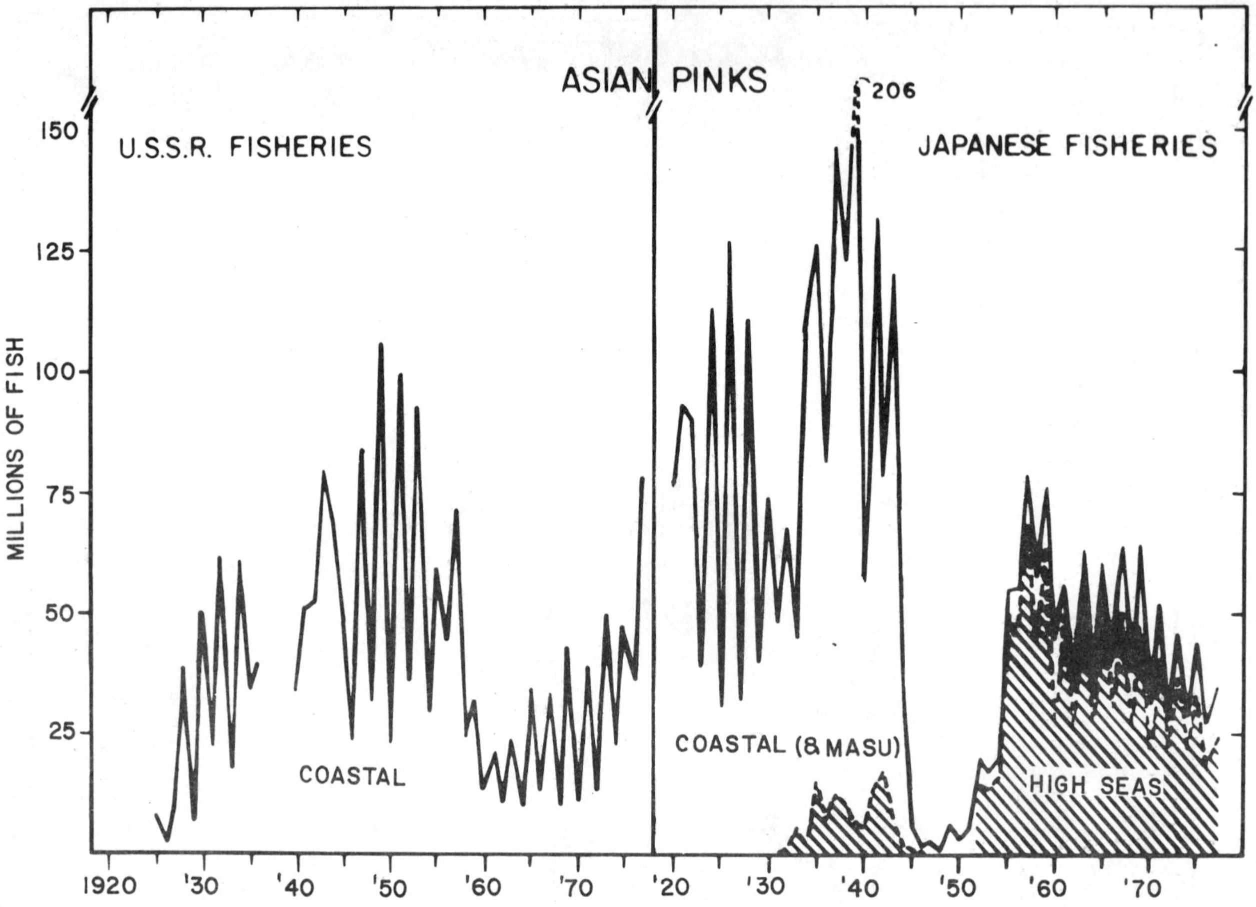

Figure 7. Catch of Asian pink salmon by commercial fisheries of the U.S.S.R. and Japan, 1920-77. (Japanese coastal catch includes masu. Data are from Table 2.)

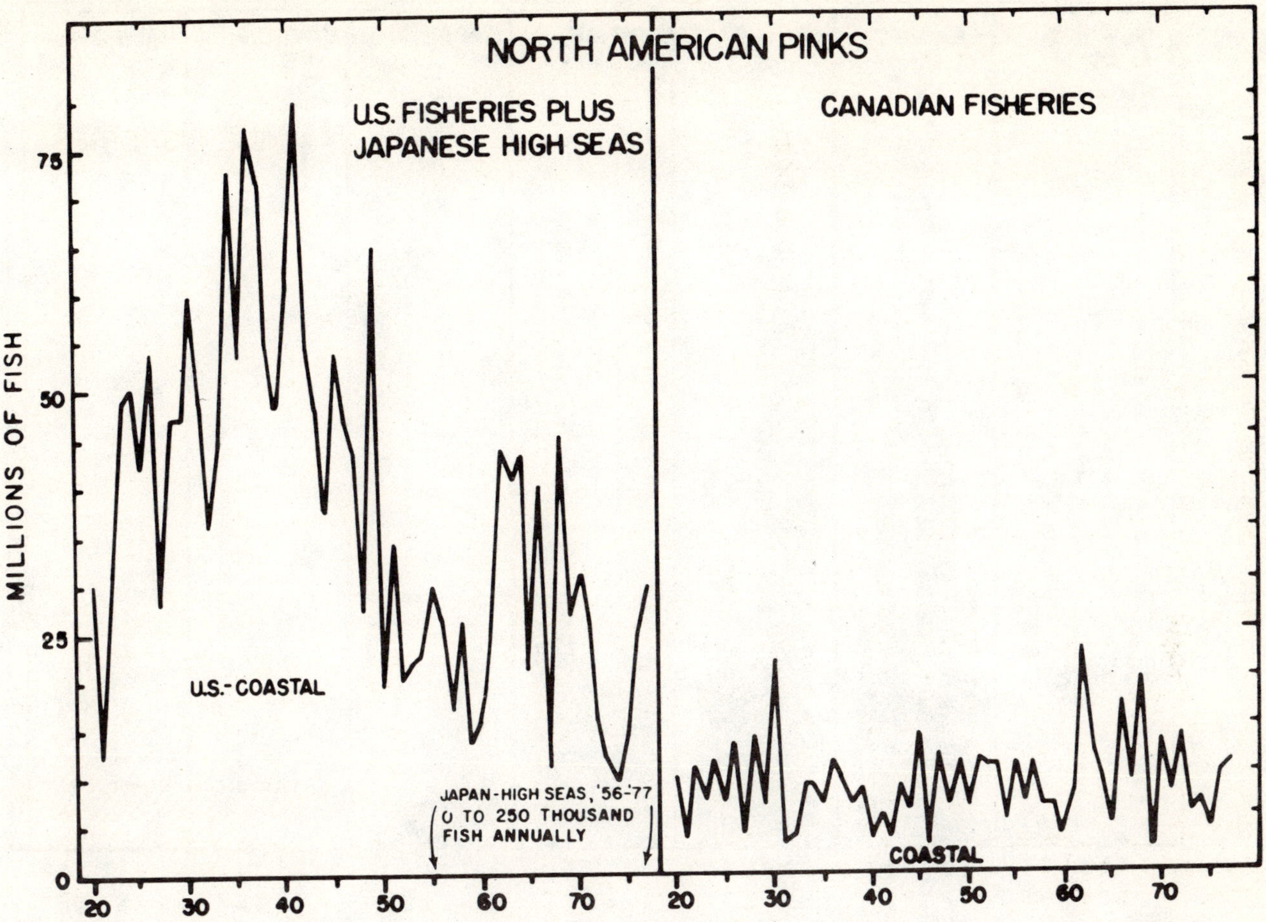

Figure 8. Catch of North American pink salmon by commercial fisheries of Japan, U.S. and Canada, 1920-77. (Data are from Table 2.)

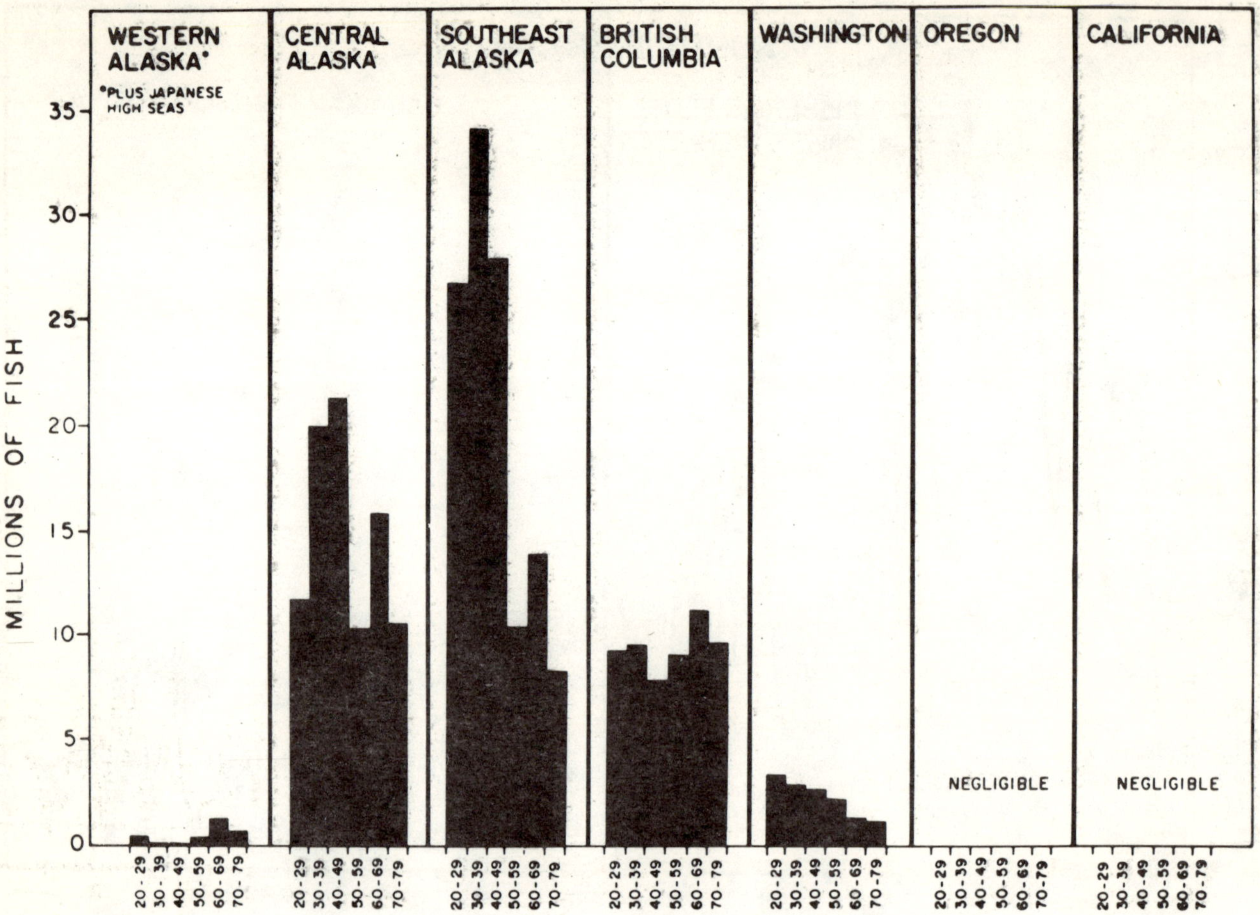

Figure 9. Average annual commercial catch of North American pink salmon, by region and decade, 1920-29 to 1970-77. (Data are from Table 2.)

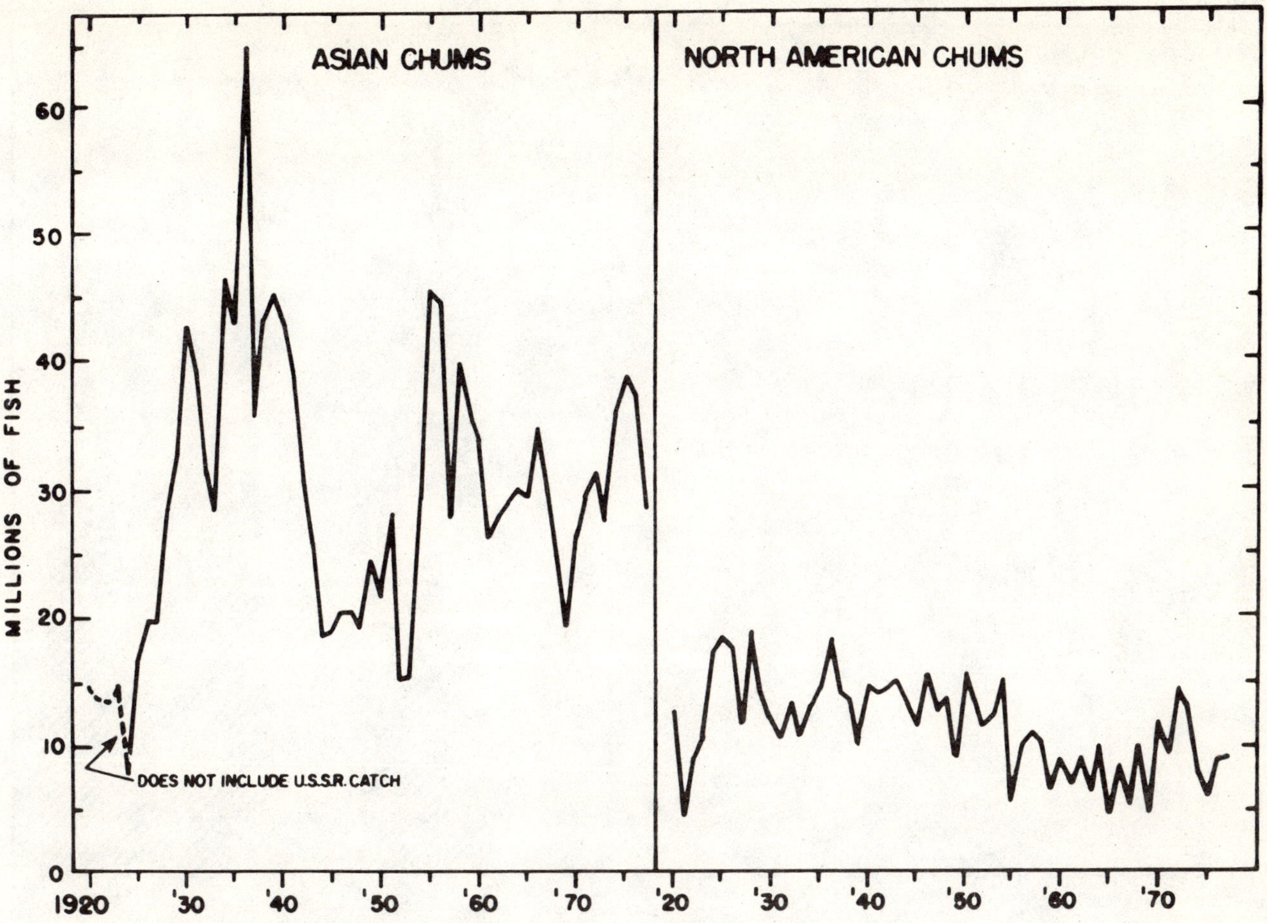

Figure 10. Catch of chum salmon of Asian and North American origins by commercial fisheries of the U.S.S.R., Japan, U.S. and Canada, 1920-77. (Data are from Table 3.)

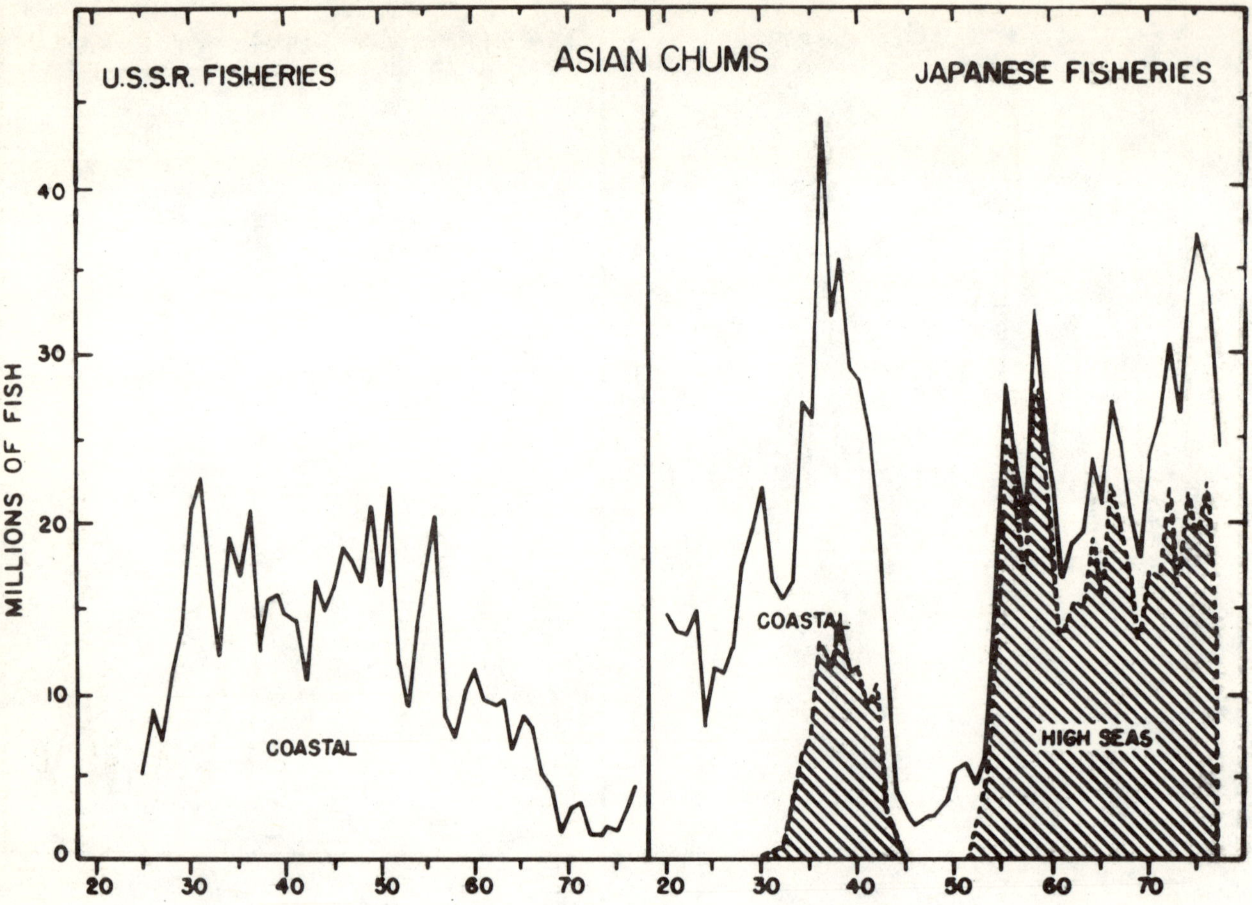

Figure 11. Catch of Asian chum salmon by commercial fisheries of the U.S.S.R. and Japan, 1920-77. (Data are from Table 3.)

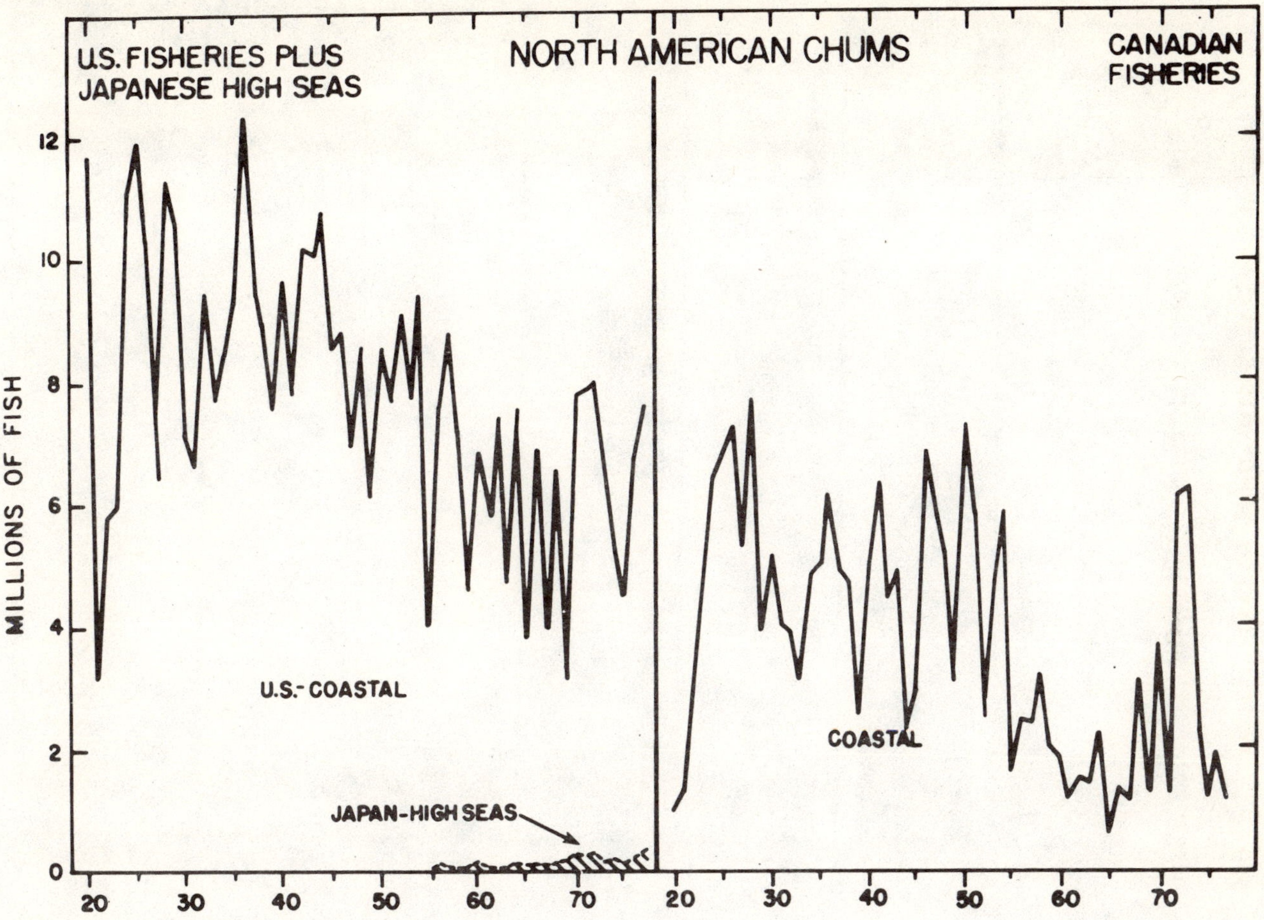

Figure 12. Catch of North American chum salmon by commercial fisheries of Japan, U.S. and Canada, 1920-77. (Data from Table 3.)

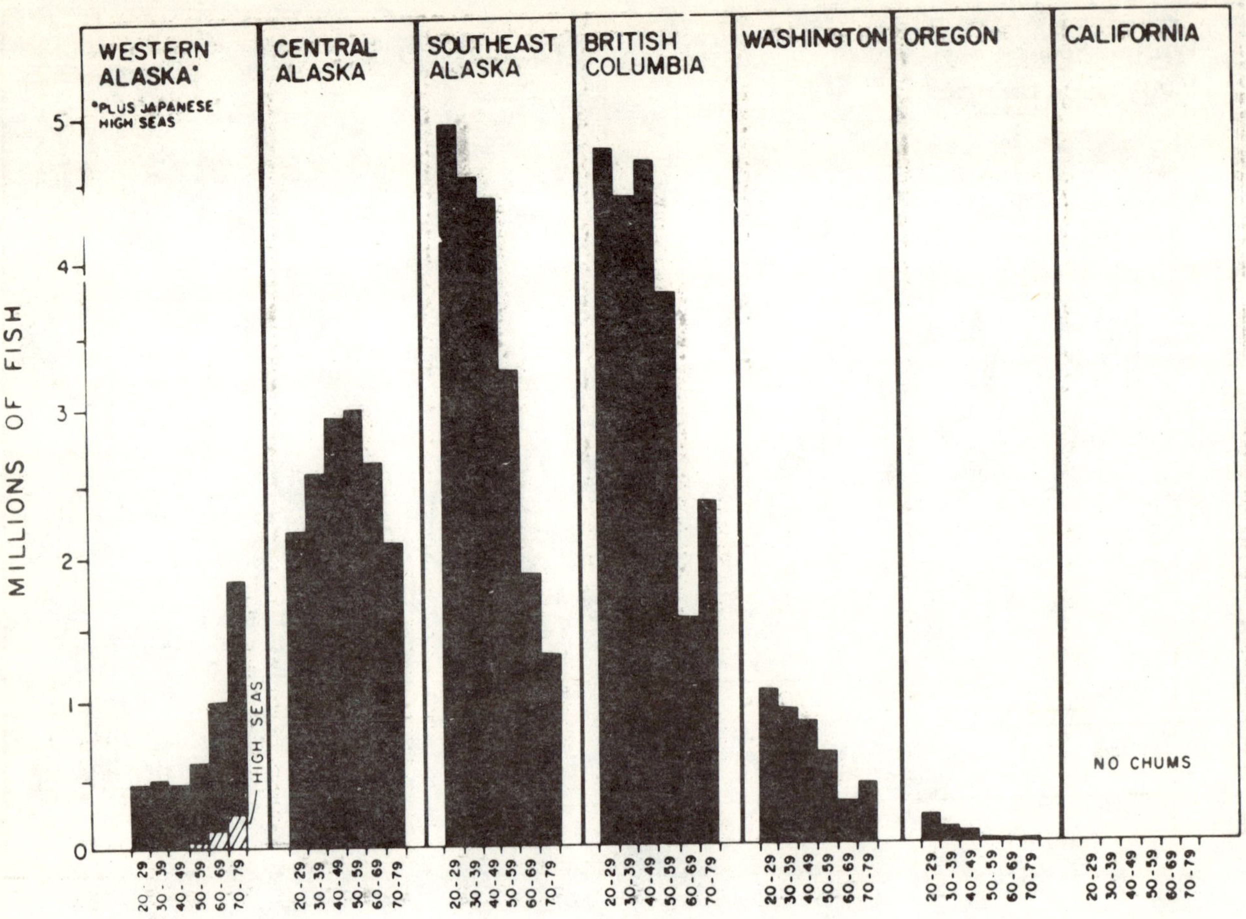

Figure 13. Average annual commercial catch of North American chum salmon, by region and decade, 1920-29 to 1970-77. (Data are from Table 3.)

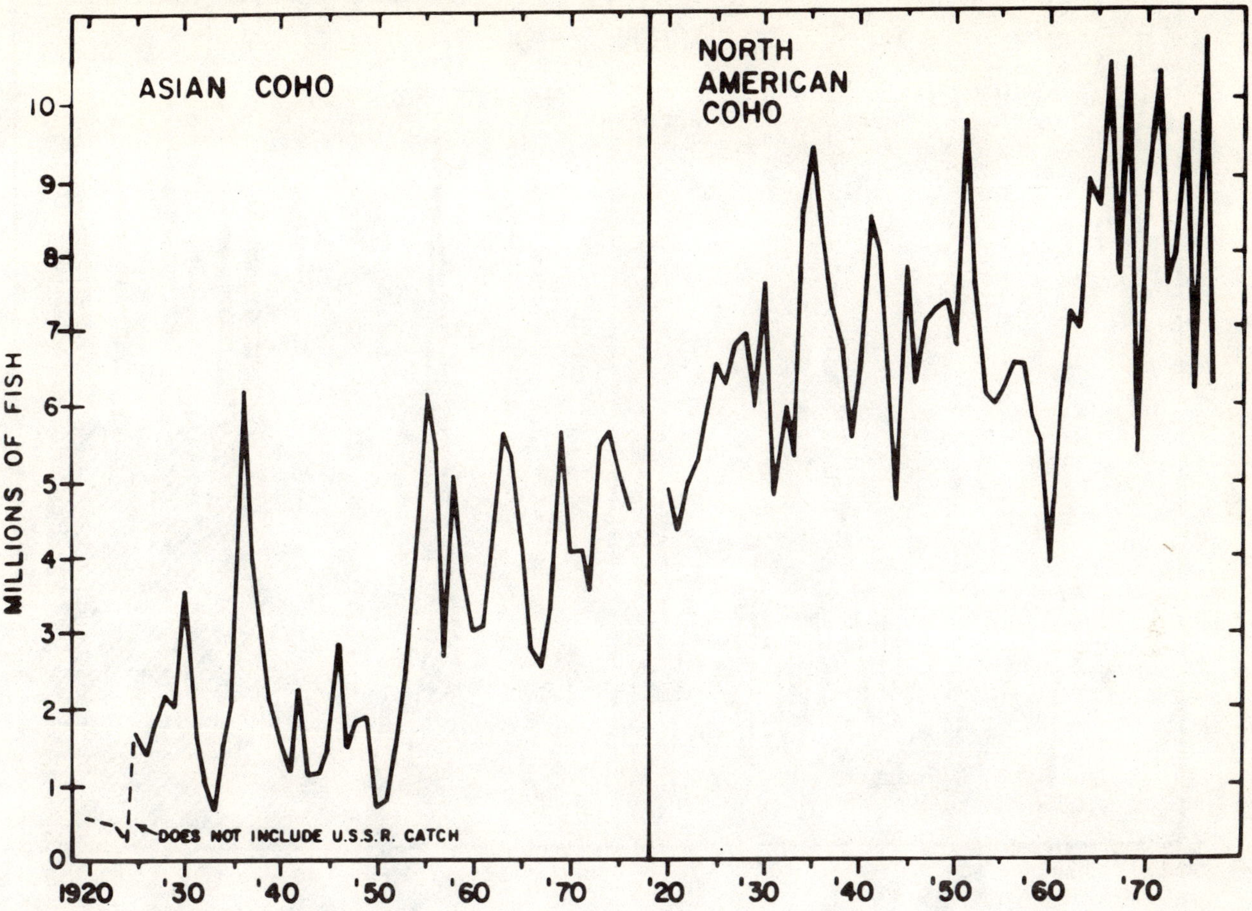

Figure 14. Catch of coho salmon of Asian and North American origins by commercial fisheries of the U.S.S.R., Japan, U.S. and Canada, 1920-77. (Data from Table 4.)

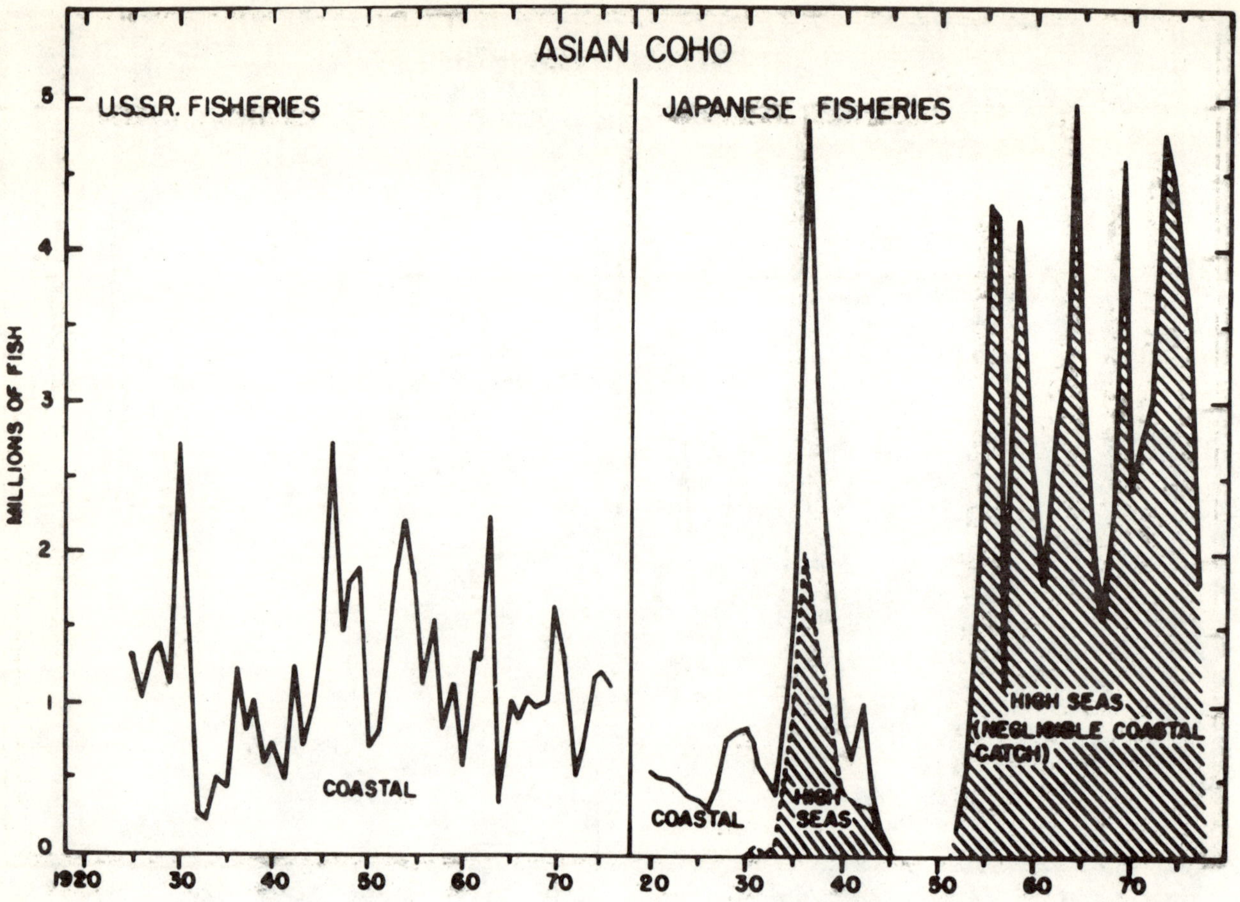

Figure 15. Catch of Asian coho salmon by commercial fisheries of the U.S.S.R. and Japan, 1920-77. (Data are from Table 4.)

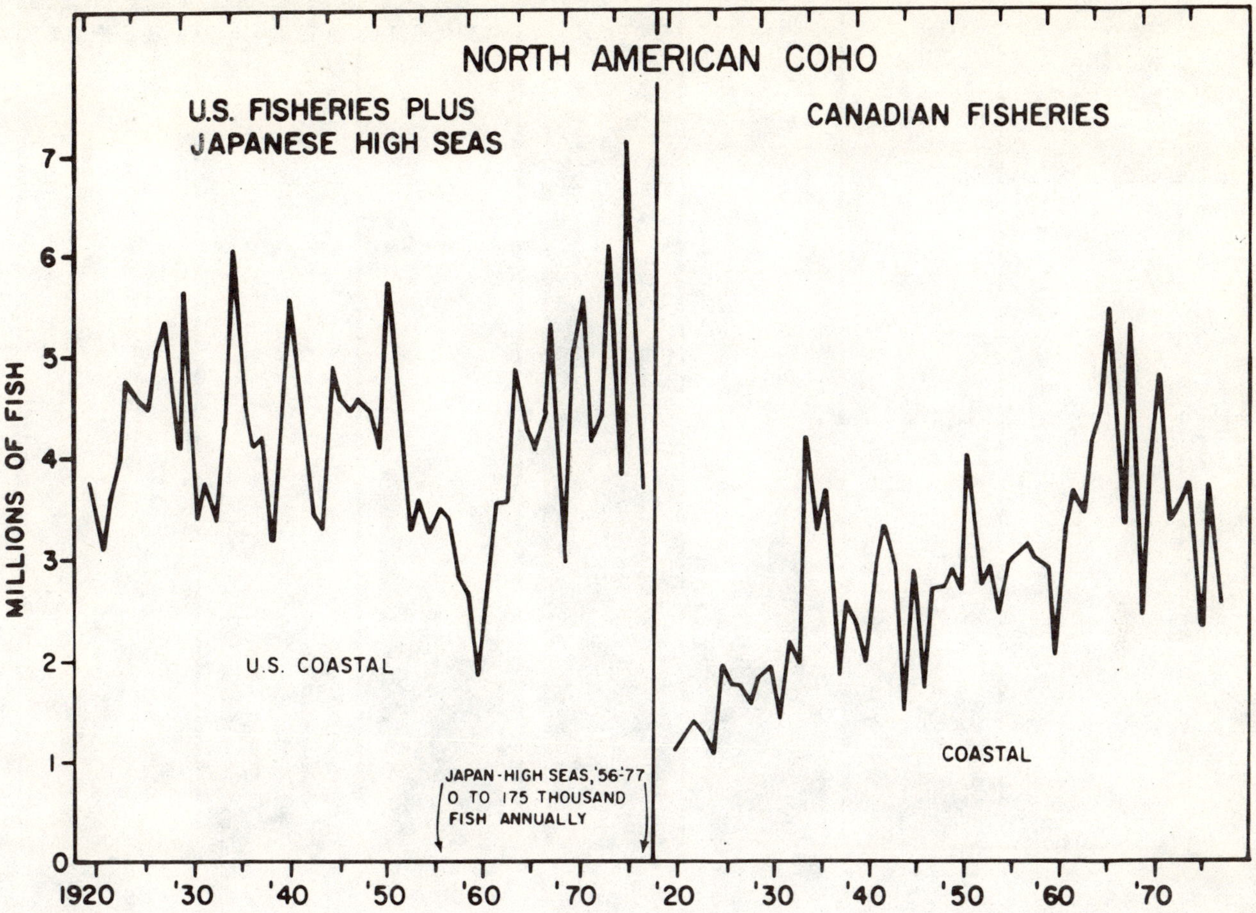

Figure 16. Catch of North American coho salmon by commercial fisheries of Japan, U.S. and Canada, 1920-77. (Data are from Table 4.)

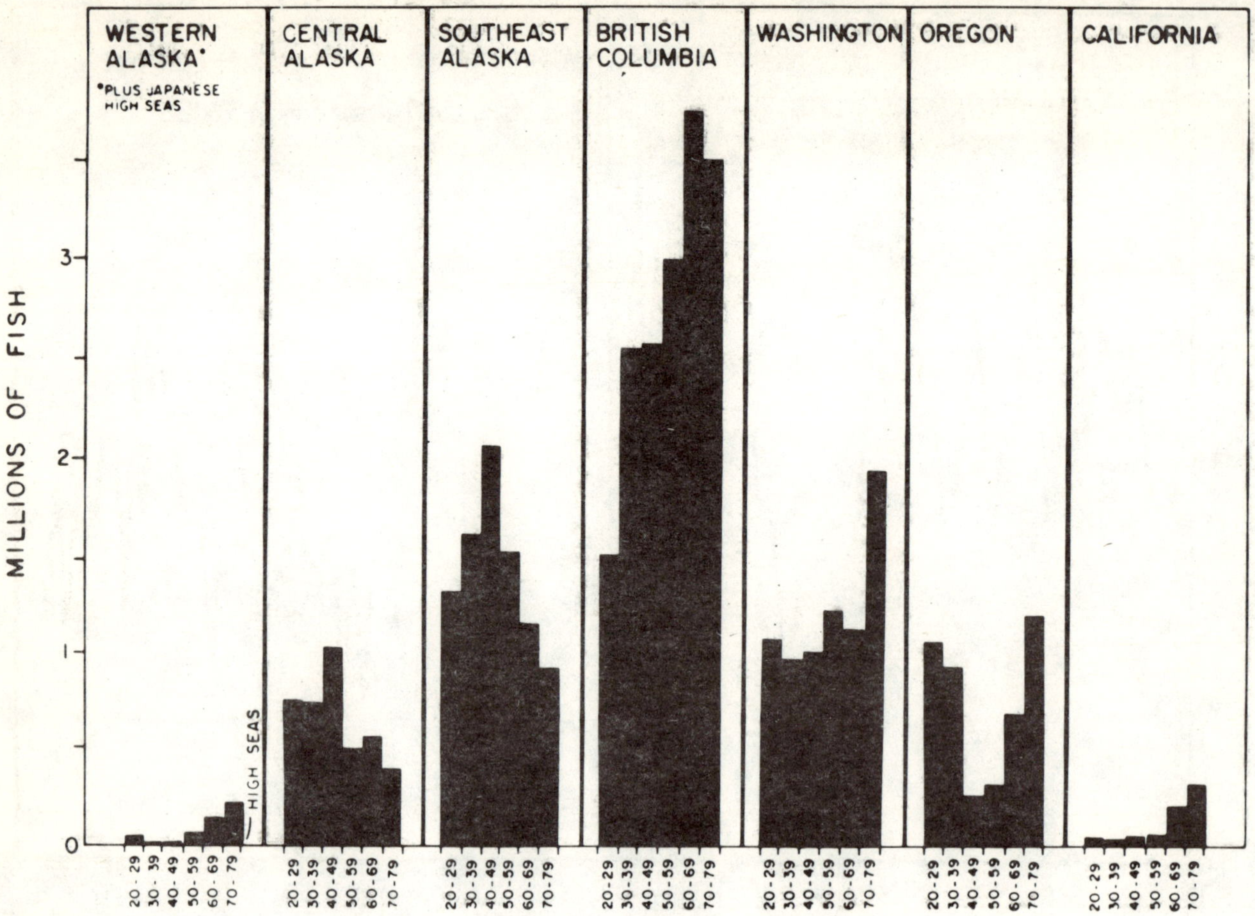

Figure 17. Average annual commercial catch of North American coho salmon, by region and decade, 1920-29 to 1970-77. (Data are from Table 4.)

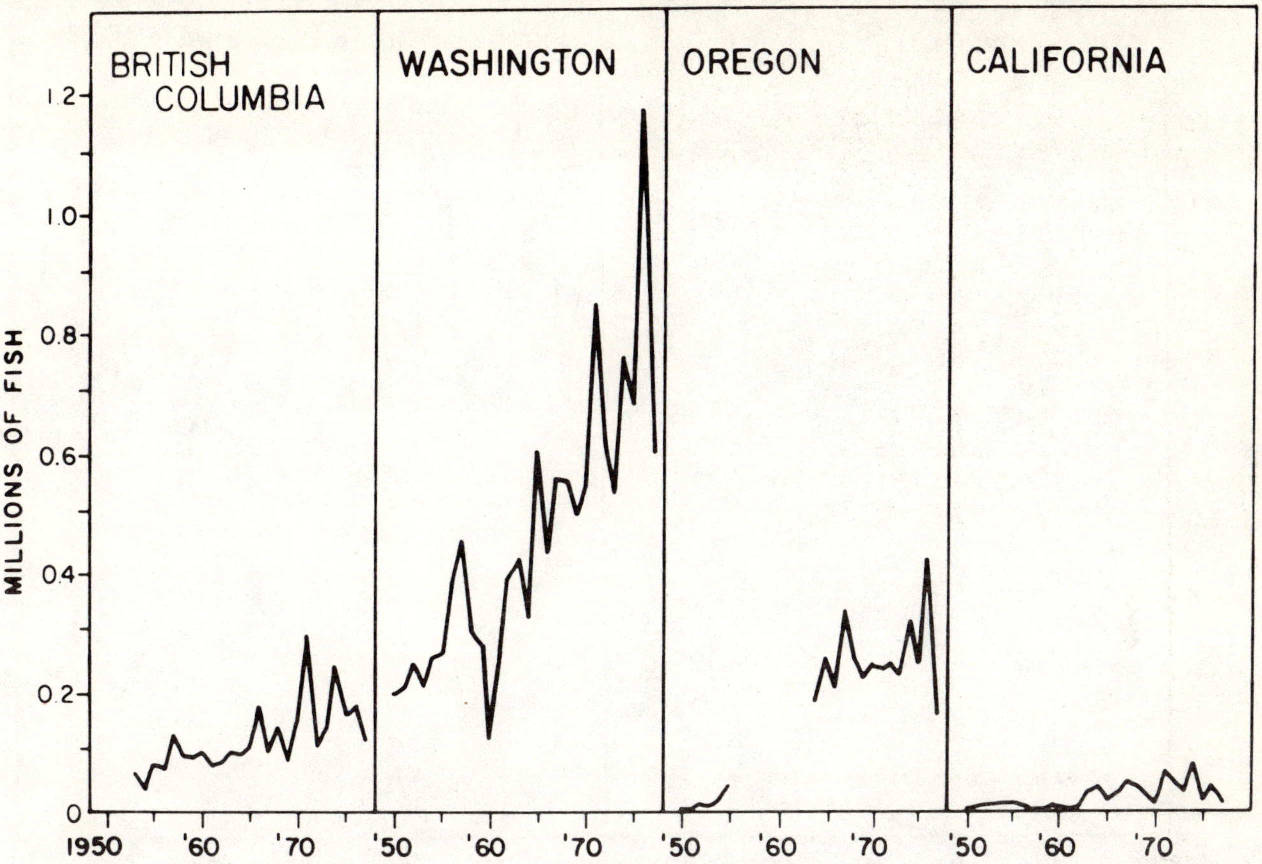

Figure 18. Catch of coho salmon by sport fisheries of British Columbia, Washington, Oregon and California, 1950-77. (Data are from Table 6.)

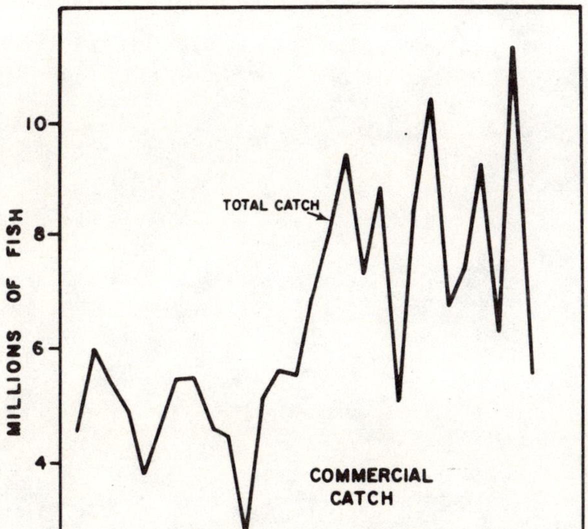

Figure 19. Catch of coho salmon by commercial and marine sport fisheries of British Columbia, Washington, Oregon and California combined, 1950-77. (Data are from Tables 4 and 6.)

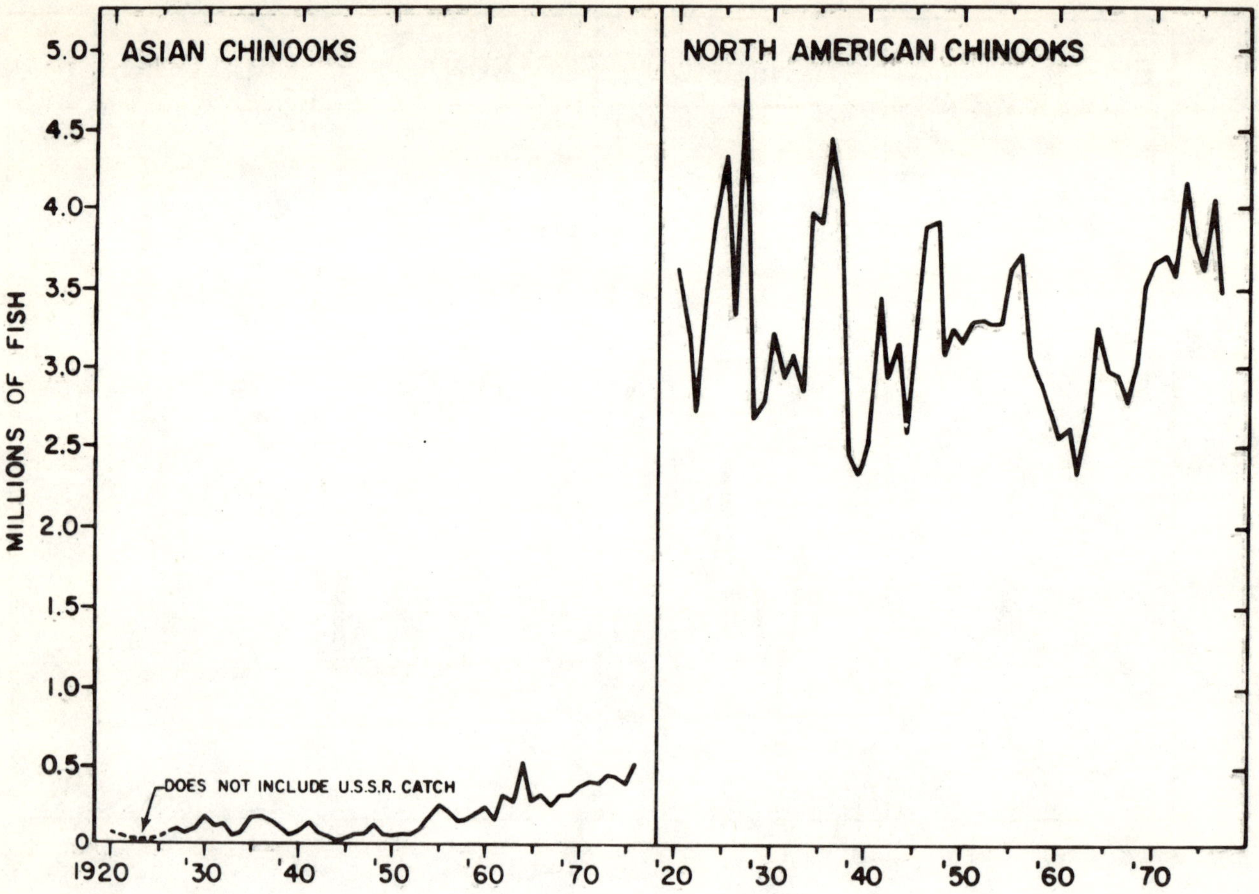

Figure 20. Catch of chinook salmon of Asian and North American origins by commercial fisheries of the U.S.S.R., Japan, U.S. and Canada, 1920-77. (Data are from Table 5.)

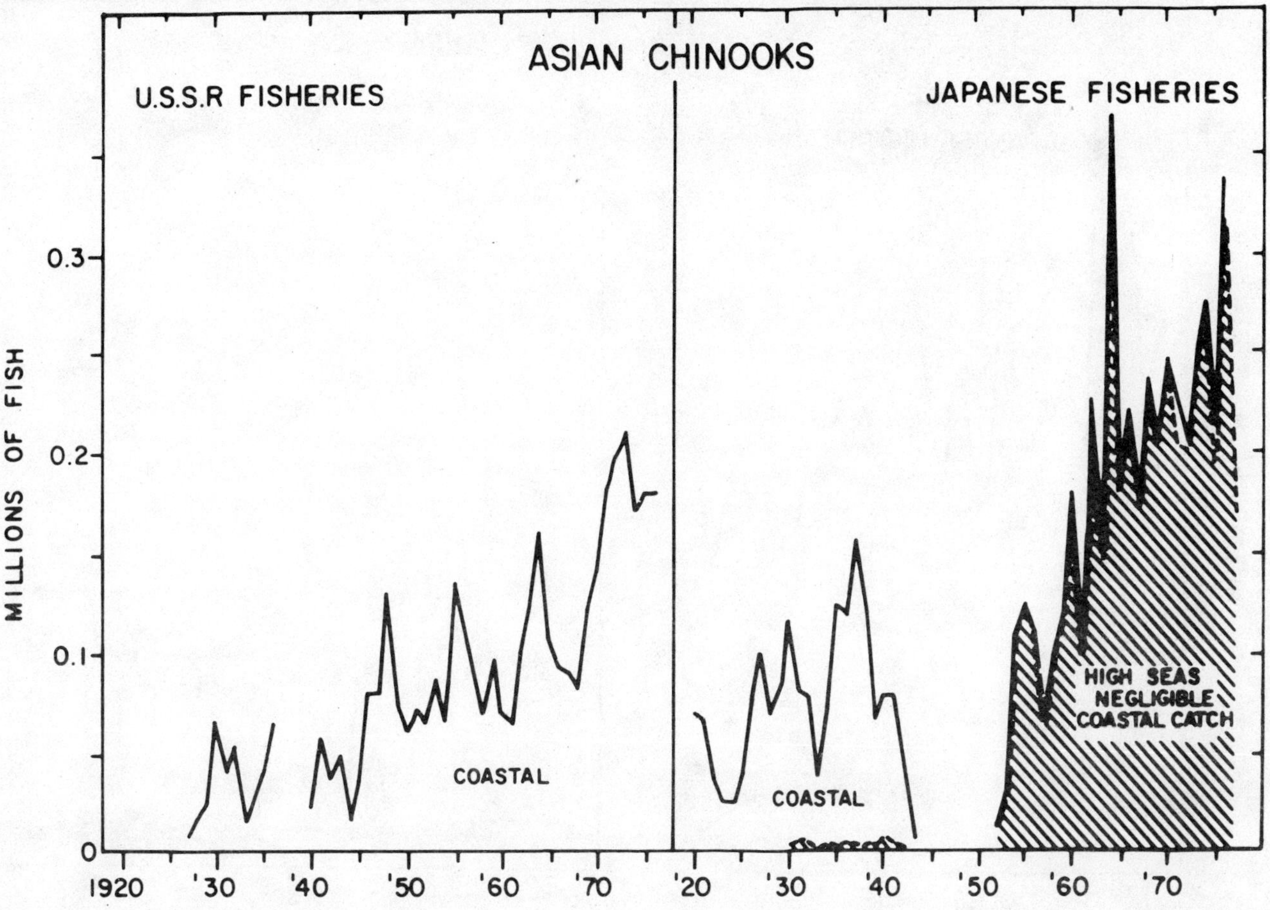

Figure 21. Catch of Asian chinook salmon by commercial fisheries of the U.S.S.R. and Japan, 1920-77. (Data are from Table 5.)

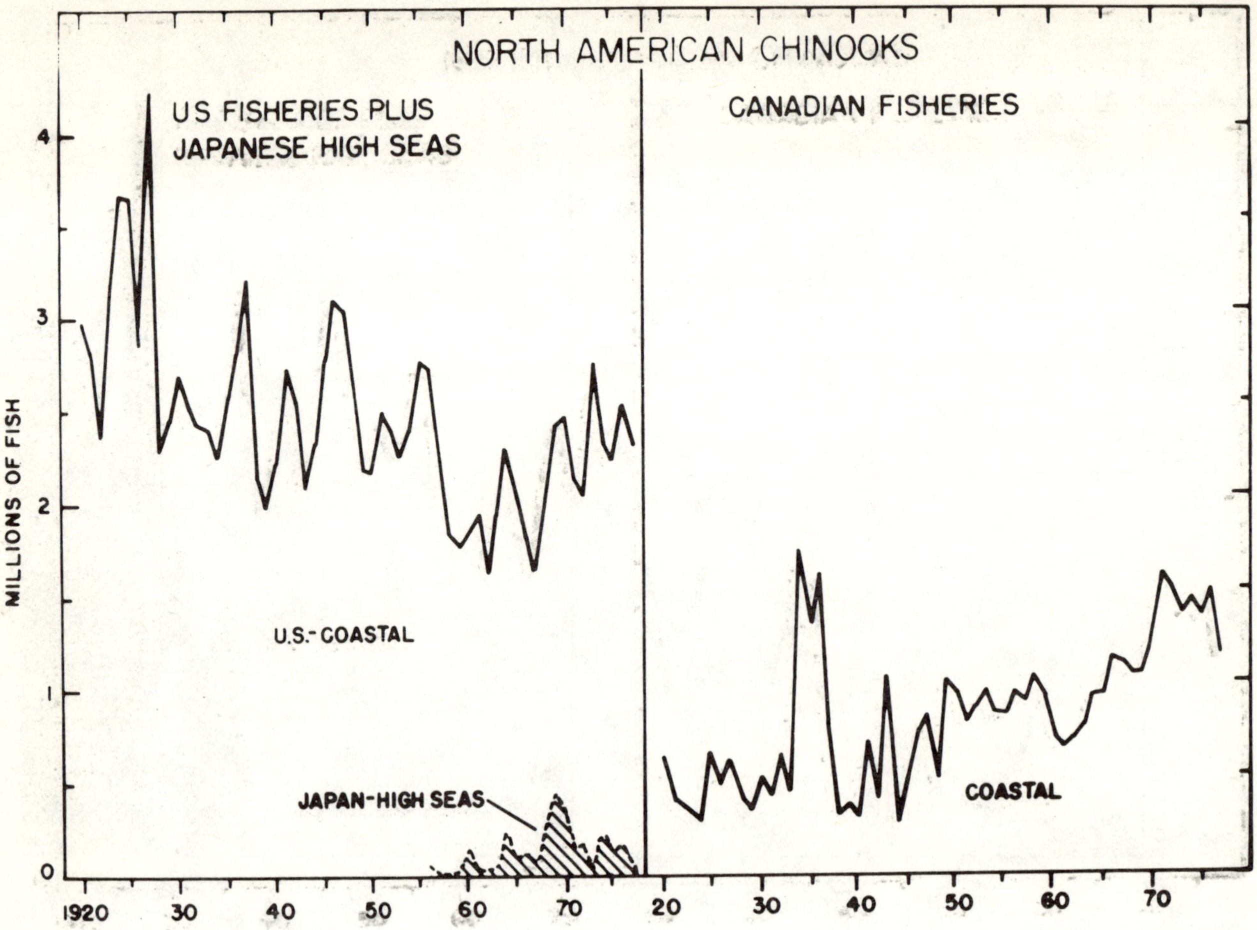

Figure 22. Catch of North American chinook salmon by commercial fisheries of Japan, U.S. and Canada, 1920-77. (Data are from Table 5.)

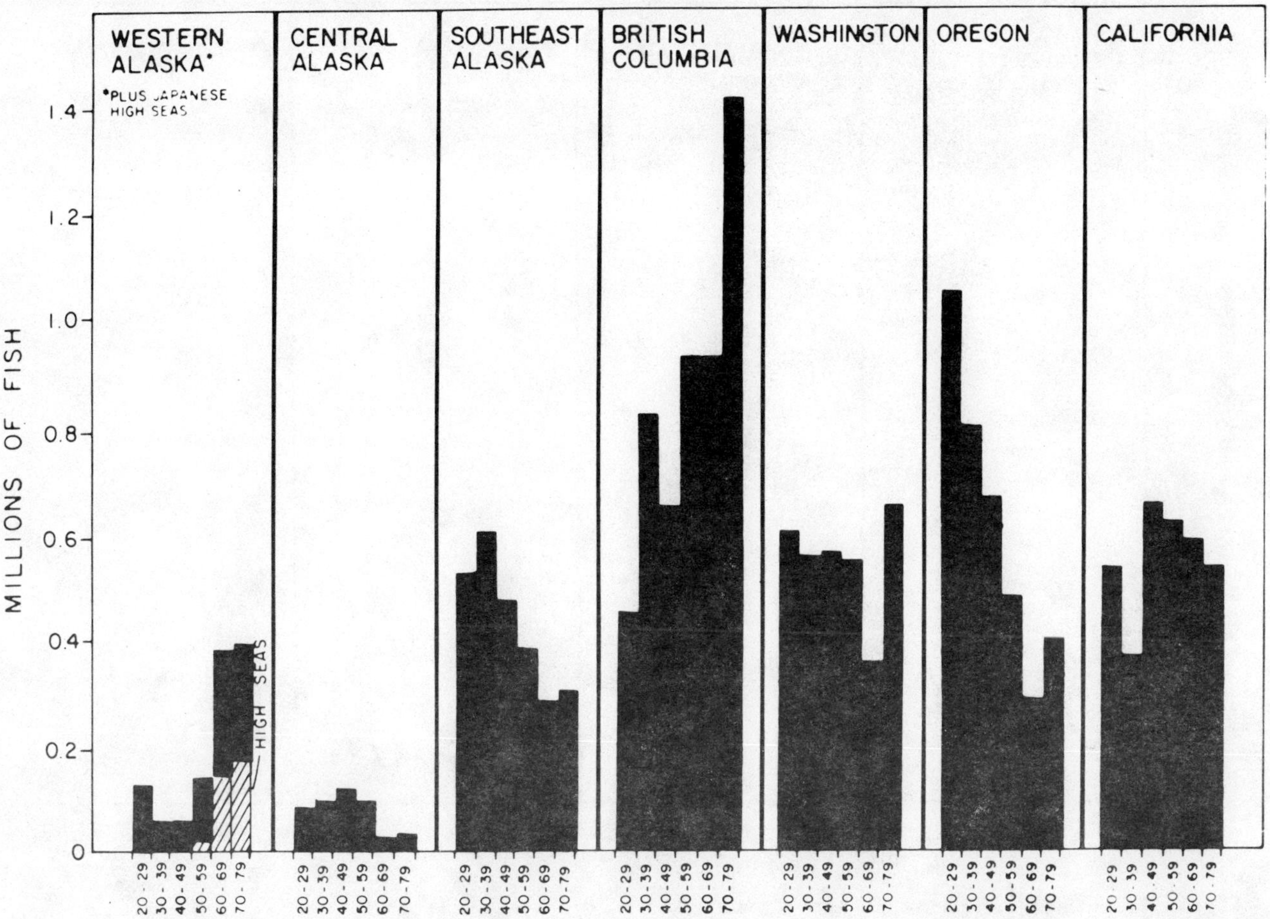

Figure 23. Average annual commercial catch of North American chinook salmon, by region and decade, 1920-29 to 1970-77. (Data are from Table 5.)

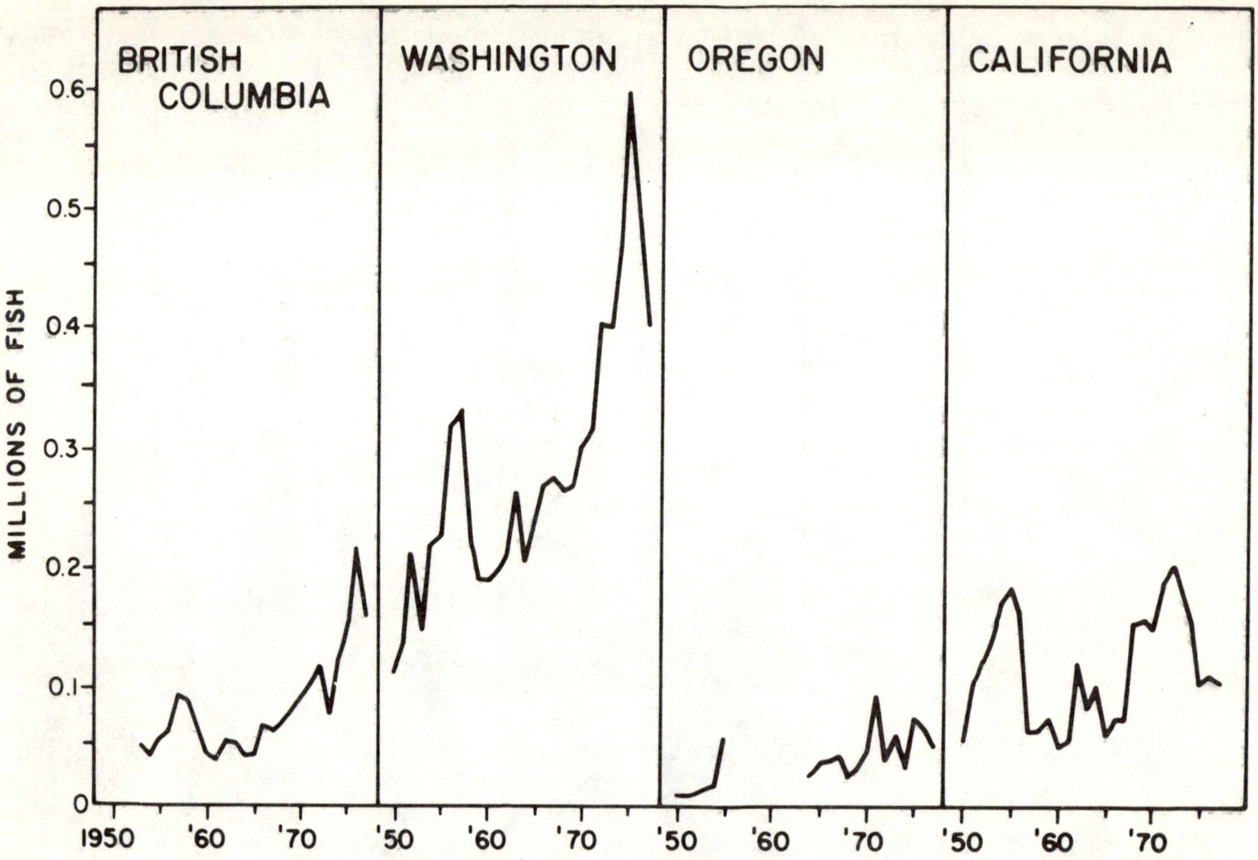

Figure 24. Catch of chinook salmon by sport fisheries of British Columbia, Washington, Oregon and California, 1950-77. (Data are from Table 6.)

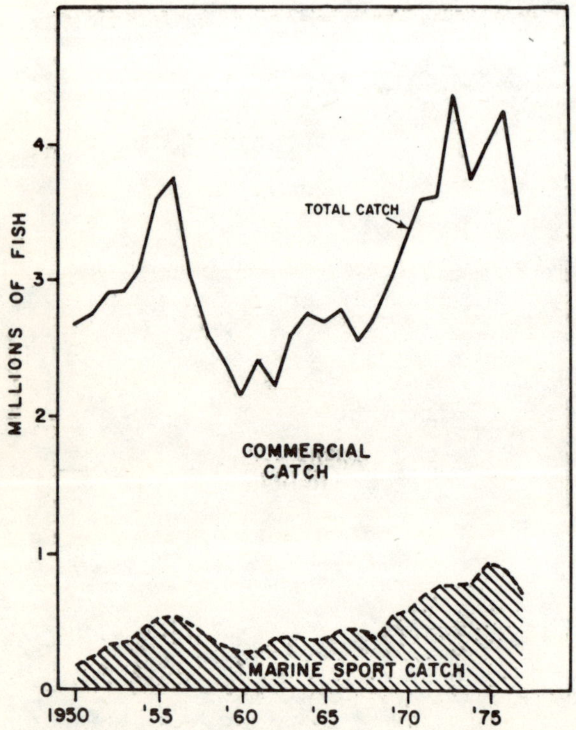

Figure 25. Catch of chinook salmon by commercial and marine sport fisheries of British Columbia, Washington, Oregon and California combined, 1950-77. (Data are from Tables 5 and 6.)

LITERATURE CITED

Department of Fisheries and Environment. 1977. Annual summary of British Columbia catch statistics, 1976. Dep. Fish. Environ., Can., Fish. Manage., Vancouver, B.C. 22 pp.

Fredin, R. A., R. L. Major, R. G. Bakkala, and G. Tanonaka. 1977. Pacific salmon and the high seas salmon fisheries of Japan. Natl. Ocean. Atmos. Admin., Natl. Mar. Fish. Serv., Northwest and Alaska Fish. Center., Seattle, Wash., Proc. Rep. 324 pp.

Holland, G. A. 1977. Information on coho and chinook salmon stocks originating in Puget Sound and Washington coastal streams north of the Columbia River. Pages 23-25 *in* Additional information on the exploitation, scientific investigation, and management of salmon stocks on the Pacific coast of the United States in relation to the abstention provisions of the North Pacific Fisheries Convention. Int. North Pac. Fish. Comm. Bull. 36.

Kasahara, H. 1963. Catch statistics for North Pacific salmon. Pages 7-82 *in* Salmon of the North Pacific Ocean. Part I. Int. North Pac. Fish. Comm. Bull. 12.

Secretariat, International North Pacific Fisheries Commission. 1974. Historical catch statistics for salmon of the North Pacific Ocean (2nd draft). Int. North Pac. Fish Comm., Vancouver, B.C. 28 pp. 92 tables. (Processed)

APPENDIX

Symbols Used in Tables 1-6

* Not available
\- No catch
\+ Less than 500 fish

a/ MS-Mothership fishery; LBDN-Landbased driftnet fishery; LBLL-Landbased longline fishery.

b/ A-Asian origin; NA-North American origin.

c/ The mothership fishery operated just outside the territorial seas off the Kamchatka Peninsula from 1929-1942. Its catch during that period is considered to have consisted of Asian fish only. The mothership fishery began operating in the North Pacific Ocean and Bering Sea in 1952 and operated in the Okhotsk Sea during 1955-1958. Estimates of interceptions of North American salmon by the mothership fishery have been made since 1956. It is assumed that the fishery caught a relatively small number of North American salmon during 1952-1955.

d/ The landbased driftnet fishery operated up to 50 miles offshore from the northern Kurile Islands during 1933-1945. Its catch during that period is considered to have consisted of Asian fish only. The fishery began operating in the North Pacific Ocean in 1952. Estimates of interceptions of North American salmon by that fishery have not been made. All fish taken are provisionally categorized as being of Asian origin.

e/ The landbased longline fishery operated in the North Pacific Ocean from 1952 through 1971. All fishing was carried out west of 160°E except in 1971 when the fleet operated as far east as 165°E. All fish taken are considered in the coastal landings.

f/ The coastal fishery operated on the U.S.S.R. coast, the Kurile Islands and Sakhalin Island, as well as in Japan, until World War II. Since then, the fishery has operated only in Hokkaido and Honshu. Catches in the Sea of Japan are included in the coastal landings.

g/ The 1922-1931 average was used for 1920-1921.

h/ Incomplete for the years from which U.S.S.R. catch data are not available

i/ Preliminary.

Table 1. Catch of Sockeye Salmon of Asian and North American Origins by Commercial Fisheries of the U.S.S.R., Japan, U.S. and Canada, in Thousands of Fish, 1920-77.

Nation	Fishery[a/]	Origin of fish caught[b/]	1920	1921	1922	1923	1924	1925	1926	1927	1928	1929
U.S.S.R.	Coastal	A	*	*	*	*	*	*	*	2836	3374	3210
Japan	High Seas-MS[c/]	NA	-	-	-	-	-	-	-	-	-	-
	" " "	A	-	-	-	-	-	-	-	-	-	5
			-	-	-	-	-	-	-	-	-	5
	" " -LBND	A[d/]	-	-	-	-	-	-	-	-	-	-
	" " -LBLL	"[e/]	-	-	-	-	-	-	-	-	-	-
	" " -Total	NA + A	-	-	-	-	-	-	-	-	-	5
	Coastal[f/]	A	4176	5657	10593	8347	8205	4326	6561	6943	11996	7975
	Total	NA + A	4176	5657	10593	8347	8205	4326	6561	6943	11996	7975
U.S.	Coastal -W. AK	NA	10280	17427	24320	18914	11030	8334	20088	11230	20566	13067
	C. AK	"	7334	7285	7361	5648	5729	5047	9003	4245	5237	4980
	SE. AK	"	2744	1499	1900	2410	2506	1844	2045	1445	1351	1901
			20358	26211	33581	26972	19265	15225	31136	16920	27154	19948
	WA	"	680	1174	640	687	776	1274	537	1018	642	1314
	OR[g/]	"	165	165	268	592	125	101	231	68	44	93
	GA[g/]	"	-	-	-	-	-	-	-	-	-	-
	Total		21203	27550	34489	28251	20166	16600	31904	18006	27840	21355
Canada	Coastal -N. Dist.	NA	*	*	*	*	*	*	*	*	*	*
	S. "	"	*	*	*	*	*	*	*	*	*	*
	Fraser Area	"	*	*	*	*	*	*	*	*	*	*
	Total		4281	2160	3959	4474	4983	4990	4232	4072	2447	3604
Total		A[h/]	4176	5657	10593	8347	8205	4326	6561	9779	15370	11190
		NA	25484	29710	38448	32725	25149	21590	36136	22078	30287	24959
		Total[h/]	29660	35367	49041	41072	33354	25916	42697	31857	45657	36149

Table 1. Catch of Sockeye Salmon of Asian and North American Origins by Commercial Fisheries of the U.S.S.R., Japan, U.S. and Canada, in Thousands of Fish, 1920-77 -- continued.

Nation	Fishery[a/]	Origin of fish caught[b/]	1930	1931	1932	1933	1934	1935	1936	1937	1938	1939
U.S.S.R.	Coastal	A	4557	3561	2849	1645	4184	1533	2479	*	*	*
Japan	High Seas - MS[c/]	NA	-	-	-	-	-	-	-	-	-	-
	" " "	A	307	596	1271	2034	4860	2496	3940	5782	4268	5601
			307	596	1271	2034	4860	2496	3940	5782	4268	5601
	" " - LBND	A[d/]	-	-	-	848	2618	867	2273	3455	3937	3070
	" " - LBLL	"[e/]	-	-	-	-	-	-	-	-	-	-
	" " - Total	NA+ A	307	596	1271	2882	7478	3363	6213	9237	8205	8671
	Coastal[f/]	A	7891	6117	6157	4287	8366	3292	6115	6727	8441	5491
	Total	NA + A	8198	6713	7428	7169	15844	6655	12328	15964	16646	14162
U.S.	Coastal - W. AK	NA	4429	13552	15917	24059	21692	3502	21198	22119	25709	14079
	C. AK	"	3377	5486	7832	6113	7260	4854	10757	6811	6458	8239
	SE. AK	"	2587	1823	1652	1010	1242	1893	2403	2187	2535	2487
			10393	20861	25401	31182	30194	10249	34358	31117	34702	24805
	WA	"	3285	743	835	1611	2692	717	528	1001	1566	592
	OR[g/]	"	80	35	26	30	59	12	81	71	85	60
	CA[g/]	"	-	-	-	-	-	-	-	-	-	-
	Total		13758	21639	26262	32823	32945	10978	34967	32189	36353	25457
Canada	Coastal - N. Dist.	NA	*	*	*	*	*	*	*	*	*	*
	S. "	"	*	*	*	*	*	*	*	*	*	*
	Fraser Area	"	*	*	*	*	*	*	*	*	*	*
	Total		5343	3070	3608	3562	4584	4607	4878	4593	5069	4082
Total		A[h/]	12755	10274	10277	8814	20028	8188	14807	15964	16646	14162
		NA	19101	24709	29870	36385	37529	15585	39845	36782	41422	29539
	Total[h/]		31856	34983	40147	45199	57557	23773	54642	52746	58068	43701

Table 1. Catch of Sockeye Salmon of Asian and North American Origins by Commercial Fisheries of the U.S.S.R., Japan, U.S. and Canada, in Thousands of Fish, 1920-77 -- continued.

Nation	Fishery[a/]	Origin of fish caught[b/]	1940	1941	1942	1943	1944	1945	1946	1947	1948	1949
U.S.S.R.	Coastal	A	1190	1610	1994	2729	3289	4211	4617	3575	2161	3054
Japan	High Seas - MS[c/]	NA	-	-	-	-	-	-	-	-	-	-
	" " "	A	3218	4298	1329	-	-	-	-	-	-	-
			3218	4298	1329	-	-	-	-	-	-	-
	" " - LBND	A[d/]	1742	1632	2309	3908	1314	236	-	-	-	-
	" " - LBLL	"[e/]	-	-	-	-	-	-	-	-	-	-
	" " - Total	NA+ A	4960	5930	3638	3908	1314	236	-	-	-	-
	Coastal[f/]	A	3981	4132	4356	2656	1428	18	-	-	-	-
	Total	NA + A	8941	10062	7994	6564	2742	254	-	-	-	-
U.S.	Coastal - W. AK	NA	5406	7645	6343	17898	11960	7695	8749	19000	15022	6586
	C. AK	"	4770	5268	4949	7970	5609	5632	4596	6243	4333	4761
	SE. AK	"	1507	1674	1566	1124	1722	1683	888	746	525	490
			11683	14587	12858	26992	19291	15010	14233	25989	19880	11837
	WA	"	990	2149	3105	290	384	791	3628	289	1180	1318
	OR[g/]	"	76	102	41	26	14	2	30	166	21	4
	CA[g/]	"	-	-	-	-	-	-	-	-	-	-
	Total		12749	16838	16004	27308	19689	15803	17891	26444	21081	13159
Canada	Coastal - N. Dist.	NA	*	*	*	*	*	*	*	*	*	*
	S. "	"	*	*	*	*	*	*	*	*	*	*
	Fraser Area	"	*	*	*	*	*	*	*	*	*	*
	Total		4222	5349	7910	2067	2004	3771	7104	3905	2828	2959
Total		A[h/]	10131	11672	9988	9293	6031	4465	4617	3575	2161	3054
		NA	16971	22187	23914	29375	21693	19574	24995	30349	23909	16118
		Total[h/]	27102	33859	33902	38668	27724	24039	29612	33924	26070	19172

Table 1. Catch of Sockeye Salmon of Asian and North American Origins by Commercial Fisheries of the U.S.S.R., Japan, U.S. and Canada, in Thousands of Fish, 1920-77 -- continued.

Nation	Fishery[a/]	Origin of fish caught[b/]	1950	1951	1952	1953	1954	1955	1956	1957	1958	1959
U.S.S.R.	Coastal	A	3231	2673	3191	1768	1389	1079	1984	1217	349	1400
Japan	High Seas - MS[c/]	NA	-	-	c/	c/	c/	c/	3335	6454	399	650
	" " "	A	-	-	736	1580	3816	12494	6955	13642	11627	8475
			-	-	736	1580	3816	12494	10290	20096	12026	9125
	" " - LBND	A[d/]	-	-	89	191	344	511	271	488	903	845
	" " - LBLL	"[e/]	-	-	+	+	+	+	1	5	9	9
	" " - Total	NA+ A	-	-	825	1771	4160	13005	19562	20589	12938	9979
	Coastal[f/]	A	-	-	-	-	-	-	-	-	-	-
	Total	NA + A	-	-	825	1771	4160	13005	10562	20589	12938	9979
U.S.	Coastal - W. AK	NA	7285	4697	11664	6653	5014	5148	10252	6631	3460	5249
	C. AK	"	6537	4136	4346	3763	3406	2815	3616	2189	1636	1937
	SE. AK	"	552	820	919	1376	1220	747	915	1071	1008	891
			14374	9653	16929	11792	9640	8710	14783	9891	6104	8077
	WA	"	1359	1220	1182	2068	4941	1118	1068	1754	5344	1890
	OR[g/]	"	37	32	128	22	45	18	46	51	144	139
	CA[g/]	"	-	-	-	-	-	-	-	-	-	-
	Total		15770	10905	18239	13882	14626	9846	15897	11696	11592	10106
Canada	Coastal - N. Dist.	NA	*	2656	3213	3109	1757	1457	2159	986	2697	1185
	S. "	"	*	720	693	1489	2788	849	516	1405	7499	1177
	Fraser Area	"	*	1038	957	1329	2165	529	582	645	1848	898
	Total		4549	4414	4863	5927	6710	2835	3257	3036	12044	3260
Total		A[h/]	3231	2673	4016	3539	5549	14084	9211	15352	12888	10729
		NA	20319	15319	23102	19809	21336	12681	22489	21186	24035	14016
		Total[h/]	23550	17992	27118	23348	26885	26765	31700	36538	36923	24745

Table 1. Catch of Sockeye Salmon of Asian and North American Origins by Commercial Fisheries of the U.S.S.R., Japan, and Canada, in Thousands of Fish, 1920-77 -- continued.

Nation	Fishery[a]	Origin of fish caught[b]	1960	1961	1962	1963	1964	1965	1966	1967	1968	1969
U.S.S.R.	Coastal	A	1402	2721	1425	1237	969	1506	1151	971	836	587
Japan	High Seas - MS[c]	NA	3948	5949	906	990	1095	6495	1588	884	1652	1754
	" " "	A	8931	7049	9684	7913	6002	5543	5666	7203	4721	4181
			12879	12998	10590	8903	7097	12038	7254	8087	6373	5935
	" " - LBND	A[d]	1627	1192	154	18	108	159	703	2566	2769	2495
	" " - LBLL	"[e]	49	26	6	2	6	1	1	1	1	1
	" " - Total	NA+ A	14555	14216	10750	8923	7211	12198	7958	19654	9143	8431
	Coastal[f]	A	-	-	-	-	-	-	-	-	-	-
	Total	NA + A	14555	14216	10750	8923	7211	12198	7958	10654	9143	8431
U.S.	Coastal - W. AK	NA	14411	12307	4990	3101	5844	24457	9562	4556	3039	6955
	C. AK	"	2835	3030	3534	2437	3198	4229	4458	3049	4260	3650
	SE. AK	"	588	744	772	678	924	1085	1054	972	831	812
			17834	16081	9296	6216	9966	29771	15074	8577	8130	11417
	WA	"	1264	1428	781	1414	541	1048	1402	2160	988	1668
	OR[g]	"	89	27	9	8	15	3	2	33	17	19
	CA[g]	"	-	-	-	-	-	-	-	-	-	-
	Total		19187	17536	10086	7638	10522	30822	16478	10770	9135	13104
Canada	Coastal - N. Dist.	NA	1265	2542	2409	1134	2873	1718	1817	3310	4662	1824
	S. "	"	896	1307	552	454	281	436	1574	2554	772	1485
	Fraser Area	"	697	716	538	498	465	866	629	878	831	957
	Total		2858	4565	3499	2086	3619	3020	4020	6742	6265	4266
Total		A[h]	12009	10988	11269	9170	7085	7209	7521	10741	8327	7264
		NA	25993	28050	14491	10714	15236	40337	22086	18396	17052	19124
		Total[h]	38002	39038	25760	19884	22321	47546	29607	29137	25379	26388

Table 1. Catch of Sockeye Salmon of Asian and North American Origins by Commercial Fisheries of the U.S.S.R., Japan, U.S. and Canada, in Thousands of Fish, 1920-77 -- continued.

Nation	Fishery[a/]	Origin of fish caught[b/]	1970	1971	1972	1973	1974	1975	1976	1977
U.S.S.R.	Coastal	A	1612	750	336	667	437	500	400	650
Japan	High Seas - MS[c/]	NA	4663	1430	921	882	954	864	1007	640
	" " - "	A	2281	2124	2263	1731	1328	1307	1270	868
			6944	3554	3184	2613	2282	2171	2277	1508
	" " - LBND	A[d/]	2966	3026	3718	3316	3158	2982	3531	1289
	" " - LBLL	"[e/]	51	2	-	-	-	-	-	-
	" " - Total	NA+ A	9961	6582	6902	5929	5440	5153	5808	2797
	Coastal[f/]	A	-	-	-	-	-	-	-	-
	Total	NA + A	9961	6582	6902	5929	5440	5153	5808	2797
U.S.	Coastal - W. AK	NA	20946	9945	2600	939	1637	5170	6275	5143
	C. AK	"	6020	3611	3073	2540	2545	2039	4912	5991
	SE. AK	"	668	623	917	1011	687	245	594	995
			27634	14179	6590	4490	4869	7454	11781	12129
	WA	"	1374	3129	1227	2678	2512	1693	1319	1709
	OR[g/]	"	12	44	43	2	-	-	-	-
	CA[g/]	"	-	-	-	-	-	-	-	-
	Total		29020	17352	7860	7170	7381	9147	13100	13838
Canada	Coastal - N. Dist.	NA	1216	1949	1748	4345	2706	1066	1896	2589
	S. Dist.	"	2288	3040	1298	2192	3805	778	2358	2583
	Fraser Area	"	575	1317	527	1063	714	428	558	1176
	Total		4079	6306	3573	7600	7225	2272	4812	6348
Total		A[h/]	6910	5902	6317	5714	4923	4789	5201	2807
		NA	37762	25088	12354	15652	15560	12283	18919	20826
		Total[h/]	44672	30990	18671	21366	20483	17072	24120	23633

DATA SOURCES (FOR STATISTICS ON TOTAL CATCHES)*

U.S.S.R.	1927-36.	As indicated in footnote 2, Table 3-I, Secretariat, INPFC (1974).
	1940-61.	INPFC ms. Catches in metric tons by district (Table 33) converted to numbers of fish using the average weight (by district) in 1962-1973 (Tables 33 and 34).
	1962-75.	As indicated in Appendix 1, Fredin et al. (1977).
	1976-77.	Catch in metric tons (Foreign Fisheries Information Releases 77-3 and 78-5, National Marine Fisheries Service, National Oceanic and Atmospheric Administration) converted to numbers of fish using average weight in 1966-1975 (Table 6.2, Fredin et al. 1977).
Japan	1920-45.	As indicated in footnotes 3-5, Table 3-1, INPFC ms.
	1952-75.	As indicated in Appendix 1, Fredin et al. (1977).
	1976-77.	Personal communication, Fishery Agency of Japan.
U.S.	1920-51.	As indicated in footnotes 2-6, Table 57, INPFC ms. Note: Oregon catches in metric tons during 1922-1937 converted to numbers of fish using 1.583 kg/fish, the average weight during 1938-1971 (Table 92, INPFC ms.).
	1952-75.	As indicated in Appendix 1, Fredin et al. (1977).
	1976-77.	Personal communications, state fisheries agencies of Alaska, California, Oregon and Washington.
Canada	1920-50.	Table 62, INPFC ms.
	1951-75.	Table 64, INPFC ms., as updated.
	1976-77.	Department of Fisheries and Environment, 1977-1978.

*Estimates of interceptions of North American sockeye salmon by the Japanese mothership salmon fishery during 1956-1975 are from Table 5.1C, Fredin et al. (1977). The estimate for 1976 is from D. D. Worlund and H. Gangmark (personal communication) and the estimate for 1977 is by Fredin (*unpublished*).

Table 2. Catch of Pink Salmon of Asian and North American Origins by Commercial Fisheries of the U.S.S.R., Japan, U.S. and Canada, in Thousands of Fish, 1920-77.

Nation	Fishery[a]	Origin of fish caught[b]	1920	1921	1922	1923	1924	1925	1926	1927	1928	1929
U.S.S.R.	Coastal	A	*	*	*	*	*	8067	1639	10164	38406	7109
Japan	High Seas - MS[c]	NA	-	-	-	-	-	-	-	-	-	-
	" " "	A	-	-	-	-	-	-	-	-	-	+
			-	-	-	-	-	-	-	-	-	+
	" " - LBND	A[d]	-	-	-	-	-	-	-	-	-	-
	" " - LBLL	"[e]	-	-	-	-	-	-	-	-	-	-
	" " - Total	NA+ A	-	-	-	-	-	-	-	-	-	+
	Coastal[f]	A	76262	92919	90393	39953	112923	30868	126522	31564	110530	41034
	Total	NA + A	76262	92919	90393	39953	112923	30868	126522	31564	110530	41034
U.S.	Coastal - W. AK	NA	2045	1	290	+	777	4	810	335	47	-
	C. AK	"	10145	163	6941	3568	19529	7196	20743	13316	11084	17965
	SE. AK	"	18121	7735	24002	39879	30029	28246	32193	8163	36051	21848
			30311	7899	31233	43447	50335	35446	53746	21814	47182	39813
	WA	"	20	4413	32	5534	100	6665	24	6526	9	7433
	OR[g]	"	-	-	-	10	-	-	-	-	-	-
	CA[g]	"	-	-	-	-	-	-	-	-	-	-
	Total		30331	12312	31265	48991	50435	42111	53770	28340	47191	47246
Canada	Coastal - N. Dist.	NA	*	*	*	*	*	*	*	*	*	*
	S. Dist.	"	*	*	*	*	*	*	*	*	*	*
	Fraser Area	"	*	*	*	*	*	*	*	*	*	*
	Total		102C4	3834	10716	7917	11754	7634	13604	4155	14334	7552
Total		A[h]	76262	92919	90393	39953	112923	38935	128161	41728	148936	48143
		NA	40535	16146	41981	56908	62189	49745	67374	32495	61525	54798
		Total[h]	116797	109065	132374	96861	175112	88680	195535	74223	210461	102941

Table 2. Catch of Pink Salmon of Asian and North American Origins by Commercial Fisheries of the U.S.S.R., Japan, U.S. and Canada, in Thousands of fish, 1920-77 -- continued.

Nation	Fishery[a]	Origin of fish caught[b]	1930	1931	1932	1933	1934	1935	1936	1937	1938	1939
U.S.S.R.	Coastal	A	50209	22638	62116	18748	61470	34325	40103	*	*	*
Japan	High Seas - MS[c]	NA	-	-	-	-	-	-	-	-	-	-
	" " "	A	197	67	1346	1793	1251	5970	1308	1230	1483	2047
			197	67	1346	1793	1251	5970	1308	1230	1483	2047
	" " - LBND	A[d]	-	-	-	3512	877	9490	6844	11376	9825	5221
	" " - LBLL	"[e]	-	-	-	-	-	-	-	-	-	-
	" " - Total	NA+ A	197	67	1346	5305	2128	15460	8152	12606	11308	7268
	Coastal[f]	A	73747	49037	65850	40313	105608	110190	73737	133694	11546	199229
	Total	NA + A	73944	49104	67196	45618	107736	125650	81889	146300	122854	206497
U.S.	Coastal - W. AK	NA	249	-	172	+	34	+	527	+	-	+
	C. AK	"	15613	13340	13045	13157	22439	19589	26150	31117	24187	20417
	SE. AK	"	43483	27244	22606	25783	50328	30248	50747	35166	30288	23722
			59345	40584	35823	38940	72801	49837	77424	66283	54475	44139
	WA	"	17	8132	9	6040	13	5825	24	5086	8	4067
	OR[g]	"	-	-	-	1	-	-	-	+	-	-
	CA[g]	"	-	-	-	-	-	-	-	-	-	-
	Total		59362	48716	35832	44981	72814	55662	77448	71369	54483	48206
Canada	Coastal - N. Dist.	NA	*	*	*	*	*	*	*	*	*	*
	S. Dist.	"	*	*	*	*	*	*	*	*	*	*
	Fraser Area	"	*	*	*	*	*	*	*	*	*	*
	Total		23873	3273	4025	9368	9289	7405	12114	10324	6991	8832
Total		A[h]	124153	71742	129312	64366	169206	159975	121992	146300	122854	206497
		NA	83235	51989	39857	54349	82103	63067	89562	81693	61474	57038
		Total[h]	207388	123731	169169	118715	251309	223042	211554	227993	184328	263535

Table 2. Catch of Pink Salmon of Asian and North American Origins by Commercial Fisheries of the U.S.S.R., Japan, U.S. and Canada, in THousands of Fish, 1920-77 -- continued.

Nation	Fishery[a/]	Origin of fish caught[b/]	1940	1941	1942	1943	1944	1945	1946	1947	1948	1949
U.S.S.R.	Coastal	A	33832	51727	52920	79759	70049	48154	24187	84022	33007	105223
Japan	High Seas - MS [c/]	NA	-	-	-	-	-	-	-	-	-	-
	" " "	A	2495	3923	4236	-	-	-	-	-	-	-
			2495	3923	4236	-	-	-	-	-	-	-
	" " - LBND	A[d/]	7221	9451	13167	8319	2206	1020	-	-	-	-
	" " - LBLL	"[e/]	-	-	-	-	-	-	-	-	-	-
	" " - Total	NA+ A	9716	13374	17403	8319	2206	1020	-	-	-	-
	Coastal[f/]	A	47511	118288	61239	112332	26358	4371	1107	2991	730	6103
	Total	NA + A	57227	131662	78642	120651	28564	5391	1107	2991	730	6103
U.S.	Coastal - W. AK	NA	258	3	172	1	58	3	41	+	53	+
	C. AK	"	31414	17776	21310	29209	18339	26126	21781	20408	12845	13086
	SE. AK	"	29094	60061	33127	18039	19344	21655	24822	14042	14352	43921
			60766	77840	54609	47249	37741	47784	46644	34450	27250	57007
	WA	"	26	2021	2	973	4	5895	+	9372	+	7605
	OR[g/]	"	-	-	-	-	-	-	-	-	-	-
	CA [g/]	"	-	-	-	-	-	-	-	-	-	-
	Total		60792	79861	54611	48222	37745	53679	46644	43822	27250	64612
Canada	Coastal - N. Dist.	NA	*	*	*	*	*	*	*	*	*	*
	S. Dist.	"	*	*	*	*	*	*	*	*	*	*
	Fraser Area	"	*	*	*	*	*	*	*	*	*	*
	Total		3795	6154	3913	9474	6636	14454	3435	12334	7013	10854
Total		A[h/]	91059	183389	131562	200410	98613	53545	25294	87013	33737	111326
		NA	64587	86015	58524	57696	44381	68133	50079	56156	34263	75466
		Total[h/]	155646	269404	190086	258106	142994	121678	75373	143169	68000	186792

Table 2. Catch of Pink Salmon of Asian and North American Origins by Commercial Fisheries of the U.S.S.R., Japan, U.S. and Canada, in Thousands of Fish, 1920-77 -- continued.

Nation	Fishery[a/]	Origin of fish caught[b/]	1950	1951	1952	1953	1954	1955	1956	1957	1958	1959
U.S.S.R.	Coastal	A	23589	99274	36992	92493	29767	59273	45382	71535	24487	32437
Japan	High Seas - MS[c/]	NA	-	-	c/	c/	c/	c/	16	-	-	-
	" " "	A	-	-	699	3065	5804	25917	17357	27881	15546	18856
			-	-	699	3065	5804	25917	17373	27881	15546	18856
	" " - LBND	A[d/]	-	-	13344	10270	9193	22396	21534	30353	24833	35129
	" " - LBLL	"[e/]	-	-	723	475	956	886	5856	10892	5621	9100
	" " - Total	NA+ A	-	-	14766	13810	15953	49199	44763	69126	46000	63085
	Coastal[f/]	A	3129	5168	5223	3172	4066	6077	10926	9544	18141	12781
	Total	NA + A	3129	5168	19989	16982	20019	55276	55689	78670	64141	75866
U.S.	Coastal - W. AK	NA	32	21	47	88	688	32	126	4	1809	22
	C. AK	"	10518	6185	10012	10602	13293	15459	12567	6892	14452	3057
	SE. AK	"	9424	22220	9803	4981	8909	9334	13728	6858	9838	7851
			19974	28426	19862	15671	22890	24825	26421	13754	26099	10930
	WA	"	10	5700	2	6045	+	5208	1	3184	5	2616
	OR[g/]	"	-	-	-	1	-	2	-	-	-	1
	CA[g/]	"	-	-	-	-	-	2	-	-	-	+
	Total		19984	34126	19864	21717	22890	30037	26422	16938	26104	13547
Canada	Coastal - N. Dist.	NA	*	6632	8303	1798	4870	3842	6235	4205	5501	1948
	S. "	"	*	4382	2931	8243	574	6561	1115	6190	1404	4345
	Fraser Area	"	*	805	1	1069	+	844	2	916	3	483
	Total		6876	11819	11235	11110	5444	11247	7352	11311	6908	6776
Total		A[h/]	26718	104442	56981	109475	49786	114549	101055	150205	88628	108303
		NA	26860	45945	31099	32827	28334	41284	33790	28249	33012	20323
		Total[h/]	53578	150387	88080	142302	78120	155833	134845	178454	121640	128626

Table 2. Catch of Pink Salmon of Asian and North American Origins by Commercial Fisheries of the U.S.S.R., Japan, U.S. and Canada, in Thousands of Fish, 1960-77 -- continued.

Nation	Fishery[a/]	Origin of fish caught[b/]	1960	1961	1962	1963	1964	1965	1966	1967	1968	1969
U.S.S.R.	Coastal	A	13240	21531	10780	24571	10627	34674	14012	33086	10084	42773
Japan	High Seas - MS[c/]	NA	23	2	25	+	15	2	23	4	27	3
	" " "	A	1857	3261	1114	6732	2266	4427	2530	7777	3796	6969
			1885	3263	1139	6732	2281	4429	2553	7781	3823	6972
	" " - LBND	A[d/]	20129	34559	14021	31255	17247	29142	16032	23051	15899	23610
	" " - LBLL	"[e/]	6068	11279	11907	17647	7755	18129	11876	18286	7104	17661
	" " - Total	NA+ A	28082	49101	27067	55634	27283	51700	30461	49118	26826	48243
	Coastal[f/]	A	17178	7213	7558	9160	5571	9455	9633	14778	11529	16711
	Total	NA + A	45260	56314	34625	64794	32854	61155	40094	63896	38355	64954
U.S.	Coastal - W. AK	NA	782	132	2981	154	1765	3	2585	39	2977	332
	C. AK	"	12313	8736	29297	14977	24945	9464	17028	3409	16664	20565
	SE. AK	"	2925	12638	11585	19145	18581	10880	20438	3111	25085	4870
			16020	21506	43863	34276	45291	20347	40051	6559	44726	25767
	WA	"	2	753	7	6302	10	729	30	4077	5	1002
	OR[g/]	"	-	-	-	5	-	42	1	268	1	65
	CA[g/]	"	-	-	-	2	-	11	+	30	1	5
	Total		16022	22259	43870	40585	45301	21129	40082	10934	44733	26839
Canada	Coastal - N. Dist.	NA	3743	5937	22569	5494	8748	3780	13654	1540	15714	932
	S. "	"	355	2252	859	6124	880	1210	3607	7746	3923	1385
	Fraser Area	"	+	117	+	584	+	110	+	532	+	255
	Total		4098	8306	23428	12202	9628	5100	17261	9818	19637	2572
Total		A[h/]	58372	77843	45380	89365	43466	95827	54083	96978	48412	107724
		NA	20148	30567	67323	52787	54944	26231	57366	20756	64397	29414
		Total[h/]	78620	108410	112703	142152	98410	122058	111449	117734	112809	137138

Table 2. Catch of Pink Salmon of Asian and North American Origins by Commercial Fisheries of the U.S.S.R., Japan, U.S. and Canada, in Thousands of Fish, 1920-77 -- continued.

Nation	Fishery[a]	Origin of fish caught[b]	1970	1971	1972	1973	1974	1975	1976	1977
U.S.S.R.	Coastal	A	11116	38785	13579	49867	23310	47700	36700	78600
Japan	High Seas - MS[c]	NA	60	5	29	4	253	4	224	5
	" " "	A	1666	8197	3766	12014	7303	14650	6991	9095
			1726	8202	3795	12018	7556	14654	7215	9100
	" " - LBND	A[d]	13403	16977	17578	24227	14424	19049	12232	15041
	" " - LBLL	"[e]	7538	14573	-	-	-	-	-	-
	" " - Total	NA+ A	22667	39752	21373	36245	21980	33703	19447	24141
	Coastal[f]	A	8968	12355	6943	9853	7315	10503	8500	11000
	Total	NA + A	31635	52107	28316	46098	29295	44206	27947	35141
U.S.	Coastal - W. AK	NA	1228	51	177	52	1160	35	1165	52
	C. AK	"	19263	14133	3344	3295	3810	8922	18284	14588
	SE. AK	"	10657	9345	12400	6455	4889	4027	5329	12697
			31148	23529	15921	9802	9859	12984	24778	27337
	WA	"	9	2499	4	2419	1	1408	+	2223
	OR[g]	"	+	2	-	3	+	-	-	96
	CA[g]	"	+	1	+	19	+	3	+	5
	Total		31157	26031	15925	12243	9860	14395	24778	29661
Canada	Coastal - N. Dist.	NA	10851	2105	13359	2023	5640	1485	6185	5158
	S. "	"	2750	5725	858	4215	1734	2793	4160	4906
	Fraser Area	"	1	626	+	284	+	356	1	44
	Total		13602	8456	14217	6522	7374	4634	10346	10358
Total		A[h]	42691	90887	41866	95961	52352	91902	64423	113736
		NA	44819	34492	30171	18769	17487	19033	35348	40024
		Total[h]	87510	125379	72037	114730	69839	110935	99771	153760

DATA SOURCES (FOR STATISTICS ON TOTAL CATCHES)*

U.S.S.R.	1925-36.	As indicated in footnote 2, Table 4-I, Secretariat, INPFC (1974).
	1940-61.	INPFC ms. Catches in metric tons by district (Table 35) converted to numbers of fish using average weight (by district) in 1962-1973 (Tables 35 and 36).
	1962-75.	As indicated in Appendix 1, Fredin et al. (1977).
	1976-77.	Catch in metric tons (Foreign Fisheries Information Releases 77-3 and 78-5, National Marine Fisheries Service, National Oceanic and Atmospheric Administration) converted to numbers of fish using average weight during 1966-1975 (Table 6.2, Fredin et al. 1977).
Japan	1920-51.	As indicated in footnotes 3-8, Table 4-I, INPFC ms.
	1952-75.	As indicated in Appendix 1, Fredin et al. (1977).
	1976-77.	Personal communication, Fishery Agency of Japan**
U.S.	1920-51.	As indicated in footnotes 2-7, Table 58, INPFC ms. Note: Oregon catches in metric tons during 1922-1937 converted to numbers of fish using 2.35 kg/fish, which was the average weight during 1970-1971 (Table 92, INPFC ms.).
	1952-75.	As indicated in Appendix 1, Fredin et al. (1977).
	1976-77.	Personal communications, state fisheries agencies of Alaska, California, Oregon and Washington.
Canada	1920-50.	Table 62, INPFC ms.
	1951-75.	Table 64, INPFC ms., as updated.
	1976-77.	Department of Fisheries and Environment, 1977-1978.

*Estimates of interceptions of North American pink salmon by the Japanese mothership salmon fishery during 1956-1975 are from Table 5.3, Fredin et al. (1977). Estimates for 1976-1977 are by Fredin (*unpublished*).

**1977 coastal catch not available. Catch assumed (by Fredin) to have been approximately the same as the average of 1971, 1973 and 1975.

Table 3. Catch of Chum Salmon of Asian and North American Origins by Commercial Fisheries of the U.S.S.R., Japan, U.S. and Canada in Thousands of Fish, 1920-77.

Nation	Fishery[a]	Origin of fish caught[b]	1920	1921	1922	1923	1924	1925	1926	1927	1928	1929
U.S.S.R.	Coastal	A	*	*	*	*	*	5045	8862	7075	10991	13382
Japan	High Seas - MS[c]	NA	-	-	-	-	-	-	-	-	-	-
	" " "	A	-	-	-	-	-	-	-	-	-	2
			-	-	-	-	-	-	-	-	-	2
	" " - LBND	A[d]	-	-	-	-	-	-	-	-	-	-
	" " - LBLL	"[e]	-	-	-	-	-	-	-	-	-	-
	" " - Total	NA+ A	-	-	-	-	-	-	-	-	-	2
	Coastal[f]	A	14522	13602	13372	14789	7922	11545	11076	12664	17659	19556
	Total	NA + A	14522	13602	13372	14789	7922	11545	11076	12664	17659	19558
U.S.	Coastal - W. AK	NA	627	499	559	211	341	295	405	283	480	767
	C. AK	"	1500	376	911	800	1981	2227	2475	2604	3835	5089
	SE. AK	"	8613	1847	3729	3913	7266	8188	6106	2220	4871	2627
			10740	2722	5199	4924	9588	10710	8986	5107	9186	8483
	WA	"	756	305	606	914	1277	1010	1262	1026	1635	1925
	OR[g]	"	185	185	12	99	272	212	74	334	476	202
	CA[g]	"	-	-	-	-	-	-	-	-	-	-
	Total		11681	3212	5817	5937	11137	11932	10322	6467	11297	10610
Canada	Coastal - N. Dist.	NA	*	*	*	*	*	*	*	*	*	*
	S. "	"	*	*	*	*	*	*	*	*	*	*
	Fraser Area	"	*	*	*	*	*	*	*	*	*	*
	Total		981	1292	3180	4908	6388	6715	7283	5347	7675	3928
Total		A[h]	14522	13602	13372	14789	7922	16590	19938	19739	28650	32940
		NA	12662	4504	8997	10845	17525	18647	17605	11814	18972	14538
		Total[h]	27184	18106	22369	25634	25447	35237	37543	31553	47622	47478

Table 3. Catch of Chum Salmon of Asian and North American Origins by Commercial Fisheries of the U.S.S.R., Japan, U.S. and Canada, in Thousands of Fish, 1920-77 -- continued.

Nation	Fishery[a/]	Origin of fish caught[b/]	1930	1931	1932	1933	1934	1935	1936	1937	1938	1939
U.S.S.R.	Coastal	A	20795	22605	15909	11885	19070	16817	20584	12209	15446	15829
Japan	High Seas - MS[c/]	NA	-	-	-	-	-	-	-	-	-	-
	" " "	A	169	441	662	1792	2649	2912	3023	2802	3856	3818
			169	441	662	1792	2649	2912	3023	2802	3856	3818
	" " - LBND	A[d/]	-	-	-	1636	3449	4383	9879	8693	10196	7327
	" " - LBLL	"[e/]	-	-	-	-	-	-	-	-	-	-
	" " - Total	NA+ A	169	441	662	3428	6098	7295	12902	11495	14052	11145
	Coastal[f/]	A	21770	16104	14827	13052	21288	18903	31035	19993	21656	18046
	Total	NA + A	21939	16545	15489	16480	27386	26198	43937	31488	35708	29191
U.S.	Coastal - W. AK	NA	321	691	965	272	345	106	278	368	580	1017
	C. AK	"	2827	2063	1520	2174	3088	3208	3237	2352	2630	2666
	SE. AK	"	2709	2851	5594	4547	3806	5084	7606	5556	4556	3389
			5857	5605	8079	6993	7239	8398	11121	8276	7766	7072
	WA	"	1172	888	1344	634	1057	844	1037	983	1009	441
	OR[g/]	"	53	140	59	107	88	104	170	143	92	58
	CA[g/]	"	-	-	-	-	-	-	-	-	-	-
	Total		7032	6633	9482	7734	8384	9351	12328	9402	8867	7571
Canada	Coastal - N. Dist.	NA	*	*	*	*	*	*	*	*	*	*
	S. "	"	*	*	*	*	*	*	*	*	*	*
	Fraser Area	"	*	*	*	*	*	*	*	*	*	*
	Total		5114	4033	3941	3120	4841	5016	6074	4882	4664	2552
Total		A[h/]	42734	39150	31398	28365	46456	43015	64521	43697	51154	45020
		NA	12196	10666	13423	10854	13225	14367	18402	14284	13531	10123
		Total[h/]	54930	49816	44821	39219	59681	57382	82923	57981	64685	55143

Table 3. Catch of Chum Salmon of Asian and North American Origins by Commercial Fisheries of the U.S.S.R., Japan, U.S. and Canada, in Thousands of Fish, 1920-77 -- continued.

Nation	Fishery[a]	Origin of fish caught[b]	1940	1941	1942	1943	1944	1945	1946	1947	1948	1949
U.S.S.R.	Coastal	A	14492	14177	10454	16657	14629	16252	18267	17779	16573	20910
Japan	High Seas - MS[c]	NA	-	-	-	-	-	-	-	-	-	-
	" " "	A	4589	3162	2392	-	-	-	-	-	-	-
			4589	3162	2392	-	-	-	-	-	-	-
	" " - LBND	A[d]	6969	5938	8199	3194	1822	140	-	-	-	-
	" " - LBLL	"[e]	-	-	-	-	-	-	-	-	-	-
	" " - Total	NA+ A	11558	9100	10591	3194	1822	140	-	-	-	-
	Coastal[f]	A	16969	15913	9483	5921	2200	2475	2183	2709	2661	3653
	Total	NA + A	28527	25013	20074	9115	4022	2615	2183	2709	2661	3653
U.S.	Coastal - W. AK	NA	359	554	169	427	473	970	272	291	658	310
	C. AK	"	3837	2928	3140	2208	3027	3704	2925	2571	2736	2456
	SE. AK	"	4625	2959	5442	6833	6881	3318	4010	3360	4004	2894
			8821	6441	8751	9468	10381	7992	7207	6222	7398	5660
	WA	"	741	1184	1197	563	396	521	1514	707	1100	537
	OR[g]	"	58	203	252	43	12	29	38	25	54	23
	CA[g]	"	-	-	-	-	-	-	-	-	-	-
	Total		9620	7828	10200	10074	10789	8542	8759	6954	8552	6220
Canada	Coastal - N. Dist.	NA	*	*	*	*	*	*	*	*	*	*
	S. "	"	*	*	*	*	*	*	*	*	*	*
	Fraser Area	"	*	*	*	*	*	*	*	*	*	*
	Total		5030	6336	4416	4922	2289	2965	6826	5749	5138	3035
Total		A[h]	43019	39190	30528	25772	18651	18867	20450	20488	19234	24563
		NA	14650	14164	14616	14996	13078	11507	15585	12703	13690	9255
		Total[h]	57669	53354	45144	40768	31729	30374	36035	33191	32924	33818

Table 3. Catch of Chum Salmon of Asian and North American Origins by Commercial Fisheries of the U.S.S.R., Japan, U.S. and Canada, in Thousands of Fish, 1920-77 -- continued.

Nation	Fishery[a/]	Origin of fish caught[b/]	1950	1951	1952	1953	1953	1955	1956	1957	1958	1959
U.S.S.R.	Coastal	A	16185	22135	11650	8969	13902	17355	20285	8472	7274	10062
Japan	High Seas - MS[c/]	NA	-	-	[c/]	[c/]	[c/]	[c/]	146	40	-	22
	" " "	A	-	-	626	2707	9404	21978	20245	11868	18787	12837
			-	-	626	2707	9404	21978	20391	11908	18787	12859
	" " - LBND	A[d/]	-	-	684	1476	2659	3949	2094	3770	9155	9045
	" " - LBLL	"[e/]	-	-	8	2	+	28	151	823	1338	1480
	" " - Total	NA+ A	-	-	1318	4185	12063	25955	22636	16501	29280	23384
	Coastal[f/]	A	5437	5943	2109	2272	3196	2273	1754	3044	3227	1827
	Total	NA + A	5437	5943	3427	6457	15259	28228	24390	19545	32507	25211
U.S.	Coastal - W. AK	NA	364	454	522	619	820	342	754	548	613	886
	C. AK	"	2259	2034	3437	3131	3570	1783	3968	4621	3244	1908
	SE. AK	"	4779	4123	4179	3542	4244	1528	2701	3413	2787	1291
			7402	6611	8138	7292	8634	3653	7423	8582	6644	4085
	WA	"	1088	1043	931	506	740	375	200	210	560	519
	OR[g/]	"	29	23	15	13	17	7	2	2	5	2
	CA[g/]	"	-	-	-	-	-	-	-	-	-	-
	Total		8519	7677	9084	7811	9391	4035	7625	8794	7209	4606
Canada	Coastal - N. Dist.	NA	*	2576	755	1598	1796	609	1179	1449	1266	424
	S. "	"	*	2701	1071	2686	3510	830	1201	838	1729	1357
	Fraser Area	"	*	528	655	386	539	129	78	126	197	234
	Total		7273	5805	2481	4670	5845	1568	2458	2413	3192	2015
Total		A[h/]	21622	28078	15077	15426	29161	45583	44529	27977	39781	35251
		NA	15792	13482	11565	12481	15236	5603	10229	11247	10401	6643
		Total[h/]	37414	41560	26642	27907	44397	51186	54758	39224	50182	41894

Table 3. Catch of Chum Salmon of Asian and North American Origins by Commercial Fisheries of the U.S.S.R., Japan, U.S. and Canada, in Thousands of fish, 1920-77 -- continued.

Nation	Fishery[a]	Origin of fish caught[b]	1960	1961	1962	1963	1964	1965	1966	1967	1968	1969
U.S.S.R.	Coastal	A	11283	9513	9242	9449	6472	8521	7873	5113	4446	1457
Japan	High Seas - MS[c]	NA	174	61	35	53	130	91	107	82	152	172
	" " "	A	10343	6067	6337	5805	8511	5945	8455	6755	7955	7549
			10517	6128	6372	5858	8641	6036	8562	6837	8107	7721
	" " - LBND	A[d]	8684	6104	7577	7538	8956	8330	11848	11078	8457	4908
	" " - LBLL	"[e]	1474	1317	1436	1810	1739	1188	1880	1219	717	295
	" " - Total	NA+ A	20675	13549	15385	15206	19336	15554	22290	19134	17281	12924
	Coastal[f]	A	1851	3315	3476	4340	4491	5577	4911	5508	2988	5091
	Total	NA + A	22526	16864	18861	19546	23827	21131	27201	24642	20269	18015
U.S.	Coastal - W. AK	NA	1923	991	1128	635	1174	535	609	646	601	747
	C. AK	"	3682	2080	4024	2350	4160	1635	2574	1198	2837	1644
	SE. AK	"	1019	2559	1996	1479	1936	1474	3273	1810	2644	561
			6624	5630	7148	4464	7270	3644	6456	3654	6082	2952
	WA	"	206	174	238	318	286	211	428	292	482	200
	OR[g]	"	1	1	3	1	2	+	1	1	+	+
	CA[g]	"	-	-	-	-	-	-	-	-	-	-
	Total		6831	5805	7389	4783	7558	3855	6885	3947	6564	3152
Canada	Coastal - N. Dist.	NA	682	658	1088	1007	1977	572	1221	907	2178	666
	S. "	"	1087	511	360	403	214	49	76	174	702	555
	Fraser Area	"	68	49	48	54	61	12	14	46	202	89
	Total		1837	1218	1496	1464	2252	633	1311	1127	3082	1310
Total		A[h]	33635	26316	28068	28942	30169	29561	34967	29673	24563	19300
		NA	8842	7084	8920	6300	9940	4579	8303	5156	9798	4634
		Total[h]	42477	33400	36988	35242	40109	34140	43270	34829	34361	23934

Table 3. Catch of Chum Salmon of Asian and North American Origins by Commercial Fisheries of the U.S.S.R., Japan, U.S. and Canada, in Thousands of Fish, 1920-77 -- continued.

Nation	Fishery[a]	Origin of fish caught[b]	1970	1971	1972	1973	1974	1975	1976	1977
U.S.S.R.	Coastal	A	3092	3460	1382	1402	1888	1800	2730	4500
Japan	High Seas - MS[c]	NA	290	251	336	129	206	126	216	300
	" " "	A	9348	9717	13037	7728	9077	7241	10228	5696
			9638	9968	13373	7857	9283	7367	10444	5996
	" " - LBND	A[d]	6585	6250	8981	7837	12531	11759	11933	6230
	" " - LBLL	"[e]	943	531	-	-	-	-	-	-
	" " - Total	NA+ A	17166	16749	22354	15694	21814	19126	22377	12226
	Coastal[f]	A	6596	9386	8040	10673	12185	17900	12060	12000
	Total	NA + A	23762	26135	30394	26367	33999	37026	34437	23926
U.S.	Coastal - W. Ak	NA	1445	1417	1396	2045	2189	2312	2707	3196
	C. AK	"	3609	4317	2726	2143	861	1315	2241	3391
	SE. AK	"	2446	1946	2942	1832	1683	687	1014	632
			7500	7680	7064	6020	4733	4314	5962	7219
	WA	"	270	183	895	607	462	216	758	373
	OR[g]	"	+	-	1	1	-	-	+	-
	CA[g]	"	-	-	-	-	-	-	-	-
	Total		7770	7863	7960	6628	5195	4530	6730	7592
Canada	Coastal - N. Dist.	NA	2484	1011	2446	2683	1506	447	630	852
	S. "	"	1017	230	3372	3362	603	625	1096	223
	Fraser Area	"	179	22	256	191	93	73	174	14
	Total		3680	1263	6074	6236	2202	1145	1900	1089
Total		A[h]	26564	29344	31440	27640	35681	38700	36951	28426
		NA	11740	9377	14370	12993	7603	5801	8836	8981
		Total[h]	38304	38721	45810	40633	43284	44501	45787	37407

DATA SOURCES (FOR STATISTICS ON TOTAL CATCHES)*

U.S.S.R.	1925-30.	As indicated in footnote 2, Table 5-I, Secretariat, INPFC (1974). Note: 95 percent of the 1920-1931 catches reported under chum salmon in Table 24 of the INPFC ms. are assumed to have been chum while the remaining 5 percent were assumed to be coho salmon. These percentages are based on chum and coho catch data for 1932-1944.
	1940-61.	INPFC ms. Catches in metric tons by district (Table 37) converted to numbers of fish using the average weight (by district) in 1962-1973 (Tables 37 and 38).
	1962-75.	As indicated in Appendix 1, Fredin et al. (1977).
	1976-77.	Catch in metric tons (Foreign Fisheries Information Releases 77-3 and 78-5, National Marine Fisheries Service, National Oceanic and Atmospheric Administration) converted to numbers of fish using the average weight in 1966-1975 (Table 6.2, Fredin et al. 1977).
Japan	1920-51.	As indicated in footnotes 3-8, Table 5-I, INPFC ms. See note above regarding Table 24 of the ms.
	1952-75.	As indicated in Appendix 1 of Fredin et al.(1977).
	1976-77.	Personal communication, Fishery Agency of Japan.**
U.S.	1920-51.	As indicated in footnotes 2-6, Table 59, INPFC ms. Note: Oregon catches in metric tons during 1922-1937 converted to numbers of fish using 5.0 kg/fish, the average weight used by Kasahara (1963).
	1952-75.	As indicated in Appendix 1, Fredin et al. (1977).
	1976-77.	Personal communications, state fisheries agencies of Alaska, California, Oregon and Washington.
Canada	1920-50.	Table 62, INPFC ms.
	1951-75.	Table 64, INPFC ms., as updated.
	1976-77.	Department of Fisheries and Environment, 1977-1978.

*Estimates of interceptions of North American chum salmon by the Japanese mothership salmon fishery during 1956-1975 are from Table 5.5, Fredin et al. (1977). Estimates for 1976-1977 are by Fredin (*unpublished*).

**1977 coastal catch data not available. Catch assumed (by Fredin) to have been approximately the same as the average catch during 1972-1976.

Table 4. Catch of Coho Salmon of Asian and North American Origins by Commercial Fisheries of the U.S.S.R., Japan, U.S. and Canada, in Thousands of Fish, 1920-77.

Nation	Fishery[a]	Origin of fish caught[b]	1920	1921	1922	1923	1924	1925	1926	1927	1928	1929
U.S.S.R.	Coastal	A	*	*	*	*	*	1308	1036	1321	1358	1166
Japan	High Seas - MS[c]	NA	-	-	-	-	-	-	-	-	-	-
	" " "	A	-	-	-	-	-	-	-	-	-	-
			-	-	-	-	-	-	-	-	-	-
	" " - LBND	A[d]	-	-	-	-	-	-	-	-	-	-
	" " - LBLL	"[e]	-	-	-	-	-	-	-	-	-	-
	" " - Total	NA+ A	-	-	-	-	-	-	-	-	-	-
	Coastal[f]	A	540	514	507	442	254	366	306	436	780	810
	Total	NA + A	540	514	507	442	254	366	306	436	780	810
U.S.	Coastal - W. AK	NA	156	86	160	9	48	17	13	+	5	59
	C. AK	"	689	91	346	366	721	743	1062	1435	1470	700
	SE. AK	"	1143	912	1259	1342	1215	1212	1185	1285	2159	1368
			1993	1089	1765	1717	1984	1972	2260	2720	3634	2127
	WA	"	742	980	1161	1257	1257	1095	946	1262	961	1008
	OR[g]	"	1000	1000	626	960	1468	1464	1258	1003	739	944
	CA[g]	"	50	50	61	52	74	71	45	48	33	37
	Total		3785	3119	3613	3986	4783	4602	4509	5033	5367	4116
Canada	Coastal - N. Dist.	NA	*	*	*	*	*	*	*	*	*	*
	S. Dist.	"	*	*	*	*	*	*	*	*	*	*
	Fraser Area	"	*	*	*	*	*	*	*	*	*	*
	Total		1123	1230	1393	1247	1077	1949	1780	1748	1579	1832
Total		A[h]	540	514	507	442	254	1674	1342	1757	2138	1976
		NA	4908	4349	5006	5233	5860	6551	6289	6781	6946	5948
		Total[h]	5448	4863	5513	5675	6114	8225	7631	8538	9084	7924

Table 4. Catch of Coho Salmon of Asian and North American Origins by Commercial Fisheries of the U.S.S.R., Japan, U.S. and Canada, in Thousands of Fish, 1920-77 -- continued.

Nation	Fishery[a]	Origin of fish caught[b]	1930	1931	1932	1933	1934	1935	1936	1937	1938	1939
U.S.S.R.	Coastal	A	2706	1033	259	230	506	433	1317	815	1009	588
Japan	High Seas - MS[c]	NA	-	-	-	-	-	-	-	-	-	-
	" " "	A	4	63	+	6	180	163	400	299	219	183
			4	63	+	6	180	163	400	299	219	183
	" " - LBND	A[d]	-	-	-	55	330	747	1641	1216	1004	610
	" " - LBLL	"[e]	-	-	-	-	-	-	-	-	-	-
	" " - Total	NA+ A	4	63	+	61	510	910	2041	1515	1223	793
	Coastal[f]	A	834	536	517	337	461	756	2810	1642	906	719
	Total	NA + A	838	599	517	398	971	1666	4851	3157	2129	1512
U.S.	Coastal - W. AK	NA	34	1	5	16	12	11	24	2	6	+
	C. AK	"	1613	814	665	694	715	496	881	579	666	426
	SE. AK	"	1999	1152	1389	1223	1956	1760	1800	1400	2200	1122
			3646	1967	2059	1933	2683	2267	2705	1981	2872	1548
	WA	"	805	684	581	727	649	1902	1118	1069	860	1226
	OR[g]	"	1162	703	1107	705	988	1890	700	1031	437	406
	CA[g]	"	45	40	35	34	32	42	37	44	28	20
	Total		5658	3394	3782	3399	4352	6101	4560	4125	4197	3200
Canada	Coastal - N. Dist.	NA	*	*	*	*	*	*	*	*	*	*
	C. "	"	*	*	*	*	*	*	*	*	*	*
	Fraser Area	"	*	*	*	*	*	*	*	*	*	*
	Total		1935	1427	2169	1968	4199	3271	3653	1841	2564	2382
Total		A[h]	3544	1632	776	628	1477	2099	6168	3972	3138	2100
		NA	7593	4821	5951	5367	8551	9372	8213	5966	6761	5582
		Total[h]	11137	6453	6727	5995	10028	11471	14381	9938	9899	7682

Table 4. Catch of Coho Salmon of Asian and North American Origins by Commercial Fisheries of the U.S.S.R., Japan, U.S. and Canada, in Thousands of Fish, 1920-77 -- continued.

Nation	Fishery[a/]	Origin of fish caught[b/]	1940	1941	1942	1943	1944	1945	1946	1947	1948	1949
U.S.S.R.	Coastal	A	724	491	1241	698	960	1465	2838	1459	1804	1893
Japan	High Seas - MS[c/]	NA	-	-	-	-	-	-	-	-	-	-
	" " "	A	126	29	31	-	-	-	-	-	-	-
			126	29	31	-	-	-	-	-	-	-
	" " - LBND	A[d/]	262	220	328	326	143	38	-	-	-	-
	" " - LBLL	"[e/]	-	-	-	-	-	-	-	-	-	-
	" " - Total	NA+ A	388	249	359	326	143	38	-	-	-	-
	Coastal[f/]	A	467	401	653	90	33	5	-	-	-	-
	Total	NA + A	855	650	1012	416	176	43	-	-	-	-
U.S.	Coastal - W. AK	NA	25	35	29	2	25	16	52	10	29	52
	C. AK	"	1129	1386	1660	709	921	882	1239	925	782	648
	SE. AK	"	1838	2515	2211	1680	1306	2588	2367	1546	2146	2280
			2993	3936	3900	2391	2252	3486	3658	2481	2957	2980
	WA	"	1071	1256	589	871	699	1068	606	1355	1260	1248
	OR[g/]	"	414	331	156	122	268	267	215	527	232	202
	CA[g/]	"	49	28	49	49	76	99	102	85	101	48
	Total		4527	5551	4694	3433	3295	4920	4581	4448	4550	4478
Canada	Coastal - N. Dist.	"	*	*	*	*	*	*	*	*	*	*
	C. "	"	*	*	*	*	*	*	*	*	*	*
	Fraser Area	"	*	*	*	*	*	*	*	*	*	*
	Total		2001	2952	3321	2931	1486	2875	1714	2696	2698	2888
Total		A[h/]	1579	1141	2253	1114	1136	1508	2838	1459	1804	1893
		NA	6528	8503	8015	6364	4781	7795	6295	7144	7248	7366
		Total[h/]	8107	9644	10268	7478	5917	9303	9133	8603	9052	9259

Table 4. Catch of Coho Salmon of Asian and North American Origins by Commercial Fisheries of the U.S.S.R., Japan, U.S. and Canada, in Thousands of Fish, 1920-77 -- continued.

Nation	Fishery[a/]	Origin of fish caught[b/]	1950	1951	1952	1953	1954	1955	1956	1957	1958	1959
U.S.S.R.	Coastal	A	691	800	1372	1914	2212	1852	1136	1546	846	1142
Japan	High Seas - MS[c/]	NA	-	-	c/	c/	c/	c/	7	-	-	-
	" " "	A	-	-	24	340	1398	3576	3875	442	3393	1423
			-	-	24	340	1398	3576	3882	442	3393	1423
	" " - LBND	A[d/]	-	-	126	272	490	728	386	695	803	1204
	" " - LBLL	"[e/]	-	-	+	+	+	+	1	19	15	16
	" " - Total	NA + A	-	-	150	612	1888	4304	4269	1156	4211	2643
	Coastal[f/]	A	-	-	+	+	+	+	+	+	+	+
	Total	NA + A	-	-	150	612	1888	4304	4269	1156	4211	2643
U.S.	Coastal - W. AK	NA	67	76	70	31	59	27	72	87	193	76
	C. AK	"	698	645	562	385	703	489	525	325	459	332
	SE. AK	"	1652	3310	1744	1164	1771	1338	917	1218	955	1024
			2417	4031	2376	1580	2533	1854	1514	1630	1607	1432
	WA	"	1407	1323	1852	1270	722	1043	1434	1108	995	988
	OR[g/]	"	176	329	440	308	249	299	476	571	208	187
	CA[g/]	"	92	69	92	102	64	56	66	89	13	35
	Total		4092	5752	4760	3260	3568	3252	3490	3398	2823	2642
Canada	Coastal - N. Dist.	"	*	1618	976	788	884	1104	1170	1133	968	722
	C. Dist.	"	*	2244	1684	2021	1425	1785	1752	1954	1945	2115
	Fraser Area	"	*	136	86	86	136	79	128	49	77	60
	Total		2666	3998	2746	2895	2445	2968	3050	3136	2990	2897
Total		A[h/]	691	800	1522	2526	4100	6156	5398	2702	5057	3785
		NA	6758	9750	7506	6155	6013	6220	6547	6534	5813	5539
		Total[h/]	7449	10550	9028	8681	10113	12376	11945	9236	10870	9324

Table 4. Catch of Coho Salmon of Asian and North American Origins by Commercial Fisheries of the U.S.S.R., Japan, U.S. and Canada, in Thousands of Fish, 1920-77 -- continued.

Nation	Fishery[a]	Origin of fish caught[b]	1960	1961	1962	1963	1964	1965	1966	1967	1968	1969
U.S.S.R.	Coastal	A	571	1344	1298	2234	332	1034	902	1043	977	1014
Japan	High Seas - MS[c]	NA	-	-	+	-	175	-	-	-	3	+
	" " "	A	962	284	1532	1895	3360	1177	469	226	895	1306
			962	284	1532	1895	3535	1177	469	226	898	1306
	" " - LBND	A[d]	1376	1486	1289	1492	1624	1913	1458	1329	1421	3328
	" " - LBLL	"[e]	113	3	5	8	4	1	1	1	3	2
	" " - Total	NA+ A	2451	1773	2826	3395	5163	3091	1928	1556	2322	4636
	Coastal[f]	A	+	+	+	+	+	+	+	+	+	+
	Total	NA + A	2451	1773	2826	3395	5163	3091	1928	1556	2322	4636
U.S.	Coastal - W. AK	NA	66	67	124	121	105	57	119	172	333	263
	C. AK	"	618	357	692	627	866	393	574	450	875	274
	SE. AK	"	721	889	1223	1275	1588	1548	1227	866	1543	596
			1405	1313	2039	2023	2559	1998	1920	1488	2751	1133
	WA	"	341	1035	1142	881	1128	1496	1681	1213	1275	920
	OR[g]	"	126	356	343	510	723	858	990	1306	925	709
	CA[g]	"	18	79	48	162	247	217	445	414	363	193
	Total		1890	2783	3572	3576	4657	4569	5036	4421	5314	2955
Canada	Coastal - N. Dist.	NA	796	1025	1455	1417	1779	1576	2032	1004	2066	742
	S. Dist.	"	1176	2240	2102	1968	2236	2824	3343	2265	3109	1649
	Fraser Area	"	57	32	68	36	134	43	37	34	82	23
	Total		2029	3297	3625	3421	4149	4443	5412	3303	5257	2414
Total		A[h]	3022	3117	4124	5629	5320	4125	2830	2599	3296	5650
		NA	3919	6080	7197	6997	8981	9012	10448	7724	10574	5369
		Total[h]	6941	9197	11321	12626	14301	13137	13278	10323	13870	11019

Table 4. Catch of Coho Salmon of Asian and North American Origins by Commercial Fisheries of the U.S.S.R., Japan, U.S. and Canada, in Thousands of Fish, 1920-77 -- continued.

Nation	Fishery[a]	Origin of fish caught[b]	1970	1971	1972	1973	1974	1975	1976	1977
U.S.S.R.	Coastal	A	1624	1273	511	705	1164	1200	1100	1400
Japan	High Seas - MS[c]	NA	20	24	-	13	149	4	33	+
	" " "	A	160	430	614	976	936	352	794	79
			180	454	614	989	1085	356	827	79
	" " - LBND	A[d]	2259	2373	2423	3796	3560	3556	2753	1722
	" " - LBLL	"[e]	1	2	-	-	-	-	-	-
	" " - Total	NA+ A	2440	2829	3037	4785	4645	3912	3580	1801
	Coastal[f]	A	+	+	5	-	-	-	-	-
	Total	NA + A	2440	2829	3042	4785	4645	3912	3580	1801
U.S.	Coastal - W. AK	NA	121	47	70	282	259	192	177	431
	C. AK	"	647	487	252	338	318	395	433	414
	SE. AK	"	759	914	1509	836	1278	427	816	748
			1527	1448	1831	1456	1855	1014	1426	1593
	WA	"	1870	2002	1253	1672	2117	1837	3031	1590
	OR[g]	"	1412	1695	925	937	1328	772	1936	467
	CA[g]	"	183	442	158	348	636	204	695	54
	Total		4992	5587	4167	4413	5936	3827	7088	3704
Canada	Coastal - N. Dist.	NA	1349	1098	1613	948	886	553	924	577
	S. Dist.	"	2496	3619	1662	2529	2781	1732	2760	2718
	Fraser Area	"	100	70	81	54	57	45	17	44
	Total		3945	4787	3356	3531	3724	2330	3701	3339
Total		A[h]	4044	4078	3553	5477	5660	5108	4647	3201
		NA	8957	10398	7523	7957	9809	6161	10822	7043
		Total[h]	13001	14476	11076	13434	15469	11269	15469	10244

DATA SOURCES (FOR STATISTICS ON TOTAL CATCHES)*

U.S.S.R.	1925-39.	As indicated in footnote 2, Table 6-I, Secretariat, INPFC (1974). Note: 5 percent of the 1920-1931 catches reported under chum salmon in Table 24 of the INPFC ms. are assumed to have been coho salmon while the remaining 95 percent are assumed to be chum salmon. These percentages are based on chum and coho catch data for 1932-1944.
	1940-61.	INPFC ms. Catches in metric tons by district (Table 39) converted to numbers of fish using average weight (by district) in 1962-1973 (Tables 39 and 40).
	1962-75.	As indicated in Appendix 1, Fredin et al. (1977).
	1976-77.	Catch in metric tons (Foreign Fisheries Information Releases 77-3 and 78-5, National Marine Fisheries Service, National Oceanic and Atmospheric Administration) converted to numbers of fish using the average weight in 1966-1975 (Table 6.2, Fredin et al. 1977).
Japan	1920-45.	As indicated in footnotes 3-5, Table 6-I, INPFC ms. See note above regarding Table 24 of the ms.
	1952-75.	As indicated in Appendix 1, Fredin et al. (1977).
	1976-77.	Personal communication, Fishery Agency of Japan.
U.S.	1920-51.	As indicated in footnotes 2-6, Table 60, INPFC ms. for Alaska, Washington and Oregon. Table 70, Kasahara (1963) for California with the combined catch of coho salmon and chinook based on coho and chinook catches in 1922-1947 and 1949 which are assumed to consist of 5 percent coho and 95 percent chinook. (These percentages are based on coho and chinook catches in 1948 and from 1950-1961.) Catch of coho in metric tons(from 1922-1951) were then converted to numbers of fish using 3.053 kg/fish, which was the average weight reported by Kasahara (1963).
	1952-75.	As indicated in Appendix 1, Fredin et al. (1977).
	1976-77.	Personal communications with state fisheries agencies of Alaska, California, Oregon and Washington.
Canada	1920-50.	Table 62, INPFC ms.
	1951-75.	Table 64, INPFC ms., as updated.
	1976-77.	Department of Fisheries and Environment 1977-1978.

*Estimates of interceptions of North American coho salmon by the Japanese mothership salmon fishery during 1956-1975 are from Table 5.7, Fredin et al. (1977). Estimates for 1976-1977 are by Fredin (*unpublished*).

Table 5. Catch of Chinook Salmon of Asian and North American Origins by Commercial Fisheries of the U.S.S.R., Japan, U.S. and Canada, in Thousands of Fish, 1920-77.

Nation	Fishery[a]	Origin of fish caught[b]	1920	1921	1922	1923	1924	1925	1926	1927	1928	1929
U.S.S.R.	Coastal	A	*	*	*	*	*	*	*	8	18	26
Japan	High Seas - MS[c]	NA	-	-	-	-	-	-	-	-	-	-
	" " "	A	-	-	-	-	-	-	-	-	-	-
			-	-	-	-	-	-	-	-	-	-
	" " - LBND	A[d]	-	-	-	-	-	-	-	-	-	-
	" " - LBLL	"[e]	-	-	-	-	-	-	-	-	-	-
	" " - Total	NA + A	-	-	-	-	-	-	-	-	-	-
	Coastal[f]	A	70	65	36	25	24	40	77	98	69	83
	Total	NA + A	70	65	36	25	24	40	77	98	69	83
U.S.	Coastal - W. AK	NA	240	184	110	97	129	110	99	100	71	155
	C. AK	"	79	27	50	48	50	88	110	147	130	131
	SE. AK	"	352	446	345	790	836	600	452	627	385	455
			671	657	505	935	1015	798	661	874	586	741
	WA	"	771	599	392	535	581	702	630	710	548	591
	OR[g]	"	1022	1022	843	1157	1307	1428	1093	1142	800	730
	CA[g]	"	522	522	628	541	764	727	464	497	342	385
	Total		2986	2800	2368	3168	3667	3655	2848	3223	2276	2447
Canada	Coastal - N. Dist.	NA	*	*	*	*	*	*	*	*	*	*
	C. "	"	*	*	*	*	*	*	*	*	*	*
	Fraser Area	"	*	*	*	*	*	*	*	*	*	*
	Total		642	407	373	343	307	667	490	614	415	345
Total		A[h]	70	65	36	25	24	40	77	106	87	109
		NA	3628	3207	2741	3511	3974	4322	3338	3837	2691	2792
		Total[h]	3698	3272	2777	3536	3998	4362	3415	3943	2778	2901

Table 5. Catch of Chinook Salmon of Asian and North American Origins by Commercial Fisheries of the U.S.S.R., Japan, U.S. and Canada, in Thousands of Fish, 1920-77 -- continued.

Nation	Fishery[a/]	Origin of fish caught[b/]	1930	1931	1932	1933	1934	1935	1936	1937	1938	1939
U.S.S.R.	Coastal	A	67	40	53	15	25	41	65	*	*	*
Japan	High Seas - MS[c/]	NA	-	-	-	-	-	-	-	-	-	-
	" " "	A	3	5	3	1	3	3	2	2	3	2
			3	5	3	1	3	3	2	2	3	2
	" " - LBND	A[d/]	-	-	-	-	-	+	2	1	-	-
	" " - LBLL	"[d/]	-	-	-	-	-	-	-	-	-	-
	" " - Total	NA + A	3	5	3	1	3	3	4	3	3	2
	Coastal[f/]	A	112	75	75	36	69	120	115	152	117	65
	Total	NA + A	115	80	78	37	72	123	119	155	120	67
U.S.	Coastal - W. AK	NA	117	58	86	59	73	18	44	45	66	47
	C. AK	"	115	101	128	98	109	97	113	113	79	84
	SE. AK	"	610	460	627	427	339	641	681	879	757	652
			842	619	841	584	521	756	838	1037	902	783
	WA	"	602	556	519	593	581	563	610	652	435	469
	OR[g/]	"	791	927	711	875	807	821	968	1071	516	527
	CA[g/]	"	458	414	359	349	329	432	383	451	292	209
	Total		2693	2516	2430	2401	2238	2572	2799	3211	2145	1988
Canada	Coastal - N. Dist.	NA	*	*	*	*	*	*	*	*	*	*
	C. "	"	*	*	*	*	*	*	*	*	*	*
	Fraser Area	"	*	*	*	*	*	*	*	*	*	*
	Total		539	425	646	448	1737	1341	1634	814	331	361
Total		A[h/]	182	120	131	52	97	164	184	155	120	67
		NA	3232	2941	3076	2849	3975	3913	4433	4025	2476	2349
		Total[h/]	3414	3061	3207	2901	4072	4077	4617	4180	2596	2416

Table 5. Catch of Chinook Salmon of Asian and North American Origins by Commercial Fisheries of the U.S.S.R., Japan, U.S. and Canada, in Thousands of Fish, 1920-77 -- continued.

Nation	Fishery[a]	Origin of fish caught[b]	1940	1941	1942	1943	1944	1945	1946	1947	1948	1949
U.S.S.R.	Coastal	A	22	58	37	49	17	37	80	80	129	75
Japan	High Seas - MS[c]	NA	-	-	-	-	-	-	-	-	-	-
	" " - "	A	8	3	1	-	-	-	-	-	-	-
			8	3	1	-	-	-	-	-	-	-
	" " - LBND	A[d]	-	-	-	-	-	-	-	-	-	-
	" " - LBLL	"[e]	-	-	-	-	-	-	-	-	-	-
	" " - Total	NA + A	8	3	1	-	-	-	-	-	-	-
	Coastal[f]	A	71	76	46	5	+	+	-	-	-	-
	Total	NA + A	79	79	47	5	+	+	-	-	-	-
U.S.	Coastal - W. AK	NA	34	61	41	69	31	46	55	101	85	88
	C. AK	"	80	134	130	156	108	116	98	127	115	126
	SE. AK	"	428	613	558	399	328	438	568	496	480	480
			542	808	729	624	467	600	721	724	680	694
	WA	"	588	657	608	487	497	546	643	637	573	508
	OR[g]	"	565	968	695	454	576	614	698	805	760	500
	CA[g]	"	509	289	505	502	785	1021	1042	876	569	491
	Total		2204	2722	2537	2067	2325	2781	3104	3042	2582	2193
Canada	Coastal - N. Dist.	NA	*	*	*	*	*	*	*	*	*	*
	S. "	"	*	*	*	*	*	*	*	*	*	*
	Fraser Area	"	*	*	*	*	*	*	*	*	*	*
	Total		326	713	416	1076	276	502	780	874	514	1046
Total		A[h]	101	137	84	54	17	37	80	80	129	75
		NA	2530	3435	2953	3143	2601	3283	3884	3916	3096	3239
		Total[h]	2631	3572	3037	3197	2618	3320	3964	3996	3225	3314

Table 5. Catch of Chinook Salmon of Asian and North American Origins by Commercial Fisheries of the U.S.S.R., Japan, U.S. and Canada, in Thousands of Fish, 1920-77 -- continued.

Nation	Fishery[a]	Origin of fish caught[b]	1950	1951	1952	1953	1954	1955	1956	1957	1958	1959
U.S.S.R.	Coastal	A	61	72	65	87	65	134	111	90	70	97
Japan	High Seas - MS[c]	NA	-	-	c/	c/	c/	c/	56	15	5	28
	" " "	A	-	-	1	8	74	76	81	16	41	40
			-	-	1	8	74	76	137	31	46	68
	" " - LBND	A[d]	-	-	8	17	31	46	24	44	45	70
	" " - LBLL	"[e]	-	-	+	+	+	+	1	5	8	9
	" " - Total	NA + A	-	-	9	25	105	122	162	80	99	147
	Coastal[f]	A	-	-	2	2	3	2	2	3	3	2
	Total	NA + A	-	-	11	27	108	124	164	83	102	149
U.S.	Coastal - W. AK	NA	38	102	92	102	124	135	133	158	182	195
	C. AK	"	188	213	115	112	86	76	84	60	45	47
	SE. AK	"	379	474	528	498	398	374	239	300	325	365
			655	789	735	712	608	585	456	518	552	607
	WA	"	480	595	662	630	570	651	519	586	482	375
	OR[g]	"	524	560	481	376	385	642	674	515	419	245
	CA[g]	"	518	540	512	536	829	884	1026	492	375	514
	Total		2177	2484	2390	2254	2392	2762	2675	2111	1828	1741
Canada	Coastal - N. Dist.	NA	*	176	227	206	156	151	145	126	204	193
	S. "	"	*	483	542	626	534	573	712	674	688	579
	Fraser Area	"	*	149	136	182	185	148	127	139	176	176
	Total		971	808	905	1014	875	872	984	939	1068	948
Total		A[h]	61	72	76	114	173	258	219	158	167	218
		NA	3148	3292	3295	3268	3267	3634	3715	3065	2901	2717
		Total[h]	3209	3364	3371	3382	3440	3892	3934	3223	3068	2935

Table 5. Catch of Chinook Salmon of Asian and North American Origins by Commercial Fisheries of the U.S.S.R., Japan, U.S. and Canada, in Thousands of Fish, 1920-77 -- continued.

Nation	Fishery[a]	Origin of fish caught[b]	1960	1961	1962	1963	1964	1965	1966	1967	1968	1969
U.S.S.R.	Coastal	A	69	64	100	124	160	107	93	91	83	122
Japan	High Seas - MS[c]	NA	135	14	30	41	256	106	112	70	227	438
	" " "	A	45	17	92	46	154	79	96	58	135	116
			180	31	122	87	410	185	208	128	362	554
	" " - LBND	A[d]	111	77	124	102	195	93	112	110	88	83
	" " - LBLL	"[e]	20	8	5	3	13	9	6	5	9	5
	" " - Total	NA + A	311	116	251	192	618	287	326	243	459	642
	Coastal[f]	A	2	3	4	4	4	5	5	5	3	5
	Total	NA + A	313	119	255	196	622	292	331	248	462	647
U.S.	Coastal - W. AK	NA	196	243	213	208	260	263	208	284	259	288
	C. AK	"	41	31	42	35	22	31	24	26	20	38
	SE. AK	"	310	230	206	258	357	287	308	301	332	314
			547	504	461	501	639	581	540	611	611	640
	WA	"	288	371	330	397	349	317	365	356	394	450
	OR[g]	"	316	284	248	328	363	300	231	270	249	339
	CA[g]	"	540	774	556	662	687	705	553	338	472	551
	Total		1691	1933	1595	1888	2038	1903	1689	1575	1726	1980
Canada	Coastal - N. Dist.	NA	187	164	195	207	299	307	316	357	351	321
	S. "	"	426	424	410	478	501	580	749	650	625	691
	Fraser Area	"	130	98	115	118	167	93	99	124	106	88
	Total		743	686	720	803	967	980	1164	1131	1082	1100
Total		A[h]	247	169	325	279	526	293	312	269	318	331
		NA	2569	2633	2345	2732	3261	2989	2965	2776	3035	3518
		Total[h]	2816	2802	2670	3011	3787	3282	3277	3045	3353	3849

Table 5. Catch of Chinook Salmon of Asian and North American Origins by Commercial Fisheries of the U.S.S.R., Japan, U.S. and Canada, in Thousands of Fish, 1920-77 -- continued.

Nation	Fishery[a/]	Origin of fish caught[b/]	1970	1971	1972	1973	1974	1975	1976	1977
U.S.S.R.	Coastal	A	141	183	197	210	172	180	180	220
Japan	High Seas - MS[c/]	NA	346	144	170	47	289	109	171	70
	" " "	A	91	62	91	72	72	53	114	23
			437	206	261	119	361	162	285	93
	" " - LBND	A[d/]	101	134	107	165	188	137	201	146
	" " - LBLL	"[e/]	47	5	-	-	-	-	-	-
	" " - Total	NA + A	585	345	368	284	549	299	486	239
	Coastal[f/]	A	5	22	5	11	14	20	20	*
	Total	NA + A	590	367	373	295	563	319	506	*
U.S.	Coastal - W. AK	NA	291	283	224	177	180	126	242	296
	C. AK	"	33	45	42	30	29	28	49	40
	SE. AK	"	322	334	287	344	347	301	242	310
			646	662	553	551	556	455	533	646
	WA	"	542	565	486	684	612	681	1060	636
	OR[g/]	"	405	315	324	658	340	391	302	467
	CA[g/]	"	517	434	492	816	527	579	465	469
	Total		2110	1976	1855	2709	2035	2106	2360	2218
Canada	Coastal - N. Dist.	NA	343	356	453	371	408	423	363	320
	S. "	"	737	1097	974	954	972	896	1096	1078
	Fraser Area	"	133	140	123	100	87	94	82	99
	Total		1213	1593	1550	1425	1467	1413	1541	1497
Total		A[h/]	385	406	400	458	446	390	515	389
		NA	3669	3713	3575	4181	3791	3628	4072	3785
		Total[h/]	4054	4119	3975	4639	4237	4018	4587	4174

DATA SOURCES (FOR STATISTICS ON TOTAL CATCHES)*

U.S.S.R.	1927-36.	As indicated in footnote 2, Table 7-I, Secretariat, INPFC (1974).
	1940-61.	INPFC ms. Catches in metric tons by district (Table 41) converted to numbers of fish using average weight (by district) in 1962-1973 (Tables 41 and 42).
	1962-75.	As indicated in Appendix 1, Fredin et al. (1977).
	1976-77.	Catch in metric tons (Foreign Fisheries Information Releases 77-3 and 78-5, National Marine Fisheries Service, National Oceanic and Atmospheric Administration) converted to numbers of fish using average weight in 1966-1975 (Table 6.2, Fredin et al. 1977).
Japan	1920-45.	As indicated in footnotes 3-5, Table 7-I, INPFC ms.
	1952-75.	As indicated in Appendix 1, Fredin et al. (1977).
	1976-77.	Personal communication, Fishery Agency of Japan.
U.S.	1920-51.	As indicated in footnotes 2-6, Table 61, INPFC ms. for Alaska, Washington and Oregon. For California: Table 70, Kasahara (1963) with the combined catch of chinook and coho salmon in 1922-1947 and 1949 assumed to consist of 95 percent chinook and 5 percent coho. (These percentages are based on coho and chinook catches in 1948 and 1950-1961.) Catch of chinook in metric tons (in 1922-1951) then converted to numbers of fish using 5.648 kg/fish, which was the average weight during 1959-1971 (Tables 80 and 81, INPFC ms.).
	1952-75.	As indicated in Appendix 1, Fredin et al. (1977).
	1976-77.	Personal communications, state fisheries agencies of Alaska, California, Oregon and Washington.
Canada	1920-50.	Table 62, INPFC ms.
	1951-75.	Table 64, INPFC ms., as updated.
	1976-77.	Department of Fisheries, 1977-1978.

*Estimates of interceptions of North American chinook salmon by the Japanese mothership salmon fishery during 1956-1975 are from Table 5.9, Fredin et al. (1977). Estimates for 1976-1977 are by Fredin (*unpublished*).

Table 6. Catch of Coho and Chinook Salmon by Marine Sport Fisheries of British Columbia, Washington, Oregon and California, in Thousands of Fish, 1950-77.

	Coho				Chinook			
Year	British Columbia a/	Washington	Oregon	California	British Columbia a/	Washington	Oregon	California
1950	*	201	3	5	*	115	11	57
1951	*	211	6	8	*	139	10	103
1952	*	250	16	10	*	213	11	123
1953	62	212	13	11	52	149	15	141
1954	36	254	22	14	43	218	17	171
1955	78	266	44	15	55	228	59	184
1956	71	386	*	13	64	318	*	163
1957	126	454	*	5	93	331	*	64
1958	96	301	*	5	89	230	*	65
1959	90	280	*	8	66	189	*	76
1960	98	123	*	6	46	189	*	50
1961	79	271	*	4	41	199	*	55
1962	83	386	*	13	56	212	*	120
1963	100	424	*	33	52	263	*	84
1964	98	329	188	40	42	207	28	101
1965	106	601	258	21	43	241	36	60
1966	175	434	206	32	69	268	39	74
1967	104	557	336	50	63	274	43	73
1968	140	552	258	40	71	263	26	154
1969	87	498	226	28	80	267	32	156
1970	154	540	244	15	91	300	47	148
1971	296	846	240	68	101	313	93	188
1972	108	616	247	44	118	401	39	201
1973	119	533	232	32	116	400	61	198
1974	247	757	318	78	122	470	33	156
1975	163	680	252	21	155	597	76	104
1976	175	1,167 b/	502 b/	40 b/	216	478 b/	66 b/	110 b/
1977	120 c/	600 b/	165 b/	12 b/	160 d/	400 b/	52 b/	104 b/

* Not available.

a/ Excluding grilse (salmon weighing 1.36 kg or less).
b/ Preliminary data.
c/ Assumed to have been approximately 2/3 of the number caught in 1976 (same as the ratio of the 1977:1976 commercial catches).
d/ Assumed to have been approximately 3/4 of the number caught in 1976 (same as the ratio of the 1977:1976 commercial catches).

DATA SOURCES (FOR CATCH STATISTICS)

1950-71. Secretariat, INPFC (1974). (Table 83-British Columbia; Table 84-Washington; Table 85-Oregon; and Table 86-California.)

1972-77. Department of Fisheries and Environment 1977-1978 (1972-1976 only); Washington Department of Fisheries; Oregon Department of Fish and Wildlife; and California Department of Fish and Game. (Provided to INPFC for updating the manuscript referred to above, and personal communication.)

Marine Mammal-Salmonid Interactions: A Review

Clifford H. Fiscus

INTRODUCTION

The relationship between marine mammals and commercially important fish stocks has been the subject of many reports since the time that the first fisherman had to defend his gear and catch from marine mammals.

Marine mammal interaction with North Pacific salmon stocks *(Oncorhynchus* sp.) are described in this paper. With our present limited knowledge, it is a difficult and perhaps impossible task to document the precise degree and effect of marine mammal predation on a particular stock.

DISCUSSION

Salmon are preyed upon from the moment a female lays her eggs until maturing adults return from the ocean to repeat the spawning cycle. Man is the most significant salmon predator. In his attempts to obtain the largest portion of mature stocks, he has devised ways to eliminate as many other predators as possible. Predatory fish have been systematically removed from spawning streams. At times, predatory birds such as grebes, mergansers, loons, herons and eagles have been killed or harrassed. Terrestrial mammals such as bears, raccoons and several of the mustelids compete with man for a share of salmon along with marine mammals.

Salmon first encounter marine mammals in the lower reaches of streams as smolts migrating through estuaries and bays. Once smolts move into the ocean, and during their years in the sea, a variety and number of marine mammals seek salmon as prey. On their return to coastal waters, adult salmon caught in fishing gear probably are most subject to marine mammal predation.

Salmonid Distribution

Member nations of the International North Pacific Fisheries Commission (INPFC) have studied salmon since 1956. Consequently, the life history of salmon in the ocean is fairly well known. The INPFC has published papers on the distribution and origin of sockeye *(O. nerka)*, coho *(O. kisutch)* and chum *(O. keta)* salmon (French et al. 1976, Godfrey et al. 1975, Neave et al. 1976). Papers on pink *(O. gorbuscha)*, chinook *(O. tshawytscha)* and masu *(O. masou)* salmon are being prepared.

Salmon range across the North Pacific from the subarctic water boundary north into arctic waters. The several species vary considerably in their seasonal north-south and onshore-offshore movements. Most salmon are found near the surface and occasionally to depths of 60 m. Most ascend toward the surface at night. Because salmon are concentrated in the surface layer of water, marine mammals inhabiting the same layer have ready access to them.

Marine Mammal Distribution and Numbers

Thirty-four marine mammal species commonly spend some portion of each year in the North Pacific and contiguous seas where salmon are abundant. A few more species could be added to the list by including those from warmer waters that infrequently enter subarctic waters in the summer. Of the 34 species which frequent salmon waters, 15 are known to prey on salmon. However, several of these take salmon only incidentally to other, preferred prey.

Marine mammals known to inhabit salmon waters are listed in Table 1 according to three categories: 1) salmon identified in stomach contents; 2) capable of preying on salmon; and 3) not known to prey on salmon.

Table 1. Marine Mammal Inhabitants of Salmonid Waters.

Marine mammal: Common name	Scientific name	Salmonids identified in stomach contents	Capable of predation on salmonids	Not known to prey on salmonids
Walrus	*Odobenus rosmarus*	-	-	X
California sea lion	*Zalophus californianus*	X	-	-
Northern sea lion	*Eumetopias jubatus*	X	-	-
Northern fur seal	*Callorhinus ursinus*	X	-	-
Sea otter	*Enhydra lutris*	-	-	X
Harbor seal	*Phoca vitulina richardsi*	X	-	-
	P. vitulina stejnegeri	X	-	-
Larga seal	*P. largha*	X	-	-
Ringed seal	*P. hispida*	-	-	X
Ribbon seal	*P. fasciata*	-	X	-
Bearded seal	*Erignathus barbatus*	-	-	X
Northern elephant seal	*Mirounga angustirostris*	-	-	X
Gray whale	*Eschrichtius robustus*	-	-	X
Minke whale	*Balaenoptera acutorostrata*	-	-	X
Sei whale	*B. borealis*	-	-	X
Fin whale	*B. physalus*	X[1]	-	-
Blue whale	*B. musculus*	-	-	X
Humpback whale	*Megaptera novaeangliae*	X[1]	-	-
Right whale	*Eubalaena glacialis*	-	-	X
Bowhead whale	*Balaena mysticetus*	-	-	X
Saddleback dolphin	*Delphinus delphis*	-	X	-
Pacific whiteside dolphin	*Lagenorhynchus obliquidens*	X	-	-
Northern right whale dolphin	*Lissodelphis borealis*	-	X	-
Whitehead grampus	*Grampus griseus*	-	-	X
False killer whale	*Pseudorca crassidens*	X[1]	-	-
Shortfin pilot whale	*Globicephala macrorhynchus*	-	-	X
Killer whale	*Orcinus orca*	X	-	-
Harbor porpoise	*Phocoena phocoena*	X	-	-
Dall's porpoise	*Phocoenoides dalli*	X	-	-
Belukha	*Delphinapterus leucas*	X	-	-
Sperm whale	*Physeter catodon*	X[1]	-	-
North Pacific giant bottlenose whale	*Berardius bairdii*	-	-	X
Goosebeak whale	*Ziphius cavirostris*	-	-	X
Bering seal beaked whale	*Mesoplodon stejnegeri*	-[2]	-	-
Archbeak whale	*Mesoplodon carlhubbsi*	-[2]	-	-

[1] Most certainly an accidental taking

[2] Food preferences unknown

I will not describe the range and seasonal distribution of all the marine mammals listed in the table. Rice (1977) gives current nomenclature and a brief description of these species' ranges. More detailed information can be found in Marine Mammal Protection Act of 1972, Annual Report: 1 April 1977 to 31 March 1978 (National Marine Fisheries Service 1978).

Generally, coastal marine mammals do not make long seasonal migrations. The arctic migrants winter in the Bering Sea and return to the Arctic through Bering Strait in the spring. Marine mammals probably remain in various parts of the Okhotsk Sea depending on the season, although some do migrate into the Pacific and Sea of Japan. Some marine mammals range north into the Bering Sea in summer, and return to the North Pacific in winter. Others may move entirely south of subarctic waters in winter.

Brief descriptions of the distribution and numbers of marine mammals known to prey on salmon follow:

California sea lion *(Zalophus californianus)* remains mostly south of the subarctic waters that salmon inhabit during the May-July breeding season. A northward migration, consisting almost entirely of males, begins in August. Males reach their northern limit in British Columbia waters in late autumn, and return south in late winter. A few males may linger in the northern portion of the range during the breeding season. Although the majority of the population remains in California waters from early autumn to early spring, as many as 4,000 animals inhabit waters north of California (Mate 1976).

Northern sea lion *(Eumetopias jubatus)* is found almost entirely within the range of salmon. Populations are concentrated in continental shelf waters. Recent estimates indicate that there are less than 250,000 northern sea lions, including about 2,000 in California; 2,000 in Oregon, 500 in Washington; 5,000 in British Columbia; 200,000 in Alaska (this estimate may be high); 3,500 on the Commander Islands; 1,000 in Kamchatka; 20,000 on the Kuril Islands; and 2,700 in the Okhotsk Sea (Fiscus *in press*).

Northern fur seal *(Callorhinus ursinus)* ranges throughout the same subarctic waters of the North Pacific as salmon do. It breeds on several islands in the Bering and Okhotsk seas, on several of the Kuril Islands, and on San Miguel Island and Castle Rock off California. When away from its breeding grounds, it rarely approaches land and seldom enters coastal areas that sea lions and harbor seals frequent. Population estimates by breeding ground location indicated that there are 1,300,000 fur seals on the Pribilof Islands; 2,000 on San Miguel Island; 265,000 on the Commander Islands; 165,000 on Robben Island; and 33,000 on the Kuril Islands.

Harbor seal *(Phoca vitulina)* inhabits the same waters that salmon do except in the northern portion of the Bering Sea and the Arctic where the larga seal *(Phoca largha)* replaces it. Harbor seals are found primarily in coastal and estuarine waters. They regularly move into rivers, and occasionally move offshore over the continental shelf. There are currently two recognized subspecies in the North Pacific (Shaughnessy and Fay 1977). One, *richardsi*, inhabits the eastern North Pacific in waters north and west to the Aleutian Islands, and in the southern Bering Sea. The other, *stejnegeri*, inhabits the western North Pacific from Hokkaido northward through the Kuril Islands and along the Kamchatka coast into the southern Bering Sea.

Estimates place the population of both subspecies in the North Pacific at 312,000 to 317,500 (Advisory Committee on Marine Resources Research 1976) with 2,000 in California; 2,000 in Oregon; 2,500 in Washington; 35,000 in British Columbia; 260,000 in Alaska; and 10,000-15,000 in the western Pacific.

Larga seal *(Phoca largha)* inhabits the northern portion of the Okhotsk and Bering seas. Small populations also inhabit the Sea of Japan and the PoHai Sea (Shaughnessy and Fay 1977). When ice forms, the larga seal leaves shore to winter on the pack ice. It returns to shore after the breeding and pupping season when the ice leaves. In summer, some larga move north through the Bering Strait into the Chukchi Sea, and return in late autumn. Popov (1976) estimated the Okhotsk Sea population at 13,000 animals, and the Bering Sea population at 135,000.

Pacific whiteside dolphin *(Lagenorhynchus obliquidens)* ranges across subarctic waters of the North Pacific. It is not known to enter the Bering Sea. In the eastern North Pacific it resides year-round from California north to British Columbia, and has been sighted in the Gulf of Alaska in April, May and November. In the western North Pacific, the Pacific whiteside dolphin has been found from Honshu north to the Kuril Islands.

Although regularly seen over the continental shelf and slope, it occasionally ranges far offshore. This dolphin apparently moves northward in summer and to the south in winter. A portion of the population always remains south of salmonid waters. No population estimates are available for the eastern North Pacific. In the western North Pacific, the population resides year-round off Japan and numbers between 30,000-50,000 (National Marine Fisheries Service 1978). This species and the Dall's porpoise *(Phocoenoides dalli)* are probably the most numerous small cetaceans found in offshore waters where salmon range.

False killer whale *(Pseudorca crassidens)* rarely ranges north to Washington or the Aleutian Islands in the eastern North Pacific. In the western North Pacific, it rarely ranges north to Honshu. It is probably more common south of salmonid waters. No population estimates are available, and it can be listed as a rare or occasional visitor to subarctic waters.

Killer whale *(Orcinus orca)* ranges throughout all oceans, but is more abundant in higher latitudes. Populations reside in the inside waters of Washington and British Columbia. Nonresident pods sometimes join them for brief periods of time. There are probably other resident populations in other parts of the species' range.

Killer whales are frequently seen over the continental shelf and slope, and are uncommon offshore. There are no population estimates for the North Pacific where their range and that of salmon overlap. In the Strait of Georgia, however, the resident population consists of about 70 animals. Japanese whalers took 1,439 killer whales in the western North Pacific from 1948-74 (National Marine Fisheries Service 1978).

Harbor porpoise *(Phocoena phocoena)* primarily has a coastal distribution in the North Pacific, almost entirely restricted to waters that salmon inhabit. The harbor porpoise ranges from California around the Pacific Rim to Japan and moves into the Arctic seasonally. Some offshore movements occur because it has been observed in the Pribilof Island area of the Bering Sea 333 km from the mainland.

Harbor porpoise are most frequently seen in harbor entrances and river mouths of estuarine and inside waters. There it usually forages for prey in the tide rips. There are no population estimates available for the North Pacific.

Dall's porpoise *(Phocoenoides dalli)* is the most abundant small cetacean occurring in the range of salmonids. It displays a seasonal north-south movement, at least in the eastern North Pacific. Some animals move south out of salmonid waters into the California bight and northern Baja California waters in winter. Some Dall's porpoise move out of the Bering Sea in winter. No population estimates are available.

Belukha *(Delphinapterus leucas)* is found in the Okhotsk Sea and Bering Sea. Considerable numbers winter there, and then move into arctic waters in spring as leads open in the ice. A small population resides in Cook Inlet (500) and Bristol Bay (1,000-1,500) (National Marine Fisheries Service 1978).

River, Estuarine and Nearshore Interactions

Salmon smolts begin migrating downstream soon after they emerge from the redd in the case of pink and chum salmon or after a stay of several months or years in freshwater in the case of other species. Seaward migration generally lasts several weeks, and usually occurs in spring. Marine mammals sometimes prey on smolts concentrated in the lower reaches of rivers and in estuaries. Smolts disperse when they reach the ocean making it more difficult for marine mammals to catch them.

Belukha, harbor porpoise, harbor seal, larga (spotted) seal, northern sea lion and perhaps the California sea lion are possible predators on smolts. The belukha is known to prey on smolts in the Kvichak and Nushagak rivers, Bristol Bay, Alaska. Investigations that the Alaska Department of Fish and Game (ADF&G) conducted (1954, 1955) showed belukha in the Kvichak River feed almost exclusively for about two weeks each year on sockeye smolts migrating to saltwater. The belukha feed on smelt prior to the smolt run, and on returning adult salmon later in the summer. The ADF&G regarded the loss of adults as relatively unimportant, but deemed the smolt loss serious. Based on the preliminary work of Fish and Vania (1971), ADF&G used recordings of killer whale sounds to frighten belukha in the Kvichak River. Belukha may also prey on smolts or adults in other estuaries of western Alaska and in the U.S.S.R.

There is no documented report of seal or sea lion predation on smolts; however, these animals and the harbor porpoise may be present in estuaries during the smolts' seaward migration. Puustinen (1975) reports that in the Columbia River, gillnetters noted that harbor seals seem to stop taking adult salmon from gillnets and concentrate on seaward migrating smolts.

Adult salmon, returning to nearshore and estuarine areas, are exposed to commercial troll, gillnet and purse seine fisheries as well as Indian traps, beach seines and set nets. An extensive sport fishery exists in some localities. Two types of marine mammal-salmon interactions occur: predation on free-swimming fish, and predation on fish caught by fishing gear.

In the eastern North Pacific, six marine mammal species are known or suspected predators on free-swimming fish: the harbor seal; northern sea lion; California sea lion; belukha; harbor porpoise; and killer whale. With the exception of the California sea lion, these species and the larga seal also inhabit the western North Pacific and Bering Sea. Although predation has been observed on free-swimming adult fish, there is little evidence of major predation.

Scheffer and Sperry (1931) give one of the early accounts of predation by harbor seals. Stomachs from 100 seals killed in Puget Sound and Willapa Bay, Washington, and the Columbia River were examined. Remains of salmon were found in only two stomachs. Pitcher (1977) examined stomachs from 196 harbor seals taken in Prince William Sound and the Copper River Delta, Alaska. Salmon remains were found in only five stomachs. Gadids, mostly pollock *(Theragra chalcogramma)* and eulachon *(Thaleichthys pacificus)* were the major food items. Pitcher's findings agree with those of Imler and Sarber (1947) who examined the stomach contents of 67 harbor seals taken from the Copper River Delta and found salmon remains in only two stomachs and eulachon remains in 64 stomachs.

The previously cited papers demonstrated the minimal effects of harbor seal predation on free-swimming salmonids as they approach their natal stream. There are exceptions. In certain circumstances, harbor seal predation on free-swimming salmon may seriously affect a particular run. For example, Mate *(in press)* noted that in 1976 a returning run of mature chum salmon was probably delayed in the estuarine waters of Netarts Bay, Oregon, by low water in the river leading to the hatchery. The hatchery reported a return of about 700 fish from the run. About 75 harbor seals in the bay were observed to be taking chum salmon. Mate concluded that if each seal took one fish per day during an estimated 6 week delay of the run, the effect would have been significant.

Walter Kirkness (personal communication, Program Director, Pribilof Islands Program, National Marine Fisheries Service, Seattle, Washington), while conducting stream surveys for the ADF&G, observed harbor seals taking free-swimming salmon in the lower reaches of the Taku and Stikine Rivers in southeastern Alaska. The seals were also taking summer and fall pink salmon at the mouths of creeks and in estuaries in southeastern Alaska, Prince William Sound and Kodiak Island.

Fishermen have made numerous reports of harbor seals damaging fish and gear in gillnet fisheries. I will cite only two reports from major salmon producing areas. Hirose *(in press)* conducted studies on the Columbia River downstream from Bonneville Dam for the Oregon Department of Fish and Wildlife from 1972-1976. The incidence of seal-damaged salmon in the annual gillnet catch ranged from 1.0-2.4 percent based on an examination of over 13,000 salmon. The greatest damage occurred in the winter and spring fisheries, and the least damage during the autumn fisheries. The gillnet fisheries take most salmon in early and late autumn. No spring gillnet fishery has been allowed since 1974. The winter fishery continues. Hirose estimated that seals damaged less than 8,000 fish annually from 1972-1976. The total annual catch of salmon in those years ranged from 317,000 to 558,000.

Matkin (1978) collected information on marine mammal-fishery interactions during the 1977 gillnet fishery in the Copper River Delta area of Alaska. He found that 8.3 percent of the salmon catch surveyed was damaged during the study period. He stated that harbor seals caused about one-quarter of the damage (2.1 percent). Sea lions caused three-quarters (6.2 percent) of the damage. Most harbor seal damage occurred inside river bars. Most northern sea lion damage occurred outside in deeper water. The northern sea lion has only recently become a predator here with the beginning of an outside salmon fishery. The inside fishery is exposed primarily to harbor seal predation, as it was in the mid-1940s when Imler and Sarber (1947) concluded that seal damage to gillnetted fish did not exceed 2 percent.

There are few published records of sea lions feeding on free-swimming salmonids in nearshore, estuarine, or lower reaches of rivers. Jameson and Kenyon (1977) spent 14.25 hours observing sea lions feeding at the mouth of the Rogue River, Oregon, in 1974 (July), 1975 (March, June and July), and 1976 (June). During the observation periods, they seldom saw more than two California sea lions (adult males), or four to six northern sea lions (both sexes from yearlings to adults) in the feeding area simultaneously. They observed 93 instances of feeding behavior: 76 (82 percent) on lampreys, 2 (2 percent) on salmonids, and 15 (16 percent) on unidentified organisms. Salmonid feedings were most easily recognized because several minutes were required to break up and consume the fish, whereas lampreys were consumed in a few seconds. Jameson and Kenyon also reported an earlier observation of two northern sea lions shot in the Klamath River, California whose stomachs contained only lampreys. Robert DeLong (marine mammologist, National Marine Mammal Laboratory, National Marine Fisheries Service, Seattle, Washington, personal communication) observed an adult male California sea lions take and eat two free-swimming salmon on 24 September 1979 in Puget Sound, off Bush Point, Whidbey Island, Washington.

Northern and California sea lions damage catches of the nearshore commercial and sport troll fisheries. However, this damage is generally not considered a problem. Briggs and Davis (1972) studied predation on hooked salmon in Monterey Bay, California from 14 April to 22 September 1969. They made 58 day-trips on vessels in the fishery. Only California sea lions were involved. On the vessels sampled, the researchers recorded a 4 percent loss of caught fish. According to them, the report must be regarded as preliminary because the sampling level in the fleet was only 0.21 percent. Sea lions were seen capturing seven free-swimming salmon, but there was some question that the salmon might have been injured.

Previously mentioned ADF&G studies (1954, 1955) document that the belukha switched prey from seaward migrating smolts to returning adult salmon in the Kvichak and Nushigak rivers. The ADF&G reported that belukha consumed 2.7 percent of the sockeye runs in 1954, and 1.0 percent in 1955. Belukha were not at that time entering other Bristol Bay salmon streams. In the western North Pacific, Kleinenberg et al. (1964) said that belukha principally prey on salmon, herring and saffron cod. Of the salmon species, belukha prey on mature pink, chum and coho. These prey species point out that this marine mammal may have difficulty catching single fish as the percentage of empty stomachs increases in the absence of migratory fishes that school.

No direct evidence exists that the harbor porpoise preys on returning adult salmon. However, Tomilin (1957) recorded predation on small salmon in the North Atlantic. The prey species Tomilin noted were principally small fishes that school. It may be that adult Pacific salmon species are too large for harbor porpoise to consume regularly.

Killer whales regularly frequent nearshore areas and occasionally wander into estuaries and rivers. Scheffer and Slipp (1948) reported salmon in the stomach of a killer whale taken in the San Juan Islands area of Washington in June 1929. Fraser *(in press)* stated that the killer whale preys on salmon. He supported this statement by reference to a Canadian investigator who observed a killer whale harrassing a salmon school and feeding on it. Fraser also mentioned a film, taken by a Washington Department of Fisheries biologist, of a killer whale tossing a salmon into the air and then catching it. Dale W. Rice (Northwest and Alaska Fisheries Center, National Marine Fisheries Service, Seattle, Washington, personal communication) observed killer whales chasing salmon. Balcomb et al. (1979) indicated that when salmon appear in abundance in certain Puget Sound areas, there is a high correlation with the occurrence of killer whales. They summarized what is known about the feeding habits of Puget Sound's killer whales and concluded that during salmon runs the killer whales predominantly eat salmon.

Offshore Interactions

We know little about the offshore interactions between salmon and marine mammals in the North Pacific in comparison to our knowledge of river, estuarine and nearshore interactions.

Eight marine mammal species prey on salmon in offshore waters. They are northern sea lion, northern fur seal, fin whale, humpback whale, Pacific whiteside dolphin, false killer whale, Dall's porpoise and sperm whale.

The northern sea lion is regularly encountered over the continental shelf and slope. Fiscus and Baines (1966) examined the stomach contents of 22 sea lions taken in offshore areas. They found salmon in only one stomach. Mathisen et al. (1962)

examined 114 sea lion stomachs taken from a rookery in the Shumagin Islands, Alaska from May to July 1958. Mature salmon pass through the area at this time, but only one stomach contained remains of a salmon. Calkins and Pitcher (1977) took 83 sea lions in the Gulf of Alaska in 1975-1976. The stomachs of 68 contained food, but no salmon were found. Walleye (Alaska) pollock *(Theragra chalcogramma)* was the major food item and represented 83.4 percent of the stomach contents collected.

More data exists on the food and feeding habits of the northern fur seal than any other marine mammal that inhabits the same waters as salmon. The fur seal seldom approaches land, the exception being bays and areas of southeastern Alaska and British Columbia where it enters protected waters during winter to feed primarily on Pacific herring *(Clupea harengus pallasi)*. Member countries of the North Pacific Fur Seal Commission (Canada, Japan, Soviet Union and the U.S.) have taken fur seals from all waters of the North Pacific since 1958. Data on food and feeding habits are available in commission reports (1962, 1969, 1971, 1975). Salmon have never formed a significant portion of the fur seal diet when fur seals were taken from January to May off California (three occurrences in 2,566 stomachs containing food) and Oregon (one occurrence in 103 stomachs containing food).

From 1967-72, Canada and the United States intensively studied fur seals off Washington. Seals were collected from November through June, but primarily from January to May. The five principal prey species in the 2,907 seals collected were salmon, Pacific herring, northern anchovy *(Engraulis mordax mordax)*, and several species of rockfish and squid (Appendix 1). By percent stomach volume, salmon species comprised 20 percent of the food consumed, but squid were the most important prey item by frequency of occurrence (North Pacific Fur Seal Commission 1975). Fur seals primarily took salmon in the waters of the continental shelf and slope, and the salmon were mostly immature pink, coho and chinook. Salmon were also found in fur seal stomachs taken in May and June in the Gulf of Alaska and western Alaskan waters, and during July and August in the Bering Sea. However, salmon only formed a small portion of the prey in these areas.

In 1958, 382 fur seals with stomachs containing food were taken. There were 14 occurrences of salmon representing 1.2 percent of the total food volume. In 1969, 876 seals were taken with food in their stomachs. Salmon occurred 22 times forming 1.6 percent of the total food volume.

U.S. scientists confined their sampling of fur seals to the eastern Bering Sea and Unimak Pass from May to October 1962 and July to September 1963. The principal food items in 1962 were capelin *(Mallotus villosus)*, gonatid squid, walleye pollock and Atka mackerel. In 1963, bathylagid deep sea smelt replaced Atka mackerel on the list. The collections were made where major runs of mature salmon were present. In 1962, 616 seals with stomachs containing food were taken. Salmon occurred nine times and formed 2.4 percent of the food volume. In 1963, 816 seals were examined. Salmon occurred 14 times and formed 2.6 percent of the total food volume.

Salmon have also been found in the stomachs of fur seals taken in the Sea of Japan, Okhotsk Sea, Bering Sea and in the western North Pacific mostly off Hokkaido, Japan; however, fur seals do not prey as extensively on salmon in these waters as they do off the Washington and British Columbia coasts.

Chum salmon up to 60 cm long have been recorded from the stomach of a fin whale taken in the western North Pacific (Tomilin 1957). This report most certainly represents an incidental taking. Fin whale in the North Pacific and adjacent seas feed mostly on krill, but also take a variety of smaller schooling fishes. In the Antarctic, euphausiids comprise their entire diet.

Pink and chum salmon have been recorded from the stomachs of humpback whales. Tomlin (1957) described an incident off the Kuril Islands where fishermen observed two humpback whales chasing pink salmon. A whaling vessel killed one of the animals soon after the incident. The whale's stomach contained mostly walleye pollock, although some pink salmon were present. Humpback whale predation on salmon is, nevertheless, believed to be incidental.

Salmon predation by the Pacific white-side dolphin has been reported (Stroud et al. *in press*). Ten dolphins were collected on 25 and 26 February 1968. They were caught west of the Columbia River mouth from pods of 25-30 animals. Eight of the 10 stomachs contained salmon. Nine pink and coho were identified in one stomach. These fish were 26-34 cm long.

A few vertebrae were found in the stomach of a false killer whale killed in Puget

Sound in May 1937 (Scheffer and Slipp 1948). These whales feed on large epipelagic fishes, and salmon would be a likely prey item when they enter subarctic waters.

The Dall's porpoise preys on a variety of small schooling fishes and on squid. It probably does not consume mature salmon. Mizue et al. (1966) examined 148 Dall's porpoise taken in the Japanese high seas salmon gillnet fishery in 1964 and 1965. Squid were the most abundant prey item found in the stomachs, followed by small fish and shrimp. One sockeye salmon was found, but no mention was made of its size or maturity.

The sperm whale principally preys on cephalopods; however, this whale also takes some benthic and demersal fish. Tomilin (1957) noted that a salmon was found in the stomach of one whale from a sample of 110 whales taken from the Kuril Islands region, which suggests that sperm whales eat salmon only incidentally.

SUMMARY AND CONCLUSIONS

Thirty-four species of marine mammals inhabit waters of the North Pacific and adjacent seas. Fifteen species reportedly eat salmon. The most serious and perhaps best documented cases of marine mammal predation on salmon have occurred in the lower reaches of rivers, in estuaries and in nearshore waters.

The belukha preys on out-migrating smolts in the estuaries of the Kvichak and Nushgak rivers, Bristol Bay, Alaska. The species also preys on smolts in other streams. No other marine mammal is known to prey on smolts, although circumstantial evidence exists that the harbor seal may. Other marine mammals that enter river and estuarine waters to prey on small fishes that school, such as smolts, include the harbor porpoise, larga seal, northern sea lion and California sea lion.

The belukha, harbor seal, larga seal, northern sea lion, California sea lion and killer whale capture and consume free-swimming adult salmon in nearshore, estuarine and river areas. The evidence suggests, however, that only the belukha consistently takes significant numbers of these mature fish. Local exceptions for other species have been reported. The intensive harbor seal predation on the 1976 run of chum salmon in Netarts Bay, Oregon is an example.

Some marine mammal species may prey on smaller and more vulnerable fish than salmon. In some instances, they may also benefit salmon by feeding on salmon predators or parasites. An example of this is in the Rogue River, Oregon where northern and California sea lions were observed eating lamprey.

Most marine mammal predation on salmon occurs after the fish are caught in near-shores estuarine and river areas on commercial and sport trolling gear, or in gillnets. Most conflicts between fishermen and marine mammals take place in these areas.

Eight marine mammal species that consume salmon inhabit offshore waters of the North Pacific. Four of these--fin whales, humpback whale, false killer whale and sperm whale--are not believed to be serious salmon predators. These four cetaceans prey on other species, and reports of them preying on salmon are based on scattered incidents.

The northern fur seal, the most abundant pinniped of the open sea, preys on salmon throughout its range; however, only off Washington in the eastern Pacific did salmon form an appreciable portion of its diet. It also preys on fish caught in the high seas gillnet fishery and the offshore troll fishery.

The northern sea lion inhabits waters over the continental shelf and occasionally seaward. A few instances of it preying on free-swimming fish have been recorded. It causes the greatest loss to both fish and gear to fishermen in nearshore and estuarine areas.

The Pacific whiteside dolphin ranges across the southern portion of the salmon range, and may regularly take immature salmon. This species probably does not consume adult salmon. Dolphins taken off the Columbia River had taken only immature salmon.

The Dall's porpoise inhabits most of the salmon range except the extreme northern portions. It is probably one of the most abundant small cetaceans in the region although no population estimates have been made. From the many stomachs examined, only one salmon has been reported. Squid was its principal prey item.

Most conflicts between marine mammals and fishermen over salmon occur in near-shore and estuarine areas, and in the lower reaches of rivers. Marine mammal species most frequently involved include the harbor seal, northern sea lion, belukha and California sea lion. Most salmon predation, except that of the belukha, occurs on caught fish with attendant gear damage.

Detailed information of marine mammal predation on immature and maturing salmon during the ocean portions of their life history are not available except for northern fur seal feeding habit studies. Additional information is needed to evaluate the impact of marine mammals on salmon stocks. Investigations to provide this needed information include: studies of local or regional marine mammal-fishery conflicts in specific fisheries to determine the extent of fish loss and damage to fishing gear; the species and numbers of marine mammal(s) involved; and the actions needed to solve marine mammal-fishery problems. For example, killer whale-salmon interactions must be studied to properly evaluate and document killer whale predation on salmon stocks. Predation on smolts migrating to the sea is not documented except in the case of the belukha in Bristol Bay, Alaska. Investigations are needed in other areas to identify marine mammal predators and determine if such predation jeopardizes specific migrations.

The harbor porpoise, present in the entrances to most estuaries and streams that produce salmon, preys on small fishes that school. Studies of this porpoise are needed to determine if it preys on salmon. If it does, we need to know the extent of its predation. The incidental take of harbor porpoise in gillnets or other fishing gear should be studied to determine possible adverse effects of the fishery on local stocks of this marine mammal.

LITERATURE CITED

Advisory Committee on Marine Resources Research. 1976. Mammals in the seas: ad hoc group III on seals and marine otters. Food Agric. Organ. U.N., Adv. Comm. Mar. Resour. Res. FAO ACMRR/MM/SC4. 182 pp.

Alaska Fisheries Board and Alaska Department of Fisheries. 1954. 1954 Annual Report. Rep. 6. 92 pp.

Alaska Fisheries Board and Alaska Department of Fisheries. 1955. 1955 Annual Report. Rep. 7. 152 pp.

Balcomb, K. C., J. R. Boran, R. W. Osborne, and N. J. Haenel. 1979. Observation of killer whales *(Orcinus orca)* in Greater Puget Sound, State of Washington. A report of field research conducted by the Moclips Cetological Society, Inc., 1 April 1976 through 31 December 1979. 27 pp.

Briggs, K. T., and C. W. Davis. 1972. A study of predation by sea lions on salmon in Monterey Bay. California Fish Game 58:37-43.

Calkins, D., and K. Pitcher. 1977. Population assessment, ecology and trophic relationships of Steller sea lions in the Gulf of Alaska. Pages 432-502 *in* Environmental assessment of the Alaskan continental shelf: annual reports of principal investigators for the year ending March 1977. Vol. 1, Receptors - Mammals. Dep. Natl. Ocean. Atmos. Admin., ERL, Boulder, Col.

Fiscus, C.H. 1979. Interactions of marine mammals with Pacific hake *(Merluccius productus)* and the hake fishery: A summary. Mar. Fish. Rev. 39(10)1-9.

Fiscus, C. H., and G. A. Baines. 1966. Food and feeding behavior of Steller and California sea lions. J. Mammal. 47: 195-200.

Fish, J. H., and J. S. Vania. 1971. Killer whale, *Orcinus orca*, sounds repel white whales, *Delphinapterous leucas*. U.S. Fish. Bull. 62:531-535.

Fraser, J. *In press*. Washington state marine mammal fisheries interaction. *In* Mate, B. R. (ed.), Workshop report for marine mammal-fishery interactions in the eastern North Pacific. Mar. Mammal Comm.

French, R., H. Bilton, M. Osako, and A. Hartt. 1976. Distribution and origin of sockeye salmon *(Oncorhynchus nerka)* in offshore waters of the North Pacific Ocean. Int. North Pac. Fish. Comm. Bull. 34. 113 pp.

Godfrey, H., K. A. Henry, and S. Machidori. 1975. Distribution and abundance of coho salmon in offshore waters of the North Pacific Ocean. Int. North Pac. Fish. Comm. Bull. 31. 80 pp.

Hirose, P. *In press*. Incidence of seal-damaged salmonids sampled from the lower Columbia River gillnet fishery, 1972-76. *In* Mate, B. R. (ed.), Workshop report for marine mammal-fishery interactions in the eastern North Pacific. Mar. Mammal. Comm.

Imler, R. H., and H. R. Sarber. 1947. Harbor seals and sea lions in Alaska. U.S. Fish. Wildl. Serv. Spec. Sci. Rep. 28. 23 pp.

Jameson, R. J., and K. W. Kenyon. 1977. Prey of sea lions in the Rogue River, Oregon. J. Mammal. 58:622.

Kleinenberg, S. E., A. V. Yablokov, B. M. Bel'kovich, and M. N. Tarasevich. 1964. Belukha. Opyt monograficheskogo issledovaniya vida (Beluga, *(Delphinapterus leucas)* investigation of the species). Izd. "Nauka", Moscow. In Russian. (Transl. by Israel Program Sci. Transl., 1969). 376 pp. (Available from U.S. Dep. Commer., Natl. Tech. Inf. Serv., Springfield, Va. as TT 67-51345.)

Mate, B. R. 1976. History and present status of the California sea lion, *Zalophus californianus*. Food Agric. Organ. U.N., Adv. Comm. Mar. Resour. Res. FAO ACMRR/MM/SC/39. 6 pp.

Mate, B. R. *In press*. Newly initiated studies look at the impact of harbor seals on aquacultural efforts to enhance chum salmon resources. *In* Mate, B. R. (ed.), Workshop report for marine mammal-fishery interactions in the eastern North Pacific. Mar. Mammal. Comm.

Mathisen, O. A., R. T. Baade, and R. J. Lopp. 1962. Breeding habits, growth, and stomach contents of the Steller sea lion in Alaska. J. Mammal. 43:469-477.

Matkin, C. O. 1978. A preliminary evaluation of fishery depredations by marine mammals on the Copper River Flats, Alaska, spring 1977. Page 6276 *in* Matkin, C. O. and F. H. Fay (eds.), Marine mammal-fishery interactions on the Copper River and in Prince William Sound, Alaska, 1978. Mar. Mammal Comm.

Mizue, K., K. Yoshida, and A. Takemura. 1966. On the ecology of the Dall's porpoise in Bering Sea and the North Pacific Ocean. Fac. Fish., Nagasaki Univ., Bull 21. 21 pp. In Japanese (English summ.).

National Marine Fisheries Service. 1978. The Marine Mammal Protection Act of 1972; Annual Report: 1 April 1977 to 31 March 1978. U.S. Dep. Commer., Natl. Ocean. Atmos. Admin., Natl. Mar. Fish. Serv. 183 pp.

Neave, F., T. Yonemori, and R. G. Bakkala. 1976. Distribution and origin of chum salmon in offshore waters of the North Pacific Ocean. Int. North Pac. Fish. Comm. Bull. 35. 79 pp.

North Pacific Fur Seal Commission. 1962. North Pacific Fur Seal Commission report on investigations from 1958 to 1961. North Pac. Fur Seal. Comm. 183 pp.

North Pacific Fur Seal Commission. 1969. North Pacific Fur Seal Commission report on investigations from 1964 to 1966. North. Pac. Fur Seal Comm. 161 pp.

North Pacific Fur Seal Commission. 1971. North Pacific Fur Seal Commission report on investigations from 1962 to 1963. North Pac. Fur Seal Comm. 96 pp.

North Pacific Fur Seal Commission. 1975. North Pacific Fur Seal Commission report on investigations from 1967 to 1972. North Pac. Fur Seal Comm. 212 pp.

Pitcher, K. W. 1977. Population productivity and food habits of harbor seals in the Prince William Sound - Copper River Delta area, Alaska. Mar. Mammal. Comm. Rep. MMC 75/03. 36 pp.

Popov, L. A. 1976. Status of main ice forms of seals inhabiting waters of the U.S.S.R. and adjacent to the country marine areas. Food Agric. Organ. U.N., Adv. Comm. Mar. Resour. Res. FAO/ACMRR/MM/SC/51. 17 pp.

Puustinen, W. W. 1975. Seals and sea lions cause loss to fishery. Columbia River Gillnetter 6(3):7-8.

Rice, D. W. 1977. A list of the marine mammals of the world. U.S. Dep. Commer., Natl. Ocean. Atmos. Admin., Natl, Mar. Fish. Serv. NOAA Tech. Rep. NMFS SSRF-711. 15 pp.

Scheffer, T. H., and C. C. Sperry. 1931. Food habits of the Pacific harbor seal, *Phoca richardii*. J. Mammal. 12:214-226.

Scheffer, W. B., and J. W. Slipp. 1948. The whales and dolphins of Washington State with a key to the cetaceans of the West Coast of North America. Am. Wildl. Nat. 39:257-337.

Shaughnessy, P. D., and F. H. Fay. 1977. A review of the taxonomy and nomenclature of North Pacific harbor seals. J. Zool. London 182:385-419.

Stroud, R. K., C. H. Fiscus, and H. Kojimura. *In press*. Food of the Pacific whitesided dolphin, *Lagenorhynchus obliquidens*; Dall's porpoise, *Phocoenoides dalli*; and northern fur seal, *Collorhinus ursinus*, off California and Washington. U.S. Fish. Bull.

Tomilin, A. G. 1957. Kitoobraznye (Cetaceans). Vol. IX. Zveri SSSR i prilezhashchikh stram (Mammals of the U.S.S.R. and adjacent countries). Izd. Akad. Nauk SSSR, Moskva, 756 pp. In Russian. (Transl. by Israel Program Sci. Transl., 1967. 717 pp. (Available U.S. Dep. Commerce, Natl. Tech. Inf. Serv., Springfield, Va. as TT 65-50086.)

APPENDIX

The United States and Canada planned and carried out joint studies on the distribution and feeding habits of the northern fur seal in the waters off Washington from 1967 through 1972.

The northern fur seal begins to appear in Washington waters during its southbound migration in November and is found in varying abundance until its northbound migration ends in early June. It is most abundant in these waters from January to May.

During the research period, 2,907 fur seals were taken. Of these, the stomachs of 1,666 contained food. Over 30 species of fish and squids have been identified from the stomachs of fur seals taken off Washington; however, only about one-third form most of the food of the fur seal in the area. Appendix Tables A-1 and A-2 rank the principal prey species found in the stomachs of fur seals taken off Washington during the study period by percentage of total stomach volume (A-1) and by frequency of occurrence (A-2).

In these tables all five species of eastern North Pacific salmon and steelhead are combined under one heading - Salmonid. The nine squids are combined under one heading - Squid.

Most salmonids were taken from the stomachs of fur seals that were collected over the continental shelf and slope. Squids were most abundant offshore of the continental shelf and slope except for the squid *Loligo opalescens*, which is a shelf species.

Combining the six years of data, salmonids formed 20 percent of the prey consumed during the period; however, if the values for 1967 (35 percent of volume) and 1968 (36.6 percent of volume) are dropped, the percentage of salmonids consumed drops to 12 percent for the period 1969-72. On an annual basis, the percentages of salmonids consumed were: 1969 (15.1 percent), 1970 (9.9 percent), 1971 (17.2 percent) and 1972 (6.6 percent).

Salmonids ranked first by percentage of food volume only in 1967 and 1968 when most seals were taken in winter as opposed to succeeding collections when most seals were taken in spring. The data suggest that salmonids may be more available to fur seals off Washington in January and February than during the months of March, April, and May. These data are being analyzed by North Pacific Fur Seal Commission biologists, and the results will be published.

Table A-1. The Five Principal Prey Species Found in Fur Seal Stomachs Taken off Washington, 1967-72, by the United States and Canada, Ranked by Percentage of Food Volume.

	Species										Number of stomachs with food						
											Winter			Spring			
Year	Salmonid[1]	Herring	Anchovy	Rockfish	Smelt	Capelin	Shad	Black Cod	Hake	Squid[2]	Dec.	Jan.	Feb.	Mar.	Apr.	May	Winter/spring combined[3]
1967	1	2	4	3	-	-	5	-	-	-		89			32		121
1968	1	-	2	3	4	5	-	-	-	-		248			93		341
1969	4	3	2	1	-	5	-	-	-	-		78			261		339
1970	4	3	2	-	-	-	-	5	-	1		60			280		340
1971	3	1	2	-	-	5	-	-	4	-		62			233		295
1972	5	1	3	4	-	-	-	-	-	2		44			162		206

[1] All species of salmonids combined under Salmonids, most being *Oncorhynchus* species.

[2] All species of cephalopods - decapoda combined under Squid.

[3] Stomachs from 24 seals taken in November or June not included in totals.

Table A-2. The Five Principal Prey Species Found in Fur Seal Stomachs Taken Off Washington, 1967-72, by the United States and Canada, Ranked by Frequency of Occurrence.

	Species									Number of stomachs with food						
										Winter			Spring			
Year	Salmonid[1]	Herring	Anchovy	Rockfish	Smelt	Capelin	Squid[2]	Lanternfish	Saury	Dec.	Jan.	Feb.	Mar.	Apr.	May	Winter/Spring combined[3]
1967	2	3	4	5	-	-	1	-	-		89			32		121
1968	2	-	4	-	5	3	1	-	-		248			93		341
1969	5	3	2	-	-	4	1	-	-		78			261		339
1970	5	3	2	-	-	-	1	4	5		60			280		340
1971	4	3	2	-	-	5	1	-	-		62			233		295
1972	4	2	3	5	-	-	1	5	-		44			162		206

[1] All species of salmonids combined under Salmonid, most being *Oncorhynchus* species.

[2] All species of cephalopods - decapoda combined under Squid.

[3] Stomachs from 24 seals taken in November or June not included in totals.

Trends in Puget Sound and Columbia River Coho Salmon

Stephen B. Mathews

INTRODUCTION

Catches of chinook *(Oncorhynchus tshawytscha)* and coho *(O. kisutch)* from California, Oregon and Washington have increased during the past two decades due to substantial increases in hatchery production. This paper compares broad trends of coho production, catch, survival and size of adult fish in two major hatchery production areas: Puget Sound and the Columbia River. The purposes are to seek possible evidence of producing and releasing too many hatchery salmon to the ocean and/or overfishing, and to provide analyses that may improve harvest and enhancement strategies.

DISCUSSION

Puget Sound Coho

The total pounds of reared fish released is perhaps the best single measure of hatchery production. This figure can be obtained by multiplying the total number of fish by average size per migrant. Survival tends to increase with the size at release (Mathews and Buckley 1976). However, timing of the release also influences survival. Test groups released later than the natural out-migration time in early May tend to survive better than groups released earlier (Mathews and Buckley 1976, Seidel and Mathews 1977*a*). Olson (1978) estimated average release dates for Puget Sound coho from planting records. He found that release dates have steadily advanced from about mid-January to early May between 1960 and the present.

Total poundage from all Puget Sound hatcheries increased about ten-fold from 1960 to the present (Fig. 1). The Washington Department of Fisheries (WDF) released most of these fish. With a trend toward a later and more favorable release time, it is not surprising that the catch has also increased.

The nature of the ocean troll and sport fisheries makes it difficult to calculate the total catch of coho produced in Puget Sound hatcheries. These fisheries harvest portions of several stocks. During 1967 to 1969, about 45 percent of the catch and escapement of marked hatchery fish was taken in the U.S. and Canadian ocean fisheries (Senn 1970*a*, 1970*b*; Senn and Satterthwaite 1971). A preliminary compilation of more recent marking studies (Morbrand et al. 1977), indicates that the ocean fisheries presently take about 55-60 percent.

To approximate catch trends for Puget Sound coho, I added the catches of all significant fisheries from which marking studies indicate major Puget Sound hatchery contribution. These include the commercial net and sport fisheries in Puget Sound and Strait of Juan de Fuca, the Canadian Strait of Juan de Fuca net fishery, the Washington Cape Flattery area troll and sport fisheries and the Canadian troll fishery in the southern Vancouver Island area and northern Washington coast. Salmon stocks from coastal rivers in Washington, southern Canadian rivers and the Columbia River support these fisheries; however, Puget Sound hatcheries contribute the majority of hatchery fish. The Columbia River provides only about one-seventh as many coho per pound released to the Canadian troll fishery as Puget Sound hatcheries do (Godfrey 1971). Production in Washington's coastal hatcheries has been relatively small. Thus, any major upward trends in the combined catch from these fisheries should primarily be due to Puget Sound hatchery production.

Contribution No. 524, College of Fisheries, University of Washington, Seattle, Washington.

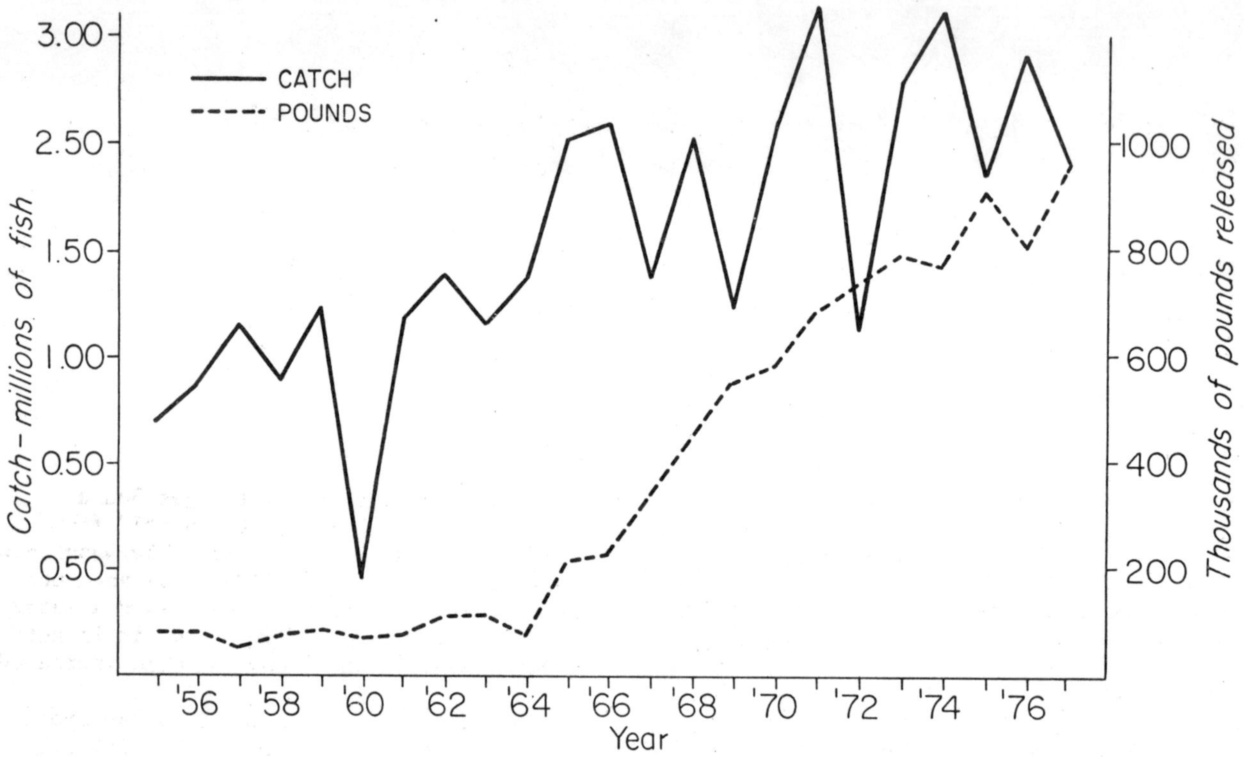

Figure 1. Total catch in fisheries taking significant numbers of coho of Puget Sound origin and total pounds released from Puget Sound hatcheries in years i-1.

Figure 1 indicates that fisheries taking Puget Sound coho have increased their catch from about 1.5 million annually during the 1950s and early 1960s to a present level of about 2.5 million. (Data sources are various published and unpublished documents of the Washington Department of Fisheries, Canadian Department of Environment and Pacific Fisheries Management Council.) Catch values for 1976 and 1977 are in many instances preliminary. (Specific data are available from the author.) Most of the increase occurred during the 1960s. Recently, the catch has remained relatively constant in spite of increasing hatchery production.

Has the catch levelled off because survival of hatchery fish has decreased? Probably not. Table 1 lists survival rate estimates (total catch + escapement ÷ number released) from marking studies. Senn (1970*a*, 1970*b*) and Senn and Satterthwaite (1971) published the 1964-66 brood year data. The 1970-73 data have not been published, but were compiled preliminarily by Morbrand et al. (1977). The 1970-73 brood experiments indicate survival rates at least as high as the 1964-66 brood experiments.

There is another indication that Puget Sound hatchery survival has not declined. Although ocean harvest rates have apparently increased, the "inside" runs in Puget Sound have increased steadily to 1977 (Fig. 2). Inside runs include total Washington net catch (all gear types) plus hatchery escapement. Estimates of non-hatchery escapement are available only for recent years and therefore not included.

The favorable survival trend of Puget Sound hatchery coho may be due to proper release timing and to the recent construction of several high flow-through release ponds by WDF (personal communication, Harry Senn, Washington Department of Fisheries, Olympia, Washington).

The above factors--increased hatchery production and apparently favorable hatchery survival, yet recently stable total catch levels--imply that non-hatchery stock contribution is declining. Even before intensive hatchery output, the fisheries were yielding some 1.5 million fish that mostly were non-hatchery salmon. In my opinion, non-hatchery production has fallen somewhat in absolute numbers contributed but may still be providing some 50 percent of the catch in these fisheries.

Table 1. Survival Rates of Yearling Coho Salmon Released From Puget Sound Hatcheries.

Brood Years	Area	Release time	Survival rate
1964	South Sound	February-April	5.59%
1964	North Sound	February-April	2.87%
1964	Hood Canal	February-April	3.53%
1965	South Sound	March-April	11.07%
1965	North Sound	March-April	4.55%
1965	Hood Canal	March-April	5.58%
1966	South Sound	April	8.17%
1966	North Sound	April-May	2.89%
1966	Hood Canal	March-May	3.19%
1970-1973	South Sound	April-June	11.04%
1972, 1973	North Sound	April-May	8.55%
1971, 1973	Hood Canal	April-May	9.12%

Sources: 1964-1966 broods, Senn 1970*a*, 1970*b*; Senn and Satterthwaite 1971; 1970-1973 broods, Mobrand et al. 1977.

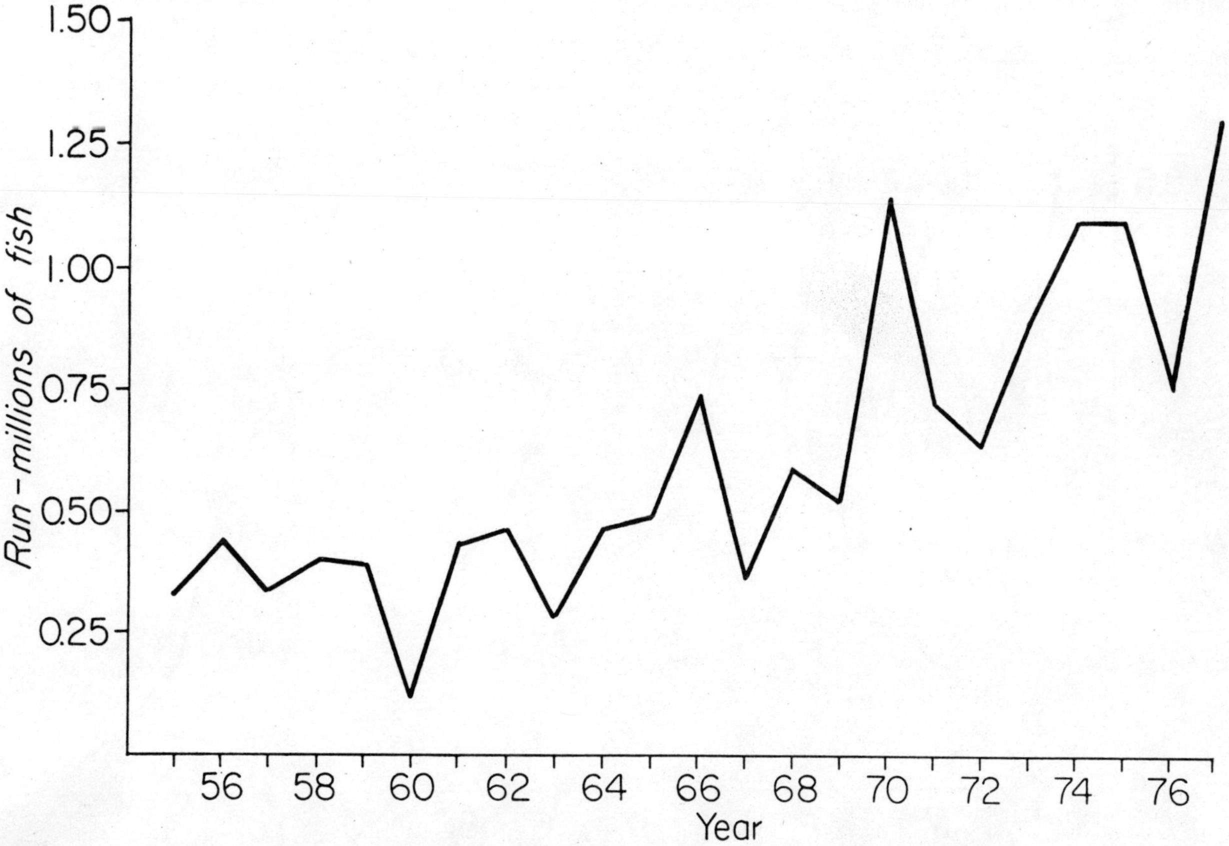

Figure 2. Puget Sound "inside" coho run (net catch plus hatchery escapement).

The likely decline in non-hatchery abundance is probably due to a combination of environmental deterioration and overfishing of weaker stocks in years of less than average survival. We could argue about the relative importance of these contributing factors to the decline; however, I contend that absolute catch levels would increase if the overall harvest rates of Puget Sound coho decreased through an orderly reduction of fishing intensity in all fisheries taking a significant number of these fish. I maintain that this would occur even if hatchery production remained constant. The fishing intensity levels of the late 1960s to early 1970s yielded similar harvests as the higher intensity levels of the most recent years, yet with less hatchery production. If these earlier fishing rates were reinstated, non-hatchery escapements would increase and resulting catch levels would similarly increase. Coupled with greater hatchery production at present, I believe that reducing the harvest rates will increase total catches in all fisheries.

The average size of adult fish is easily obtainable and could possibly indicate hatchery overplanting. The opening of the ocean-troll season for coho, traditionally June 15, is the earliest date that average adult size can be observed. Figure 3 shows the average dressed weight in pounds for coho landed at Neah Bay during June. Landings at this time of year would be mainly coho caught along the north Washington coast (Umatilla and Prarie grounds) with a relatively high component from Puget Sound. The average early-season troll weight has clearly dropped, more than a full pound in the last 20 years. Increased poaching prior to June 15 may have contributed to this trend; however, I do not regard this as a significant factor, particularly since a similar trend has not been observed throughout the Washington troll fishery. Another contributing factor may be the increased planting of "delayed release" coho (Mathews and Buckley 1976) that do not migrate far in the ocean. They apparently grow more slowly. However, the declining trend in June troll weight began well before WDF's delayed release program that began in the 1970s.

There is no apparent trend indicating that Puget Sound coho are smaller at the end of their ocean life. To index this statistic, I computed the average (round) weight of fish caught in the Canadian purse seine fishery in the Strait of Juan de Fuca. A purse seine is less selective for size than a gillnet. I chose the Canadian fishery because the Washington seine fishery was, until recently, restricted from fishing within Puget Sound during September when

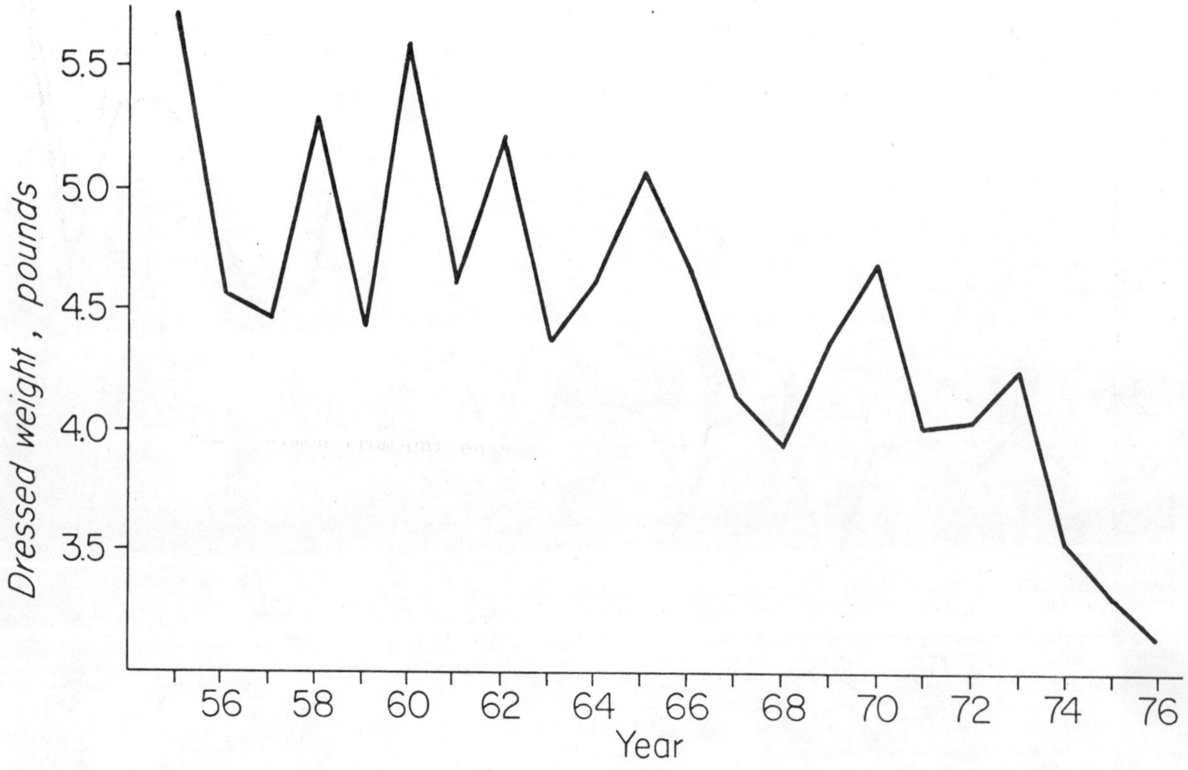

Figure 3. Average June weight of troll coho salmon landed from Cape Flattery vicinity.

peak catches occur. There has been no downward trend in average size at maturity similar to that in the June troll weight (Fig. 4). Thus, if Puget Sound coho now grow at a slower rate prior to June of their last year at sea, they must be making this up by an increased growth rate during their last summer. This trend can perhaps be seen best in Figure 5, which shows the average percent increase in weight from June to September in the Cape Flattery troll fishery.

These apparent growth trends could be due to many things, including stock changes, artifacts of fishery practices or increased population density due to hatchery production. With this multitude of possible causes for the decline in opening troll weights, one can only speculate that increased hatchery production accounts for it. However, if it is due to hatchery production, such a density-induced effect probably takes place within Puget Sound shortly after release. There is some evidence that Puget Sound coho, although migrating to saltwater in May, do not leave for the open ocean until late summer. While commercial trolling in the vicinity of Cape Flattery, I do not catch many yearling coho "shakers", or fish smaller than the legal size, on my gear until about mid-August. They are usually abundant for about two weeks then disappear around Labor Day. These observations may coincide with the out-migration of Puget Sound yearling coho. Al Hartt (personal communication, Fisheries Research Institute, University of Washington, Seattle, Washington) reported that when purse seining and tagging salmon in the outer Strait of Juan de Fuca in 1968, yearling coho were most abundant in late August. Thus, Puget Sound coho could be remaining within the relative confines of Puget Sound until about mid-August. Population densities from hatchery production would conceivably be more likely to depress growth if the fish were confined for several months than if they immediately migrated to the open ocean.

If density from increased hatchery production were negatively affecting growth, a negative correlation should exist between abundance and adult size. Such a relationship is indicated for June troll weight versus abundance (catch) (Fig. 6), but not for adult weight (Fig. 7). However, both June weight and abundance correlate with time (weight, negatively; abundance, positively).

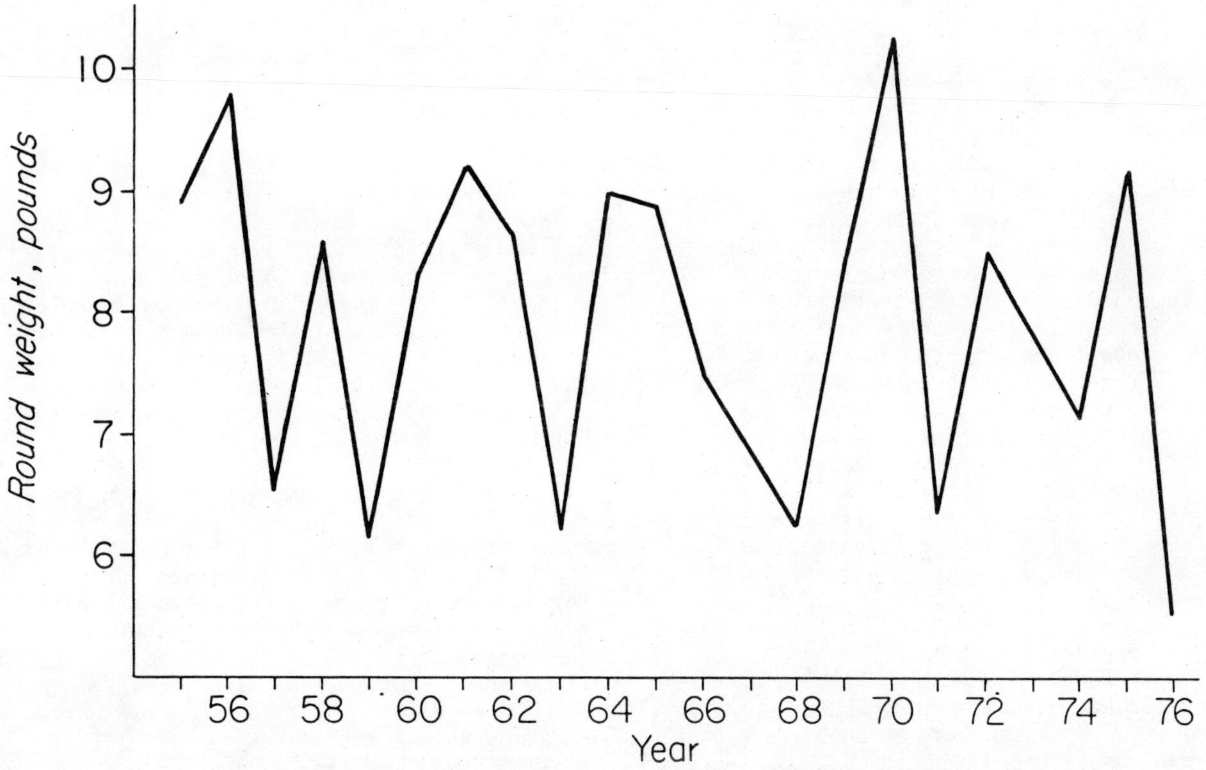

Figure 4. Weight of Puget Sound coho at maturity. Average from Canadian Strait of Juan de Fuca purse seiners.

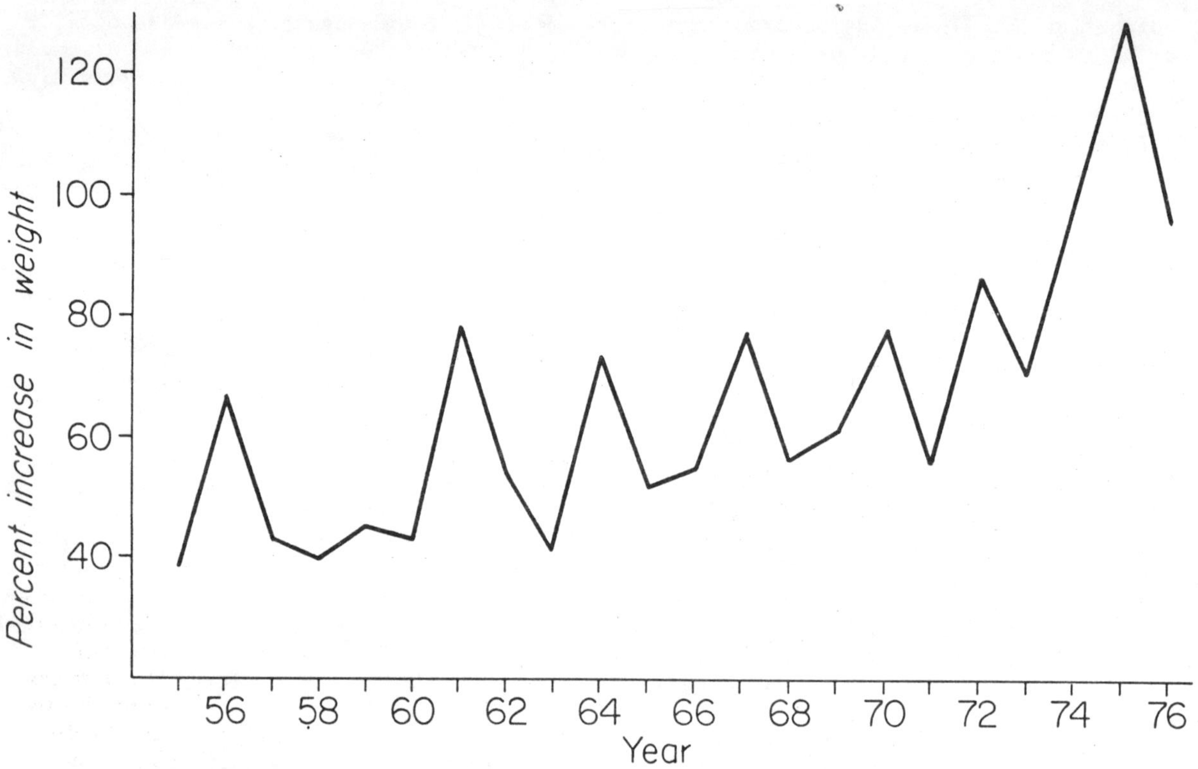

Figure 5. Growth rate of Puget Sound coho in the Cape Flattery troll fishery. $\frac{\text{Sept. wt.} - \text{June wt.}}{\text{June wt.}} \times 100$.

Columbia River Coho

Figure 8 plots total pounds of coho released from Columbia River hatcheries, and catch totals from all significant fisheries that take Columbia River coho as revealed from marking experiments. The specific fisheries for which catches were summed included California troll and ocean sport, Oregon troll and ocean sport, Washington Columbia River mouth troll and ocean sport (Ilawaco), and Columbia River net. Central and northern Washington troll catches were excluded because hatchery coho were mostly from the early Toutle stock. This stock tends to enter fisheries south of their ocean entry point (Hopley 1976). The trends resemble those of Puget Sound to a point. Catches increased directly with releases through the late 1960s and early 1970s, but then began drifting downward even with increasing hatchery production.

Hatchery production has apparently had a greater effect on these fisheries than Puget Sound production had on its fisheries, although the comparison may be obscured by the unknown loss of the non-hatchery contribution to the Puget Sound area fisheries. Catches in fisheries that Columbia River hatcheries support increased threefold. Catch levels have risen from less than 1 million a year to averages of about 2.5 million, clearly the result of hatchery output.

The recent and disturbing divergence between catch and hatchery output may have a different cause than for Puget Sound. The decline in non-hatchery contribution explains well this trend for Puget Sound. For Columbia River supported fisheries, however, the non-hatchery contribution was apparently so low before the push in hatchery production that it could have disappeared entirely without explaining the present divergence between catch and hatchery output.

The non-hatchery contribution from the river may have declined. Ocean fishery rates are high, higher than for Puget Sound. In 1968 and 1969, ocean fisheries accounted for 75 percent of the total catch and escapement of marked Columbia River hatchery coho (Wahle et al. 1974). This rate has increased since then according to the preliminary analysis by Mobrand et al. (1977) and may presently be about 85 percent. It is doubtful that runs not supplemented in some manner by hatchery production can sustain these rates without decline. However, because of rela-

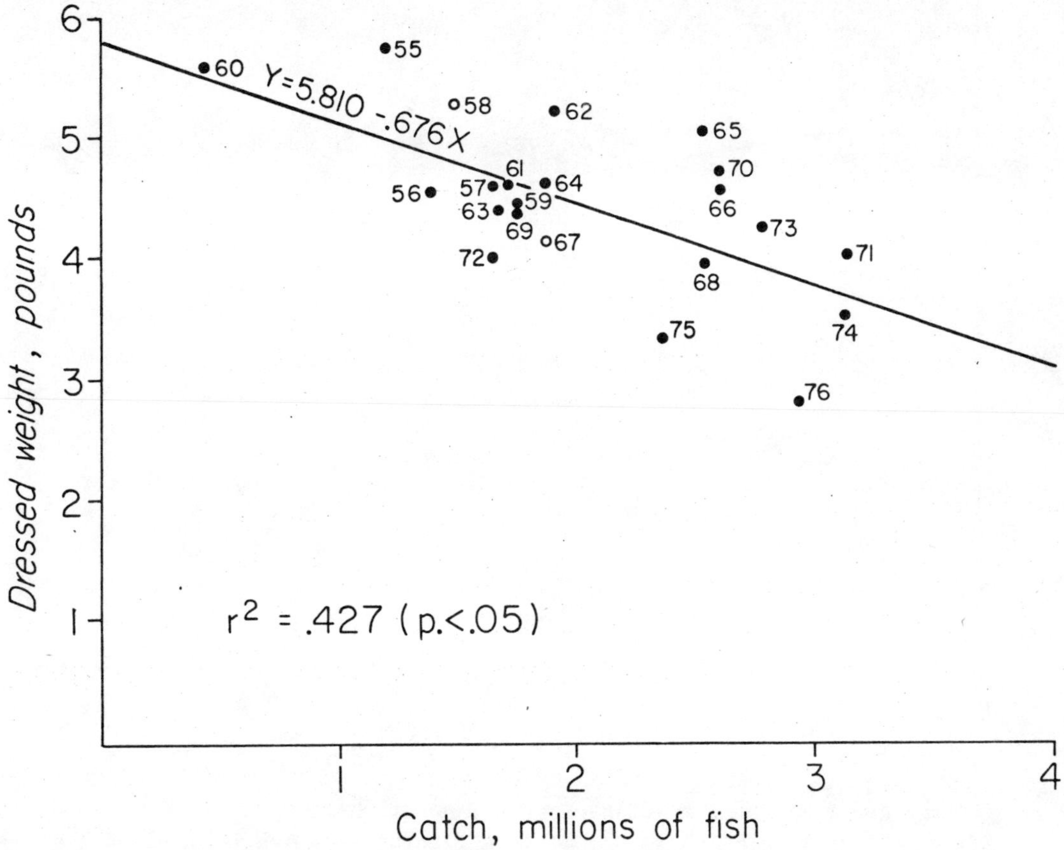

Figure 6. Cape Flattery June weight versus total catch in fisheries which take significant numbers of Puget Sound coho.

tively low non-hatchery production, the major cause for divergence between catch and hatchery production more likely lays in declining, or perhaps a better term is irregular, hatchery survival. Table 2 lists survival data for a number of Columbia River hatchery experiments by brood year. These data may not be entirely representative as agencies have conducted a number of other experiments. Their data have not yet been fully analyzed and written up.

The table reveals a declining trend in survival through time. I was able to find partial data for only one marking experiment with the 1973 brood, which constituted the high 1976 catch. This was an experiment done at the Kalama Hatchery. These incomplete recovery data indicated a probable survival rate in the 2-3 percent range, again relatively low in accordance with recent trends. However, considering the high 1976 catch level, the average survival rate of Columbia River hatchery coho that year was probably higher than this one marking experiment indicates. Despite yearly fluctuations, survival of Columbia River hatchery coho could be declining and does not appear to be as successful as Puget Sound coho. Unlike Puget Sound, catches probably cannot be increased by decreasing the overall catch rates because non-hatchery fish contribute a minor proportion to the Columbia River supported fisheries. This does not imply that natural production should be foregone. Some leveling of catch rates coupled with off-station hatchery releases are necessary to maintain production.

The gillnet and sport fisheries in the river, the Bonneville Dam count, and hatchery escapements and dam counts in the lower river also indicate a declining hatchery survival rate (Fig. 9). The ocean catch rates have increased, of course. However, the ocean survival rate (100 percent-ocean catch rate) has dropped roughly by half in the recent decade while hatchery output has about doubled. If hatchery coho survival had remained constant during this time, catches in the inside fisheries should also have remained constant.

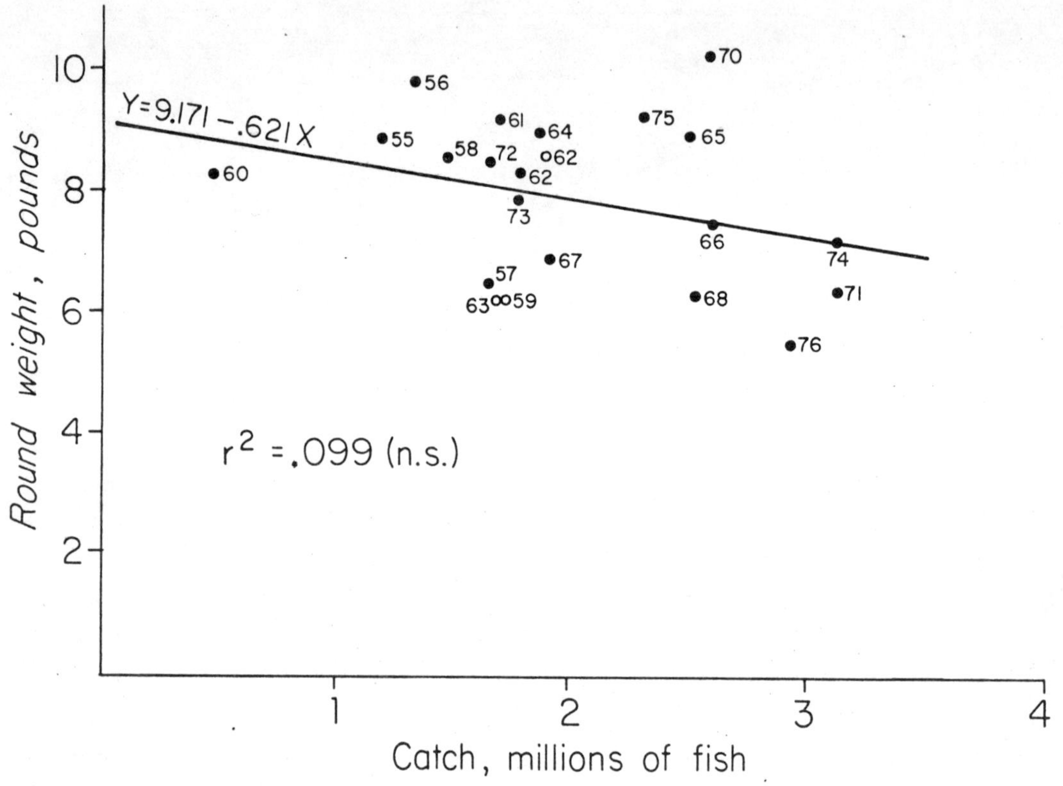

Figure 7. Average adult weight (Canadian purse seine) of Puget Sound coho versus total catch in fisheries which take significant numbers of Puget Sound coho.

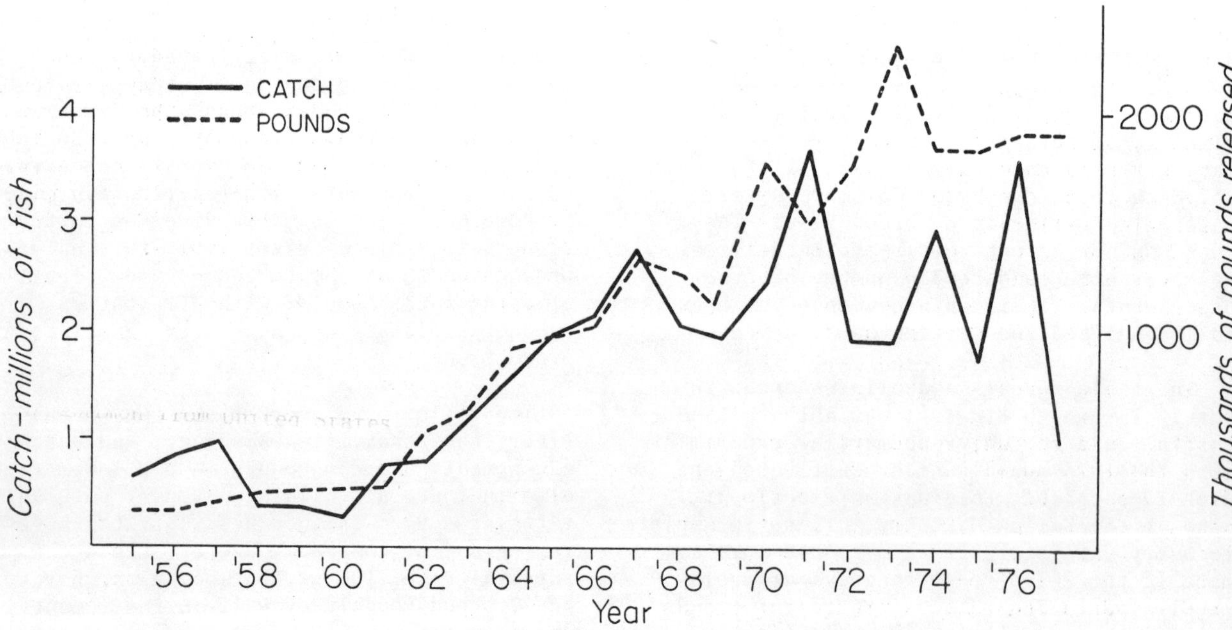

Figure 8. Total catch in fisheries taking significant numbers of coho of Columbia River origin, and total pounds released from Columbia River hatcheries in years i-1.

Table 2. Survival Rates of Yearling Coho Salmon Released from Columbia River Hatcheries.

Brood years	Stock	Hatcheries	Survival rate %	Data sources
1961	Washougal	Washougal	6.82[1]	Senn and Noble (1968)
1964	Grays	Grays	12.35[2]	Hopley and Mathews (1975)
1965-1966	-	Mid- and lower river hatcheries	8.14	Wahle et al. (1974)
1967	Early Cowlitz	Cowlitz	3.45	Hopley (1976)
1967	Late Cowlitz	Cowlitz	3.68	Hopley (1976)
1968	Early Cowlitz	Cowlitz	5.77	Hopley (1976)
1968	Late Cowlitz	Cowlitz	5.84	Hopley (1976)
1969	Early Cowlitz	Cowlitz	2.50	Hopley (1976)
1969	Late Cowlitz	Cowlitz	3.92	Hopley (1976)
1970	Late Cowlitz	Elokomin	2.50	Seidel and Mathews (1970*b*)
1970	Elokomin	Elokomin	2.00	Seidel and Mathews (1970*b*)
1972	Late Cowlitz	Klickitat	0.63[3]	Seidel and Mathews (1970*c*)
1972	Elokomin	Klickitat	0.23[3]	Seidel and Mathews (1970*c*)
1972	Toutle	Toutle	4.50[4]	Seidel and Mathews (1970*a*)
1972	Cowlitz	Klickitat	3.90	Seidel and Mathews (1970*d*)
1972	Elokomin	Klickitat	2.40	Seidel and Mathews (1970*d*)
1972	Cowlitz	Fisher Slough	1.33	Seidel and Mathews (1970*d*)

[1] For the 19/lb group fed Oregon Moist Pellet.

[2] Average of two groups incubated under different conditions.

[3] Ocean fishery recovery data were incomplete. However, survival for these groups was clearly low.

[4] This was the survival rate for groups released in April and May. Experimental groups "delayed" until June and July before release survived at much higher rates.

Apparently no trends between time and weight exist either among fish landed in the early troll fishery or returning as adults to the Columbia River (Figs. 10 and 11). There is no size-abundance relationship either (Figs. 12 and 13). If overplanting hatchery fish in the Columbia River system negatively influences survival, which is, of course, speculative, such a density factor does not appear to restrict ocean growth. Perhaps this implies that any possible negative influence on survival from overplanting takes place either in the river, or at an early time in the estuary. By the time significant ocean growth begins, compensatory mortality may have occurred and reduced abundance to a level where it does not affect growth.

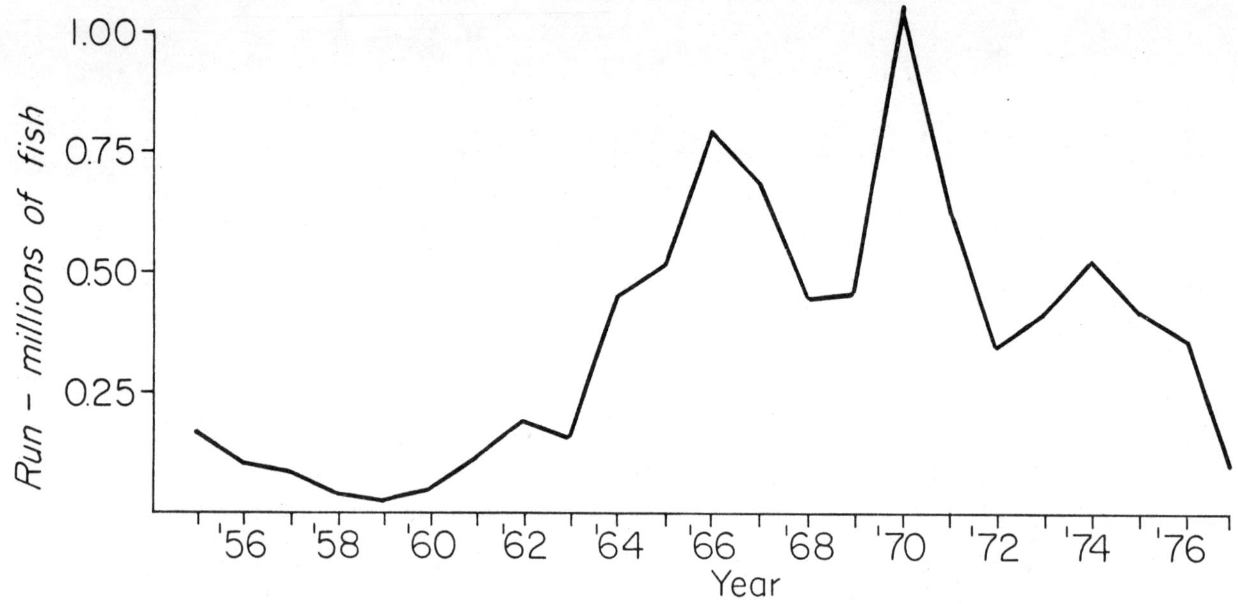

Figure 9. Columbia River "inside" coho run (river net and sport catch, Bonneville Dam count, and lower river hatchery escapement and dam count).

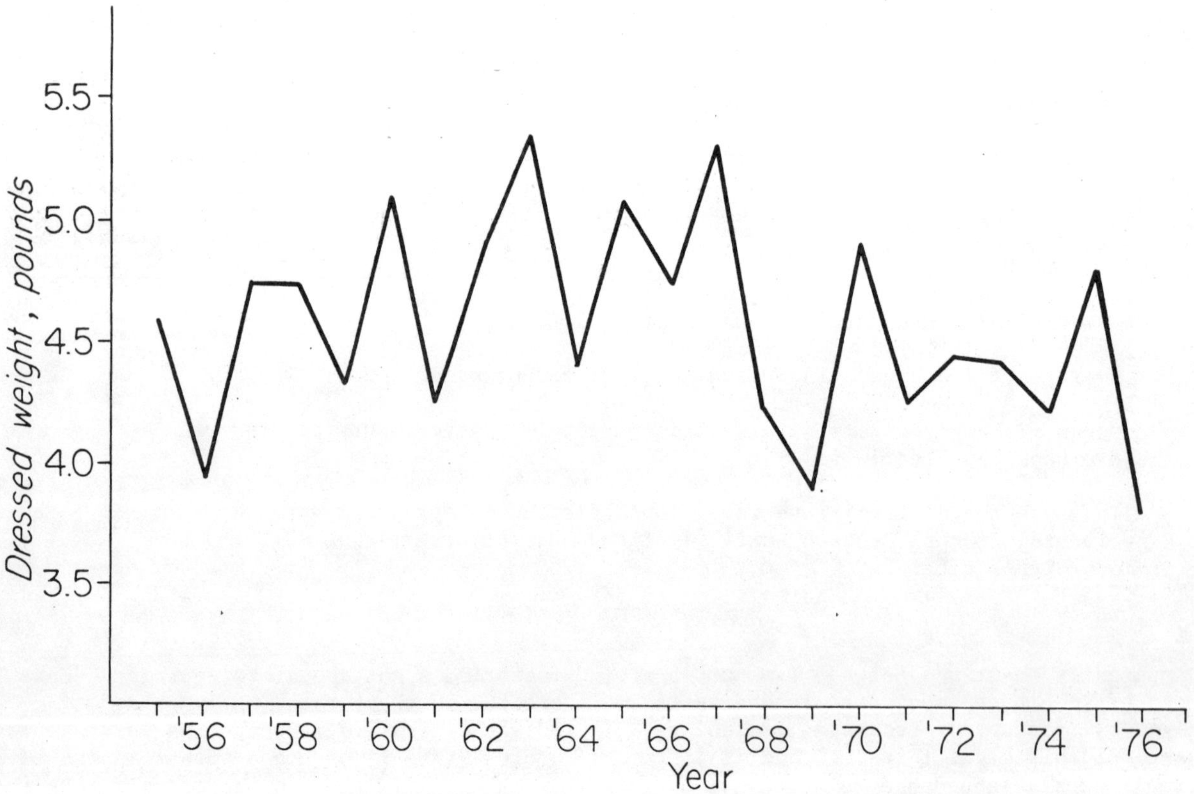

Figure 10. Average June weight of troll coho salmon landed from Tillamook Head to Willipa Harbor.

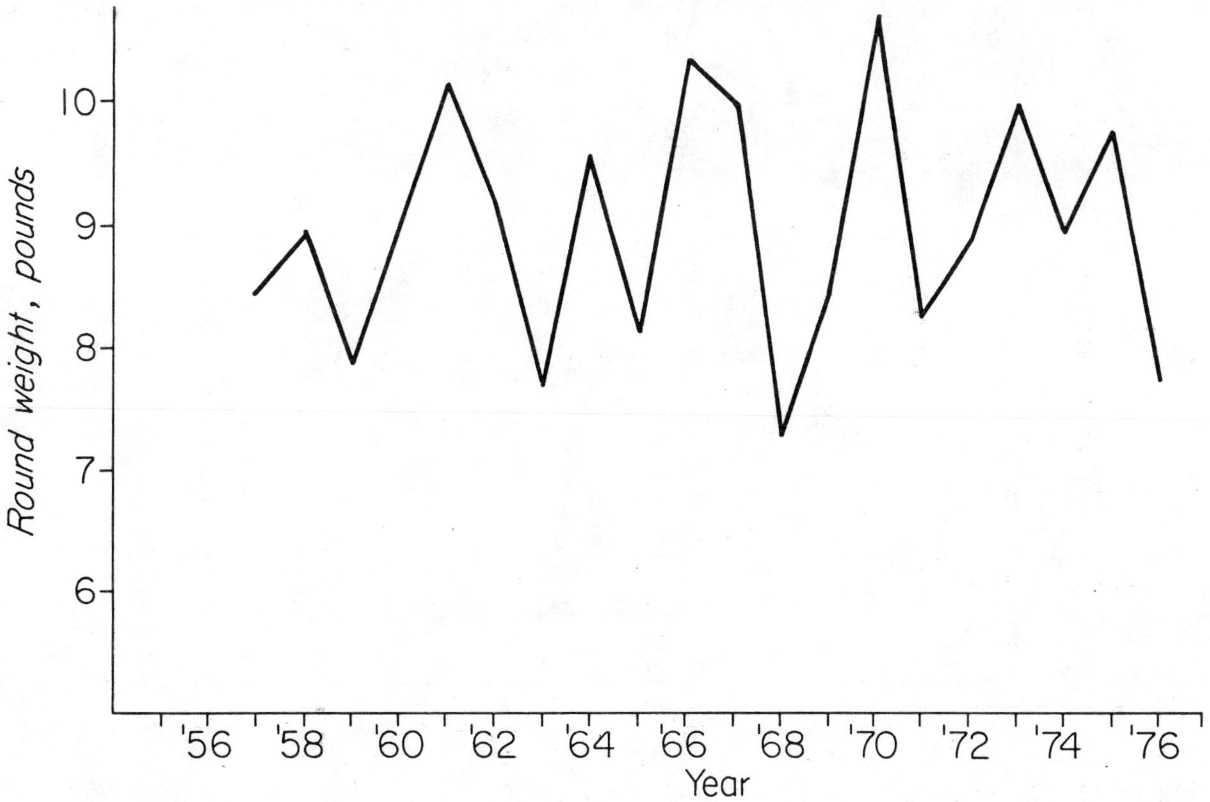

Figure 11. Weight of Columbia River coho at maturity. Average in Columbia River gillnet fishery.

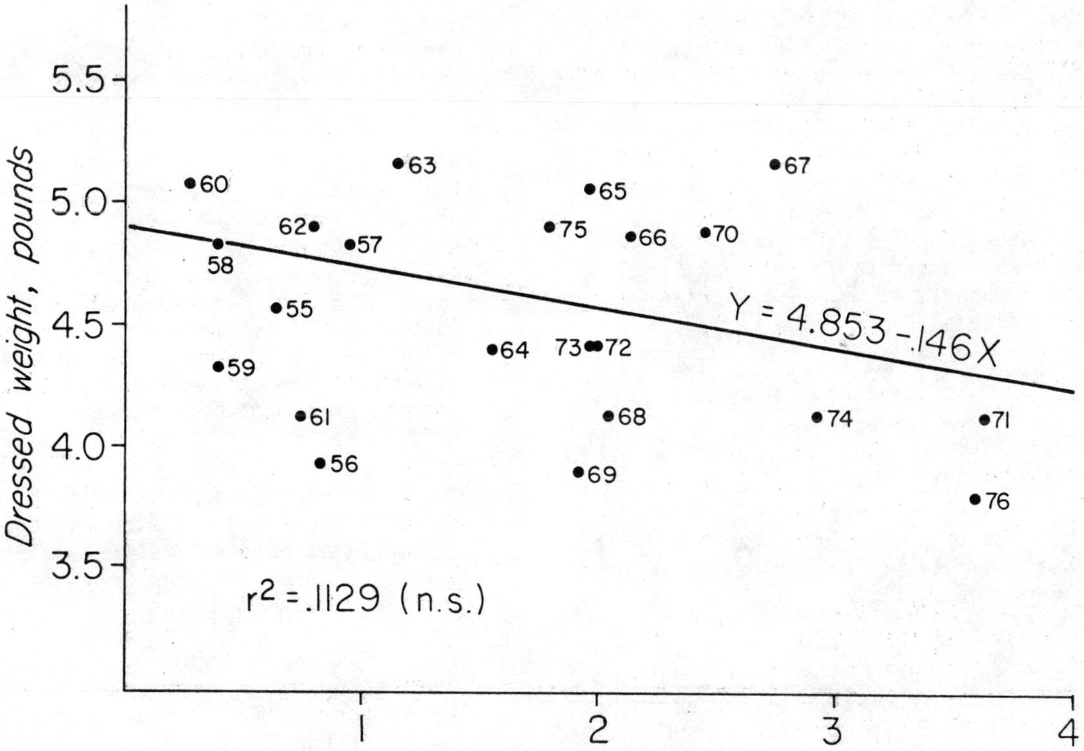

Figure 12. Columbia River average June troll weight versus total catch in fisheries which take significant numbers of Columbia River coho.

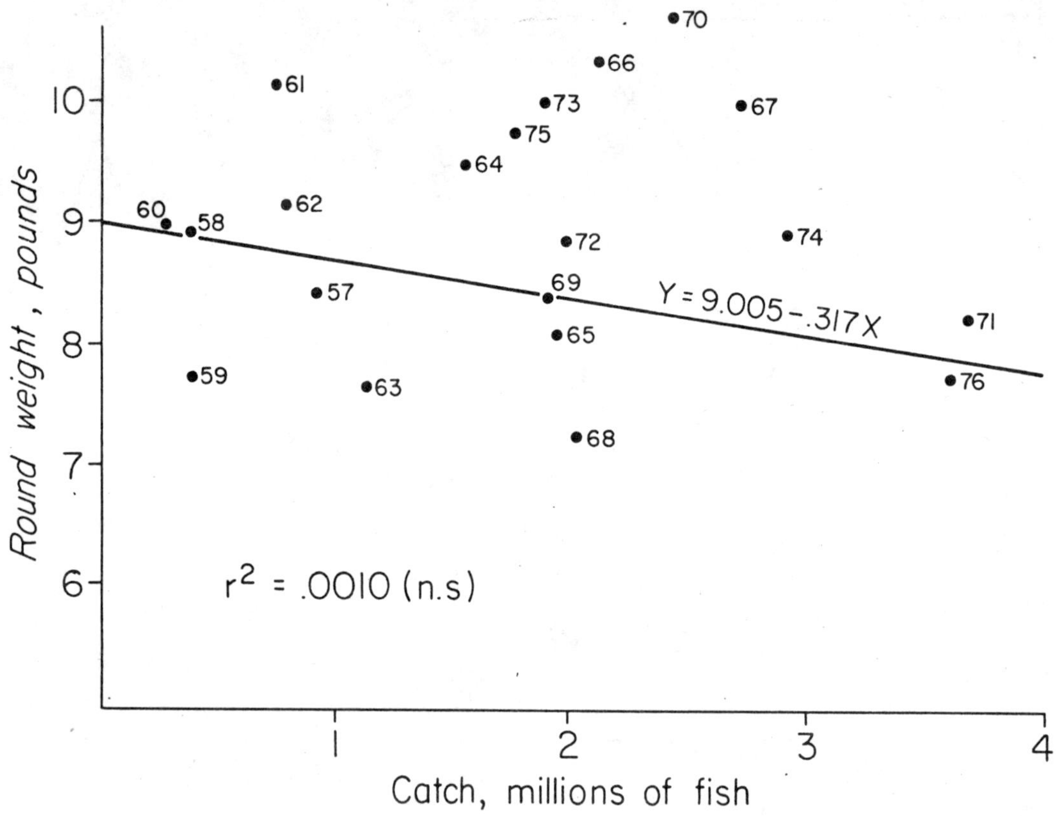

Figure 13. Average adult weight of Columbia River coho versus total catch in fisheries which take significant numbers of Columbia River coho.

We could speculate *ad infintum* from these trends. However, let me offer merely one more speculative comparison. Puget Sound coho exhibit a depressed growth rate, perhaps because of increased densities of hatchery releases. Their survival rate has not declined. These fish may spend some months in Puget Sound before leaving for the open ocean. Columbia River coho, on the other hand, probably enter the ocean relatively soon after out-migration. This indicates the importance of Puget Sound as an interim rearing area to the survival of coho.

LITERATURE CITED

Godfrey, H. 1971. The contribution of 1966 brood coho salmon from United States hatcheries to United States and Canadian fisheries in 1969. Fish. Res. Bd. Can. MS. Rep. Ser. 1153. 16 pp.

Hopley, C. W. 1976. Informal interim report on portions of 1967, 1968 and 1969 brood Cowlitz River coho stock timing evaluation. Contract rep. prepared for Wash. Dep. Fish. 18 pp. (mimeo)

Hopley, C. W., and S. B. Mathews. 1975. Grays River coho gravel incubation study. Contract rep. prepared for Wash. Dep. Fish. 12 pp. (mimeo)

Mathews, S. B., and R. Buckley. 1976. Marine mortality of Puget Sound coho salmon *(Oncorhynchus kisutch)*. J. Fish. Res. Bd. Can. 33(8):1677-1684.

Mobrand, L. E., S. B. Mathews, and F. W. Olson. 1977. A summary of survival and fishery contributions of hatchery-produced salmon in Washington State. Spec. tech. rep. to President Task Force. 15 pp. (mimeo)

Olson, F. W. 1978. An evaluation of factors affecting the survival of Puget Sound hatchery-reared coho salmon *(Oncorhynchus kisutch)*. M.S. Thesis. Univ. of Wash., Seattle. 72 pp.

Seidel, P.R., and S. B. Mathews. 1977*a*. 1972-brood Toutle River coho time/size at release study. Contract rep. prepared for Wash. Dep. Fish. 16 pp. (mimeo)

Seidel, P. R., and S. B. Mathews. 1977*b*. Comparison of 1970-brood late running Cowlitz River coho and early running Elokomin River coho. Contract rep. prepared for Wash. Dep. Fish. 8 pp. (mimeo)

Seidel, P. R., and S. B. Mathews. 1977*c*. Comparison of times of release of 1972-brood Cowlitz and Elokomin coho reared and released at Klickitat Hatchery. Contract rep. prepared for Wash. Dept. Fish.

Seidel, P. R., and S. B. Mathews. 1977*d*. 1972-brood coho conditioning experiments at Klickitat and Elokomin hatcheries. Contract rep. prepared for Wash. Dep. Fish. 8 pp. (mimeo)

Senn, H. 1970*a*. Evaluation of 1964 brood coho released from ten Puget Sound and one coastal hatcheries. Wash. Dept. Fish. Prog. Rep. 40 pp. (mimeo)

Senn, H. 1970*b*. Evaluation of 1965 brood coho released from ten Puget Sound and three coastal hatcheries. Wash. Dep. Fish. Prog. Rep. 42 pp. (mimeo)

Senn, H. G., and R. E. Noble. 1968. Contribution of coho salmon *(Oncorhynchus kisutch)* from a Columbia River watershed hatchery. Wash. Dep. Fish. Res. Paper 3(1):51-62.

Senn, H. G., and K. Satterthwaite. 1971. Evaluation of 1966 brood coho released from eleven Puget Sound and two coastal hatcheries. Wash. Dep. Fish. Prog. Rep. 42 pp. (mimeo)

Wahle, R. J., R. R. Vreeland, and R. H. Lander. 1974. Bioeconomic contribution of Columbia River hatchery coho salmon, 1965 and 1966 broods, to the Pacific salmon fisheries. Natl. Mar. Fish. Serv. Fish. Bull. 77(1):139-159.

Vulnerability of Pink Salmon Populations to Natural and Fishing Mortality

William J. McNeil

INTRODUCTION

Wild salmon production has declined steadily throughout the range of Pacific salmon species due to habitat alteration and overfishing. Pink salmon *(Oncorhynchus gorbuscha)* have historically been the most plentiful of all Pacific salmon species, but has declined the most in overall abundance. In Alaska, the pink salmon harvest is currently about one-third of historic levels (Fig. 1).

This paper reviews information on survival of pink salmon in North America, and examines the vulnerability of pink salmon populations to increased mortality from alterations of freshwater habitat and changes in rate of fishing.

DISCUSSION

Survival

Pink salmon have the simplest life history of all Pacific salmon species. They live only two years from egg fertilization to spawning. Fry migrate to sea soon after emerging from spawning beds. Reproduction and early development occur in freshwater. Growth and maturation occur mostly in saltwater.

Freshwater Survival: Egg-to-fry survival for wild populations of pink salmon has averaged 10 percent of potential egg deposition in five Alaska and British Columbia streams (Table 1). Density-dependent and density-independent mortality factors affect the survival of eggs, alevins and fry in spawning streams (McNeil 1968). Competition for space on spawning beds appears to contribute significantly to density-dependent mortality. Weather variations typically account for density-independent mortality.

Marine Survival: Tagging studies with British Columbia pink salmon, where marked fish were recovered from the fishery as well as from spawning areas, have provided information on marine survival. Parker (1968) calculated catch plus escapement of wild fish from four brood years to be 22, 5, 4 and 2 percent of fry. Bams (1972, 1974) calculated catch plus escapement of hatchery fish from two brood years to be 7 and 2 percent of fry. The average value of these six observations is 7 percent, which seems high. The average of the five lower values (excluding the 22 percent value Parker reported) is 4 percent. This value seems more consistent with what we would expect. An average marine survival of 4 percent was estimated from results of marking studies with pink salmon from five Soviet hatcheries (Rukhlov and Lubaeva 1978).

Expected marine survival (Sm) of wild pink salmon can be calculated from statistics on adults returning per spawner (R), average egg content per spawner (E) and freshwater survival (Sf). To make this calculation, I will assume that an average pink salmon female contains 1,800 eggs, and females and males occur in equal numbers. Therefore, 1,800 eggs require two spawners, and E = 900 eggs per spawner.

The number of adults returning per spawner is the product of egg content per spawner, freshwater survival and marine survival:

$$R = E \cdot Sf \cdot Sm, \text{ and}$$

$$Sm = \frac{R}{E \cdot Sf}.$$

Statistics on the number of Alaska pink salmon returning per spawner have been summarized for 158 observations by McNeil (1967*a*). The average value is R = 3.06 adults per spawner. For a population of pink salmon where E = 900 eggs per spawner and Sf = 0.10,

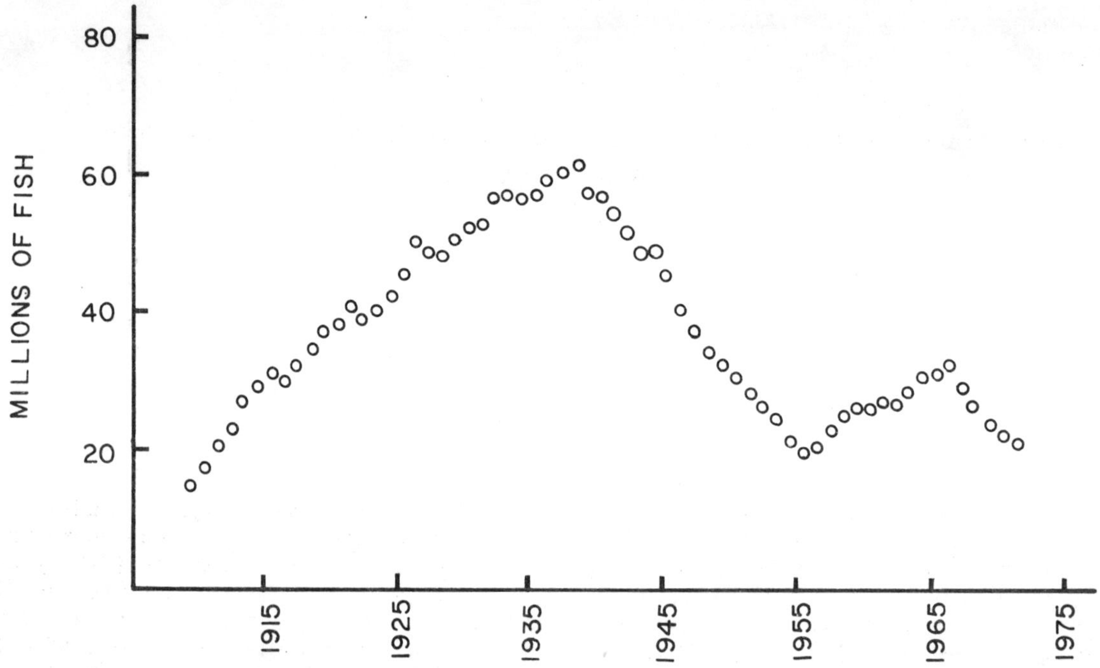

Figure 1. Ten-year moving average of catch of pink salmon in Alaska.

Table 1. Egg-to-Fry Survival of Wild Pink Salmon Populations in Five British Columbia and Alaska Spawning Streams.

Stream	No. of Observations	Mean Survival (%)	Reference
McLinton Cr., B.C.	6	14.5	Pritchard (1948)
Hooknose Cr., B.C.	14	7.3	Parker (1962)
Pleasant Bay and Whale Pass, Alaska	3	19.6	Wright (1964)
Sashin Cr., Alaska	18	9.3	Ellis (1969)
Auke Cr., Alaska	4	8.6	Taylor (1975)

average marine survival is calculated to be:

$$Sm = \frac{3.06}{(900) \quad (0.10)} = 0.034 =$$

3.4 percent

Importance of Freshwater Habitats

Conserving wild populations of pink salmon requires an average escapement onto the spawning ground of one adult fish for each parent spawner. An average return of 3.06 adults per spawner provides a generous surplus of fish for harvest, this return gives 2.06 surplus adults per spawner for a 67 percent rate of exploitation.

Freshwater survival is the key to providing this surplus for ocean fisheries, which fish mixed stocks. We will assume for purposes of modeling that natural mortality in marine waters for any given year is fairly constant--between 3 and 4 percent. We will allow freshwater survival to vary among streams (Table 1).

Figure 2 shows how the percentage of surplus fish after replacement of the spawning stock in a stream changes as freshwater survival changes. A wild stock with 10 percent freshwater survival, for example, provides a harvestable surplus of 60 to 70 percent. A hatchery stock with 80 percent freshwater survival, on the other hand, provides a harvestable surplus of 94 to 96 percent.

Changes in Rate of Fishing

Pink salmon populations become more susceptible to overexploitation by a fishery as freshwater survival declines. Observed freshwater survival rates of wild pink salmon populations range from an average of about 7 to 20 percent (Table 1). According to Figure 2, populations within this range support a maximum rate of exploitation ranging from 47 to 81 percent for Sm = 3 percent, and from 60 to 86 percent for Sm = 4 percent. If a fishery on mixed stocks is managed to harvest fully those stocks experiencing average freshwater survival, rates of exploitation are likely to approach 70 percent. With a 70 percent rate of exploitation, values of freshwater and marine survival that avoid overfishing a stock are:

Sf $\geq$ 12 percent for Sm = 3 percent

and

Sf $\geq$ 9 percent for Sm = 4 percent

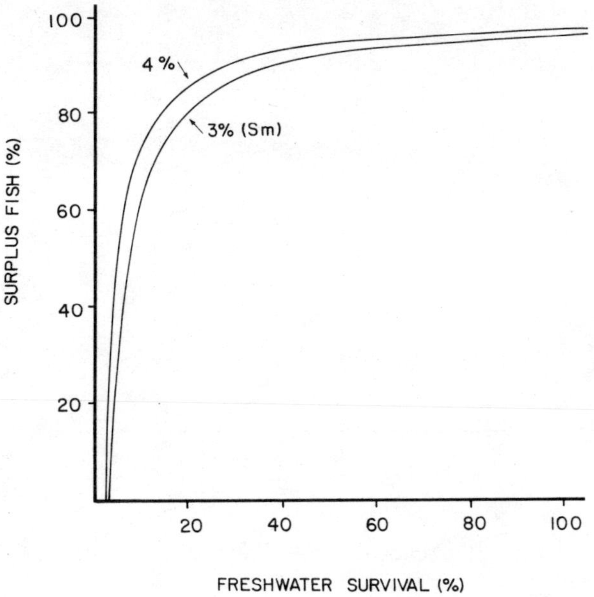

Figure 2. Percentage of surplus fish after replacement of spawning stock where marine survival (Sm) ranges from 3 to 4 percent.

Populations with relatively low freshwater survival are especially vulnerable to overfishing. This vulnerability can be illustrated by structuring a hypothetical example where a high enough exploitation rate keeps spawning escapement below the replacement level. For this exercise, I will assume that:

Sf = 0.100
Sm = 0.034
E = 900 eggs per spawner
R = 3.06 returning adults per spawner.

A population with these values for the stated parameters will not be overfished at an exploitation rate of 67.3 percent or less. How fast will the population decline, however, if the exploitation rate exceeds 67.3 percent? Declining spawner escapements, plotted in Figure 3, illustrate the average rate that the hypothetical population declines at exploitation rates of 69, 72 and 75 percent. The results show that fishing reduces the spawning population by half in 14 generations (28 years for pink salmon) at a 69 percent rate of exploitation. The spawning population is reduced by half in five generations (10 years) at a 72 percent rate of exploitation and in three generations (6 years) at a 75 percent rate of exploitation.

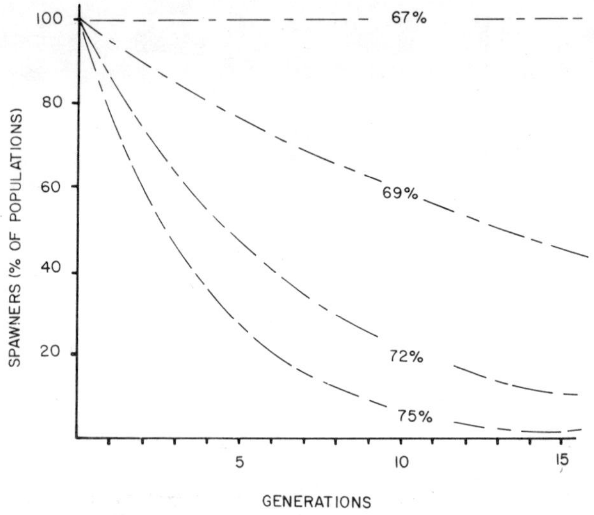

Figure 3. Change in escapement of spawners at different rates of exploitation varying from 67 to 75 percent where R = 3.06 adults returning per spawner.

CONCLUSION

The problem of managing fisheries on mixed populations is very basic to conserving wild salmon populations. Our inability to resolve this problem is possibly the key factor contributing to the depletion of pink and other salmon species in Alaska and elsewhere.

While a typical salmon fishery operates concurrently on a mixture of spawning populations, the goal of managers is to achieve an optimum level of escapement for each separate population. Healthy spawning populations provide a surplus of harvestable fish in most years, but the size of the surplus varies greatly from year to year and different spawning populations that pass through any given fishing ground have different sized surpluses. Furthermore, the rate of exploitation can vary unpredictably for each spawning population. It becomes extremely difficult, therefore, to manage fisheries on mixed spawning populations without overharvesting some and underharvesting others. Because we often regulate fishing on mixed populations based on the status of stronger populations, the weaker populations are frequently depleted over a succession of generations.

Land and water use activities that deteriorate freshwater salmon habitat likely reduce freshwater survival. In many cases, this means that the rate of exploitation must be reduced to avoid depleting wild salmon populations. Unfortunately, pressures to develop land and water resources work synchronously with pressures to increase the salmon harvest. It is not surprising, therefore, that many salmon populations have been depleted.

Major initiatives are underway to restore salmon fisheries through artificial propogation. Hatcheries and spawning channels produce more than 500 million juvenile pink salmon annually (McNeil 1976*b*). The evidence suggests that hatcheries can increase returns per spawning pink salmon by five to eight times over natural spawning (Bams 1972, 1974; Bailey et al. 1976). This implies that artificially propagated populations can produce 15 to 25 adult pink salmon returning per spawner, which would support a rate of exploitation between 93 and 96 percent. Therefore, fisheries on mixed populations of hatchery and wild fish could readily destroy wild populations unless surplus hatchery fish can be harvested in a terminal fishery or segregated from wild fish by some other means.

LITERATURE CITED

Bams, R. A. 1972. A quantitative evaluation of survival to the adult stage and other characteristics of pink salmon *(Oncorhynchus gorbuscha)* produced by a revised hatchery method which simulates optimal natural conditions. J. Fish. Res. Bd. Can. 29:1151-1167.

Bams, R. A. 1974. Gravel incubators: a second evaluation on pink salmon, *Oncorhynchus gorbuscha*, including adult returns. J. Fish. Res. Bd. Can. 31:1279-1385.

Bailey, J. E., J. J. Pella, and S. G. Taylor. 1976. Production of fry and adults of the 1972 brood of pink salmon, *(Oncorhynchus gorbuscha)*, from gravel incubators and natural spawning at Auke Creek, Alaska. Fish. Bull. 74:961-971.

Ellis, R. J. 1969. Return and behavior of adults of the first filial generation of transplanted pink salmon, and survival of their progeny, Sanshin Creek, Baranof Island, Alaska. U.S. Fish. and Wildlife Serv., Spec. Sci. Rep. - Fish. 589. 13 pp.

McNeil, W. J. 1968. Survival of pink and chum salmon eggs and alevins. Pages 101-117 *in* Northcote, T. G. (ed.), Symposium on salmon and trout in streams, H. R. MacMillan lectures in fisheries. Univ. British Columbia.

McNeil, W. J. 1976*a*. Potential for salmon aquaculture in Alaska. Pages 25-67 *in* Roseberg, D. H. (ed.), Proceedings on the Conference on Salmon Aquaculture and the Alaska Fishing Community. Univ. Alaska Sea Grant Rep. 76-2:25-67.

McNeil, W. J. 1976*b*. Review of transplantation and artificial recruitment of anadromous species. FAO Tech. Conf. on Aquaculture. Kyoto, Japan.

Parker, R. R. 1962. A concept of the dynamics of pink salmon populations. Pages 203-211 *in* Wilimovsky, N. J. (ed.), Symposium on pink salmon. H. R. MacMillan lectures in fisheries. Univ. British Columbia.

Parker, R. R. 1968. Marine mortality schedules of pink salmon of the Bella Coola River, central British Columbia. J. Fish. Res. Bd. Can. 25:757-794.

Pritchard, A. L. 1948. Efficiency of natural propagation of the pink salmon *(O. gorbuscha)* in McClinton Creek, Masset Inlet, B. C. J. Fish. Res. Bd. Can. 7:223-236.

Rukhlov, F. N., and O. S. Lubaeva. 1978. On the efficiency of the Sakhalin pink hatcheries. Pages 35-36 *in* Biology of salmons. Ministry of Fisheries, U.S.S.R.

Taylor, S. G. 1975. Production of pink salmon fry from a hatchery with gravel incubators at Auke Creek, Alaska, and return of the 1973 brood. Northwest and Alaska Fisheries Center, Auke Bay Laboratory, Biannual Progress Rep. to Alaska Dep. Fish and Game, 24 December 1975. 30 pp.

Wright, A. T. 1964. A study of the carrying capacity of pink and chum salmon spawning areas in Alaska. *In* Studies to determine optimum escapement of pink and chum salmon in Alaska. Part 9. Alaska Dep. Fish and Game. Final Summary Rep. for Contract No. 14-17-007-22 (U.S. Fish and Wildlife Serv.)

Some Features of Ocean Migrations and Timing of Pacific Salmon

Robert L. Burgner

INTRODUCTION

In this paper, I will review features of Pacific salmon migrations in the ocean, comment on some differences among species and stocks in ocean distribution patterns, document how precisely Bristol Bay sockeye salmon time the return to their home rivers and examine how predictable are variations in that timing.

Our knowledge of Pacific salmon migrations in the ocean has largely been derived from investigations beginning in 1955 by Canada, Japan and the United States. These nations belong to the International North Pacific Fisheries Commission (INPFC).

A recent series of comprehensive commission bulletins synthesizes much of the published and unpublished information on migrations and distributions of the five major salmon species: pink salmon *(Oncorhynchus gorbuscha)* (Takagi et al. *in press*), chum salmon *(O. keta)* (Neave et al. 1976), sockeye salmon *(O. nerka)* (French et al. 1976), coho salmon *(O. kisutch)* (Godfrey et al. 1975) and chinook salmon *(O. tshawytscha)* (Major et al. 1978). An additional report by Fredin et al. (1977) provides useful additional information on the five species, and Japan's high seas salmon fisheries.

The extensive catch, catch per unit effort (CPUE) and biological data collected each spring and summer since 1970 by several Japanese research vessels remains unpublished in summary form. These vessels fish consistent transect patterns with gillnets in the northwest Pacific and Bering Sea, and provide the opportunity to study annual distribution patterns of the five salmon species relative to oceanographic and climatic features. In addition, Hartt and Dell (unpublished) will provide new information of juvenile salmon migrations during their first summer at sea.

DISCUSSION

Features of Ocean Migration

Neave (1964) and Royce et al. (1968) discuss features of ocean migration patterns and guidance mechanisms of Pacific salmon. During the great distances salmon travel in the ocean, they clearly rely on inherited responses to guidance stimuli. These fish have had no opportunity to learn from parents. They migrate over broad expanses of the ocean and often travel circular routes not traversed in reverse on the return journey. During most of the migration in the open ocean, salmon swim near the surface and away from landmasses that might provide cues. For example, even on their final homeward migration, Bristol Bay sockeye apparently remain offshore until close to the bays of their natal streams (Straty 1975). Sonar observations and catch patterns in gillnets show that during feeding periods the salmon disperse rather than form defined schools. Salmon use ocean currents in their migrations (Royce et al. 1968), but the fish also migrate positively across ocean domains and major current features.

Weather during the migration is often stormy with overcast conditions. Celestial navigation appears to be an unreliable means of navigation for fish on a definite time schedule (Royce et al. 1968). Several possible migratory cues have been proposed

---

Contribution No. 488, College of Fisheries, University of Washington, Seattle, Washington 98195

including polarized light, photoperiod, pheromone or other chemical detection, and sensitivity to an electric potential established by current velocity or fish movement relative to a magnetic field. Certain factors, such as reaction to temperature, salinity or food supply do not appear to be useful to salmon as migration cues. If water is too warm or saline, or food insufficient, these alone do not appear to provide directional cues for salmon to move to a more favorable environment.

It is clear, however, that strong migration patterns are genetically determined. Salmon progeny transplanted to radically new environs, such as Okhotsk Sea pinks to the Barents Sea, stray extensively from their new "home" stream. Apparently they experience great difficulty in finding a route back in a foreign ocean environment. Under normal circumstances, however, salmon obviously retain general orientation, recognize water masses that they have traversed before and probably are cued by annual heat budget and photoperiod cycles. Regarding chum salmon, Neave et al. (1976) concluded:

> "While ocean currents must affect the direction, speed or effort expended in movements (or in resistance to movement) of the fish, they do not account for the observed distribution and migration of the various stocks."

Whatever a species' or stock's ocean migration pattern, it presumably has evolved to optimize growth and survival. The pattern provides "signposts" for salmon to follow in the ocean so they can return home on a controlled timetable. In this regard, man can possibly alter the survival benefits of certain inherited migration behaviors by intensively fishing competitor or predator species, or overprotecting others (e.g. marine mammals). In future ocean-ranching ventures, it may be desirable to choose salmon stocks and release schedules that allow them to take innate advantage of ocean pastures and migration routes, and at the same time avoid overtaxing certain ocean areas.

Distribution Patterns

Across the North Pacific and Bering Sea, ocean domains and surface current patterns (Fig. 1) clearly influence migration paths and seasonal distributions of different salmon stocks (Royce et al. 1968). In winter and early spring, stocks distribute more southerly in the western North Pacific than in the eastern North Pacific. This reflects the influence of colder water in the western subarctic gyre area (Fig. 1). Keeping this in mind, we can say that pink

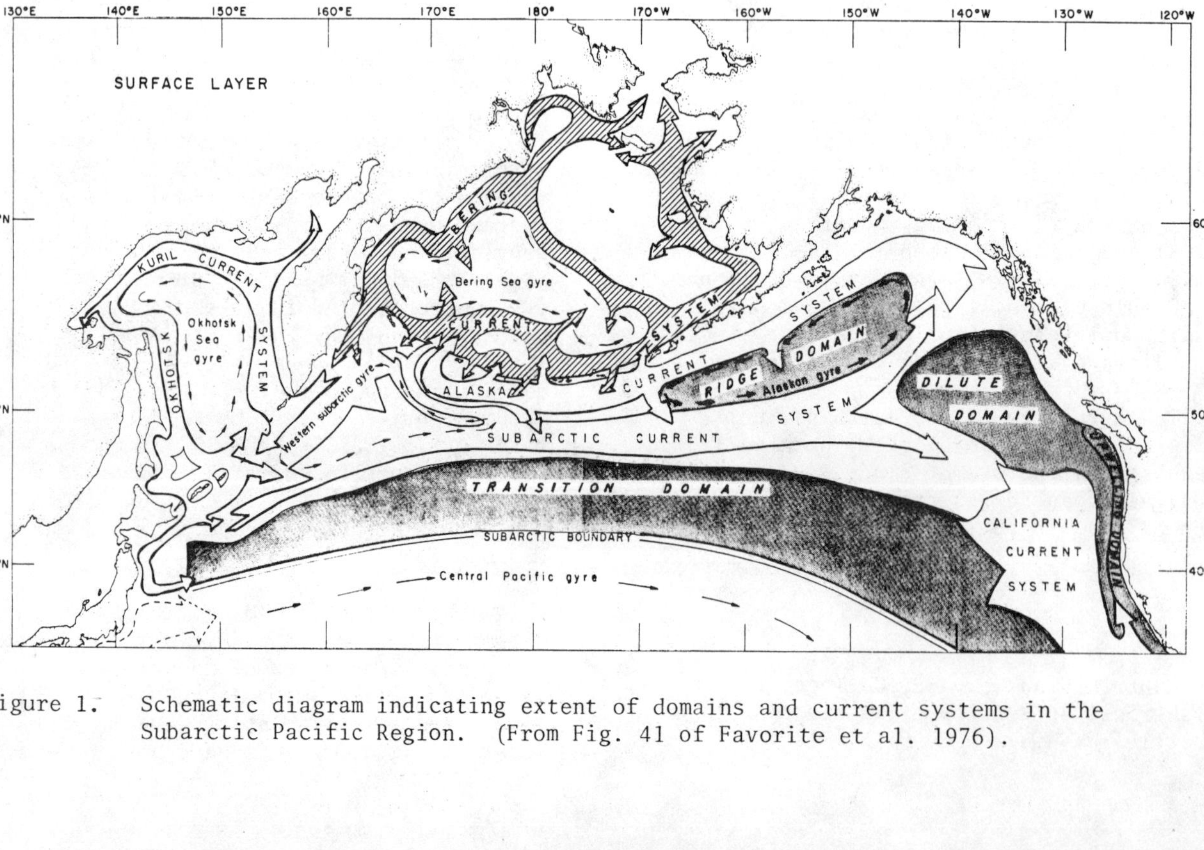

Figure 1. Schematic diagram indicating extent of domains and current systems in the Subarctic Pacific Region. (From Fig. 41 of Favorite et al. 1976).

salmon from different geographical areas use distinctly different areas of the ocean. We can also suggest different genetic adaptations of stocks originating in different regions. Pink salmon are useful in this respect because they spend only a little more than one year in the ocean. Figure 2, adapted from a series of figures in Takagi et al. (*in press*), shows the general ocean distribution of stocks from six general geographic areas on the North Pacific Rim. Distribution patterns were modified slightly to restrict them to known distributions confirmed by tagging studies.

Washington-British Columbia stocks move well northward in the Gulf of Alaska, but they do not move as far westward in their southern range as do the Alaska stocks bordering the Gulf of Alaska. Western Alaska and east Kamchatka stocks show limited overlap in the Bering Sea, but both move through the Aleutians and overlap extensively in the eastern North Pacific. The western Alaska contingent has time to intermingle extensively with other Alaskan stocks in the gulf, but not with Washington-British Columbia stocks.

In the western North Pacific, stocks bordering the Okhotsk Sea move southward and eastward to feeding grounds generally quite separate from the East Kamchatka-Karaginsky stocks. Their winter distribution extends considerably farther south than that of pink salmon in the eastern North Pacific. Finally, pinks from the west coast of Sakhalin Island, Amur River and Primore coast move into the Sea of Japan. They are much more restricted in their ocean migrations than other pink stocks.

Within these broad geographic regions, ocean migrations could undoubtedly be more finely resolved. All of the depicted stocks generally move counterclockwise in the ocean. Pink salmon populations in these different geographical areas may possibly use the gyres and surface current systems depicted in Figure 1 to minimize energy expended in their extensive migrations. From the distinct distributions shown in Figure 2 it is clear, however, that strong genetic determinants are at work.

Catches by Japanese research vessels show interesting differences in the distributions of salmon species in the western North Pacific. The 1976 early to mid-June CPUE (Fig. 3) illustrate these. Chum salmon distribute widely because many stocks and age groups intermingle. Immature fish tend to be farther south, and maturing fish migrate shoreward over a considerable

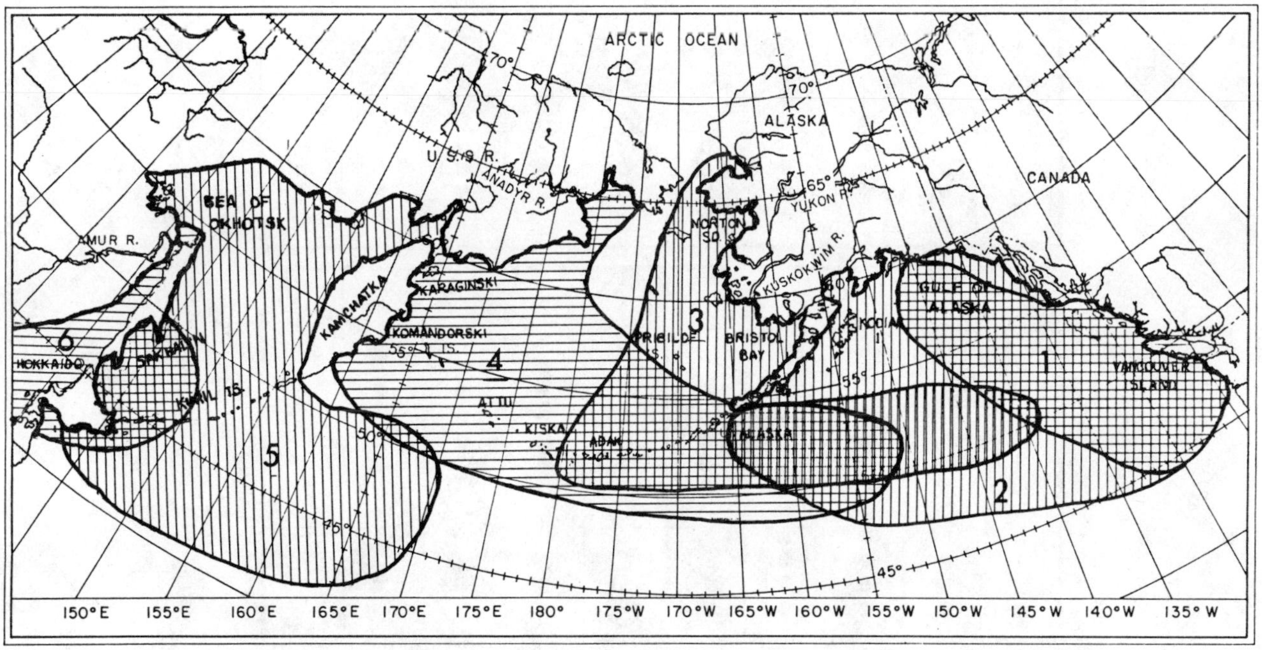

Figure 2. Primary ocean distribution of pink salmon originating in different geographical regions: (1) Washington and British Columbia; (2) Southeast, central, southwest Alaska; (3) N. side Alaska Peninsula to Kotzebue Sd; (4) E. Kamchatka to Anadyr Bay; (5) W. Kamchatka, N. Okhotsk Sea coast, E. Sakhalin Is., Kurile Is., Hokkaido Is.; (6) W. Sakhalin Is., Amur R., Primore Coast. (Modified and simplified from Takagi et al. *in press*.

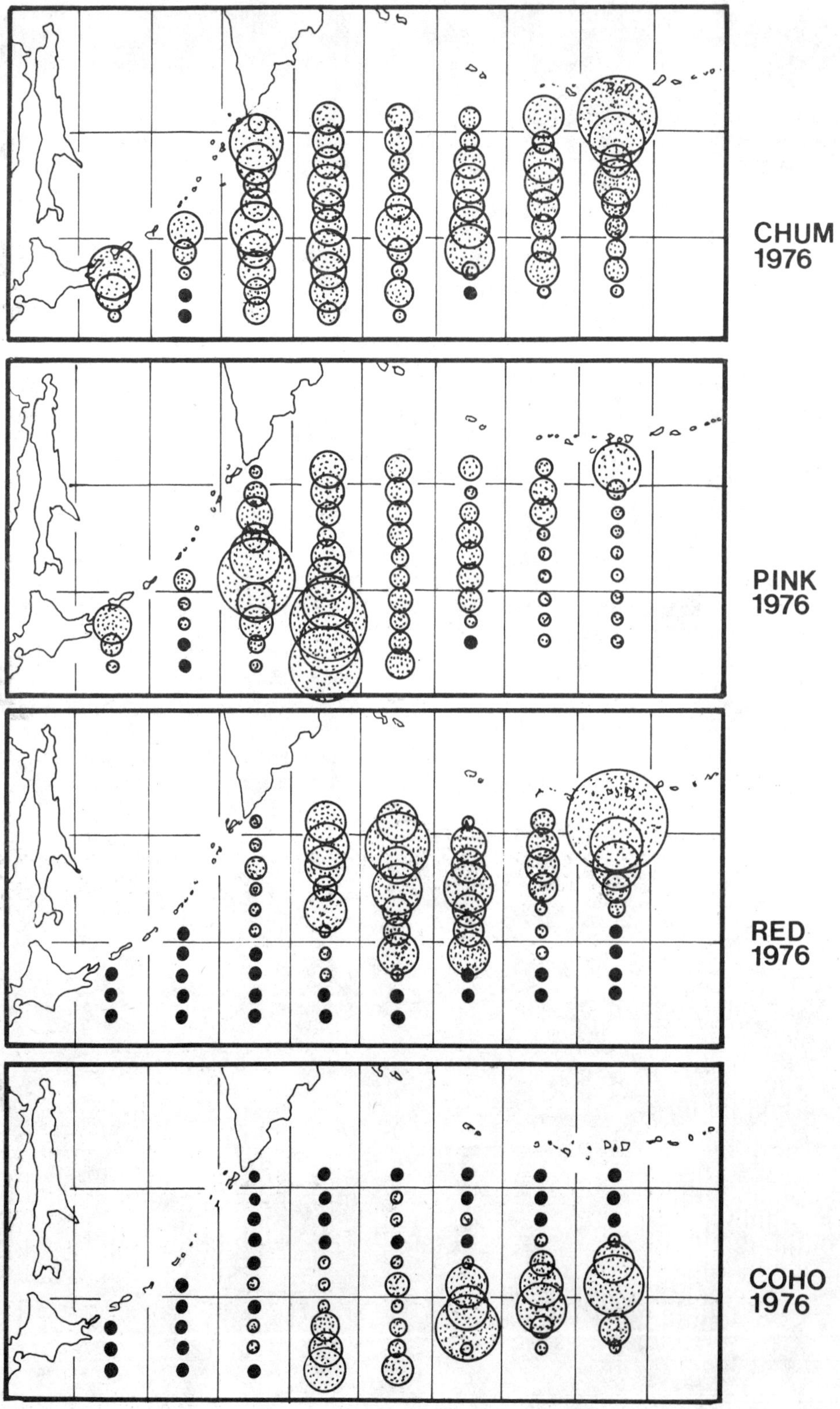

Figure 3. Patterns of salmon catch per tan in commercial-size gillnets fished by Japanese research vessels on transects in early to mid-June 1976. (Solid dot = no catch, scale of circle size differs among species. Data courtesy of Japan Fisheries Agency.)

period of time. The single year class of maturing pink salmon concentrates in the southwestern sector, and heads primarily for spawning areas bordering the Okhotsk Sea. By this time, sockeye, both matures and older immatures, have moved northward and have their own distinctive distribution which barely overlaps that of the coho salmon to the south and east. Although Asian coho only stay at sea a few months longer than pink salmon do, they enter the ocean at a larger size and range farther to the southeast. The extent that North American coho overlap into the eastern portion of this North Pacific area remains unknown.

Distinctness of Run Timing

Different species and stocks exhibit specific features of oceanic distribution. However, individuals of a species bound for a particular river system begin their homeward migrations from divergent points over vast ocean areas. Remarkably, they arrive over a short time span usually so consistently that variations in ocean currents and temperatures, or salinity and temperature gradients as Fujii (1975) proposed, apparently do not greatly affect their arrival. From 1956-1976, for example, 80 percent of the Bristol Bay run passed through the fishery during a 12.9 day period (SD 1.58 days) with a mean of 4 July (extremes 28 June 10 July) and standard deviation of only 2.92 days. Figure 4 illustrates the sockeye run timing in Bristol Bay from 1966-1975.

Because it is such a short migration period, it is important to determine if the runs are early, normal or late in order to apportion catch and spawning escapement in the Bristol Bay fishery. Oceanographic conditions the fish encounter in their final migration south of the Aleutians, through the passes and northeastward in the Bering Sea (Fig. 1) probably influence the time they enter the river fisheries. Of the possible influencing factors, only temperature data are readily available for examination. McLain and Favorite (1976) commented on the apparent effects of anomalously cold oceanic conditions in the southeastern Bering Sea on the onshore movements of Bristol Bay sockeye in 1956 and 1976. Nishiyama (1977) related mean sea surface temperature of the Bering Sea in June and the timing of the Kvichak River sockeye run for the years of 1965-1971. He concluded that the runs are earlier in warmer years than in colder years.

I have examined further the relationships between deviations in sea surface temperature and run timing. Also, information on sea surface temperature is not necessarily reliable for the region. Consequently, I have examined the relationship between deviations in regional air or river temperatures and annual migration timing of the Bristol Bay sockeye runs. As the run timing variable, I have used the date of the 50 percent point of the annual run to Bristol Bay.

Fishing Information, published monthly since 1960, provides charts of sea surface temperature deviations from the long-term mean in North Pacific waters. I utilized these charts to approximate monthly sea surface temperature deviations immediately south of the Aleutian area between the Alaska Peninsula and Adak Island. During the April-June period, Bristol Bay sockeye are expected to migrate through this general area. Similar temperature data for the Bering Sea were not available. The annual dates of the Bristol Bay runs' 50 percent point were regressed against annual anomalies in sea temperature in separate regressions using temperatures for May, June, April-May, May-June and April-June for the years 1960-1977. All regressions were significant at the 5 percent level except the one using April-May data. The regressions using May-June and May were significant at the 1 percent and 2 percent level, respectively (Figs. 5 and 6).

Air temperatures may be more accurate and are available in a more timely fashion. I examined the correlation between Adak and Cold Bay weather station air temperatures, averaged together, and the sea surface temperature deviations for the same period as above (May, June, April-May, May-June, April-June). All were significant at the 1 percent level, with the highest correlations for May ($r = .786$, significant at the 0.1 percent level) (Fig. 7). It appears valid to substitute air temperatures in lieu of good sea surface temperature data.

I correlated April-June air temperatures from Adak, Cold Bay, St. Paul and King Salmon with Bristol Bay run timing. All except June temperatures at St. Paul correlated significantly with run timing at the 5 percent level. The regression of run timing against Adak-Cold Bay air temperatures during May provided the best relationship ($r = .7501$, Fig. 8). This suggests that temperature effects directly or indirectly explain over 50 percent of the deviation in Bristol Bay run timing.

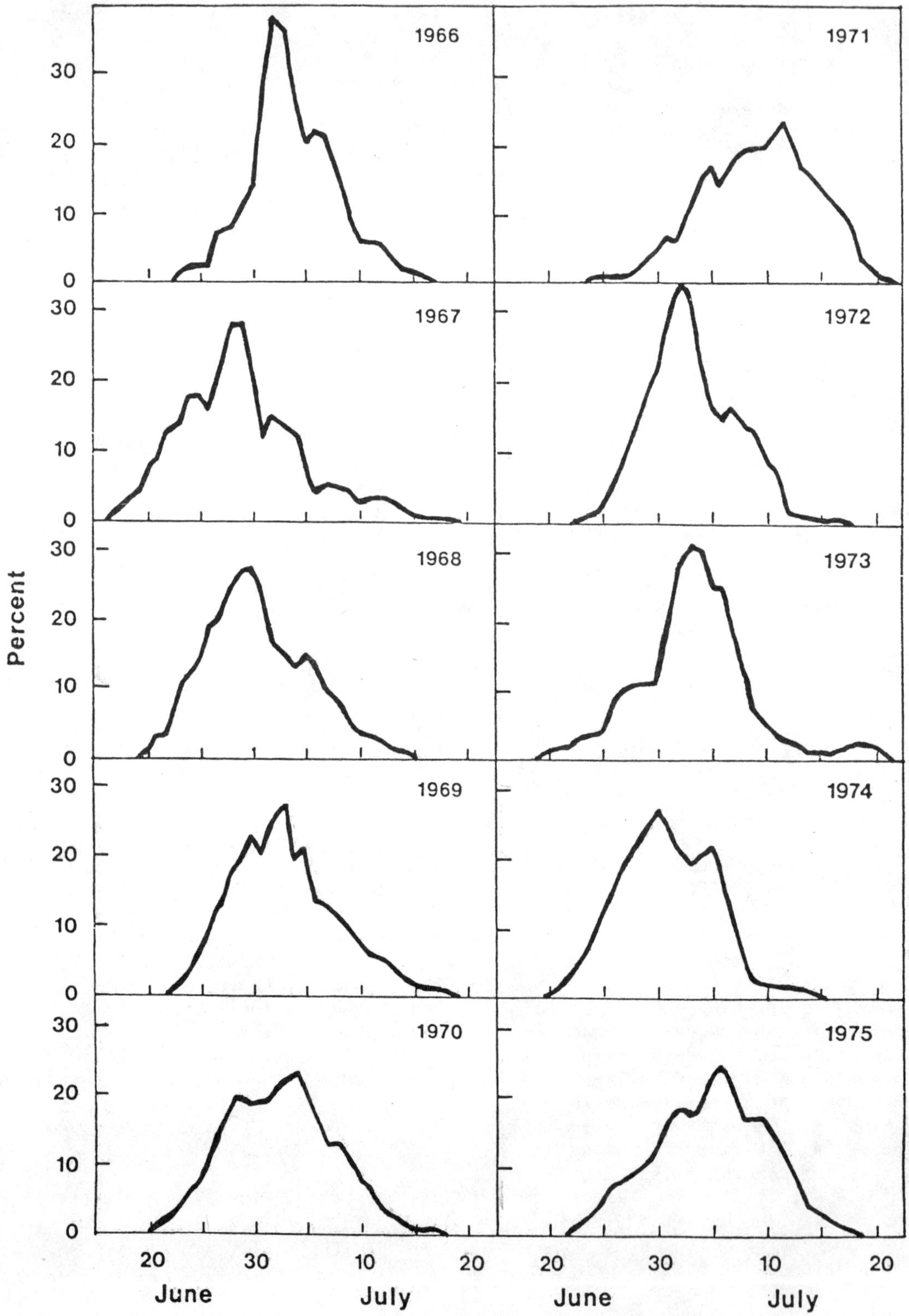

Figure 4. Timing of the sockeye salmon runs in Bristol Bay, 1966-1975.

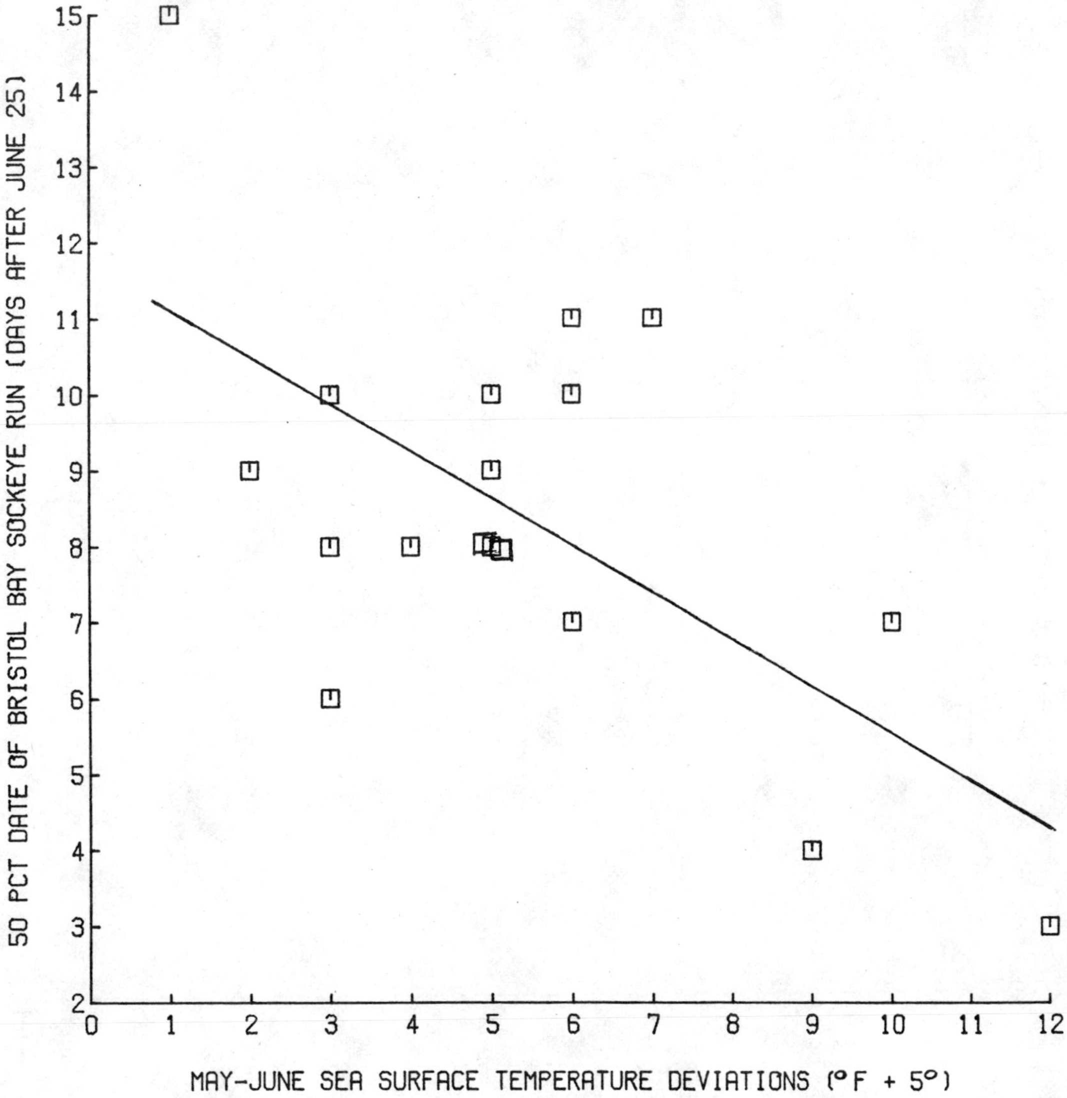

Figure 5. Regression of annual timing of sockeye run to Bristol Bay on the sum of May and June sea surface temperature anomalies immediately south of the eastern Aleutians.

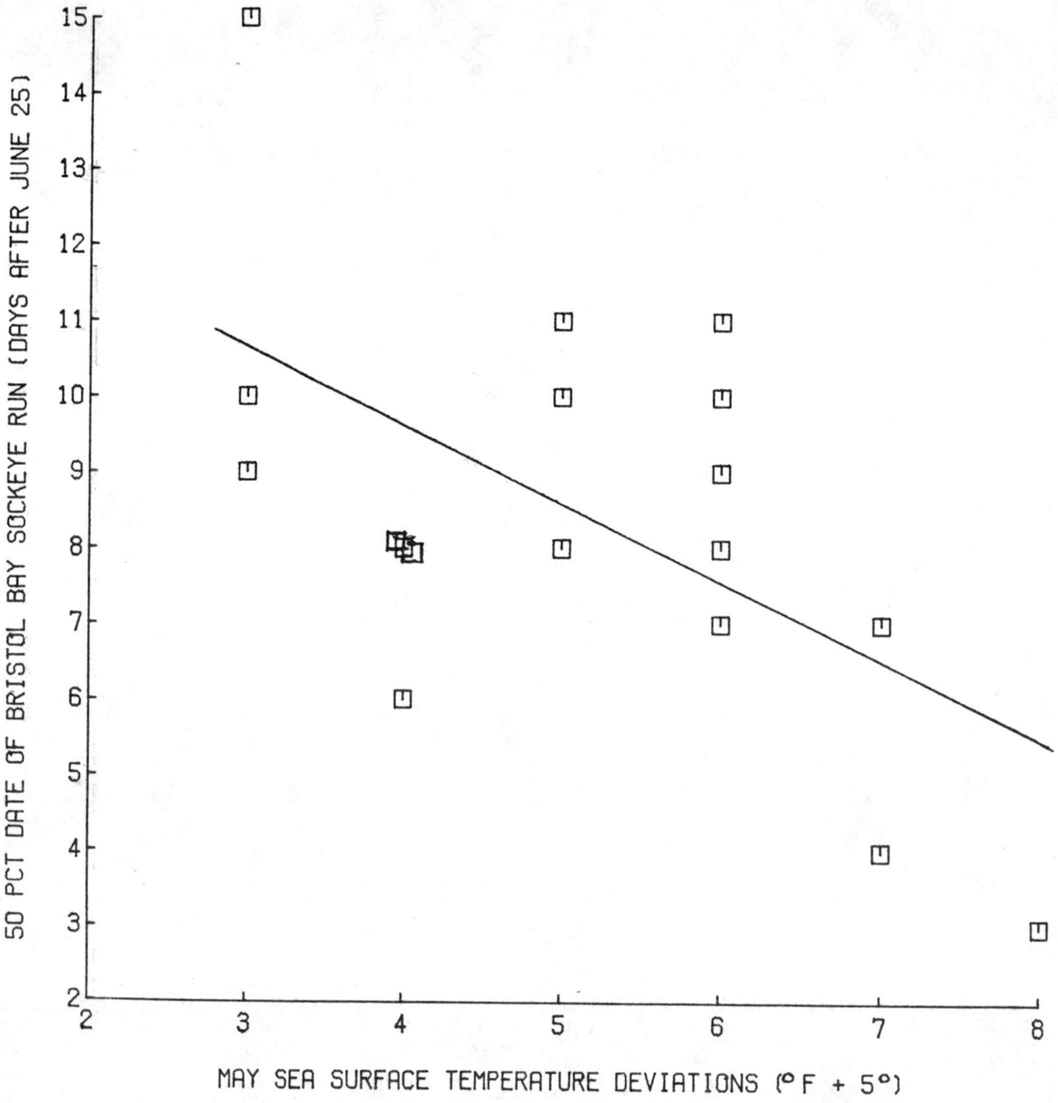

Figure 6. Regression of annual timing of sockeye run to Bristol Bay on monthly sea surface temperature anomalies for May immediately south of the eastern Aleutians.

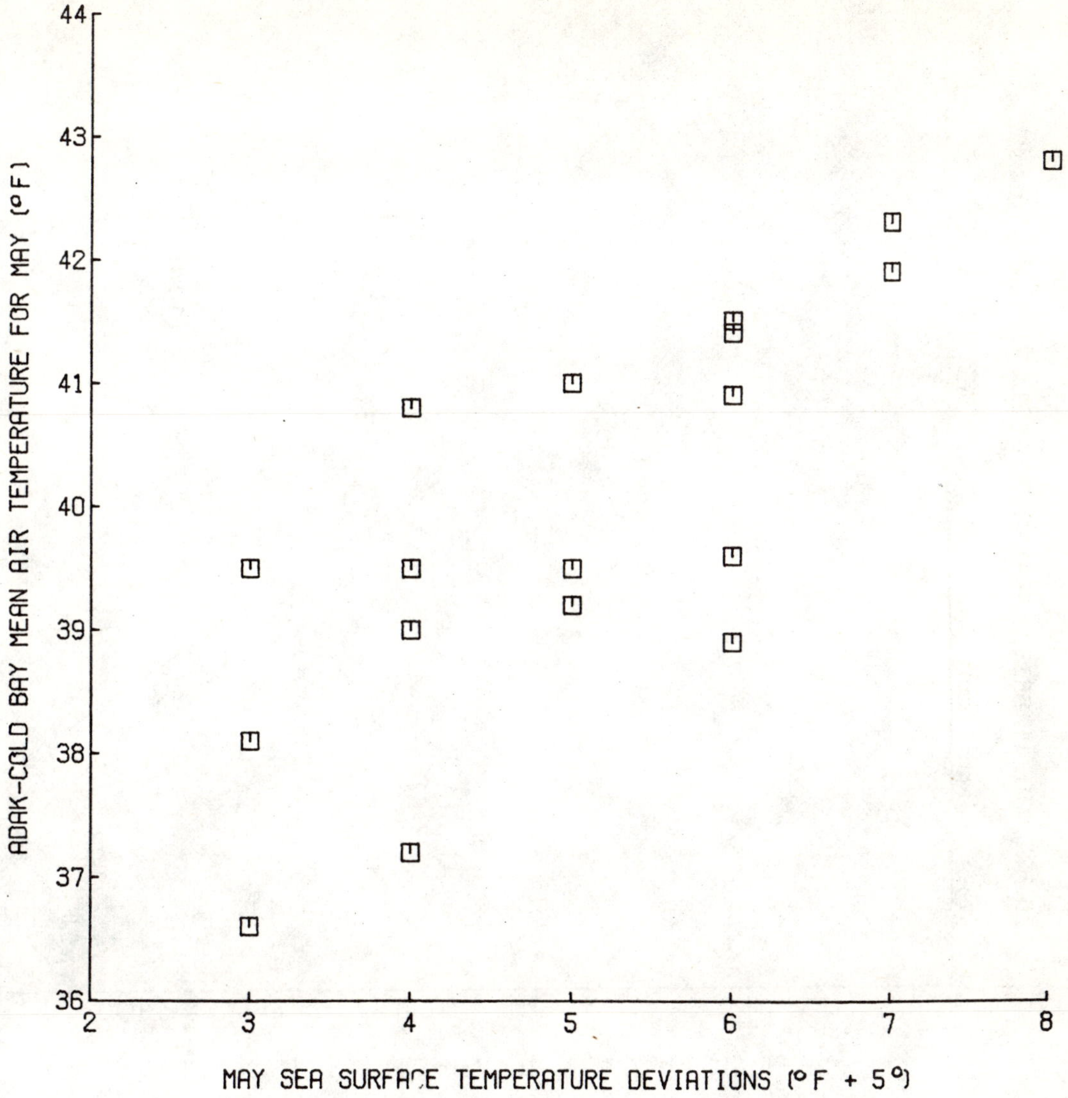

Figure 7. Correlation between monthly sea surface temperature anomalies immediately south of the eastern Aleutians and Adak-Cold Bay mean air temperatures for May.

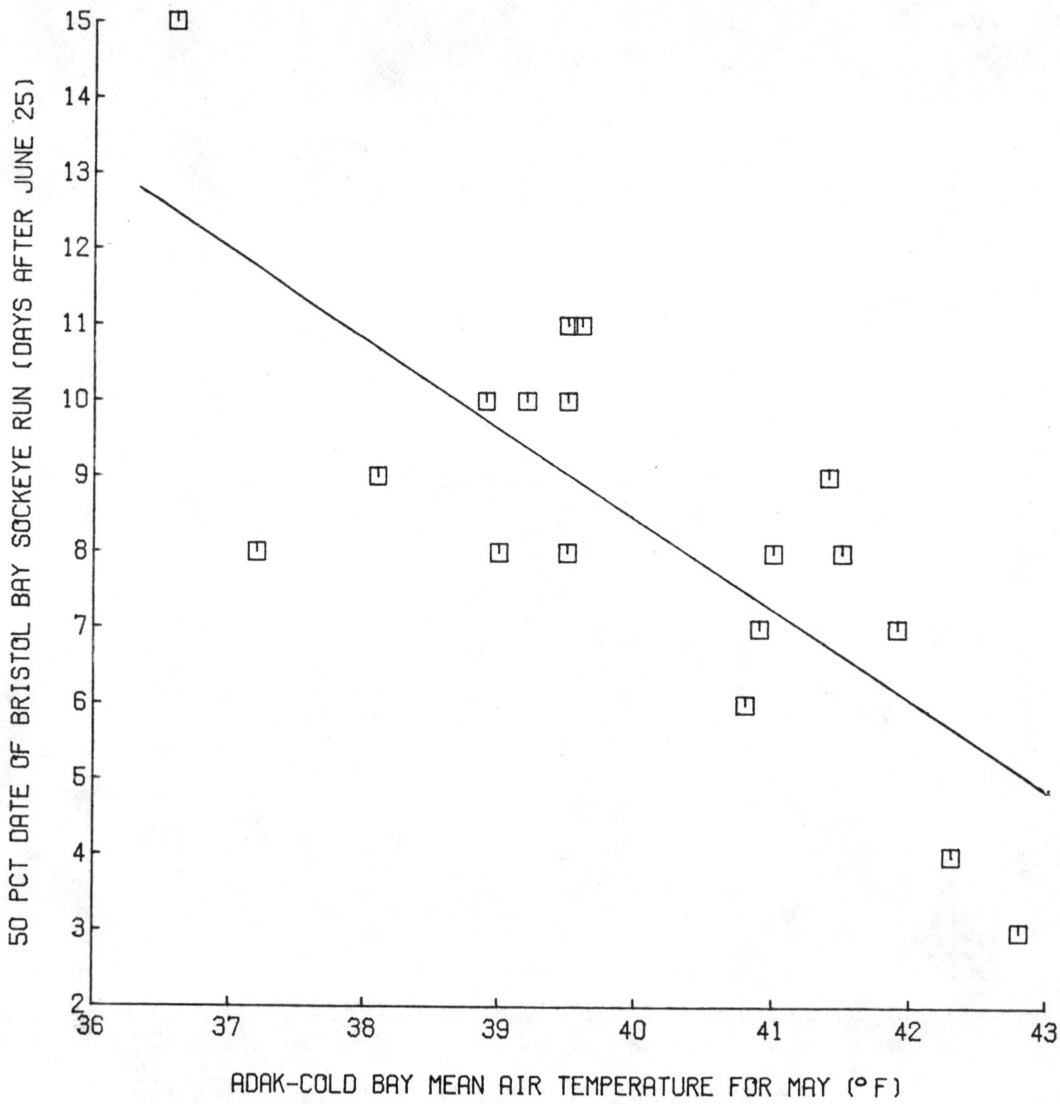

Figure 8. Regression of annual timing of sockeye run to Bristol Bay on Adak-Cold Bay mean air temperatures for May.

These correlations are sufficiently high to be useful in predicting run timing. Further, air temperature data becomes available about three weeks before the Bristol Bay sockeye run arrives. However, work is needed to determine possible cause and effect relationships, and to examine other variables that might be used to account for additional deviations in the relationship. Better sea-surface temperature data, data on sea surface conditions (storms) or data on gonad maturation of sockeye sampled en route to Bristol Bay might prove useful.

SUMMARY

We now know much about the distribution of Pacific salmon in the ocean and their migratory routes. However, the guidance mechanisms salmon use to find their way at sea remain largely unknown. Different species and stocks within species clearly have well defined, genetically influenced migratory patterns. They still disperse over broad geographic areas of the ocean. On their return to home streams, the timing of annual runs is remarkably distinct. Bristol Bay sockeye exemplify this. Annual deviations in timing of the Bristol Bay runs highly correlate with deviations in sea surface and air temperatures in late spring prior to arrival. Because accurate forecast information on run timing could be extremely valuable in regulating the Bristol Bay sockeye fishery, I suggest that additional studies be conducted to improve forecast potential.

LITERATURE CITED

Favorite, F., A. J. Dodimead, and K. Nasu. 1976. Oceanography of the subarctic Pacific regions, 1960-1971. Int. North Pac. Fish. Comm. Bull. 33. 187 pp.

Fredin, R. A., R. L. Major, R. C. Bakkala, and G. Tanonaka. 1977. Pacific salmon and the high seas salmon fisheries of Japan. U.S. Dep. Commerce, Natl. Mar. Fish. Serv., Northwest Alaska Fish. Center Proc. Rep., December 1977. 324 pp.

French, R. R., T. Bilton, M. Osako, and A. C. Hartt. 1976. Distribution and origin of sockeye salmon *(Oncorhynchus nerka)* in offshore waters of the North Pacific Ocean. Int. North. Pac. Fish. Comm. Bull. 34. 113 pp.

Fujii, T. 1975. On the relation between the homing migration of the western Alaska sockeye salmon *Oncorhynchus nerka* (Walbaum) and oceanic conditions in the eastern Bering Sea. Memoirs Faculty of Fish., Hokkaido Univ. 22(2):99-191.

Godfrey, H., K. A. Henry, and S. Machidori. 1975. Distribution and abundance of coho salmon in offshore waters of the North Pacific Ocean. Int. North Pac. Fish. Comm. Bull. 31. 80 pp.

Hartt, A. C., and M. B. Dell. Migrations and growth of juvenile Pacific salmon and steelhead trout during their first summer in the open sea. Unpublished manuscr. Fish. Res. Inst. See also this symposium: Hartt, A. C., Juvenile salmonids in the oceanic ecosystem--the critical first summer.

Major, R., J. Ito, S. Ito, and H. Godfrey. 1978. Distribution and abundance of chinook salmon *(Oncorhynchus tshawytscha)* in offshore waters of the North Pacific Ocean. Int. North Pac. Fish. Comm. Bull. 38. 54 pp.

McLain, D. R., and F. Favorite. 1976. Anomalously cold winters in the southeastern Bering Sea, 1971-75. Mar. Sci. Communications 2(5):299-334.

Neave, F. 1964. Ocean migrations of Pacific salmon. J. Fish. Res. Bd. Can. 21(5):1227-1244.

Neave, F., T. Yonemori, and R. G. Bakkala. 1976. Distribution and origin of chum salmon in offshore waters of the North Pacific Ocean. Int. North Pac. Fish. Comm. Bull. 35. 79 pp.

Nishiyama, T. 1977. Food-energy requirements of Bristol Bay sockeye salmon *Oncorhynchus nerka* (Walbaum) during the last marine lifestage. Research Institute of North Pac. Fish. Fac. of Fish., Hokkaido Univ. Spec. Vol. 1977: 289-320.

Renner, A. (ed.). 1972-1977. Fishing information. Natl. Ocean. Atmos. Admin., Natl. Mar. Fish. Serv. (monthly reports for years 1972-1977).

Robinson, M. K. 1976. Atlas of North Pacific Ocean monthly mean temperatures and mean salinities of the surface layer. U.S. Nav. Oceanogr. Off. Ref. Publ. 2. 194 pp.

Royce, W. F., L. S. Smith, and A. C. Hartt. 1968. Models of ocean migrations of Pacific salmon and comments on guidance mechanisms. Natl. Ocean. Atmos. Admin. Fish. Bull 66:441-462.

Straty, R. 1975. Migratory routes of adult sockeye salmon, *Oncorhynchus nerka*, in the eastern Bering Sea and Bristol Bay. Natl. Ocean. Atmos. Admin. Tech. Rep. Natl. Mar. Fish. Serv. SSRF-690. 32 pp.

Takagi, K., V. Aro, A. C. Hartt and M. B. Dell. *In press*. Distribution and origin of pink salmon in offshore waters of the North Pacific Ocean. Int. North Pac. Fish. Comm. Bull.

Feeding Ecology of Lake Washington Juvenile Sockeye Salmon and the Salmon Enhancement Problem

D. M. Eggers

INTRODUCTION

As biologists working with salmon-producing ecosystems, perturbations in these systems concern us. The kinds of perturbations are numerous and include logging on natal streams, pollution, urbanization and, recently, large-scale salmon enhancement projects.

Declining salmon stocks and the ramifications of impossible demands for increased catch allocations among competing fisheries forced these large-scale enhancement projects into being in this country as well as in Canada, Japan and the Soviet Union. In this paper, I will discuss some potential fallacies in enhancement from the perspective of a recently developed hypothesis (Eggers 1978) regarding the limnetic feeding behavior and predator avoidance of sockeye salmon *(Oncorhynchus nerka)* in Lake Washington. In developing the hypothesis, I described their feeding repertoire and assessed their behavior relative to the hypothesis. I substantiated the hypothesis by contrasting the feeding habits of juvenile sockeye in Lake Washington with those of other sockeye in different lakes with different environmental characteristics.

The hypothesis suggests that salmon feeding behavior is a dynamic and finely tuned response to ambient conditions that trade off the competing necessities of obtaining food and avoiding predators. Factors in the limnetic environment include seasonal and vertical gradients of prey abundance, prey species composition, light intensity and temperature. In addition, a sizeable population of northern squawfish *(Ptychocheilus oregonensis)* preys on juvenile sockeye. Owing to all of these factors, juvenile sockeye in the lake display complicated seasonal and diel patterns of feeding behavior (Woodey 1972, Eggers 1975, Doble and Eggers 1977), and a wide variation in the fitness of sockeye observed over the set of alternative behavioral decisions can be expected.

DISCUSSION

The Behavior of Juvenile Sockeye Salmon in Lake Washington

Anadromous sockeye salmon from Lake Washington spawn in the Cedar River. They then spend one year in the lake and two years in the ocean maturing before they return to reproduce (Woodey 1972). To reproduce, sockeye must obviously avoid predation and survive their residence in freshwater to attain sufficient energy reserves for migration to oceanic feeding grounds. To acquire energy reserves, juvenile sockeye must make several behavioral decisions including when and how long to feed, where to feed in the water column, what prey to eat and what behavior to adopt when not feeding.

Active feeding, however, requires active movement. Movement increases the chance that a piscivorous predator, such as the northern squawfish, will see juvenile sockeye, and also increases the chance of a moving sockeye encountering a foraging piscivore. Both sockeye, a planktivore, and northern squawfish, a piscivore, feed in lighted portions of the water column.

In open waters, planktivores have no place to hide. To avoid predators, they rely on behaviors which are inefficient in exploiting food. Therefore, the

---

Contribution No. 525, College of Fisheries, University of Washington, Seattle, Washington.

planktivore must make a trade-off between feeding and avoiding predation. This trade-off gives rise to the complicated pattern of feeding behavior of juvenile sockeye salmon in Lake Washington. In reviewing these behaviors, I found that they include diel feeding chronologies, size-selective predation, vertical migration, location in the water column, and diel and vertical patterns of school formation and dispersal (Eggers 1978). These behaviors vary while sockeye reside in the lake, and reflect varying zooplankton availability, light intensity and limnetic distribution of northern squawfish.

A process that optimizes the competing demands of feeding and avoiding predation best explains these observed behaviors among juvenile sockeye in the lake (Table 1). The juveniles feed briefly, school when light intensity is high and migrate into the upper water column where zooplankton are more abundant only during periods of low light intensity. These behaviors limit northern squawfish predation.

In addition, foraging sockeye ignore small prey items if large prey abound. This behavior further reduced the time necessary for juvenile sockeye to meet their energy requirements and thus optimizes survival, or fitness.

Two aspects of the Lake Washington situation are important. First, sockeye occur in a wide depth interval. The fish show complicated temporal and spatial patterns of school formation and dispersal, which suggests that a fairly large subset of the total possible behavioral repertoires may have comparable fitness. Second, the fish

Table 1. Outline of Sockeye Salmon Optimization Process.

I. Costs and benefits of feeding
   A. Feeding results in energy and growth
      1. Energy necessary for metabolism
      2. Energy necessary for reproduction
      3. Energy necessary for migration
      4. Size advantage for avoiding predators
   B. Feeding increases vulnerability to predators
      1. Feeding requires light and salmon are necessarily exposed to visual piscivores
      2. Feeding requires active swimming
         a. Activity increases visibility
         b. Activity increases probability of encountering moving object

II. Ways to mitigate costs of feeding and associated loss of feeding efficiency
   A. Feed briefly but ration intake is necessarily limited
   B. Select large prey to maximize energy intake rate or minimize time to ingest a given level of ration. This behavior results in efficient feeding
   C. Schooling decreases chance of being preyed upon but results in reduced feeding efficiency (provided zooplankton are randomly distributed) because of competition among school members through visual field overlap and resource depression along the school axis
   D. Occur in low light intensity but zooplankton are generally in lighter regions of the water column and less visible in low light intensity
   E. Cryptic coloration that decreases object contrast in pelagic environment (dark dorsal with silvery mirrorlike sides) but this diverts energy from reproduction and growth

III. Decisions available to sockeye salmon
   A. Timing of feeding in the diel photoperiod
   B. Duration of feeding
   C. What prey organisms to pursue upon encounter
   D. Whether to school or not
   E. Size of the school
   F. Where in the water column to feed
   G. If not feeding, what is the best way to avoid predation

alter their behavior in consistent seasonal patterns, which include vertical migration, location in the water column, diel feeding chronologies, schooling behavior and wintertime behavior. To interpret these behavioral patterns, zooplankton abundance and size, sockeye energy requirements and the distribution of northern squawfish must be considered. The interpretation suggests that feeding behavior is indeed dynamic, and ambient conditions strongly influence it.

Optimal Behavior and the Evolutionary Approach

As a consequence of natural selection, behavior should be an optimization process that operates under varying environments. An organism budgets or allocates its time, matter and energy resources among conflicting demands of feeding, avoiding predators, reproduction, migration, social behavior and so forth (Pianka 1974). Clearly, many possible strategies of resource allocation exist and many different behaviors are observable in communities of organisms.

This line of reasoning relates to an emerging ecological paradigm, or model, based on the evolutionary approach (Reiger and Rapport 1978). It evaluates behavior in terms of a control variable (e.g. net energy intake rate), which is assumed to be commensurate with fitness (McArthur and Pianka 1966, Emlen 1966, Schoener 1971, Charnov 1976, Werner and Hall 1974, Maynard-Smith 1974). The model seeks broad generalizations about how community structure has been derived from natural selection. It implicitly contains the following logic: As a consequence of natural selection, organisms possess physiological machinery that translates environmental stimuli into observed behaviors. In order for this machinery to be selected, there must be a high degree of correlation between the organism's behavior and the ecologically optimal, or most fit, behavior. Thus the organism's behavior and the ecologically optimal behavior correlate closely.

Optimal Behavior and the Tactical Approach

The organism's tactics and limitations in the face of environmental constraints and uncertainty influence optimal behavior. However, the evolutionary approach mostly ignores these factors.

In organisms that display behavioral plasticity, the tactical approach, or model, is especially important (Holling 1968). The model seeks to determine how the organism decides among alternative behaviors.

Clearly, an organism does not calculate fitness by any *a priori* model based on the current environment for each element of its behavioral repertoire. Rather, an organism possesses physiological machinery that gives a comfort stimulus in the current environment and physiological state (Balchen 1976). Organisms behave in a manner that maximizes comfort. In order for it to increase its comfort, the organism must "know" the first derivative of comfort with respect to alternative behaviors. This knowledge is either acquired from past experience, or programmed into the physiological machinery, or to some extent both. For example, when the organism "knows" it is hungry, it must initiate foraging to increase its comfort.

How do we reconcile the tactical approach with the evolutionary approach? Both conceptually involve optimization models. The tactical approach involves a short-term optimization of comfort, while the evolutionary approach involves a long-term optimization of fitness.

Both approaches involve implicit subjectivity. In general, we do not know enough about the physiological machinery to deduce observed behavior from an environmental stimulus and physiological state of the organism as the tactical approach requires. To make this approach tractable, the ecological manifestation of the comfort optimization must be devised. For example, Holling (1966) constructed an empirical model in which the attack rate of mantids preying on dead houseflies increased with increasing hunger.

The evolutionary approach has a similar pattern. The variable commensurate with fitness must be devised to predict the ecologically optimal behavior.

In practice, the ecological manifestation of comfort in the tactical approach and the control variable commensurate with fitness in the evolutionary approach are equivalent constructions. Both approaches deal with the same phenomenon. They must ultimately reach the same goal, and considering both approaches provides useful insights into behavioral mechanisms. The tactical approach implies that behavior is an immediate response to environmental alteration whereas the evolutionary approach implies that the great physiological complexity of an organism can be reduced to tractable models.

It appears that juvenile sockeye salmon in Lake Washington change their location in the water column, the timing and duration of their feeding, their pattern of school formation and dispersal, and the set of prey they pursue according to some short-term optimization of comfort. Comfort here means minimizing exposure to predators while achieving minimum ration requirements. These behaviors are also commensurate with fitness.

Ecological Considerations in Large-Scale Salmon Enhancement

Some argue that the ocean historically supported several times the present standing stocks of salmon. They say that because ocean productivity has not apparently declined, the ocean can support historic levels of salmon. What have been the consequences of drastically reducing the standing stocks of salmon in the ocean through fishery exploitation? Will the forage base formerly cropped by salmon be available to enhanced salmon stocks of the future, or are competing species now cropping this productivity?

Historical levels of salmon standing stocks indicate that salmon are probably the most efficient foragers. Salmon species should persist at lower food densities than competitor species. If competitor species now crop the productivity that salmon once did before exploitation, then enhancing salmon stocks in the future would most likely exclude these competitor species.

Man's impact on Pacific salmon has not only reduced the standing stocks of salmon but altered the spatial and temporal distribution of salmon in the ocean. This latter effect has resulted from different magnitudes of environmental degradation and fisheries exploitation among the various stocks of Pacific salmon.

There is evidence that temporal and spatial aspects of smolt migration are critical to a stock's survival. Large-scale enhancement may upset these unique temporal and spatial aspects of migration, and impact the entire marine rearing system. There is evidence that sockeye smolts enter the ocean along a latitudinal gradient that correlates closely with a latitudinal gradient in the timing of the northward progression of offshore zooplankton blooms in spring (Hall and English *in press*). Straty (1974) observed similar patterns for Bristol Bay sockeye stocks.

The five Pacific salmon species differ markedly in their marine life histories. On entering the marine environment, sockeye proceed immediately to feeding grounds on the high seas. Chum *(O. keta)* and pink *(O. gorbuscha)* salmon rear in estuaries, marshes and littoral regions before migrating to coastal pelagic or high seas feeding areas. Coho *(O. kisutch)* rear in freshwater streams and rivers. Upon entering the marine environment, they move to pelagic environments in both coastal and high seas areas where they feed primarily in the surface stratum of the water column.

The life history of chinook salmon *(O. tshawytscha)* is perhaps the most complicated. The period of freshwater residence varies widely among stocks of this species. Little is known about their early marine life history, but they are usually found in coastal zones and feed much deeper in the water column than coho.

These differences in life history between the species are perhaps the manifestation of natural selection operating to minimize competition for resources such as food, space and habitat. One can argue that where niche separation is greatest, competition is most intense and resources most limiting. For example, coho and chinook are territorial. They set up and defend feeding stations in riparian rearing areas. Thus, one can say that freshwater rearing habitat limits these species. This may explain the great success of hatchery-reared chinook and coho as these smolts bypass natural rearing.

Coho, chinook and sockeye exploit pelagic environments immediately upon entering the marine system. These species are large as smolts because they reared to some extent in freshwater. Chum and pink do not rear in freshwater and are smaller when they enter the marine system. They rear in estuaries, marshes and inshore environments before becoming pelagic. The prey organisms in marine pelagic environments may be too large for chum and pink salmon fry to exploit effectively. Also, chum and pink salmon fry have more potential predators because of their small size. Littoral environments may provide refuge where the young salmon can grow large enough to safely exploit pelagic environments.

We have little information on the relative production per unit area in littoral and pelagic environments, but the vast difference in amount of habitat available to salmon suggests that littoral environments could be more easily "filled up" by enchancement activities.

According to the hypothesized mechanism of feeding behavior developed for Lake Washington, a larger sockeye population depresses the abundance of prey. Consequently, salmon exploit their forage base aggressively. This might mean longer feeding periods or moving into alternative habitats. These behavioral alterations may result in higher vulnerability to predation and thus higher mortality.

If resources were limiting in inshore refuge areas that chum and pink salmon exploit, then resource partitioning by cohabiting pink and chum fry would be expected. Research into competitive interactions between chum and pink fry is a necessary precursor to large-scale enhancement of these species.

Available information on the feeding ecology of cohabiting salmon in the high seas shows little resource partitioning among species. Manzer (1964) observed a great deal of overlap in the vertical distribution of coho, pink, chum and sockeye salmon in the Gulf of Alaska. Chum salmon occurred deeper and pink and coho catches were small. But those caught were closer to the surface than sockeye. LeBrasseur (1966) showed a great deal of diet overlap among Pacific salmon. Observed diet differences among different salmon species seemed to reflect the effects of salmon size rather than niche separation. Larger salmon tended to eat larger prey items. The lack of observed competitive interactions among cohabiting salmon on the high seas suggests that prey resources are not limiting and can support more salmon.

SUMMARY AND CONCLUSIONS

The above view of salmon feeding ecology suggests that a diversity of methods is the best strategy of salmon enhancement. These methods should mimic the pattern of life history and the great diversity of stocks among Pacific salmon before they were heavily exploited.

The evolutionary manifestation of competition seems to be most developed for riparian rearing species. Salmon possess very plastic feeding behavior adaptive for exploiting a great number of habitats during their life history. Sockeye in Lake Washington possess elaborate behaviors for minimizing vulnerability to predators, suggesting that salmon, particularly small individuals, experience significant predation. This should not be overlooked in enhancement planning. Careful enhancement design is necessary to circumvent the predation problem. It should be noted that fisheries exploitation effectively controls predators as many potential salmon predators are themselves valuable species.

For species that rear in restricted marine littoral habitats, care should be taken to spread out releases temporally and spatially so the fish take advantage of all available habitat. Release strategies should be timed to the seasonal dynamics of the available forage base. By all means, concentrated releases of pink and chum should be avoided, because resource depression due to high population densities are potentially detrimental to these species.

LITERATURE CITED

Balchen, J. G. 1976. Principles of migration in fishes. Report STF48 A76045. Found. of Scient. and Ind. Res. (SINTEF), Trondheim, Norway.

Charnov, E. L. 1976. Optimal foraging, the marginal value theorem. Theoretical Pop. Biol. 9:129-136.

Doble, B. D., and D. M. Eggers. 1977. Diel feeding chronology, rate of gastric evacuation, daily ration and prey selectivity in Lake Washington juvenile sockeye salmon *(Oncorhynchus nerka)*. Trans. Am. Fish. Soc. 107:36-45.

Eggers, D. M. 1975. A synthesis of the feeding behavior and growth of juvenile sockeye salmon in the limnetic environment. Ph.D. Thesis. Univ. of Wash., Seattle. 217 pp.

Eggers, D. M. 1978. Limnetic feeding behavior of juvenile sockeye salmon in Lake Washington and predator avoidance. Limnol. Oceanogr. *In press*.

Emlen, J. M. 1966. The role of time and energy in food preference. Am. Natur. 100:611-617.

Hall, C. A. S., and K. English. *In press*. Computer simulation of energy losses and benefits of Pacific salmon migration. Paper presented at 41st Annual Meeting of American Society of Limnology and Oceanography. 19-22 June 1978, Victoria, British Columbia.

Holling, C. S. 1966. The functional response of invertebrate predators to prey density. Mem. Entomol. Soc. Can. 48:1-86.

Holling, C. S. 1968. The tactics of a predator. Symp. Roy. Entomol. Soc. London 4:47-58.

LeBrasseur, R. J. 1966. Stomach contents of salmon and steelhead trout in the northeastern Pacific Ocean. J. Fish. Res. Bd. Can. 23:85-100.

McArthur, R. H., and E. R. Pianka. 1966. On optimal use of a patchy environment. Am. Natur. 100:603-609.

Manzer, J. I. 1964. Preliminary observations on the vertical distribution of Pacific salmon (genus *Oncorhynchus*) in the Gulf of Alaska. J. Fish. Res. Bd. Can. 21:891-903.

Maynard-Smith, J. 1974. Models in ecology. Cambridge Univ. Press. 146 pp.

Pianka, E. R. 1974. Evolutionary ecology. Harper and Row, New York. 356 pp.

Regier, H. A., and D. J. Rapport. 1978. Ecological paradigms, once again. Bull. Ecol. Soc. Am. 59:2-6.

Schoener, T. W. 1971. Theory of feeding strategies. Ann. Rev. Ecol. Syst. 2: 369-404.

Straty, R. R. 1974. Ecology and behavior of juvenile sockeye salmon *(Oncorhynchus nerka)* in Bristol Bay and the eastern Bering Sea. Pages 285-320 *in* Hood, D. W., and E. J. Kelly (eds.), Oceanography of the Bering Sea with emphasis on renewable resources. Instit. of Marine Sci., Univ. Alaska.

Werner, E. E., and D. J. Hall. 1974. Optimal foraging and the size selection of prey by the bluegill sunfish *(Lepomis macrochirus)*. Ecol. 55:1042-1052.

Woodey, J. C. 1972. Distribution, feeding and growth of juvenile sockeye salmon in Lake Washington. Ph.D. Thesis. Univ. of Wash., Seattle. 207 pp.

Salmon Outmigration Studies in Hood Canal: A Summary—1977

Nicholas J. Bax, Ernest O. Salo, Bruce P. Snyder, Charles A. Simenstad, and William J. Kinney

INTRODUCTION

In 1977, the Fisheries Research Institute (FRI) conducted the third phase of a five-year program to study salmonid out-migrations in Hood Canal (Fig. 1). FRI also assessed the impact of pier construction and associated human activities at Bangor Annex on these migrations. The U.S. Navy is presently constructing a Trident submarine base there.

Hood Canal is an important migration route and nursery area for four salmon species, two anadromous trout species and the organisms that form their food supply. Several investigators conclude that conditions during early marine life are important to overall salmonid growth and survival (Shepard 1948, Vernon 1958, Wickett 1958, Gilhousen 1962, Manzer and Shepard 1962, Martin 1966, Hurley and Woodall 1968). This phase of the salmon life history is also a time of high natural mortality (Parker 1965, 1968).

The FRI conducted a beach seining and townetting program, and the Washington Department of Fisheries (WDF) initiated a mark-recapture program (Whitmus and Olsen *in press*). These programs confirm the migration routes and timing that previous studies suggested.

This report summarizes:

1. Sampling techniques used.
2. The out-migration patterns of juvenile chum salmon (*Oncorhynchus keta*).
3. The relationship between the catch-per-unit-of-effort (CPUE) and abundance of juvenile salmon.
4. The relationship between CPUE and the numbers and times of hatchery releases.
5. Estuarine feeding habits of juvenile chum salmon and some possible effects on prey abundance.

METHODS AND MATERIALS

Bax et al. (1978) gives detailed descriptions of the nearshore and offshore sampling techniques and the habitats sampled in this study.

Nearshore Sampling

Eight beach seine stations on the east shore, and four on the west shore, were sampled regularly from early January to late July (Fig. 2). Nighttime sampling was conducted from early April to early June.

Two sizes of beach seines were used. A 10 m x 2 m beach seine with a 6 mm stretch mesh bag was used at the beginning of the sampling until late March. A larger beach seine was also used from late January to late July. Its dimensions were 37 m with 18 m wings of a 3 cm stretch mesh. A 3 cm stretch mesh was used in the bag. The bag's dimensions were 0.6 m x 2.4 m x 2.3 m.

Visual survey-transects 0.8 km long were conducted by boat 2-15 m from the shore through late July. Polarized glasses and a mechanical counter were used to count salmonids. Accurate surveys required special environmental conditions (Schreiner et al. 1977).

Contribution No. 501, College of Fisheries, University of Washington, Seattle, Washington 98195

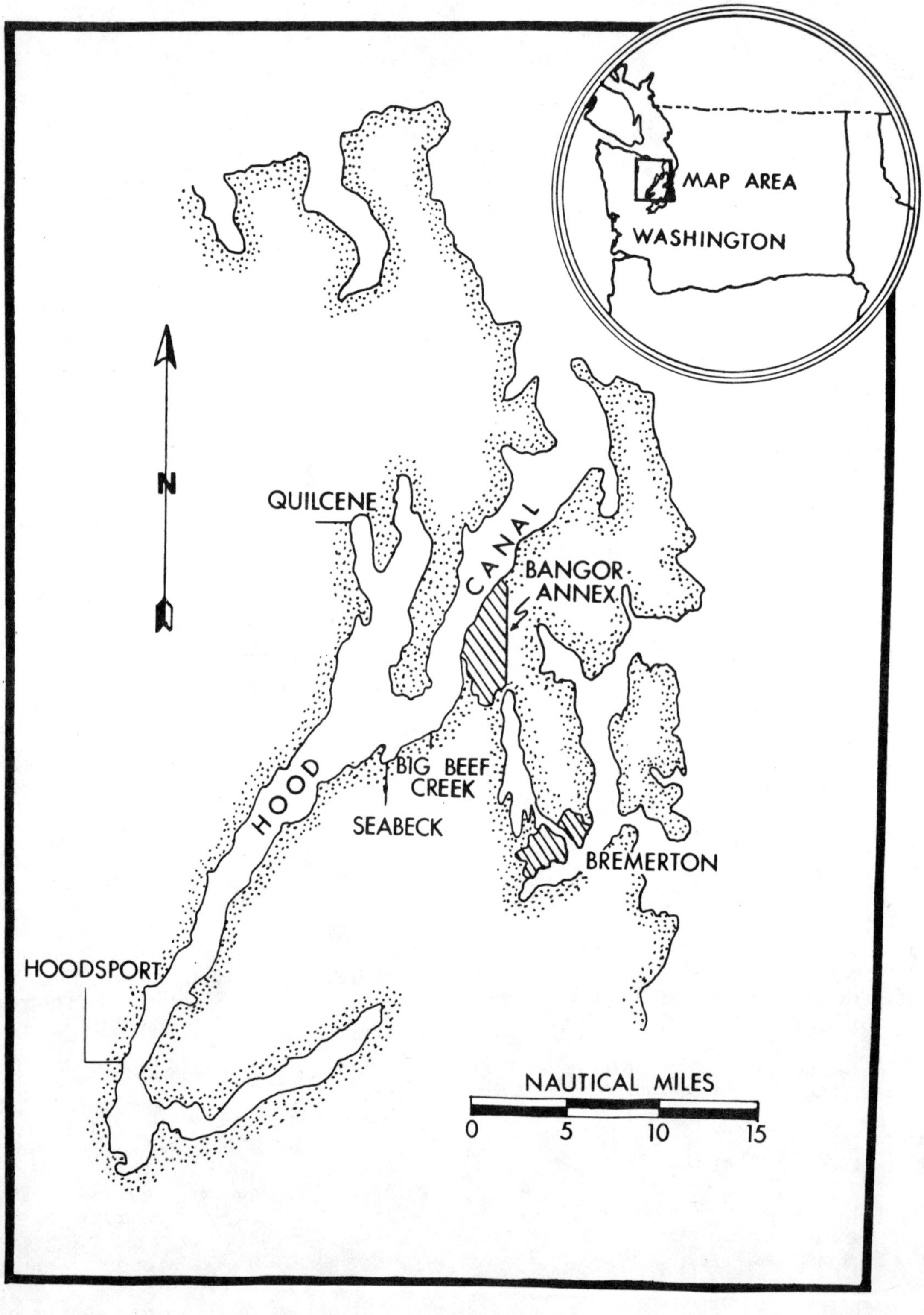

Figure 1. Location of Bangor Annex, Big Beef Creek and the Quilcene and Hood Canal (Hoodsport) fish hatcheries, Hood Canal, Washington.

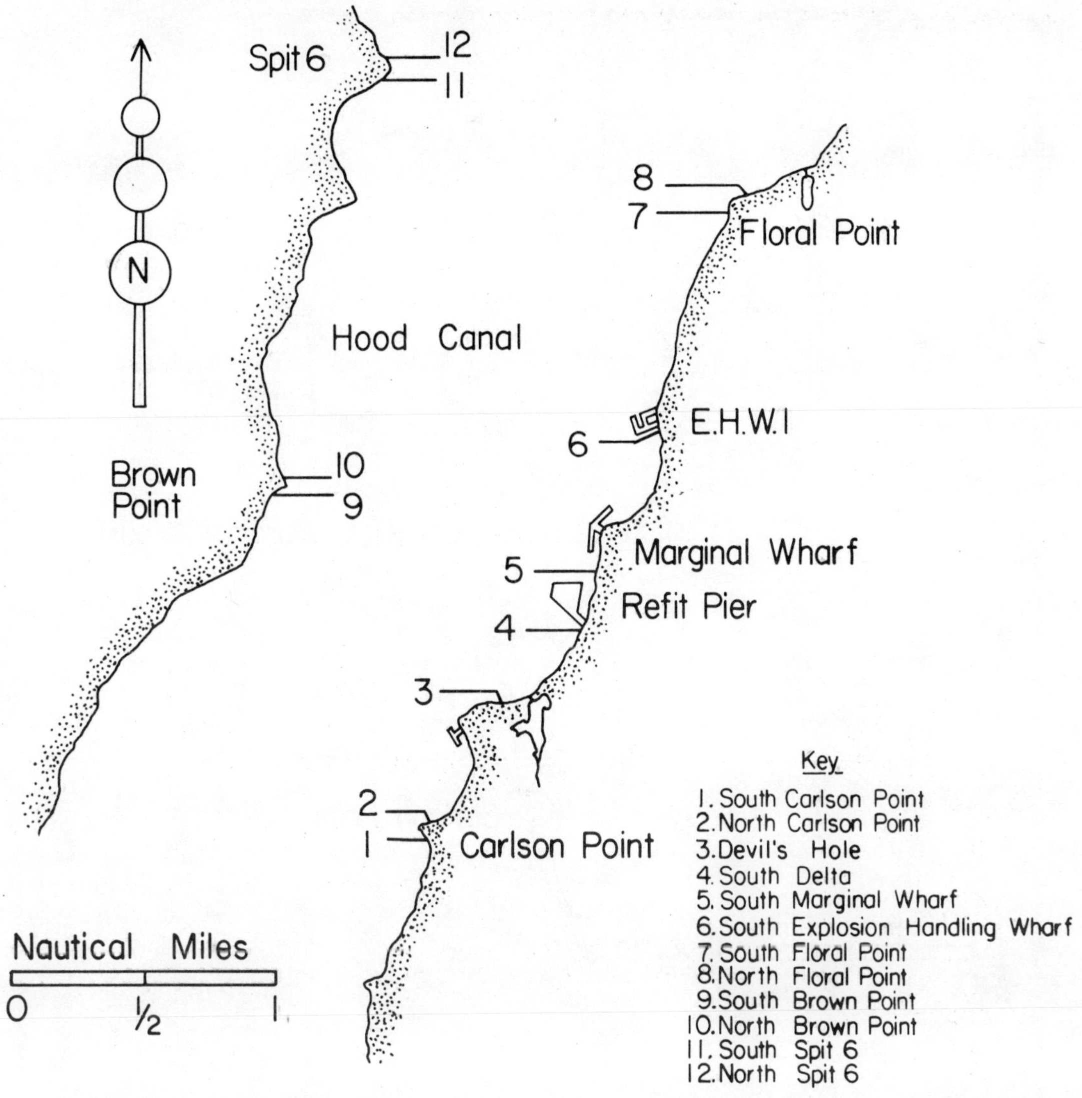

Figure 2. Beach seine sampling stations for shoreline salmonid out-migration studies, January through July 1977, Hood Canal, Washington.

Offshore Sampling

From late January to late July, surface townet transects 0.8 km long were sampled in daytime (Fig. 3). From early April until late July, nighttime sampling was conducted over the same transects. The sampling net was a surface trawl with a 3 x 6 m opening and stretch mesh sizes ranging from 76 mm at the opening to 6 mm at the bag. The net's wings were spread vertically by 3.75 cm diameter galvanized pipes that were connected with a short nylon bridle to single warps leading to each vessel. The net was towed between the *M/V Tenas*, a 38 ft. (11.6 m) diesel-powered vessel moving at a water speed of between 1.5 and 2.0 knots, and the *M/V Narwhal*, a 26 ft. (7.9 m) motorized whaler. At 10 minute intervals, two crewmen in an outboard skiff pursed the cod end of the tow-net, and removed all fish and debris. This technique allowed continuous sampling of the offshore transect pattern. Caught salmonids were transported in nontoxic 20 liter plastic buckets full of water to the *M/V Tenas* to be identified and sorted.

Fish Specimen Analysis

Usually less than 100 fish of each species were subsampled from each catch. The remainder were counted and released.

The subsamples were narcotized in MS-222 (tricaine methane sulfonate), preserved on ice and processed later that day. At regular intervals, concurrent with plankton sampling, five chum fry were preserved in a formalin

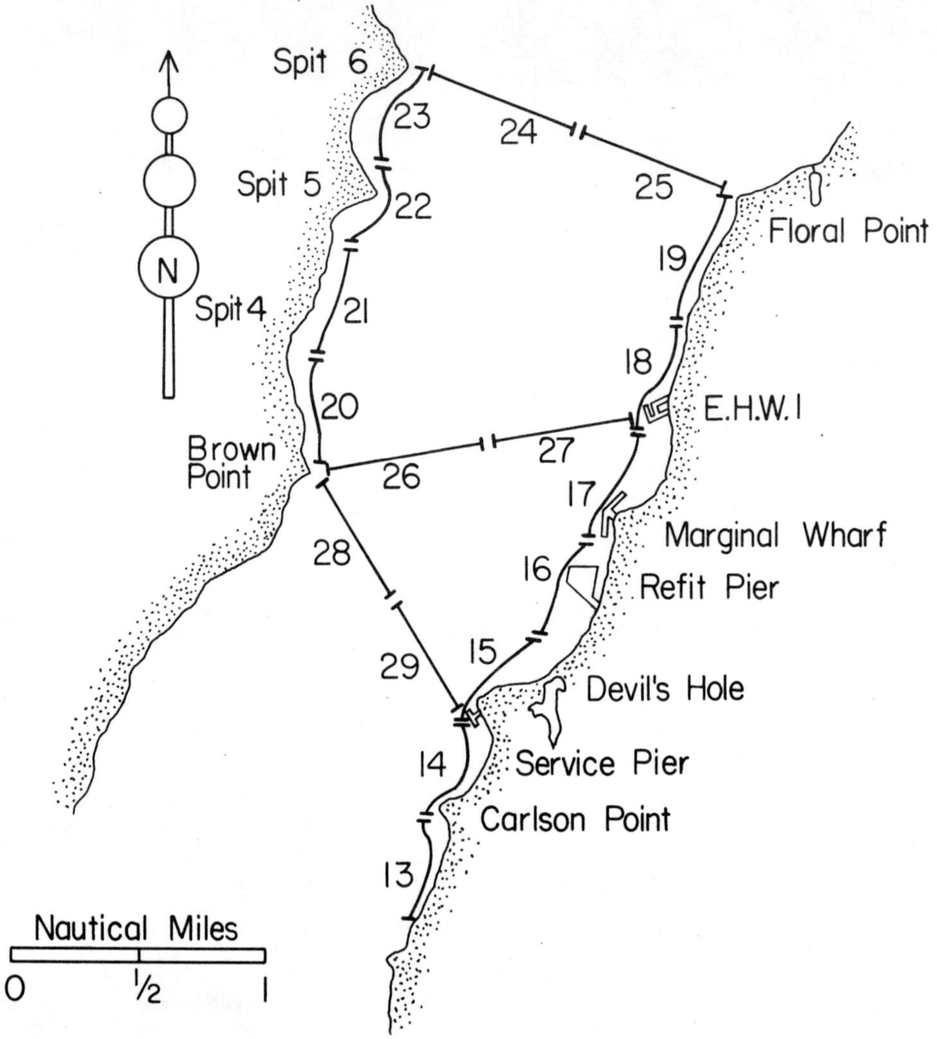

Figure 3. Townet surface trawl pattern used during salmonid out-migration studies, January through July 1977, Hood Canal, Washington.

solution for stomach analysis. Likely predators were treated in the same manner.

Environmental Data Collection

Environmental observations were taken after each beach seine set at 0.5 m - 1.0 m depths, 10 m - 15 m from shore. The dissolved oxygen concentration was measured with a Yellow Springs Instrument (YSI) Model 54 oxygen meter. Temperature was measured with a glass thermometer. Water samples were processed for total nonfilterable residues (TNFR). Weather and sea conditions were recorded at each site.

At the end of each 10 minute townetting transect, water samples and readings were taken at 1 m depths for TNFR, temperature, salinity and conductivity. Water visibility was measured with a 15 cm Secchi disk.

Epibenthic Plankton Sampling

Epifauna at four shallow, sublittoral sites in the vicinity of the Trident Submarine base (Fig. 4) was sampled in replicate using a suction-pump system. The pump system (Fig. 5) consisted of a self-priming, gas powered, 5.1 cm (2 in.) centrifugal pump. The pump drew water and associated planktors through a 25.4 cm (10 in.) conical expander into a 5.1 cm flexible plastic hose. Once through the pump, the water sample passed through a sealed-register totalizing flowmeter into a double stainless steel cylinder in which two nested conical nets were suspended. Mesh sizes of the nets were 505 μ and 209 μ with area/aspect ratios of 1:2.5 and 1:5.3, respectively. Standard net buckets with window screens of appropriate mesh size retained the epibenthic organisms.

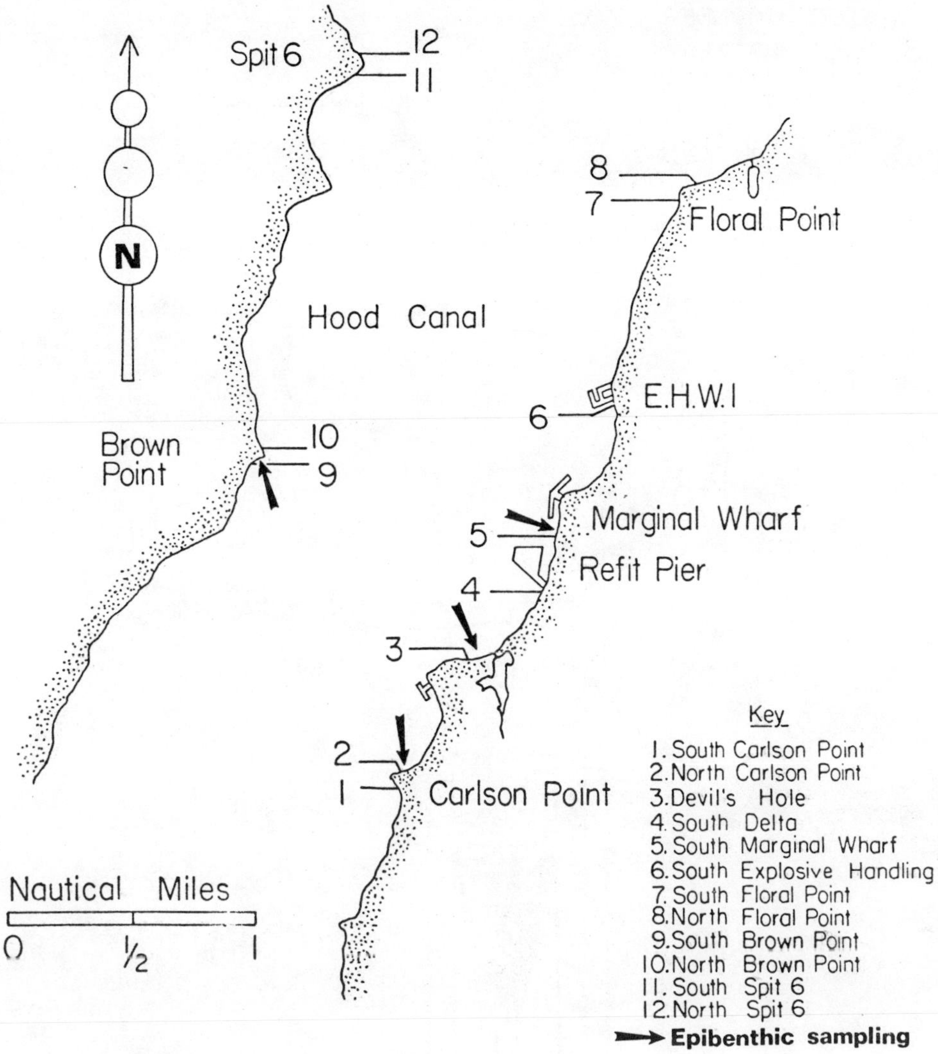

Figure 4. Beach seine and epibenthic sampling stations for shoreline salmonid outmigration studies January through July 1977, Hood Canal, Washington.

The pumping system was operated from a 26 ft. whaleboat maneuvered to stations at approximately the 0.3 m tide level and anchored. SCUBA equipped divers randomly placed a 1 m^2 round sampling cylinder on the substrate, then proceeded to vacuum the area within by moving the expander cone systematically 10 cm above the surface of the benthos. This distance was maintained by a ring extending from the expander cone. The ring did not contact the substrate. Two nested nets were dropped into place within the sampling tank and removed after 378.5 liters (100 gal.) had been filtered. Organisms retained in the nets were removed and preserved in 5 percent buffered seawater formalin in labeled PVC jars. The sampling process was repeated twice at each site after replacement of the sampling cylinder upon similar substrate nearby.

Bax et al. (1978) discusses the identification of epibenthic organisms and statistical treatments to use. Helmut Koch, Eastern Washington University, identified gammarid amphipod samples. Some harpacticoid copepod samples were identified by Beverly Kask, Pacific Marine Biological Station, Nanaimo, British Columbia. Laboratory results were recorded on MESA/EDS format forms that included prey code, life history stage, count, wet weight, total contents weight and remarks.

Approximately 100 specimens of gammarid amphipods and harpacticoid copepods from selected epibenthic and chum salmon stomach samples were measured to determine the size frequency of abundant taxa that juvenile salmonids utilize. Total length, minus antennae and setae, was measured to the

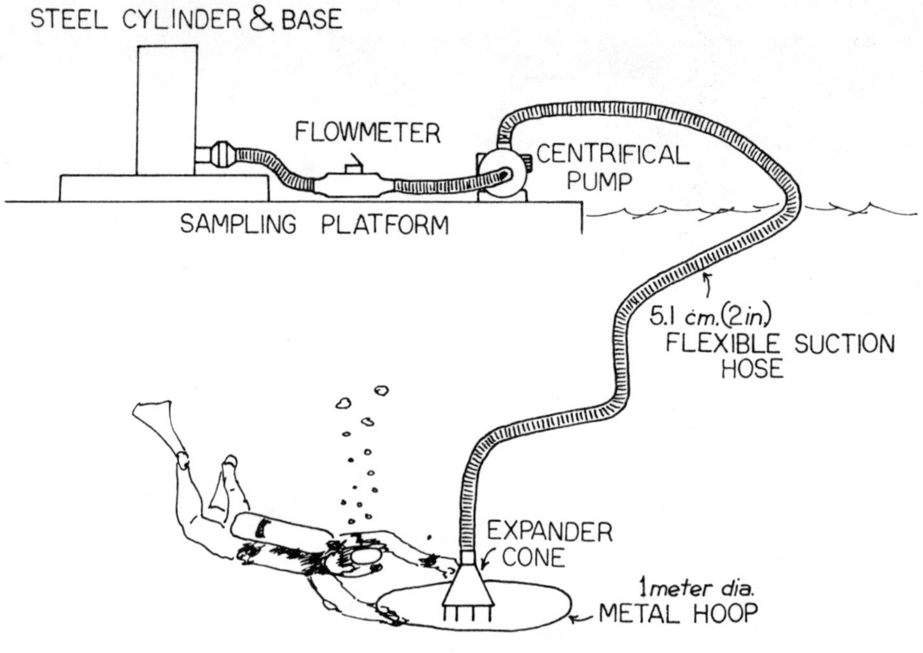

a. Epibenthic pump system components

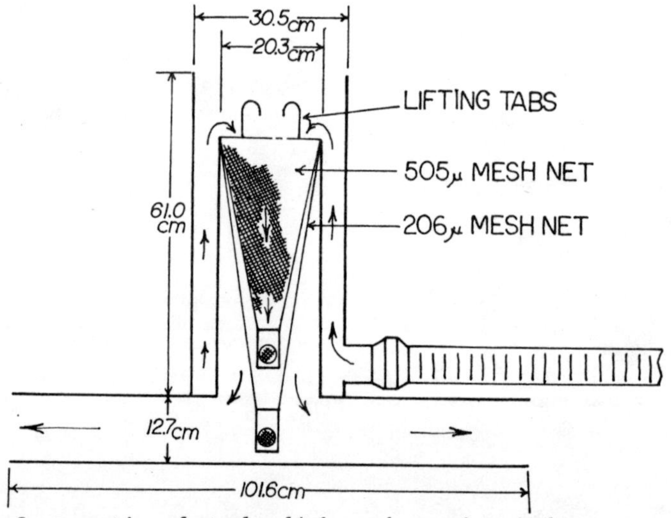

b. Cross-section of steel cylinder and nested conical nets

Figure 5. Overall system design and construction detail of epibenthic pump sampling system.

nearest 0.1 mm on amphipods, and to the nearest 0.025 mm on harpacticoids. Measurements were made using a dissecting microscope with a calibrated reticle micrometer.

Fish Stomach Analysis

Samples of chum salmon and associated fishes were collected at, or just offshore, beaches where epibenthic plankton samples were collected. Stomach content analyses included the numerical and gravimetric composition of prey organisms, the degree of fullness and the degree of digestion.

RESULTS AND DISCUSSION

Catch-Per-Unit-Effort

Catch-per-unit-effort (CPUE) computed for each salmonid species was of the form:

$$CPUE = Ci/Ei$$

C is the number of fry captured, E the intensity of sampling effort, and i is any given unit of time (Ricker 1968). Imperfect beach seine sets or townet hauls were excluded from the data analysis. The mean weekly CPUE, being the more consistent, was used as the primary statistic. Day-to-day sampling was not consistent in regard to the time of day.

Visibility, measured with a Secchi disk, dropped throughout the sampling season from 8.7 m in early March to 4.2 m in late June. Salinity in the towing area varied from 27.4 ppt in early April to 31 ppt two weeks later. The water temperature in the Bangor area rose from a minimum of 7.5°C in late February to 15.6°C in mid-June. Dissolved oxygen increased from 8.0 ppm in February to 11.0 ppm in late July. Total nonfiltrable residue measurements collected at wave-affected nearshore sites varied more than those at offshore locations, but no obvious trends were observed.

A stepwise multiple regression analysis showed the relative importance of each recorded environmental variable on the chum fry CPUE. The sampling week was entered first into the regression equation so that seasonal trends would not affect the choice of subsequent variables. Results for the beach seine indicate a highly significant positive relationship at the 0.99 level on the flood tide between the CPUE and temperature (Table 1). It was also found that the catch was higher on the sides of spits protected from the incident tide.

On both stages of the tide, a highly significant seasonal trend was noticed in the CPUE for the townet (Table 2). Weather conditions on the flood tide were significantly related to CPUE at the 0.01 level. Salinity and tide height on the ebb tide was significantly related at the 0.05 level.

Although the relationships mentioned were significant, they explained little of the variance in the CPUE. The coefficient of determination, r^2, was less than 0.10 in all cases.

Migration Periods and Peaks

Chum Salmon: The weekly CPUE with the 37 m beach seine (Fig. 6), indicated that two major peaks in abundance of chum occurred in the nearshore environment. The early February peak was most noticeable on the west shore. The later, and larger, peak was observed on both sides of Hood Canal from late May to early July. In this case CPUE values were higher on the east shore, although they peaked one week later than on the west shore. This same relation between the size and timing of east and west shore peaks was noticed for smaller peaks in mid-April and mid-March.

To find the effect of individual sampling locations on CPUE, a model 1, two-way analysis of variance was run with the logarithmic transformation of the CPUE as the dependent variable and sampling location and sampling week as the independent variables. Differences were found (Table 3). To find which

Table 1. A Comparison of Environmental Factors Affecting the CPUE of Chum Fry[1] with the 37 m Beach Seine From January to July 1977, Using Multiple Regression Analysis.

Step	Variable entered	F to enter	Significance	Partial regression coefficient	Coefficient of determination	Overall F	Significance
				Ebb Tide			
1	Week	1.23	0.27	0.02	0.01	1.23	0.27
2	Tide height	2.36	0.13	0.02	0.03	1.80	0.17
3	Weather	0.94	0.33	-0.08	0.04	1.62	0.20
4	Sea conditions	0.71	0.40	-0.05	0.04	1.31	0.27
5	Sea temperature	0.26	0.61	-0.04	0.04	1.10	0.37
				Flood Tide			
1	Week	0.91	0.34	-0.07	0.01	0.91	0.34
2	Sea temperature	8.98	0.00	0.28	0.10	4.98	0.01
3	Tide height	1.31	0.26	0.03	0.11	3.77	0.01
4	Weather	0.39	0.53	0.06	0.11	2.91	0.03

[1] Logarithmic transformation (LG_{10} (chum + 1)) used to stabilize variance and achieve normality.

Table 2. A Comparison of Environmental Factors Affecting the CPUE of Chum Fry[1] With the Surface Townet From February to July 1977, Using Multiple Regression Analysis.

Step	Variable entered	F to enter	Significance	Partial regression coefficient	Coefficient of determination	Overall F	Significance
				Ebb Tide			
1	Week	46.74	0.00	0.04	0.14	46.74	0.00
2	Salinity	5.01	0.03	0.05	0.15	26.20	0.00
3	Tide height	4.02	0.05	0.03	0.17	18.99	0.00
4	Sea state	2.30	0.13	0.05	0.17	14.88	0.00
5	Weather	0.44	0.51	0.03	0.17	11.97	0.00
6	Sea temperature	0.08	0.78	0.01	0.17	9.96	0.00
				Flood Tide			
1	Week	30.20	0.00	0.06	0.12	30.20	0.00
2	Weather	17.28	0.00	-0.19	0.18	24.83	0.00
3	Tide height	2.36	0.13	0.02	0.19	17.44	0.00
4	Salinity	0.42	0.52	0.02	0.19	13.15	0.00
5	Sea state	0.53	0.47	0.05	0.19	10.61	0.00
6	Sea temperature	0.36	0.55	-0.04	0.20	8.87	0.00

[1] Logarithmic transformation (LG_{10} (chum + 1)) used to stabilize variance and achieve normality.

particular locations were causing the differences the Student--Newman--Keuls (SNK) multiple comparison procedure was used. No one location was different from the remaining ones. Generally, those sites with an extensive littoral zone - the south side of spits, south EHW and Devil's Hole - had high CPUE values while sites with less extensive littoral zones - the north sides of spits - had lower CPUE values. South Marginal Wharf and South Refit Pier also had low CPUE values, perhaps due to construction activity in the area as the CPUE at South Marginal Wharf in 1976 was relatively high (Schreiner 1977).

Weekly CPUE with the surface townet indicated a peak in chum fry abundance from early June to mid-July (Fig. 7). An earlier, and smaller, peak was noticed in late April. The major peak in abundance measured by townet sampling started and ended later than observed with the beach seine. The peak was higher on the east shore but started one week earlier on the west side than on the east side of Hood Canal.

Data were analyzed in a manner similar to the beach seine data to find the effect of individual sampling locations on CPUE. There did not appear to be any one site, or sites, different from the rest (Table 4, Part A). Using t-tests to test our *a priori* hypothesis that there might be a difference between east and west shores, we found that the east shore mean CPUE was indeed higher than the west (Table 4, Part B). This is surprising as the majority of chum fry enter Hood Canal from rivers and hatcheries on the west shore. The other *a priori* hypothesis tested was the difference between shoreline and offshore catches. We found that the mean CPUE on townet transects decreased with distance from shore (Table 4, Part C). Thus the chum fry at this stage of their out-migration use areas predominantly close to shore.

Due to weather requirements for visual surveys of fry abundance, (Schreiner 1977), surveys were carried out only sporadically during the 1977 sampling season. The results were inconclusive (Bax et al. 1978).

Diurnal variation in catches was noticed for both the beach seine and townet (Table 5). Both the CPUE with the 37 m beach seine and its coefficient of variation decreased at nighttime when compared to daytime catches ($\alpha < .001$ and $\alpha < .0005$, respectively). For the surface townet the same decrease in the CPUE coefficient of variation was noticed at nighttime ($\alpha < .01$). The CPUE with the surface townet, in contrast to the beach seine, increased at nighttime ($\alpha < .0005$). It has been suggested that this indicated an offshore movement of fry, and the break up of schools at night.

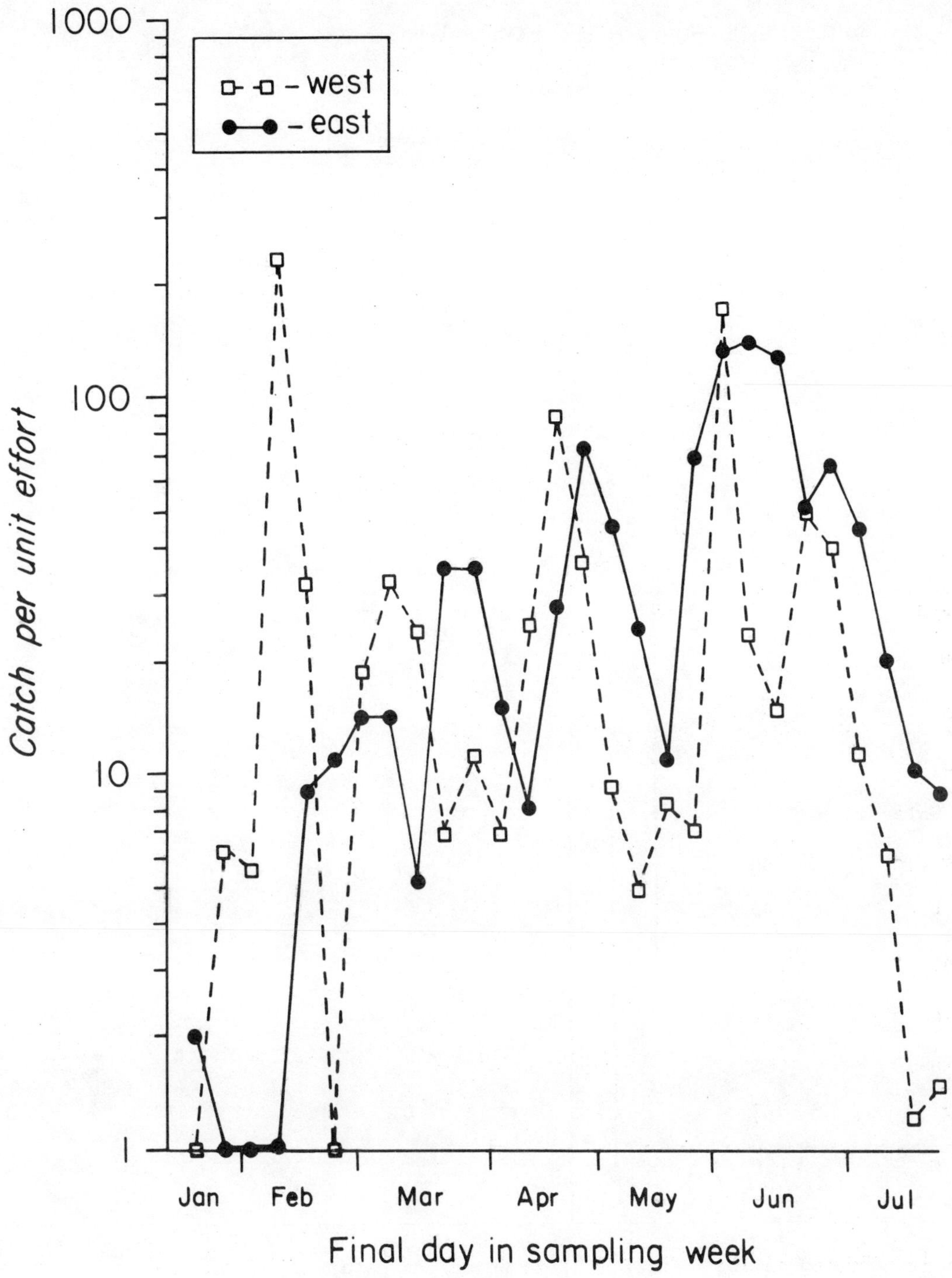

Figure 6. Comparison of CPUE on the east and west shores of Hood Canal in the Bangor Annex area, for chum fry, with the 37 m beach seine, 14 January to 28 July 1977.

Table 3. Analysis of Variance to Show the Effect of Sampling Location and Sampling Week on the CPUE[1] of Chum Salmon Fry With the 37 m Beach Seine.

ANOVA SUMMARY TABLE

Source of variation	Sum of squares	Degrees of freedom	Mean square	F	Significance of F
Totals	446.518	773	.578		
Cells	94.960	36	2.638	5.530	.001
Location	26.581	11	2.412	5.056	.001
Week	68.388	12	2.736	5.735	.001
Error	351.557	737	.477		

SNK MULTIPLE COMPARISON

Location	Mean CPUE	Groups[2]		
S. Carlson	1.20	│		
S. E.H.W.	1.14	│	│	
S. Brown	.93	│	│	
S. Floral	.92	│	│	
Devil's Hole	.90	│	│	│
N. Brown	.85	│	│	│
S. Spit 6	.82	│	│	│
N. Floral	.74	│	│	│
N. Carlson	.71		│	│
S. Marginal	.68		│	│
S. Delta	.61		│	│
N. Spit 6	.50		│	│

[1] CPUE is computed as $\log_{10}$ (CPUE + 1) to normalize data.

[2] The vertical lines enclose groups of sites not different from each other (p = .05)

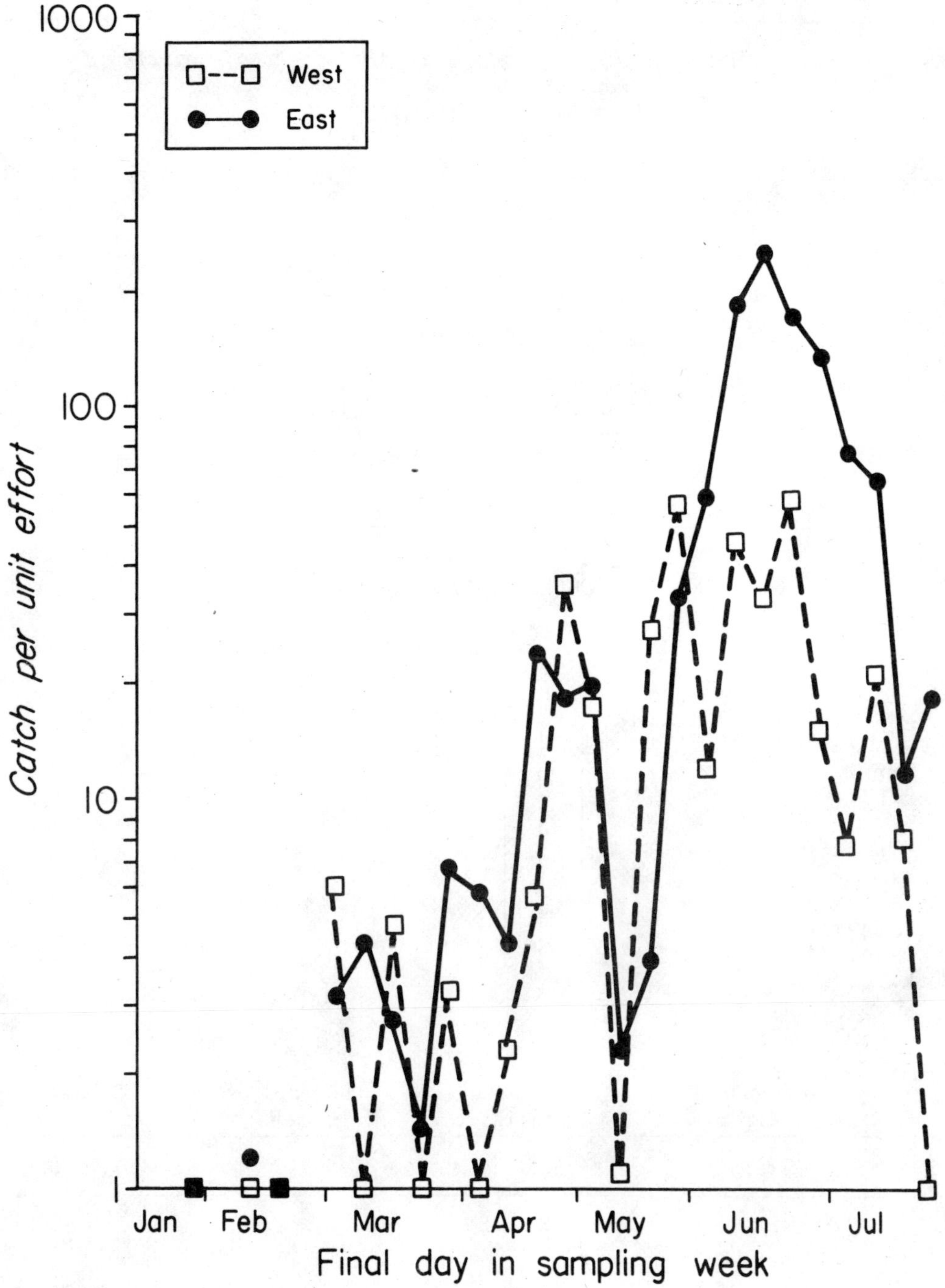

Figure 7. Comparison of CPUE on the east and west sides of Hood Canal in the Bangor Annex area, for chum fry, with the surface townet, 21 January to 28 June 1977.

Table 4. Analysis of Variance and T-Tests To Show the Effect of Sampling Location and Sampling Week on the CPUE[1] of Chum Salmon Fry With the Surface Townet.

A. Shoreline Locations

ANOVA SUMMARY TABLE

Source of variation	Sum of squares	Degrees of freedom	Mean squares	F	Significance of F
Total	359.036	662	.542		
Cells	136.836	30	4.561	12.973	.001
Location	12.557	10	1.256	3.572	.001
Week	125.662	20	6.283	17.371	.001
Error	222.199	632	.352		

SNK MULTIPLE COMPARISONS

Location	Mean CPUE	Groups[2]		
Carlson Pt. - Service Pr.	1.06	|		
King Spit - Carlson Pt.	1.03	|	|	
EHW No. 1 - Buoy B.	.89	|	|	|
Buoy B - Floral Pt.	.88	|	|	|
Marginal - EHW No. 1	.86	|	|	|
Service Pr. - Devil's Hole	.85	|	|	|
N. Spit 4 - Spit 5	.83	|	|	|
Devil's Hole - Marginal	.75		|	|
S. Spit 4 - N. Spit 4	.71		|	|
Brown Pt. - S. Spit 4	.66		|	|
Spit 5 - Spit 6	.61			|

B. East vs. West Shore

Difference	Standard error	T statistic	Degrees of freedom	Significance of T
5.0117	1.5781	3.176	976	.002

C. Shoreline vs. Offshore Transects

Difference	Standard error	T statistic	Degrees of freedom	Significance of T
27.0514	3.5549	7.610	976	.000

[1] CPUE is computed as $\log_{10}$ (CPUE + 1) to normalize data.

[2] The vertical lines enclose groups of sites not different from each other (p = .05)

Table 5. Comparison of Day-Night Catches of Chum Salmon Fry With the 37 m Beach Seine and Surface Townet From February to July 1977, Hood Canal, Washington.

Beach Seine			Surface Townet	
Day	Night	Variable	Day	Night
38	23	Mean CPUE	24	29
98	41	Standard deviation	89	85
256%	181%	Coefficient of Variation	378%	292%

Coho Salmon Smolts: Coho salmon *(O. kisutch)* smolts were caught in the beach seine and surface townet from late April through to the end of sampling in July. Smolt CPUE was higher with the 37 m beach seine than with the surface townet (Table 6). The CPUE was also higher at night than in the day, which may indicate avoidance of the net during the day (Table 7). Beach seine catches peaked in the week ending 14 July. Townet catches peaked in the week ending 30 June. The smolts may have been wild fish, or in part hatchery-reared smolts released from Hoodsport on 27 June. Both beach seine and townet catches peaked earlier in the week ending 19 May. This peak was coincident with the recapture of wild coho smolts with clipped adipose fins from Big Beef Creek that were released approximately one week prior to recapture.

Chinook Salmon: Chinook salmon *(O. tshawytscha)* smolts, yearlings and adults were caught throughout the 1977 sampling season. More were caught with the 37 m beach seine than with the surface townet. The capture of smolts and yearlings peaked slightly in May. Smolts were released from the Hood Canal Hatchery from 21 April to 18 May prior to the peak catches.

Cutthroat Trout: Juvenile and adult cutthroat trout *(Salmo clarkii)* were caught throughout the sampling season, but only in the 37 m beach seine. Peak catches were in early June.

Steelhead Trout: Eight juvenile steelhead *(S. gairdneri)* were caught in the townet. The 37 m beach seine caught one adult and four juveniles.

Hatchery Influence on CPUE

Hatchery-reared chum salmon fry are released from the west shore of Hood Canal (Fig. 1). However, beach seine and townet catch statistics from 1975 and 1976 suggest that the majority of hatchery fry crossed to the east shore prior to reaching Bangor Annex.

Catch statistics for 1977 show a more even distribution of chum fry on both shores, with the majority of townet catches still on the east shore. The observed peaks in CPUE of fry at Bangor Annex were not as clearly related to hatchery releases as in the previous two years. This was probably due to the almost continuous releases from late March to early July. In addition, a strong possibility exists that both wild and hatchery fish migrate through Hood Canal at a faster rate earlier in the season. Data from mark-recapture experiments (Whitmus and Olsen *in press*) show that an April release of marked fry from Hoodsport took less time to arrive at the Bangor Annex than a later June release.

Nearshore: The abundance of chum fry on the east shore, as indicated by the 37 m beach seine catches at South Floral Point, showed an early peak in early March before hatchery releases. Beginning in late March CPUE peaks reflected prior hatchery releases from the Hood Canal Hatchery (Fig. 8). In March, there was one week between the time of the hatchery release and subsequent CPUE peak at Bangor. By late June, the period between release and subsequent CPUE peak increased to three weeks.

West shore catches peaked a week earlier than east shore catches. Chum fry releases from Quilcene Hatchery later in the season coincided with the highest overall CPUE at Bangor Annex (Fig. 9). The time that these fry arrived is not clear from the CPUE values, which may also be influenced by a release of over three million fry from the Hood Canal on 19 May. The decrease in CPUE at the end of June and in July while fry from the Quilcene Hatchery were still being released may be because fry were no longer available to the beach seine.

Table 6. Weekly Release of Fin-clipped Coho Smolts From the Big Beef Creek Wild Out-migrations and Weekly CPUE of Coho Smolts in Hood Canal, Washington From April to July 1977.

Final day in sampling week	No. of coho smolts tagged and released in Big Beef Creek	Weekly CPUE of coho smolts					
		37 m Beach Seine			Surface townet		
		Fin-clipped	Non-clipped	No. sets	Fin-clipped	Non-clipped	No. tows
April 28	568	0	0	35	0	0	52
May 5	9454	0	0.11	37	0	0.04	56
12	10521	0.28	0.51	39	0.02	0.53	59
19	6541	3.39	2.61	41	0.17	1.21	48
26	1569	1.11	1.47	36	0.07	0.31	45
June 2	2066	0.77	1.77	39	0	0.33	55
9	335	0.68	1.61	31	0.03	0.09	32
16		0.62	2.49	37	0	0.20	45
23		0.52	2.96	23	0	1.44	25
30		0.19	2.03	36	0.02	1.49	47
July 7		0.03	4.23	40	0	0.89	47
14		0.14	4.82	22	0.03	0.89	35
21		0	2.58	24	0	0.26	39
28		0	1.92	12	0	0	12

Table 7. Comparison of Day-Night Catches of Coho Smolts From April to July 1977 With the 37 m Beach Seine and Surface Townet in Hood Canal, Washington.

37 m Beach Seine			Surface Townet	
Day	Night	Variable	Day	Night
1.89	3.64	Mean CPUE	0.28	1.17
7.47	2.96	Standard deviation	0.63	1.51
395%	81%	Coefficient of variation	226%	129%

Offshore: Hatchery releases appeared to similarly affect offshore abundance of chum fry, as indicated by the surface townet. The data suggest a two-week delay from the time fry were released from Hood Canal Hatchery until their arrival at Bangor Annex (Fig. 10). From mid-June to the end of sampling, the townet CPUE declined while fry were being released from Quilcene Hatchery (Fig. 11). This may indicate that larger fry avoided the net.

Fry Condition

At the start of the sampling season the mean fork length of captured chum fry was between 35 mm and 41 mm with small variance (Fig. 12). At the end of March and beginning of April, the mean length of chum fry captured with the townet showed a sharp peak, which was concurrent with an increase in variance (Fig. 13). This peak may be due to the arrival at Bangor Annex of Hood Canal Hatchery fry released on 11 March. The increase in mean length from 40 mm at release to as much as 52 mm at capture suggests considerable growth in the first few weeks the fry enter the marine environment. This growth rate compares with maximum growth rates of chum fry in freshwater with unlimited food available under hatchery conditions (personal communication, S. Schroeder, FRI). From mid-April, when the influence of hatchery-reared fish was detected in catches at Bangor Annex, until the end of the sampling season, chum fry showed a steady increase in mean length and variance. Length of fry at the time of release from the hatcheries showed a concurrent increase. The mean length of fry caught in the townet was consistently higher than those caught in the beach seine. A two-way model 1 analysis of variance showed that the differences were significant. The significant interaction of week and gear on mean length, together with the wide range of fork lengths among fry caught with the two gears, suggest that there is not a distinct size, or size range, when the fry move offshore and become available to the townet. Rather, an interaction of size and other factors, influenced by the season, determine the inshore/offshore distribution of the chum fry.

The mean condition factor of caught chum fry increased early in the season, but decreased toward the end. This trend was most noticeable in beach seine catches. The fry with the highest mean condition factors were caught at the period of maximum recruitment. These results are contrary to those obtained in 1975 and 1976 when the condition factor was found highest during periods of low recruitment. Thus, no drop in condition factor was found at the end of the sampling season in 1975 and 1976.

The drop in mean condition factor found at the end of the 1977 sampling season may be due to the short time fry from the Quilcene Hatchery spend in the marine environment prior to their arrival at Bangor Annex. It is thought that Quilcene Hatchery fry were the predominant fry at Bangor Annex at the end of the sampling season.

Trophic Relationships

Epibenthic Plankton Community: Thirty-four major taxa of epibenthic organisms were represented in the plankton pump collections (Bax et al. 1978). The most taxonomically diverse groups included the harpacticoid copepods, polychaete annelids, gammarid amphipods, shrimp and bivalves.

The numerically prevalent organisms were harpacticoid copepods. They were followed by gammarid amphipods; crustacean eggs; ostracods; calanoid copepods; isopods; nematodes; barnacle nauplii and cyprides; prosobranch larvae; juvenile shrimp; and cumaceans in decreasing order of overall percent composition. Of the harpacticoid copepods, the prevalent species were *Harpacticus* sp.; *Amphiascopsis cinctus* and a species of the family Laophontidae. The most common gammarids were an undescribed *Pontogeneia* sp., *Calliopiella pratti* and *Anisogammarus pugettensis*.

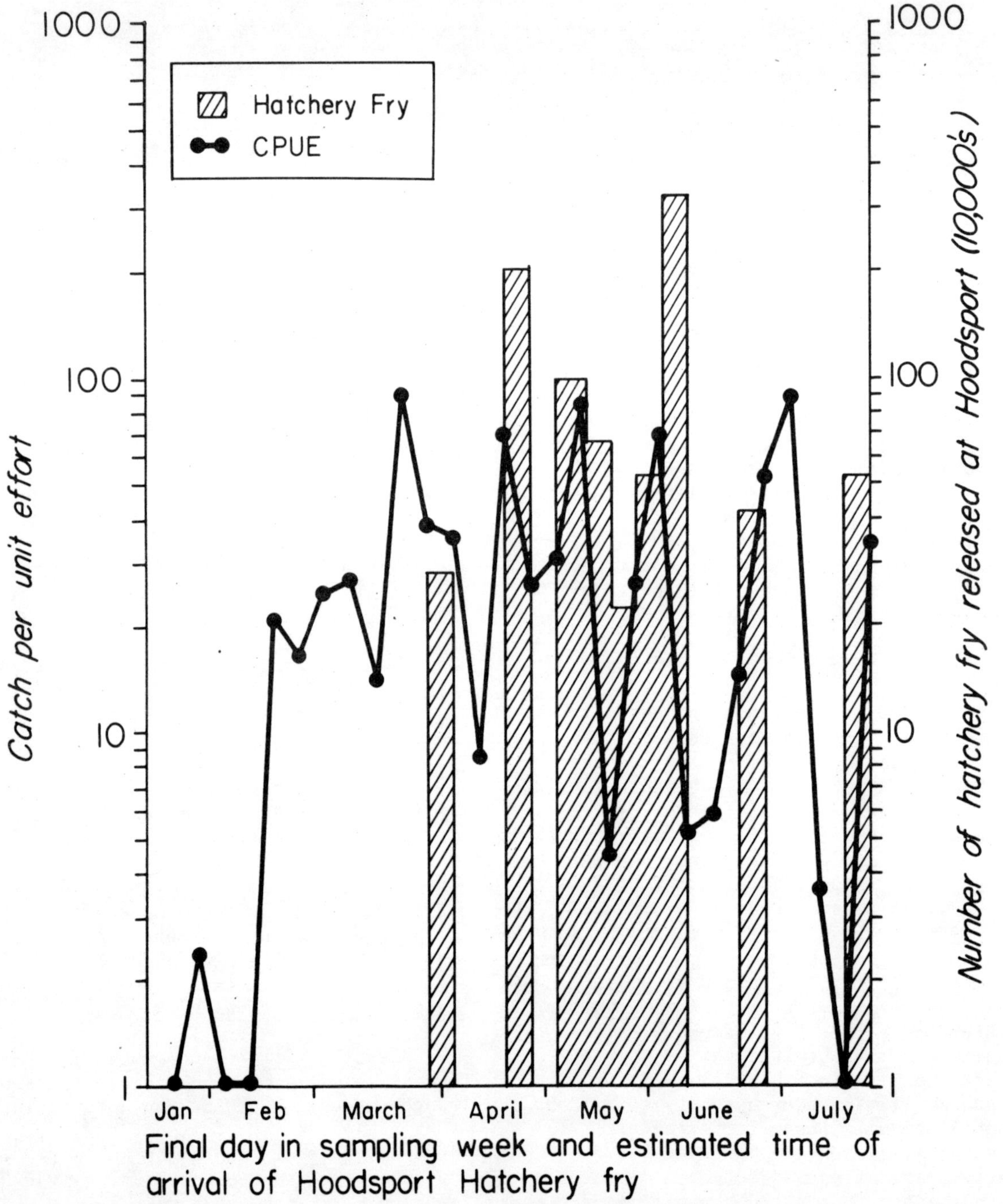

Figure 8. Comparison of estimated time of arrival of Hood Canal hatchery chum fry at Bangor (date of release + 3 weeks) and CPUE of chum fry at South Floral Point with the 37 m beach seine from 14 January to 28 July 1977.

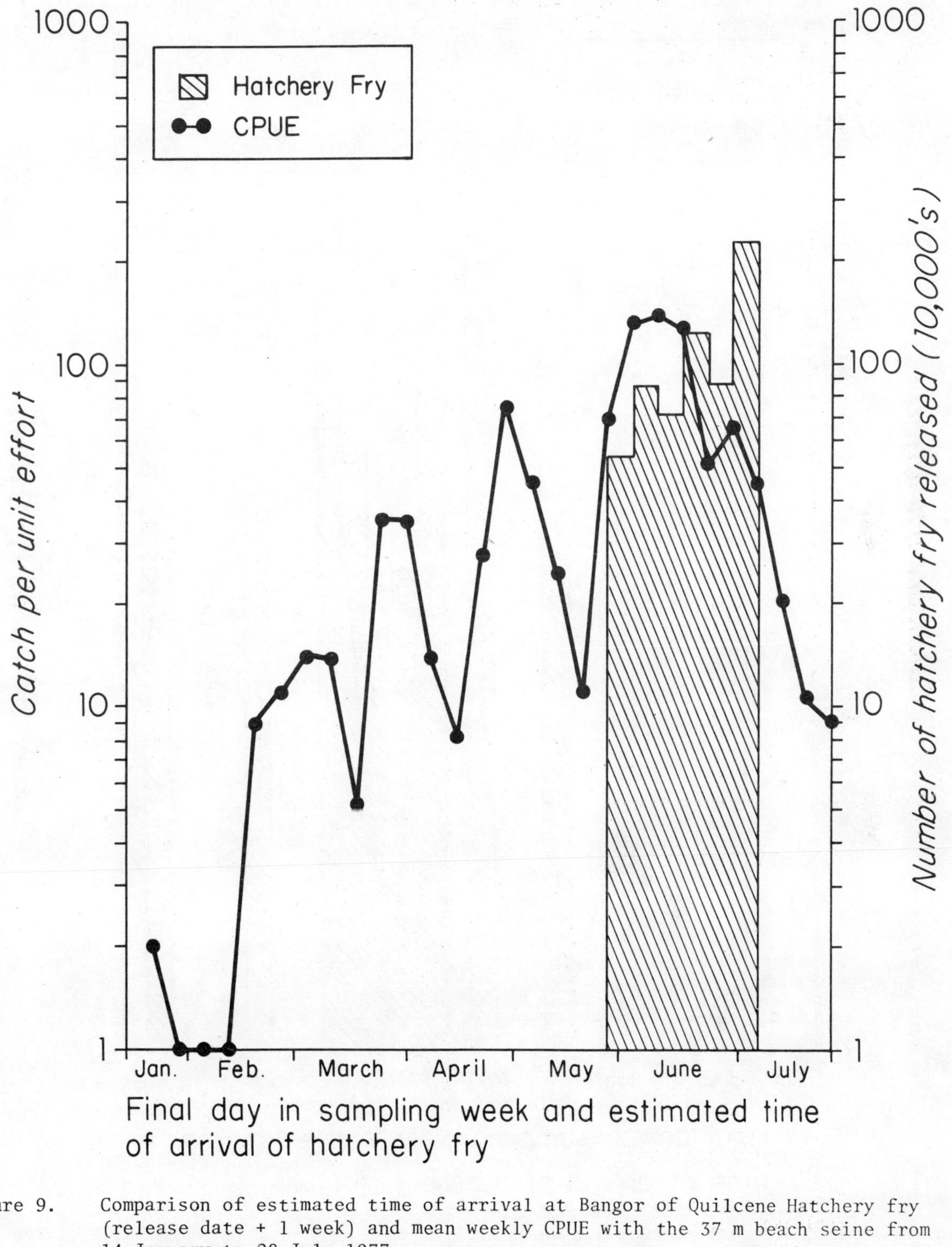

Figure 9. Comparison of estimated time of arrival at Bangor of Quilcene Hatchery fry (release date + 1 week) and mean weekly CPUE with the 37 m beach seine from 14 January to 28 July 1977.

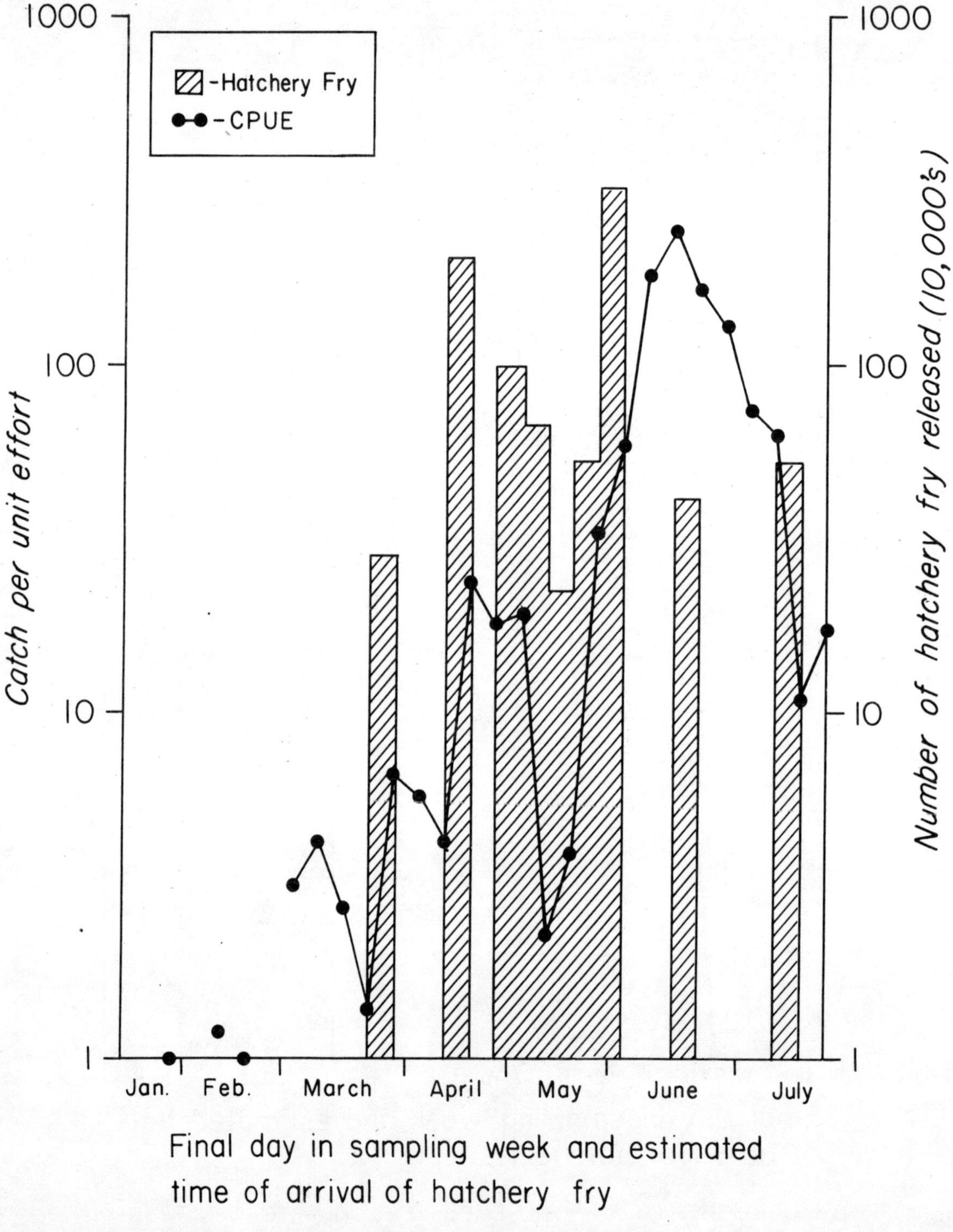

Figure 10. Comparison of estimated time of arrival at Bangor of Hood Canal Hatchery fry (release date + 2 weeks) and mean weekly CPUE with the surface townet from 21 January to 28 July 1977.

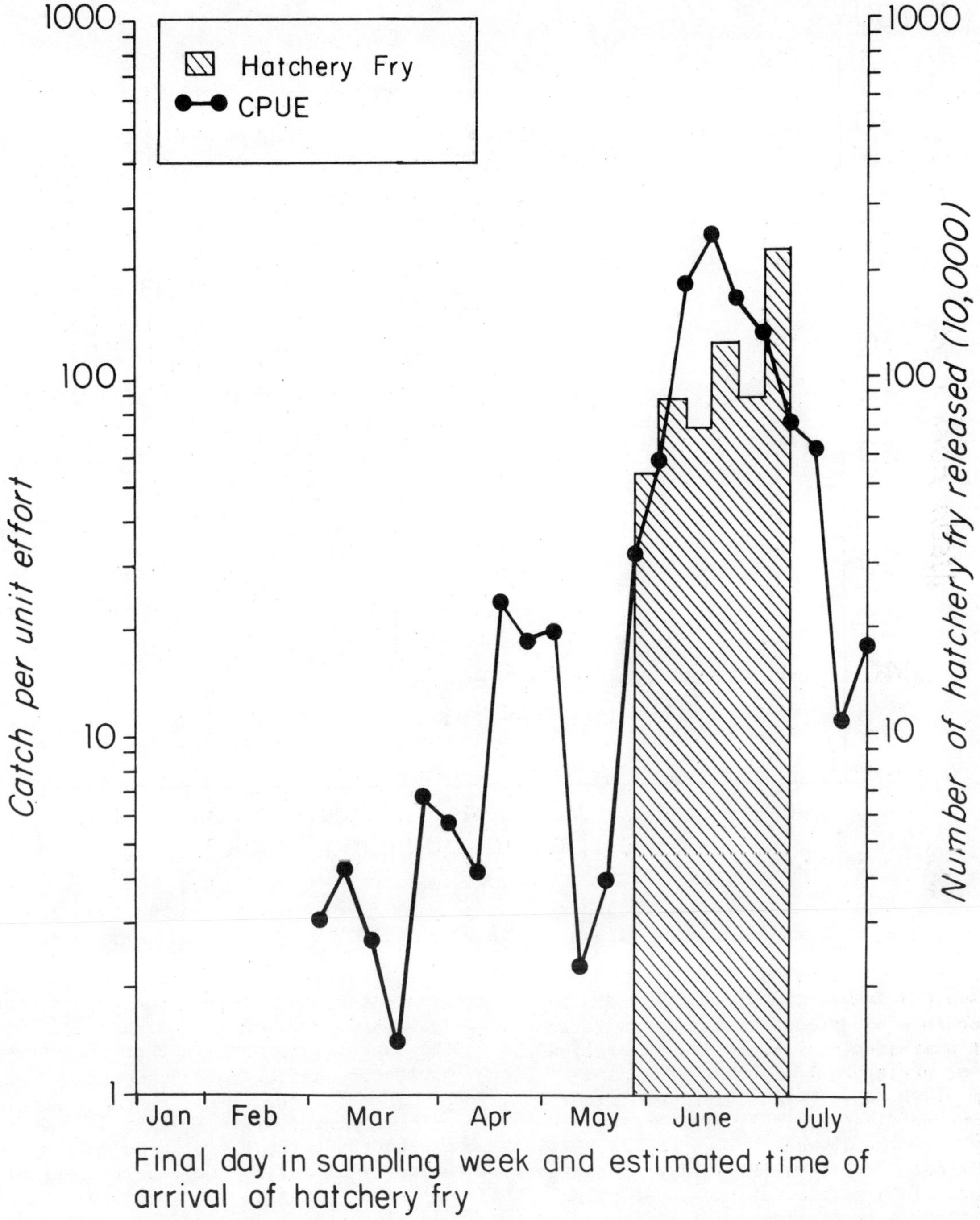

Figure 11. Comparison of estimated time of arrival at Bangor of Quilcene Hatchery fry (release date + 1 week) and mean weekly CPUE with the surface townet from 21 January to 28 July 1977.

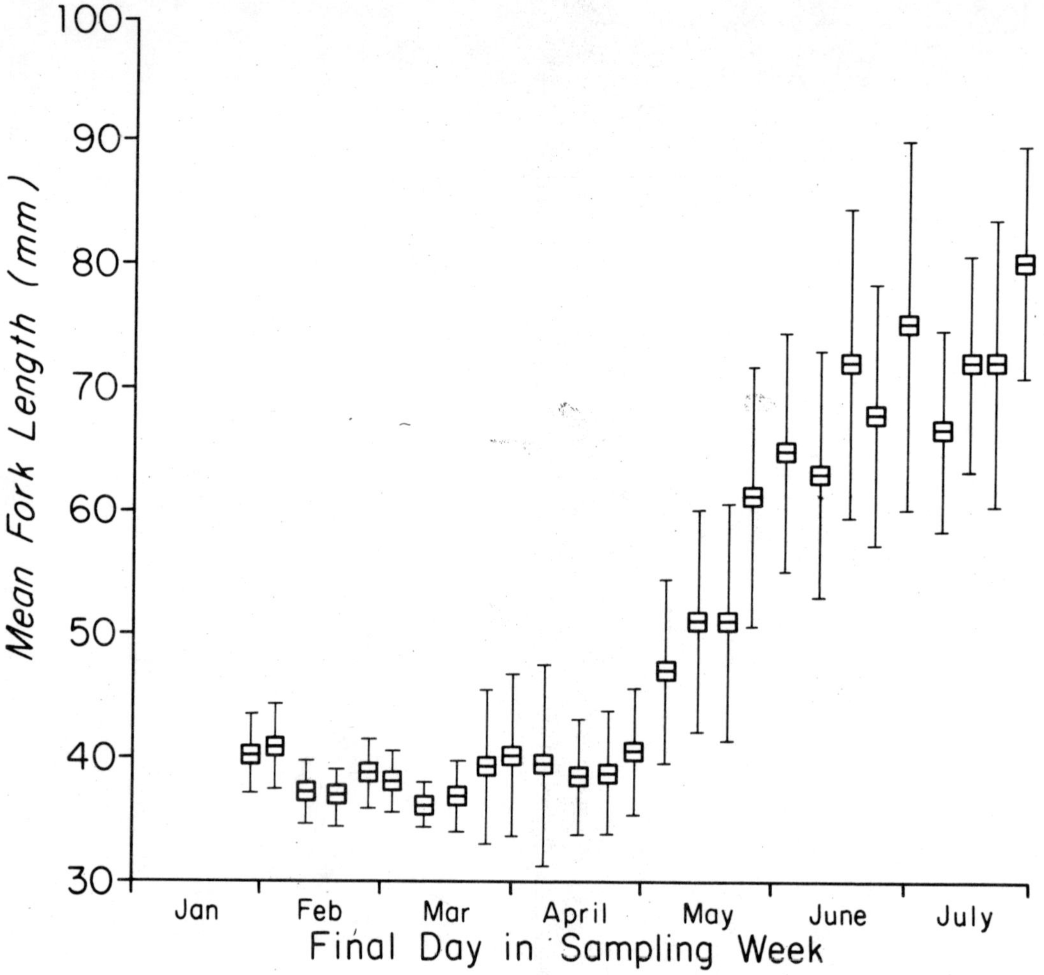

Figure 12. Weekly mean fork length and standard deviation of chum fry caught with the 37 m beach seine from 21 January to 28 July 1977 in Hood Canal, Washington.

Between 30 December 1976 and 22 July 1977, the abundance of epibenthic organisms varied between a minimum of 7,625.0 ± 2,322.1/1,000 liters occurring on 17 February 1977 and a maximum of 99,344.0 ± 15,030.4/1,000 liters occurring on 21 April 1977. The average density over the 27-week period was 28,902.0 ± 28,399.6/1,000 liters (Fig. 14). Densities of epibenthic organisms at the three principle sampling sites followed a similar seasonal trend. All showed maximum densities in mid-April to early May. Harpacticoid copepod densities at the three sites tended to be quite variable with Devil's Hole Delta typically having the highest densities. Carlson Point showed more frequent and extreme fluctuations than the other sites. Gammarid amphipod densities also varied considerably between sites, especially at Brown Point. Specific taxa composition for the dominant harpacticoids varied at three sites with different genera or species predominating at each site. Percent composition of gammarids occurring at Brown Point over time illustrated one species, the underscribed *Pontogeneia* sp., to predominate and three other species to vary in their relative importance.

The dominant organism taxa also differed between the three sites. The epibenthic community of Devil's Hole Delta typically had higher densities of veneriform bivalves, general copepods, mysids and crangonid shrimp. Carlson Point had significantly more polychaetes, principally polynoids; prosobranch larvae; unidentified crustacean larvae; barnacle larvae; spaeromatid; valveriferan and epicaridean isopods; gammarid and caprellid amphipods; hippolytid shrimp; and holothurians.

Prey Composition of Juvenile Chum Salmon: Chum salmon captured with the beach seine in shallow, sublittoral habitats had fed predominantly upon harpacticoid copepods and gammarid amphipods. Calanoid copepods, insects, hyperiid amphipods and euphausiids

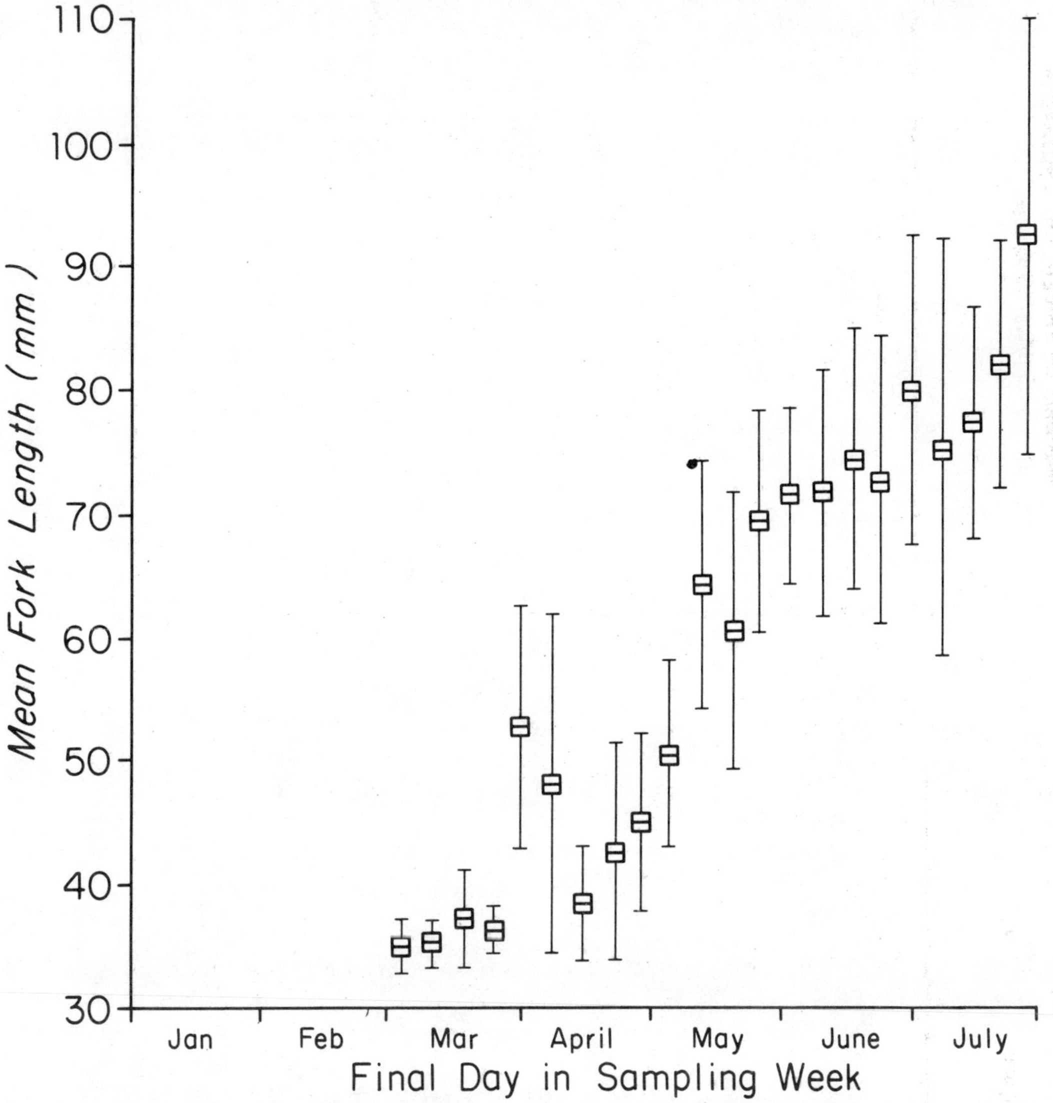

Figure 13. Weekly mean fork length and standard deviation of chum fry caught with the surface townet from 25 February to 28 July 1977 in Hood Canal, Washington.

were of secondary importance. Harpacticoid copepods and gammarid amphipods predominated throughout the out-migration period except during mid-May when a pulse of euphausiids, calanoid copepods and hyperiid amphipods appeared in the prey spectra.

The stomachs of chum fry captured in neritic waters using the townet contained both epibenthic and nearshore prey organisms including harpacticoid copepods; gammarid amphipods; crustacean larvae; and insects. They also contained pelagic forms including euphausiids; calanoid copepods; and hyperiid amphipods. Even though they were captured offshore, chum fry caught in the townet contained mostly epibenthic organisms in their stomachs until early May when prey composition shifted dramatically to pelagic organisms. Euphausiids predominated in June, eventually tapering off with the appearance of calanoid copepods and hyperiid amphipods in late June.

Harpacticoid copepods dominated the Index of Relative Importance prey spectra (IRI) of beach seined chum fry at all sites. This single index combines percent frequency of occurrence and percent numerical and gravimetric composition. Bax et al. (1978) gives a more thorough description of this index. Gammarid amphipods usually ranked second in importance except at Devil's Hole Delta where euphausiids ranked higher; at Marginal Wharf where calanoid copepods and hyperiid amphipods were more important; and at the EHW site where dipteran insects prevailed.

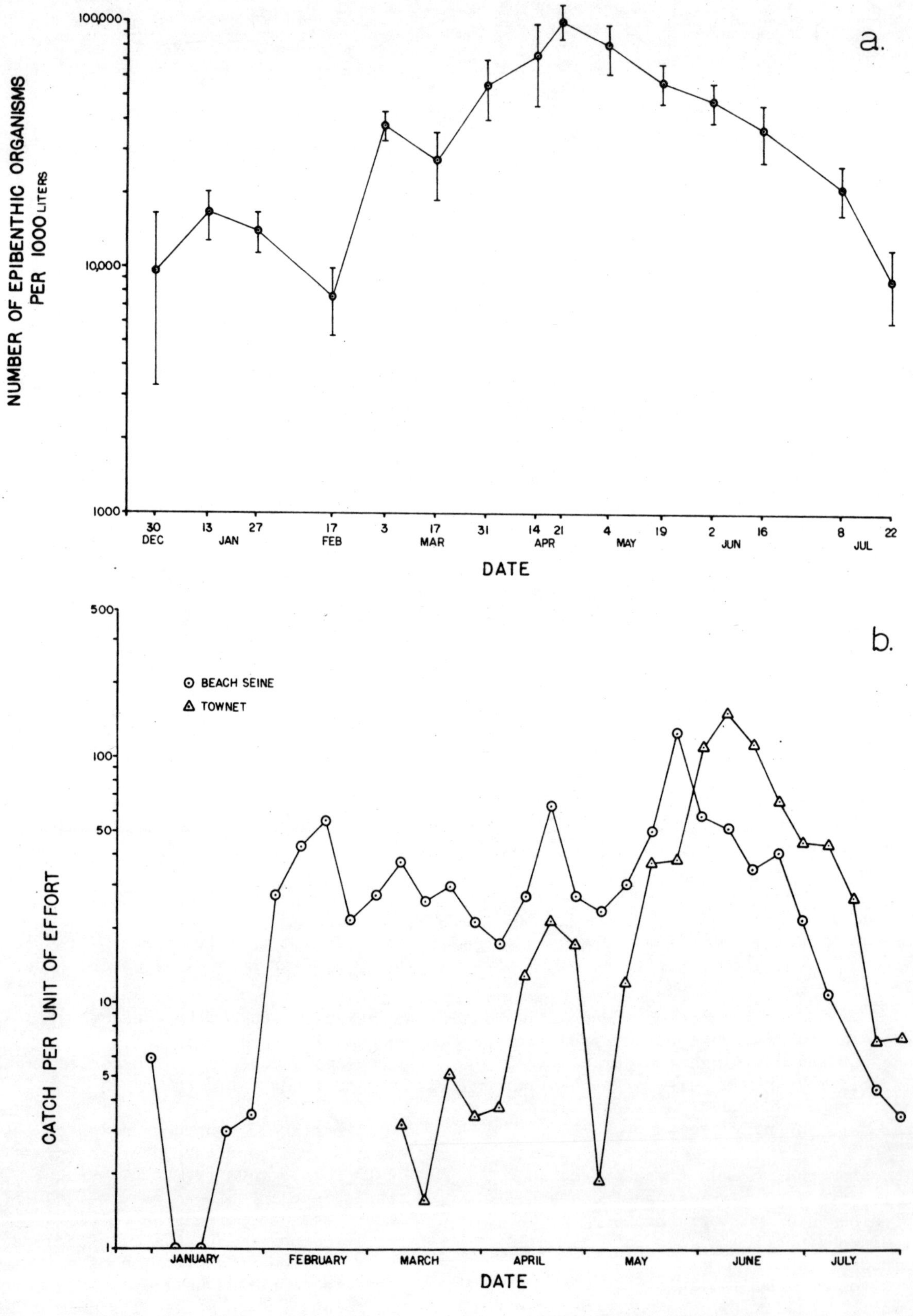

Figure 14. Total mean number of shallow sublittoral epibenthic organisms per 1,000 liters (a) and catch-per-unit-of-effort of chum salmon (b) in Hood Canal, Washington, 1977. Bars represent ± 1 S.D.

Prey composition of chum fry caught in the townet varied more from site to site than beach seined chum (Table 8). Harpacticoid copepods, hyperiid amphipods, euphausiids and calanoid copepods were the predominant prey of chum fry in neritic waters adjacent to Carlson Point. The same prey taxa predominated in chum fry at Devil's Hole Delta Point, but euphausiids were more important. The prey spectra from Brown Point was less diverse than any other site and dominated by euphausiids (87 percent of total IRI). Harpacticoid and calanoid copepods, and hyperiid amphipods were the prevalent organisms composing the prey spectra from Marginal Wharf.

Predators of Juvenile Chum and Pink Salmon: Many workers support the hypothesis that predation in estuarine habitats is a principal component of juvenile pink *(O. gorbuscha)* and chum salmon mortality. Iwamota and Salo (1977) review this hypothesis, especially as it pertains to the potential impact of shoreline docks and bulkheads.

Predation by Other Salmonids: Other salmonids--juvenile and immature coho and chinook salmon, and cutthraot and Dolly Varden trout *(Salvelinus malma)*--have been particularly implicated as predators of pink and chum salmon fry. On the basis of observational data, Sano (1966), Heiser and Finn (1970), Parker (1971), Allen (1974) and Walker (1974) suggested that coho smolts have the potential to significantly prey on chum fry. From stomach content analysis, Stober et al. (1973) suggested that coho smolts prey on chum and pink salmon fry, but the researchers provided no data.

Similarly, there is little indication that juvenile and immature chinook salmon prey on chum and pink salmon fry, except for the incomplete evidence of Stober et al. (1973). Samples from Nisqually Reach (Fresh et al. *in press*) and northern Puget Sound (Miller et al. 1977) provided little or no evidence. However, resident chinook, referred to as blackmouth, were not effectively sampled in most of these studies. They may represent a significant

Table 8. Dominant Prey Organisms, Expressed as Percent of Total IRI, Composing Prey Spectra of Juvenile Chum Salmon at Different Sites in Northern Hood Canal, Washington in 1977.

	Harpacticoid copepods	Gammarid amphipods	Insects	Calanoid copepods	Euphausiids	Hyperiid amphipods	Cumaceans	Shrimp	Barnacle larvae	Decapod larvae
Beach seine										
Carlson Point	67.1	23.6	3.8	1.7	0.5	0.5	0.1	0.1		
Devils Hole Delta	57.5	13.5	2.5	9.6	13.1	2.6	0.1	0.2		
Brown Point	58.5	30.3	3.8	1.7	1.3	0.6	1.7	0.6		
Marginal Wharf	67.3	6.2	3.8	11.4	0.7	8.9	0.0	0.0		
EHW	74.4	7.6	10.7	3.1	0.0	2.7	0.2	1.1		
Spit 6	54.2	43.5	0.4	0.8	0.0	0.0	0.0	0.2		
Floral Point	59.1	32.7	5.2	0.6	0.0	0.3	0.3	0.0		
Townet										
Carlson Point	37.6	5.9	3.0	10.6	12.6	21.0	0.4	0.1	2.4	0.4
Devils Hole Delta	11.9	2.7	2.6	19.5	38.2	19.6	0.0	0.0	0.1	1.0
Brown Point	5.7	2.9	0.8	1.2	87.0	1.2	0.1	0.1	0.3	0.0
Marginal Wharf	29.8	7.9	2.9	27.6	5.5	18.5	0.0	0.1	0.4	0.6

mortality factor, especially considering the WDF delayed release programs designed to expand these stocks.

During our sampling, Juvenile coho and chinook salmon were often caught in association with the out-migrating chum fry. Juvenile pink salmon were not common because a significant adult return to Hood Canal does not occur in even-numbered years. Sea-run cutthroat and rainbow trout (steelhead) were also caught in the beach seines and townet. In order to determine the significance of their predation on chum fry, stomachs from specimens longer than 100 mm were analyzed.

Juvenile coho (length $\bar{x}$ = 122.4 mm) fed specifically on brachyuran crab larvae and euphausiids. The latter comprised over 80 percent of the total number of prey in the sample. Larval fish, gammarid and hyperiid amphipods and calanoid copepods were also common prey items. However, they were not numerically important.

Juvenile chinook salmon (length $\bar{x}$ = 267.0 mm), including immature resident blackmouth, most frequently fed on shrimp larvae, insects and juvenile Pacific herring *(Clupea harengus pallasi)*. However, brachyuran crab larvae and juvenile Pacific sand lance larvae *(Ammodytes hexapterus)* comprised greater percentages of the total prey numbers. Two chum fry were found in the stomach of one juvenile chinook, but comprised only 0.43 percent of the total number of prey items. Our Hood Canal data showed only minimal evidence (1.7 percent total IRI) of chinook predation on other juvenile salmonids.

Sea-run cutthroat trout have also been implicated as potential predators of chum fry (Heiser and Finn 1970). Our Hood Canal sampling illustrated this. This species (length $\bar{x}$ = 327.5 mm) fed predominantly on gammarid amphipods, which alone comprised over 55 percent of the total number of prey organisms. Spaeromatid isopods *(Gnorimosphaeroma oregonensis* and *Exosphaeroma media)*, juvenile fish including several chum and other salmon fry, callianassid shrimp, cumaceans and Pacific sand lance were common, but not abundant, in the examined stomach contents. Salmon fry were fourth in frequency of occurrence of all prey taxa; however, fish overall constituted only 11.4 percent of the total IRI prey spectra.

The most common prey organisms of steelhead smolts (length $\bar{x}$ = 166.7 mm) included crab larvae; various insect taxa (scolioidea, diptera and coleoptera); gammarid amphipods; euphausiids; ostracods (Myodocopa); calanoid copepods; and crab larvae. Other ostracods and dipteran insects were numerically predominant. Only one steelhead smolt stomach contained fish remains. Those remains consisted of 15 unidentified fish larvae.

Predation By Species Other than Salmonids: Other investigators have suggested that marine fish other than salmonids potentially prey on pink and chum salmon fry. These species include numerous cottid, or sculpin, species (Heiser and Finn 1970, Beall 1972, Simenstad 1976), and walleye pollock *(Theragra chalcogramma)* (Armstrong and Winslow 1968).

During our sampling of out-migrating salmonids in the nearshore region of northern Hood Canal, we caught numerous nonsalmonid species that we considered as potential predators of juvenile chum salmon. Stomach contents of 13 species were examined. These species included spiny dogfish *(Squalus acanthias)*; Pacific herring; Pacific hake *(Merluccius productus)*: Pacific tomcod *(Microgadus proximus)*; Pacific cod *(Gadus macrocephalus)*; whitespotted greenling *(Hexagrammos stelleri)*; buffalo sculpin *(Enophrys bison)*; great sculpin *(Myoxocephalus polyacanthocephalus)*; cabezon *(Scorpaenichthys marmoratus)*; staghorn sculpin *(Leptocottus armatus)*; striped seaperch *(Embiotoca lateralis)*; shiner perch *(Cymatogaster aggregata)*; and starry flounder *(Platichthys stellatus)*.

The examinations revealed that none of the pelagic or epibenthic plankton-feeding species preyed on juvenile salmonids. Pacific herring prey predominantly on calanoid copepods and the euphausiid *Euphausia pacifica*. More facultative plankton feeders such as spiny dogfish, Pacific hake and Pacific tomcod utilized both pelagic and epibenthic plankton such as euphausiids, mysids and gammarid amphipods. The true benthic feeding fishes did not indicate any piscivorous food habits either. Starry flounder preyed on bivalves and polychaetes while cabezon and buffalo sculpin ate a variety of benthic crab species.

Facultative benthivores, however, were quite omnivorous. Fishes were found in the stomach contents of several species, and salmonids found in Pacific cod and Pacific staghorn sculpin. Pacific cod preyed principally on gammarid amphipods, but we found one chum fry in one of two stomachs that were examined. The sample size was too small to determine if this species preys significantly on chum fry.

Two of 12 Pacific staghorn sculpin stomachs we examined contained eight chum fry.

The fry comprised 4 percent of the prey items with gammarid amphipods *(Ansiogammarus pugettensis)*, euphausiids *(E. pacifica)*, unidentified crabs and fish eggs being the predominant food items. Stomach analysis of this species sampled from Nisqually Reach (Fresh et al. *in press*), northern Puget Sound (Miller at al. 1977) and the Strait of Juan de Fuca (Simenstad 1977) also did not indicate if Pacific staghorn sculpin preys significantly on juvenile salmonids.

To adequately assess the trophic role of neritic predators, as well as that of the deeper-occurring salmonids, more studies must be implemented. New sampling techniques, such as purse seining and midwater trawling, may be required to sample effectively.

Results of stomach examinations also provided interesting insights into the food habits of many species that did not prey on salmon. Three whitespotted greenling stomachs contained mostly gammarid and caprellid amphipods, and cragonid shrimp. Great sculpins fed on caligoid copepods and gammarid amphipods *(A. confervicolus)*. One striped seaperch stomach contained 1,120 gammarid amphipods while the other embiotocid, shiner perch, fed on a diverse array of isopods *(Exosphaeroma media)*, bivalves *(Mytilus edulis)*, gastropods *(Littorina scutulata)* and gammarid amphipods.

Overlap Between Epibenthic Plankton Community and Chum Prey: We measured the positive electivity of chum fry occurring in shallow sublittoral habitats for epibenthic organisms. Ivlev's coefficient, E (Fig. 15), was used. Positive electivity appears to shift from crustacean larvae, juvenile shrimp and calanoid copepods early in the out-migration period to gammarid amphipods, harpacticoid copepods, euphausiids, crustacean eggs and hyperiid amphipods during the peak out-migration period. Electivity then shifts to calanoid copepods and hyperiid amphipods as the migration ends. Insects always had positive E values primarily because the plankton pump did not sample them well.

Size selection of harpacticoid copepods and gammarid amphipods was quite apparent (Fig. 16). For example, the sizes (metasome lengths) of harpacticoid copepods characterizing the Brown Point epibenthic plankton community ranged from 0.817 ± 0.812 mm to 0.628 ± 0.144 mm, averaging 0.716 ± 0.174 mm through the 1977 out-migration period. Harpacticoids sampled from the stomach contents of juvenile chum caught in the nearshore region adjacent to the plankton pumping site ranged in size from 1.499 ± 0.132 mm to 0.781 ± 0.182 mm. In many instances the upper distributions of the harpacticoids that chum fry consumed were completely out of range of those that the plankton pump sampled. In addition, the mean harpacticoid sizes were typically larger in the stomach contents of chum fry caught with the townet in neritic waters compared to those chum fry caught in shallow sublittoral habitats with the beach seine (Fig. 17). This may also reflect size-selective predation as a function of predator size since chum caught in the townet were generally larger than those caught with seines.

There was some indication, though far from conclusive, that the intense pressures of size-specific predation depressed the mean size distributions of epibenthic harpacticoids during the peak out-migration period (Bax et al. 1978).

Chum fry migrating through the shallow sublittoral zones of different habitats in northern Hood Canal exhibited few differences in prey composition. Harpacticoid copepods appear to be the preferred food organism in all cases, accounting for 53-74 percent of the total IRI prey spectra. Harpacticoids contributed most to the diets of chum fry collected north of Devil's Hole Delta. This suggests that the shallow delta region, with sandy substrate and abundant eelgrass, may host the maximum concentrations of the prey resource. The epibenthic pump data supports this. Conversely, gammarid amphipods appear to be most common in the diets of chum fry collected from exposed spits which typically have clean gravel and coarse sand substrates in the shallow sublittoral zone (Brown Point, Floral Point and Spit Six).

SUMMARY

Our sampling indicated that chum salmon out-migrants were the most abundant of the salmonid species, with a minor peak of wild chum stock in early February and a major peak of both wild and hatchery stock from mid-May to mid-July. Coho salmon appeared in late April, with a minor peak in mid-May associated with the Big Beef Creek stock. A major peak of wild and hatchery stock occurred in mid-July. Chinook salmon were caught throughout the sampling period, with a small peak of hatchery and wild stock in May. Cutthroat trout were caught over the entire sampling period, with a peak in early June.

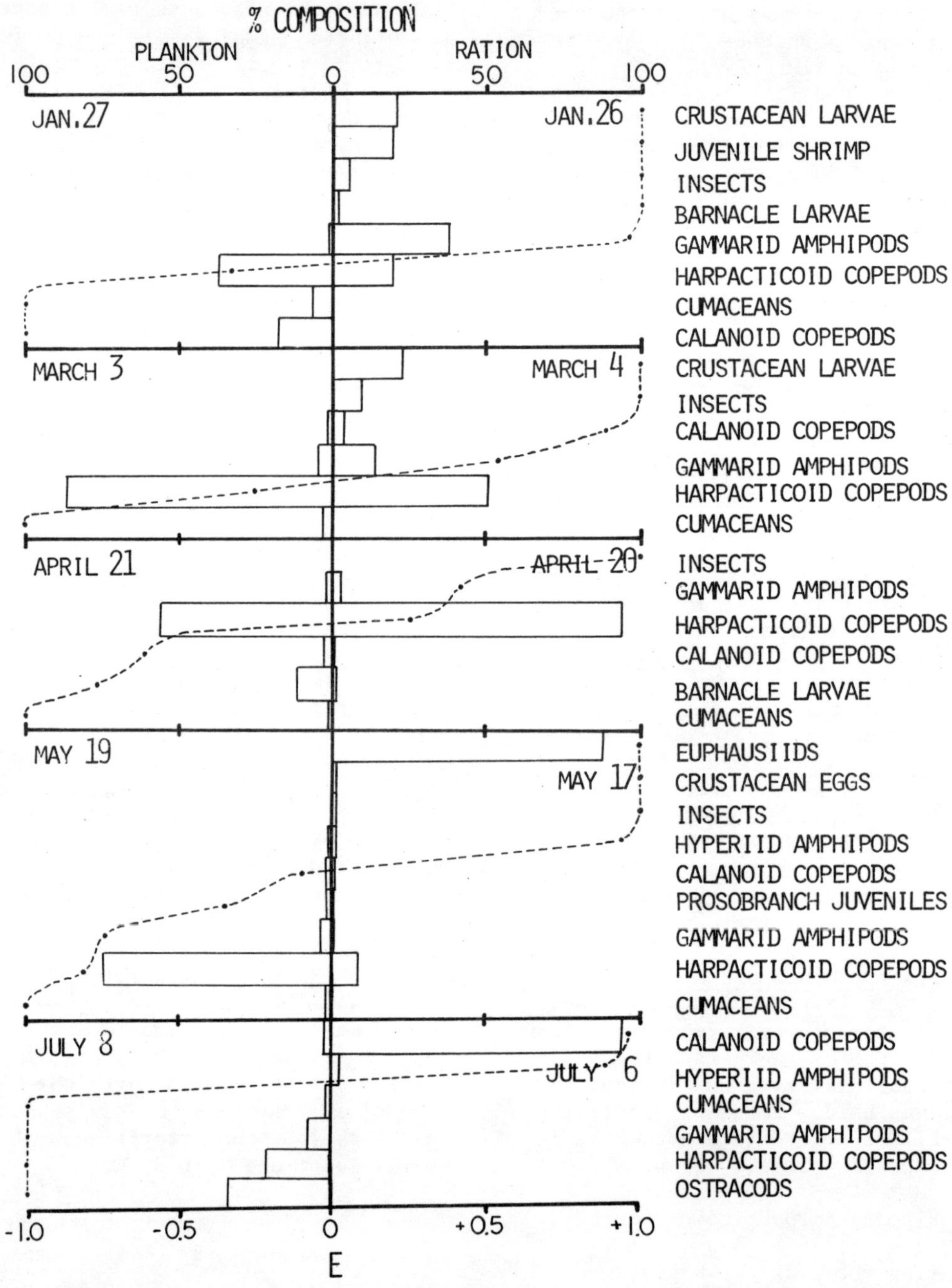

Figure 15. Electivity curves for epibenthic plankton sample composition and ration composition of juvenile chum salmon, Hood Canal, Washington, 1977.

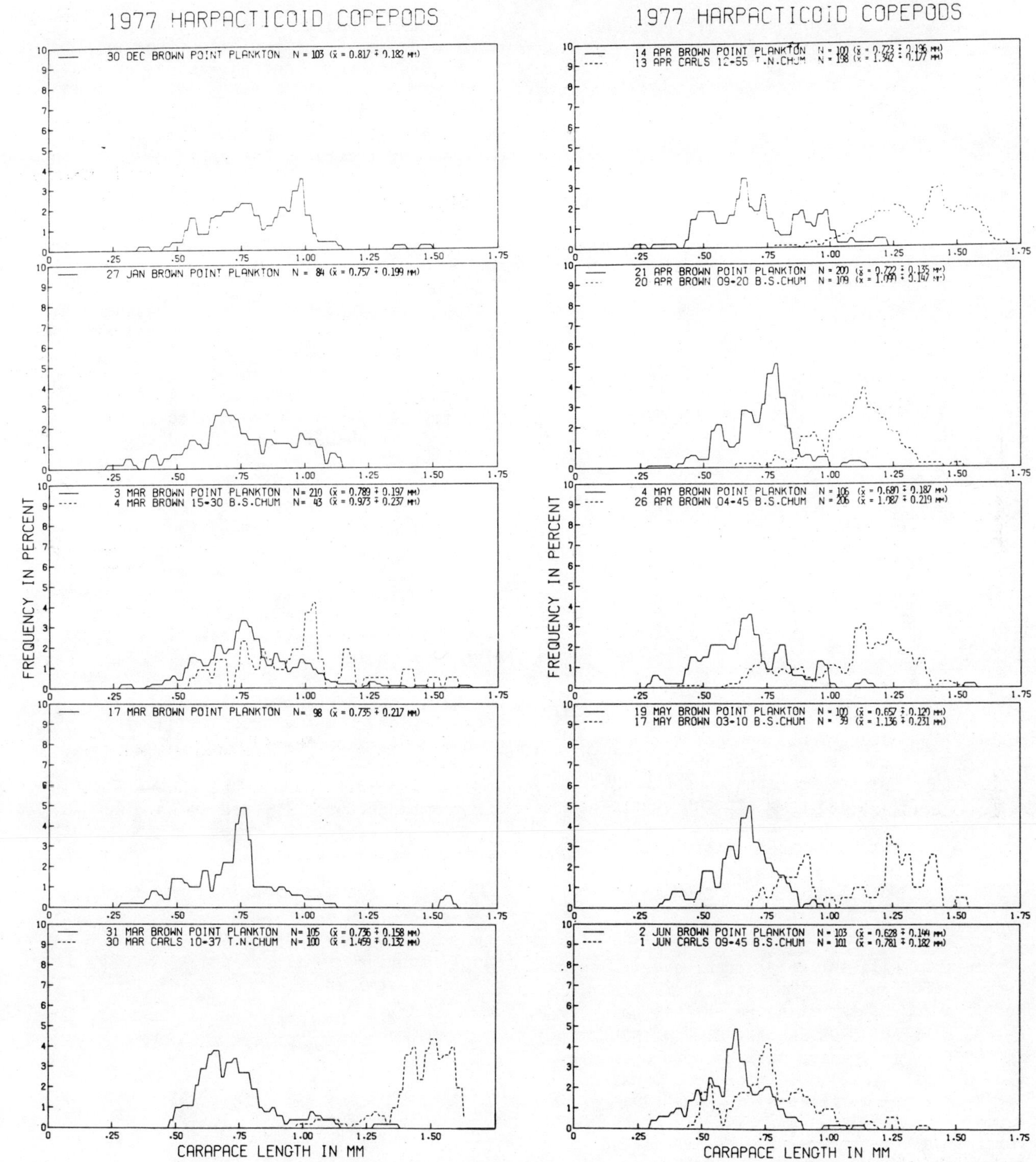

Figure 16. Harpacticoid copepod size (metasome length) distributions from epibenthic plankton community (solid line) and in stomach contents of juvenile chum salmon (dashed line) during 1977 out-migration period in Hood Canal, Washington.

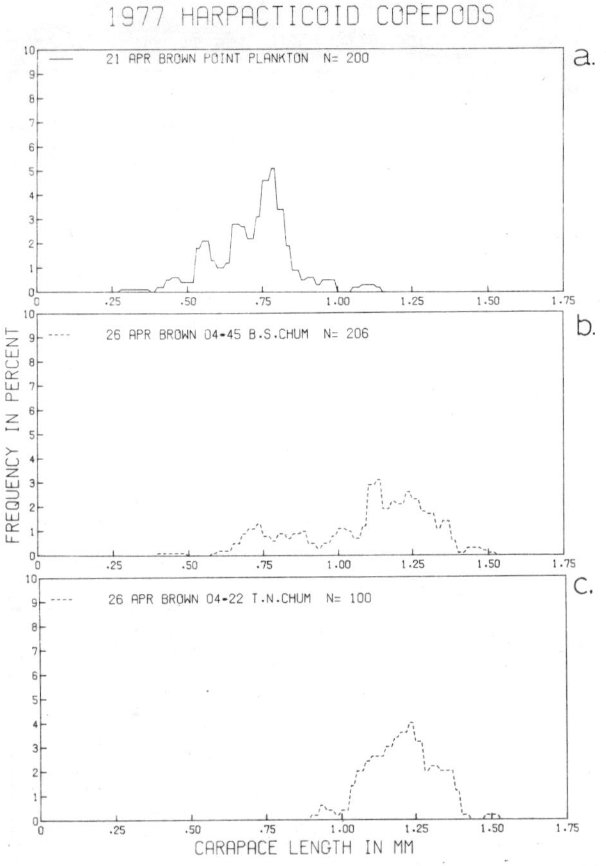

Figure 17. Harpacticoid copepod size (metasome length) distributions from epibenthic plankton community (a), and from stomach contents of juvenile chum salmon caught in shallow sublittoral (b) and neritic (c) environments in Hood Canal, Washington, late April 1977.

Some significant relationships were found between CPUE and environmental variables. And from 1977 data, it appeared that hatchery releases affect CPUE as they did in 1975 and 1976. Preliminary mark-recapture studies showed that fry released from Hood Canal and Big Beef Creek hatcheries are caught at Bangor Annex (Whitmus and Olson *in press*). Further mark-recapture studies are needed to quantify the contribution of the hatcheries to the CPUE at Bangor Annex.

The sampling results also indicated that more salmonids were caught on the west shoreline in 1977 than in the previous two years. This was especially evident in the early wild stocks. The majority of later wild and hatchery stocks was found on the east shore. We also did not notice a concentration of salmonids around piers in 1977.

The mean condition of fry that we sampled decreased at the end of the season. A decrease in mean lengths was not evident.

Our sampling of epibenthic organisms with the pump in the shallow sublittoral environment indicated the presence of harpacticoid copepods, gammarid amphipods, crustacean eggs, ostracods, calanoid copepods, isopods, nematodes, barnacle and prosobranch larvae, juvenile shrimp and cumaceans as the most prevalent organisms.

By analyzing chum fry stomach contents, we found that fry sampled in inshore waters fed predominantly on harpacticoid copepods, gammarid amphipods, calanoid copepods, insects, hyperiid amphipods and euphausiids. Chum fry sampled offshore fed on gammarid amphipods, euphausiids, calanoid copepods, crustacean larvae, hyperiid amphipods and insects. Chum fry exhibited both taxonomic and size feeding selectivity. During the early period of the out-migration, the abundance of harpacticoid copepods appeared depressed, and there was some evidence that size-specific predation may have depressed the mean size distributions of epibenthic harpacticoids during the peak of the out-migration period.

We found no evidence of significant predation on chum fry by other salmonids or marine fish species. Only chinook salmon, sea-run cutthroat trout, Pacific cod and staghorn sculpin had chum fry in their stomach contents, and in no case did chum fry constitute an important component of the overall diet.

The studies will continue in 1979 with the expectation that the numbers of hatchery produced chum salmon will be significantly increased--perhaps to 70 to 100 million juveniles.

ACKNOWLEDGMENTS

This work was sponsored by the U.S. Navy, the Washington Sea Grant Program under the National Oceanic and Atmospheric Administration, U.S. Department of Commerce, and the Washington State Department of Fisheries.

LITERATURE CITED

Allen, B. 1974. Early marine life history of Big Qualicum River chum salmon. Pages 137-148 *in* Hargind, D. R., (ed.), Proceedings of 1974 Northeast Pacific pink and chum salmon workshop. Dep. Environ., Fish., Vancouver, B.C.

Armstrong, R. H., and P. C. Winslow. 1968. An incidence of walleye pollock predation on salmon young. Trans. Amer. Fish. Soc. 97(2):202-203.

Bakshtansky, E. L. 1969. The impact of the environmental factors on survival of the far eastern young salmon during the acclimatization of the latter in the northwest part of USSR. Int. Comm. N.W. Atlantic Fish. Spec. Publ. 6:477-480.

Bax, N. J., E. O. Salo, B. P. Snyder, C. A. Simenstad, and W. J. Kinney. 1978. Salmonid out-migration studies in Hood Canal. Univ. Wash. Fish. Res. Inst. Final Report, Phase III. FRI-UW-7819. 128 pp.

Beall, E. P. 1972. The use of predator-prey tests to assess the quality of chum salmon, *Oncorhynchus keta*, fry. M.S. Thesis. Univ. of Wash., Seattle. 105 pp.

Brooks, J. L., and S. L. Dodson. 1965. Predation, body size, and composition of plankton. Sci. 150:28-35.

Burgner, R. L., K. K. Chew, J. S. Isakson, O. A. Mathisen, P. A. Lebednik, R. E. Norris, C. E. O'Clair, M. M. Peck, C. A. Simenstad, P. N. Slattery, and G. J. Tutmark. 1969. Research program on marine ecology and oceanography, Amchitka Island, Alaska. Annual progress report for July 1, 1968-June 30, 1969. Fish. Res. Inst., Univ. Wash. Battelle Mem. Inst. BMI-171-128.

Confer, J. L., and P. I. Blades. 1975. Omnivorous zooplankton and planktivorous fish. Limnol. Oceanogr. 20(4):571-579.

Feller, R. J. 1977. Life history and production of meiobenthic harpacticoid copepods in Puget Sound. Ph.D. Dissertation. Dep. Oceanogr., Univ. of Wash., Seattle. 249 pp.

Feller, R. J., and V. W. Kaczynski. 1975. Size selective predation by juvenile chum salmon *(Oncorhynchus keta)* on epibenthic prey in Puget Sound. J. Fish. Res. Bd. Can. 32(8):1419-1429.

Fiscus, G. 1969. 1968 Admiralty Inlet chum salmon tagging. Wash. Dep. Fish. Annual Report 78:13-19.

Foskett, D. R. 1951. Young salmon in the Nanaimo area. J. Fish. Res. Bd. Can. Pac. Prog. Rep. 86:18-19.

Fresh, K. L., C. A. Simenstad, D. Rabin, and E. O. Salo. *In press*. DuPont-Nisqually salmonid and non-salmonid resources. Univ. Wash. Fish. Res. Inst.

Galbraith, M. G., Jr. 1967. Size-selective predation on *Daphnia* by rainbow trout and yellow perch. Trans. Amer. Fish. Soc. 96(1):1-10.

Gerke, R. J., and V. W. Kaczynski. 1972. Food of juvenile pink and chum salmon in Puget Sound, Washington. Wash. Dep. Fish. Tech. Rep. 10. 27 pp.

Gilhousen, P. 1962. Marine factors affecting survival of Fraser River pink salmon. Pages 105-109 *in* Wilimovsky, N. J. (ed.), Symposium on pink salmon, H. R. MacMillan lectures in fish. Univ. British Columbia, Vancouver.

Harris, C. K., and A. C. Hartt. 1977. Assessment of pelagic and nearshore fish in three bays on the east and south coasts of Kodiak Island, Alaska. Univ. Wash. Fish. Res. Inst. Final Report. FRI-UW-7719. 190 pp.

Healey, M. C., R. J. LeBrasseur, J. R. Sibert, W. E. Barraclough, and J. C. Mason. 1976. Ecology of young salmon in Georgia Strait. Pages 201-207 *in* Gunstrom, G. K. (ed.), Proceedings of 1976 Northeast Pacific pink and chum salmon workshop. Alaska Dep. Fish Game, Juneau.

Heiser, D. W., and E. L. Finn, Jr. 1970. Observations of juvenile chum and pink salmon in marine and bulkheaded areas. Wash. Dep. Fish., Mgmt. Resource Div., Supple. Prog. Rep. Puget Sound stream studies. 28 pp.

Hurley, D. A., and W. L. Woodall. 1968. Responses of young pink salmon to vertical temperature and salinity gradients. Int. Pac. Salmon Fish. Comm. Prog. Rep. 19. 80 pp.

Iwamoto, R. N., and E. O. Salo. 1977. Estuarine survival of juvenile salmonids: a review of the literature. Unpublished report to Wash. Dep. Fish., Contract No. 807, Univ. Wash. Fish. Res. Inst., Seattle. 64 pp.

Jones, B. C., and G. H. Geen. 1977. Food and feeding of spiny dogfish in British Columbia waters. J. Fish. Res. Bd. Can. 34(1):2067-2078.

Kaczynski, V. W., R. J. Feller, J. Clayton, and R. J. Gerke. 1973. Trophic analyses of juvenile pink and chum salmon in Puget Sound. J. Fish. Res. Bd. Can. 30(7):1003-1008.

Kislalioglu, W., and R. N. Gibson. 1975. Field and laboratory observations on prey-size selection in *Spinachia spinachia* (L.). Pages 29-40 *in* Barnes, H. (ed.), Proceedings of 9th European marine biology symposium. Aberdeen, Scotland Univ. Press.

Koski, K. V. 1975. The survival and fitness of two stocks of chum salmon *(Oncorhynchus keta)* from egg deposition to emergence in a controlled-stream environment at Big Beef Creek. Ph.D. Dissertation. Univ. of Wash., Seattle. 212 pp.

LeBrasseur, R. J. 1969. Growth of juvenile chum salmon *(Oncorhynchus keta)* under different feeding regimes. J. Fish. Res. Bd. Can. 26:1631-1645.

Manzer, J. I., and M. P. Shepard. 1962. Marine survival, distribution and migration of pink salmon off the British-Alaska coast. Pages 113-121 *in* Wilimovsky, N. J. (ed.), Symposium on pink salmon, H. R. MacMillan lectures in fish. Univ. British Columbia, Vancouver, B.C.

Martin, J. W. 1966. Early sea life of pink salmon. Pages 111-124 *in* Sheridan, W. L. (ed.), Proceedings of 1966 Northeast Pacific pink salmon workshop. Alaska Dep. Fish Game Inform. Leafl. 87.

Mason, J. C. 1974. Behavioral ecology of chum salmon fry *(Oncorhynchus keta)* in a small estuary. J. Fish. Res. Bd. Can. 31:83-92.

Miller, B. S., C. A. Simenstad, L. L. Moulton, W. A. Karp, K. L. Fresh, F. C. Funk, and S. F. Borton. 1977. Puget Sound baseline program: Nearshore fish survey. Univ. Wash. Fish. Res. Inst. Final Report. FRI-UW-7710. 220 pp. (Also State of Washington Dep. Ecol. Baseline Study Rep. No. 10.)

Morrill, C. 1974. Pink and chum salmon predation studies. Wash. Dep. Fish. Mgmt. Res. Div., Comp. Rep., May 1974. 28 pp.

Okada S., and A. Taniguchi. 1971. Size relationship between salmon juveniles in shore waters and their prey animals. Bull. Pac. Fish., Hokkaido Univ. 22:30-36.

Parker, R. R. 1965. Estimation of sea mortality rates for the 1961 brook-year pink salmon of the Bella Coola area, British Columbia. J. Fish. Res. Bd. Can. 22(6):1523-1531.

Parker, R. R. 1968. Marine mortality schedules of pink salmon of the Bella Coola River, central British Columbia. J. Fish. Res. Bd. Can. 25:757-794.

Parker, R. R. 1971. Size selective predation among juvenile salmonid fishes in a British Columbia inlet. J. Fish. Res. Bd. Can. 28(10):1503-1510.

Pyke, C. H., H. R. Pulliam, and E. I. Charnov. 1977. Optimal foraging - A selective review of theory and tests. Quart. Rev. Biol. 52(2):137-154.

Ricker, W. E. (ed.). 1968. Methods for assessment of fish production in fresh waters. Blackwell Sci. Publ., Oxford. 313 pp.

Salo, E. O., T. E. Prinslow, R. Campbell, and B. P. Snyder (eds.). *In press*. The effects of dredging on migrations of juvenile salmonids at Bangor, Washington, 1977. Univ. Wash. Fish. Res. Inst. Final Report to U.S. Dep. Navy.

Sano, S. 1966. Salmon of the North Pacific Ocean - Part III. A review of the life history of North Pacific salmon. 3. Chum salmon in the Far East. Int. North. Pac. Fish. Comm. Bull. 10:41-57.

Schreiner, J. U. 1977. Salmonid out-migration studies in Hood Canal, Washington. M.S. Thesis. Univ. of Wash., Seattle. 111 pp.

Schreiner, J. U., E. O. Salo, B. P. Snyder, and C. A. Simenstad. 1977. Salmonid out-migration studies in Hood Canal. Univ. Wash. Fish. Res. Inst. Final Report, Phase II. FRI-UW-7715. 64 pp.

Shepard, M. P. 1948. Responses of young chum salmon, *Oncorhynchus keta* (Walbaum), to changes in the sea water of the environment. M.A. Thesis. Univ. of British Columbia, Vancouver, B.C.

Sibert, J., T. J. Brown, M.C. Healey, B. A. Kask, and R. J. Naiman. 1977. Detritus-based food webs: Exploitation by juvenile chum salmon. Sci. 196:649-650.

Simenstad, C. A. 1976. Trophic relations of juvenile chum salmon and associated salmonids in nearshore environments of northern Puget Sound. Page 186 *in* Gunstrom, G. K. (ed.), Proceedings of 1976 Northeast Pacific pink and chum salmon workshop. Alaska Dep. Fish Game, Juneau.

Simenstad, C. A. 1977. Prey organisms and prey community composition of juvenile salmonids in Hood Canal, Washington Pages 163-176 *in* Simenstad, C. A., and S. J. Lipovsky (eds.), Proceedings of 1st Pacific Northwest technology workshop on fish food habits studies, Astoria, Oregon, October 13-15, 1976. Wash. Sea Grant. WSG-W077-2.

Stober, Q. J., S. J. Walden, and D. T. Griggs. 1973. Juvenile salmonid migration through Skagit Bay. Pages 35-70 *in* Stober, Q. J., and E. O. Salo (eds.), Ecological studies of the proposed Kiket Island nuclear power plant site. Univ. Wash. Fish. Res. Inst. Final Report. FRI-UW-7304.

Terry, C. 1977. Stomach analysis methodology: Still lots of questions. Pages 87-92 *in* Simenstad, C.A., and S. J. Lipovsky (eds.), Proceedings of 1st Pacific Northwest technology workshop on fish food habits studies, Astoria, Oregon, October 13-15, 1976. Wash. Sea Grant. WSG-W077-2.

Vernon, E. H. 1958. An examination of factors affecting the abundance of pink salmon in the Fraser River. Int. Pac. Salmon Comm., Prog. Rep.

Walker, J. H. C. 1974. Mechanics of size selected predation by coho smolts on pink and chum salmon fry. Pages 114-120 *in* Harding, D. R. (ed.), Proceedings of 1974 Northeast Pacific pink and chum salmon workshop. Dep. Environ., Fish., Vancouver, B.C.

Werner, E. E., and D. J. Hau. 1974. Optimal foraging and the size selection of prey by the bluegill sun fish *(Lepomis macrochirus)*. Ecol. 55:1042-1052.

Whitmus, C. J., and S. Olsen. *In press*. The 1977 migratory behavior of juvenile chum salmon released from the Hood Canal Hatchery at Hoodsport, Washington. Univ. Wash. Fish. Res. Inst.

Wickett, W. P. 1958. Review of certain environmental factors affecting the production of pink and chum salmon. J. Fish. Res. Bd. Can. 15(5):1103-1126.

Wong, B., and F. J. Ward. 1972. Size selection of *Daphnia pulicaria* by yellow perch *(Perca flavescens)* fry in west Blue Lake, Manitoba. J. Fish. Res. Bd. Can. 29(12):1761-1764.

The Ecology of Juvenile Salmon in Georgia Strait, British Columbia

M. C. Healey

INTRODUCTION

We understand less about the first few months that Pacific salmon *(Oncorhynchus* sp.) spend in the ocean than all other phases of their life history. Mortality during these first few months at sea may be heavy and possibly critical to the numbers of returning adults (Parker 1965, 1968).

Recent attempts to forecast stock abundance in British Columbia corroborate the hypothesis that events soon after salmon enter the sea determine adult returns of some species, but the critical period appears to occur after the initial heavy mortality (personal communication, 1978, W. P. Wickett, Fisheries and Oceans Canada, Nanaimo; personal communication, 1978, D. Blackbourn, International Pacific Salmon Fisheries Commission, New Westminster). We can only speculate on the causes of this mortality, but they can be considered intimately related to growth rate and presumably reflect the carrying capacity of ocean habitats in some way.

All major salmon producing nations are currently enhancing salmon stocks. These attempts to increase production by releasing large numbers of fry or smolts may result in local or general overexploitation of food resources for salmon, or elicit increased predation in the marine environment (Peterman and Gatto 1978). As a result, there may be no net gain in production. Understanding the ecological relationships affecting young salmon soon after they enter the sea, therefore, is essential to fully understanding variations in stock abundance and assessing the ocean's capacity to produce more salmon.

In this report, I shall review information on the distribution, abundance, growth rates and feeding habits of juvenile salmon in Georgia Strait, an important marine nursery area in southern British Columbia. Young salmon enter the strait in spring and early summer. They spend a few weeks to a few months exploiting the food resources there and then leave for oceanic feeding grounds. In reviewing the available information about the ecology of juvenile salmon during their brief residence in the strait I shall try to refute the hypothesis that the major movement patterns of the fish are in response to declining food supplies. If I cannot refute the hypothesis, then food limitation shortly after juvenile salmon enter the sea may limit the potential to increase stocks by artificial enhancement.

INFORMATION SOURCES AND METHODS

Information collected by the scientific staff at the Pacific Biological Station in Nanaimo, British Columbia, and my own previously unpublished data, form the foundation of this report. I will use information from other sources as well. My principal information sources were as follows: Between April and July 1966-1969 the dispersal of juvenile salmon in the Fraser River Plume and the southern Gulf Islands was investigated (Barraclough and Phillips 1963, Phillips and Barraclough 1978, Barraclough and Fulton 1978, Barraclough et al. 1968). A two-boat trawl, 6.1 m wide and 3 m deep towed at approximately 5.5 km/hr was used for sampling. Because of the room needed to tow the trawl, these samples represent mainly deeper water some distance from shore.

In June 1966, and from September to December 1972 and 1973, bays and inlets on both sides of southern Georgia Strait were surveyed by a fine mesh purse seine

(Godfrey 1968, Godfrey and Ball 1974). The purse seine, approximately 480 m long and 48 m deep, was fished in water depths of 50 m or greater at some distance from shore. Sets were made in locations where juvenile salmon, especially coho *(O. kisutch)* and chinook *(O. tshawytscha)* were expected to be captured.

In July and August 1974, the Gulf Islands and the open waters of southern Georgia Strait were surveyed by this same purse seine (Beamish et al. 1976). This survey was extended to cover the entire strait from late July to September 1975 and from August to September 1976 (Healey 1978). The Gulf Islands region was also surveyed monthly from May to October 1976 by a slightly smaller purse seine approximately 360 m long by 36 m deep (Healey 1978). During these latter surveys, seine sets were made as close as possible to a predetermined sample location without reference to the possible presence of fish in the area.

A more detailed examination of the distribution and abundance of juvenile salmon in the Nanaimo area was undertaken from 1975 to 1977. Beach seines and a variety of small mesh purse seines were used to sample several depth zones from very shallow water near shore through the intermediate depths. The deeper water was sampled with the large purse seine and the trawl described previously. Downstream migrant chum fry *(O. keta)* in the Nanaimo River and chinook juveniles on the Nanaimo estuary were marked during these studies. The marked fish provided estimates of dispersion to various available nursery areas and early marine growth (Healey et al. 1977*a*, 1977*b*; Healey 1979, 1980).

Much of the available information on the life of salmon soon after they enter the sea is only semiquantitative. Abundance data are generally in the form of catch-per-set values, and comparisons between gear are impossible. Furthermore, catch frequencies generally do not correspond to any simple probability function, making comparisons within a gear type difficult. I shall present both arithmetic average catch values and range in catch in the tables, but these should be viewed as indices of abundance only.

Fish size and growth rate are particularly relevant to any estimate of feeding rate or food requirements. Few direct estimates of early marine growth exist. Consequently, I have inferred growth rates from temporal changes in the size of unmarked fish. LeBrasseur and Parker (1964) described difficulties in estimating growth this way.

Several studies that I cite present detailed data on diet. I will not review these data thoroughly, but will use them to describe food habits during different stages in the life of juvenile salmon in Georgia Strait. From these descriptions, I shall infer feeding preferences and variations in the availability of certain food items.

STUDY AREA

Georgia Strait, a partly enclosed body of water, lies between the southern end of Vancouver Island and the mainland shore of British Columbia (Fig. 1). It connects with the open Pacific Ocean through Juan de Fuca Strait in the south, and Johnstone and Queen Charlotte straits in the north. Approximately 220 km long by 33 km wide, the strait has a surface area of 6,900 km^2 (Waldichuk 1957). Tidal pulses entering through Juan de Fuca and Johnstone straits, winds, and the Fraser River discharge drive surface currents in the strait, which could influence the distribution of young salmon. In general, the open waters of the strait circulate counterclockwise, but local current patterns are complex and not easily predicted (Waldichuk 1957, 1958; Tully and Dodimead 1957, Giovando 1973, Giovando and Tabata 1970).

Biological productivity is greatest in the southern half of the strait. Tidal mixing, entrainment in the Fraser River plume, nutrient loading from the Fraser River, and turbidity in the surface waters (Parsons et al. 1969*a*, Stockner et al. 1979) influence productivity there.

Primary production in the strait compares with other coastal marine waters and has increased from about 120 g carbon/m^2/yr in 1965 and 1966 (Parsons et al. 1970) to about 345 g carbon/m^2/yr (Stockner et al. 1979). Surface zooplankton blooms occur between March and May in the southern half of Georgia Strait, and produce standing crops in excess of 1,000 organisms/m^3. In the central and northern parts of the strait, zooplankton bloom later and standing crops are smaller (Hutchinson and Lucas 1931, Stephens et al. 1969, Parsons et al. 1969*b*, Stockner et al. 1979).

Zooplankton biomass in the surface waters increases in March and April, peaks in May, and declines in June. In central Georgia

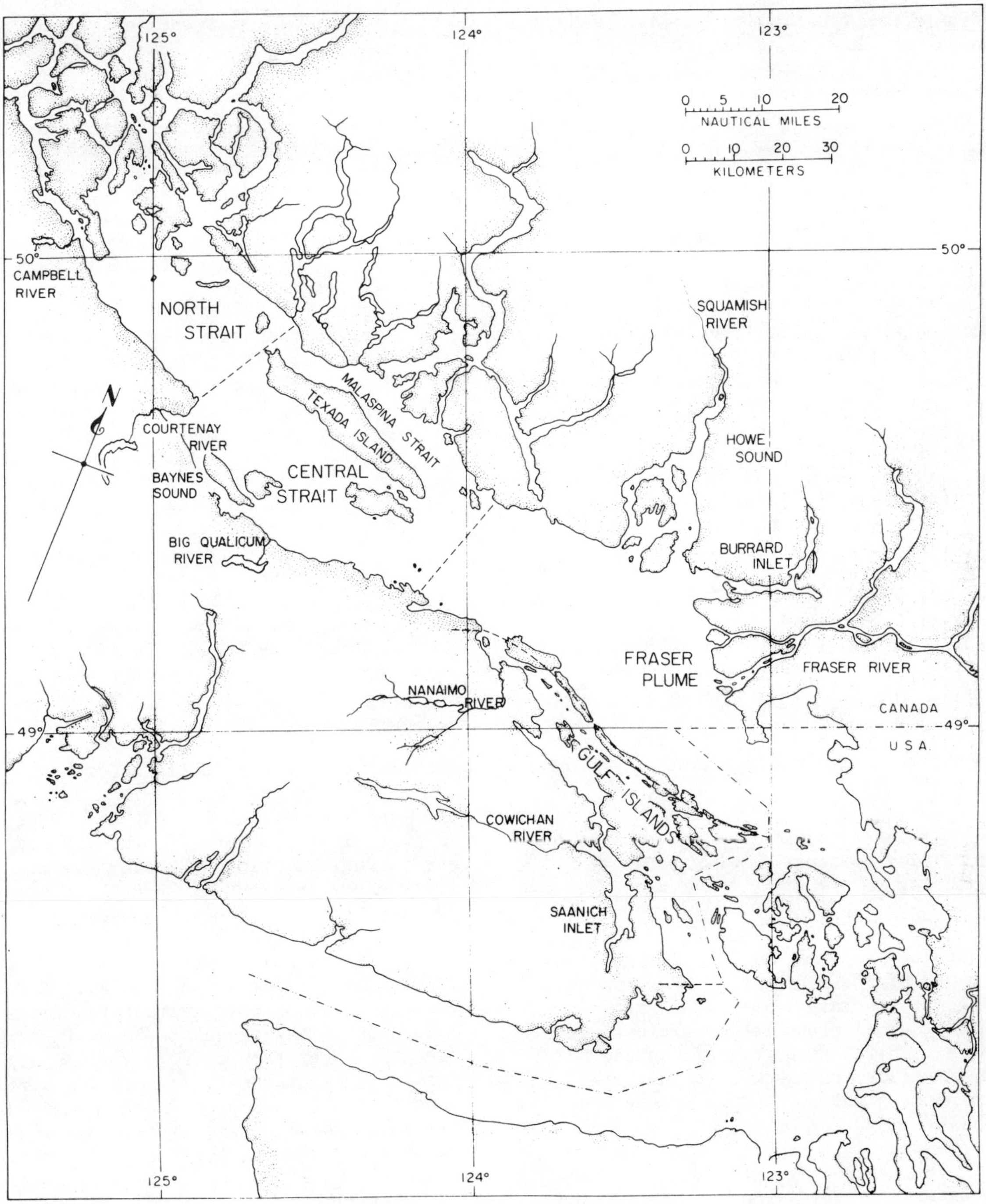

Figure 1. Map of Georgia Strait showing major salmon rivers, the four sampling regions and major landmarks referred to in the text.

Strait (Station 1, Stephens et al. 1969), zooplankton biomass in the surface 20 m may remain high throughout August and September, but small zooplankton dominate the biomass after June.

The principal salmon rivers tributary to Georgia Strait include the Fraser River and the Squamish River on the mainland shore, and the Cowichan, Nanaimo, Qualicum, Courtenay and Campbell rivers on the Vancouver Island shore (Fig. 1). The Fraser River is the greatest single salmon producing system in British Columbia. During the 11-year period from 1965 to 1975, escapements to these and other tributary rivers each year averaged about 1.2 million sockeye *(O. nerka)*, 1.8 million pink *(O. gorbuscha)* (odd-year runs only), 850,000 chum, 230,000 coho, and 100,000 chinook. Most of these fish spawn in the Fraser system. Escapements to the Fraser during this period averaged 1.2 million sockeye, 1.6 million pink, 200,000 chum, 65,000 coho, and 59,000 chinook (Aro et al. 1977). Each spring, a substantial number of juvenile salmon enter the strait--some 300 million in odd years with an additional 200 million pink fry in even years.

I shall compare catches of juvenile salmon within four broad regions of Georgia Strait (Fig. 1):

1) The Gulf Islands, a region of high productivity and considerable habitat structuring.

2) The Fraser River Plume, including the area of the open strait under greatest influence of the plume, Howe Sound and the entrance to Burrard Inlet. This area of high biological productivity exhibits less physical habitat structuring than the Gulf Islands region, but shows considerable patchiness in primary and secondary productivity.

3) The Central Strait, including the open waters of central Georgia Strait and Jervis Inlet. This is an area of low biological productivity except for Baynes Sound and Malaspina Strait.

4) The North Strait, including the area of Georgia Strait most influenced by the tidal pulse entering through Johnstone Strait. This is an area of relatively low biological productivity.

DISCUSSION

Sockeye Salmon

Timing of Migration and Areas of Concentration: Almost all sockeye smolts enter Georgia Strait from the Fraser River. The smolts move downstream mainly in April and May. In late April, they are concentrated just off the river mouth. By late May they have dispersed rapidly south and west, and are concentrated along the east and west sides of the outer Gulf Islands chain. By June, most sockeye are concentrated in the Gulf Islands region, especially along the Vancouver Island shore (Barraclough and Phillips 1978, Healey 1978). Most sockeye leave the strait in late June and July via Juan de Fuca Strait. The timing of peak catches in the Fraser Plume and Gulf Islands region suggests that the smolts take 20 to 30 days to move through Georgia Strait.

Not all sockeye smolts leave Georgia Strait in June and July. Small numbers occurred throughout the strait in August and September 1975, and relatively large catches were made in Howe Sound and off the north arm of the Fraser River in September 1976 (Table 2).

Sockeye fry apparently accompany smolts downstream in the Fraser River during spring. Fry have been found rearing in marshes and sloughs in the upper areas of the Fraser Delta during spring and early summer. Presumably, migrant fry use these areas as rearing sites, although some may be washed out to sea as indicated by the sockeye fry caught in the plume during May 1967 (Phillips and Barraclough 1978). The fry reside in the Fraser Delta for several weeks until they reach 60-70 mm in fork length. They then enter the plume and gradually disperse throughout the strait. Small sockeye captured in the plume region in July (Phillips and Barraclough 1978) probably represent the first of these fry migrants to leave the delta. Catches of underyearling sockeye throughout the strait in August and September suggest that these fish probably spend some time within the strait, possibly making use of the low salinity water in the plume to adjust to marine waters. We do not know exactly when these sockeye leave the strait, but they may emigrate in September. They are comparable in size with yearling smolts by then.

Table 1. Catch-Per-Set, Range of Catch, Average Length, and Average Weight of Juvenile Salmon Captured in the Gulf Islands Region, May to October 1976.

		Month					
		May	June	July	August	September	October
Number of sets		40	40	26	45	40	44
Species							
Chum	$\overline{X}$ catch	4.5	140.0	71.0	2.8	1.3	1.8
	Range of catch	0-40	0-2700	0-240	0-20	0-39	0-80
	$\overline{X}$ length	72	88	101	112	154	182
	$\overline{X}$ weight	3.7	7.3	11.7	16.5	48.5	85.4
Pink	$\overline{X}$ catch	1.6	660.0	7.4	11.0	23.0	0.3
	Range of catch	0-33	0-9900	0-25	0-89	0-509	0-8
	$\overline{X}$ length	36	79	85	111	147	193
	$\overline{X}$ weight	0.34	5.0	6.4	16.0	41.9	106.2
Coho	$\overline{X}$ catch	6.0	1.3	5.9	5.2	6.0	1.2
	Range of catch	0-109	0-6	0-75	0-25	0-200	0-15
	$\overline{X}$ length	116	168	187	212	263	281
	$\overline{X}$ weight	18.6	66.1	95.3	146.0	306.0	384.0
Chinook 0+	$\overline{X}$ catch	0.0	1.7	2.1	2.0	2.8	1.7
	Range of catch	-	0-26	0-11	0-14	0-20	0-13
	$\overline{X}$ length	-	132	159	139	210	215
	$\overline{X}$ weight	-	31.1	57.9	36.9	147.0	159.0
Chinook 1+, 2+	$\overline{X}$ catch	2.3	0.4	0.7	0.3	0.2	0.2
	Range of catch	0-21	0-5	0-5	0-3	0-4	0-3
	$\overline{X}$ length	328	346	341	381	399	410
	$\overline{X}$ weight	650.0	777.0	740.0	1072.0	1250.0	1370.0
Sockeye	$\overline{X}$ catch	6.3	7.3	1.4	0.0	0.02	0.1
	Range of catch	0-102	0-53	0-15	-	0-1	0-3
	$\overline{X}$ length	106	108	117	-	160	185
	$\overline{X}$ weight	12.8	13.6	17.7	-	50.1	81.1

Size and Growth Rate: Sockeye caught in April, May and June averaged about 80-120 mm in length, which is about the size of smolts captured in the river. Those caught in the Gulf Islands region averaged about 15 mm longer than those captured in the Fraser Plume. The average size of sockeye caught in both regions did not increase with time (Table 1), which indicates that the fish probably spend too little time in any region of Georgia Strait to increase significantly in size.

Sockeye smolts caught in the Fraser River Plume during July averaged only 54-79 mm in fork length (Phillips and Barraclough 1978). In late summer of 1975 and 1976, most sockeye captured were 90-110 mm in fork length although some larger fish were also caught (Fig. 2). In May, 1967, sockeye caught in the plume displayed a biomodal length frequency with one mode at 39 mm and the other at 95 mm. The presence of these small fish suggested that a stock different from the smolt sized fish caught in April and May. Examining scales from sockeye 90-110 mm long caught during late summer 1976 revealed that these fish were young-of-the-year that had migrated as fry. Sockeye 150-200 mm long caught at the same time were yearling smolts.

It is difficult clearly to determine the growth rate of juvenile sockeye because they move so quickly through the strait. Sockeye caught in the Gulf Islands region in July to October 1976 appeared to increase in length by 0.7 mm/day (Table 1). Young-of-the-year that entered the Fraser delta as 35 mm fry in May and were caught as 90-110 mm fish in September must have grown in length at a rate of about 0.6 mm/

Table 2. Catch-Per-Set and Range of Catch for Each Species in Different Regions of the Strait of Georgia In Late Summer 1975 and 1976. Average Number of Sets - 42 Per Region in 1975, 31 Per Region in 1976.

		Region							
		Gulf Islands		Fraser Plume		Central Strait		North Strait	
Species	Year	Mean	Range	Mean	Range	Mean	Range	Mean	Range
Chinook	1975	2.9	0-43	8.1	0-84	0.4	0-8	0.1	0-1
	1976	3.10	0-20	4.2	0-27	3.3	0-28	3.4	0-16
Coho	1975	10.0	0-73	6.0	0-47	5.3	0-46	7.2	0-36
	1976	7.0	0-200	2.7	0-17	3.8	0-29	5.8	0-24
Chum	1975	35.0	0-344	20.0	0-197	2.7	0-40	14.3	0-233
	1976	1.4	0-39	0.63	0-5	0.78	0-9	1.3	0-14
Sockeye	1975	1.0	0-7	0.74	0-14	0.52	0-2	1.3	0-4
	1976	0.02	0-1	3.4	0-77	0.41	0-5	0.38	0-3
Pink	1975	0		0		0		0	
	1976	27.0	0-509	0.96	0-6	0.98	0-11	1.6	0-18

day. Smolts from the Gulf Island region were 15 mm longer on the average than those from the Fraser Plume, indicating a growth rate of 0.5-0.75 mm/day over their 20 to 30 day residence in the strait. Therefore, the growth rate of sockeye appears to be about 0.6 mm/day while they reside in Georgia Strait.

The length-weight relationship for sockeye captured in Georgia Strait is $W_{(g)} = 2.41 \times 10^{-6} L^{3.32}_{(mm)}$. Given this relationship, a daily growth of 0.6 mm over the length range 80-120 mm for smolts implies an instantaneous daily growth in weight of 0.0202, or about 2 percent of body weight per day. For sockeye fry entering the strait in July, growth of 0.6 mm/day over the length range 70-100 mm implies an instantaneous daily growth in weight of 0.0237 or about 2.4 percent of body weight per day.

Diet: Data on sockeye diet are rather scattered. For the purpose of this review, I selected information where comparisons could be made between years, or between times of year. The data on sockeye smolts caught in the Fraser Plume during late April and early May (Barraclough 1967, Robinson et al. 1968) are suitable for this purpose, as are data I collected in late summer of 1975 and 1976.

Smolts captured in the Fraser Plume during 1966 fed mainly on insects, euphausiids and copepods. In 1967, they fed on copepods, insects and fish larvae (Table 3). Sockeye captured throughout Georgia Strait in late summer fed mainly on amphipods and insects. Crab larvae and fish larvae were important diet items in late summer 1975, as were copepods and ctenophores in 1976 (Table 4). The fish examined were of similar size in all samples. Therefore, I interpret major differences in diet between years and times of year to reflect differences in the availability of food items. The persistence of certain items in the sockeye's diet through time, despite fluctuations in the item's abundance, reflects feeding preference. Sockeye apparently prefer copepods and insects, but they eat amphipods, euphausiids and fish larvae when available. Of these latter food items, sockeye apparently prefer euphausiids over fish larvae otherwise it is difficult to examine the shift to euphausiids in May 1967 compared to April 1976, since fish larvae were abundant both years (Barraclough 1967, Robinson et al. 1968). The importance of ctenophores as a food item for sockeye during September 1976 probably indicates an abundance of ctenophores rather than a preference for them.

Pink Salmon

Timing of Migration and Areas of Concentration: During even-numbered years, large numbers of pink salmon are present in Georgia Strait. Most come from the Fraser River, although some rivers that flow into the north end of the strait contribute moderate numbers.

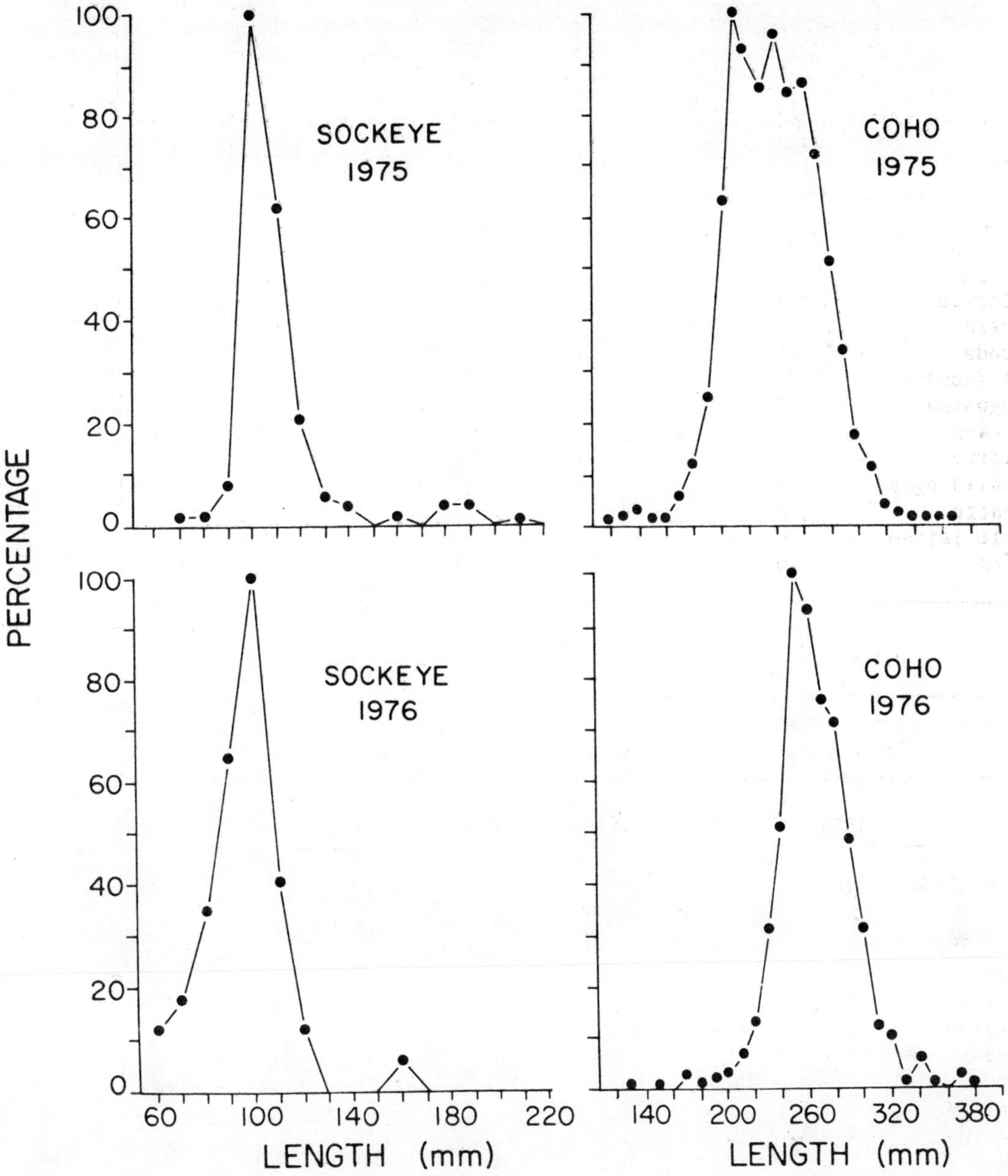

Figure 2. Length-frequency distributions for sockeye and coho captured in Georgia Strait in late summer 1975 and 1976. In each distribution the number of fish at each size is expressed as a percentage of the number of fish at the modal size.

Pink fry migrate down the Fraser River mainly in March and April. They disperse quickly across the strait to occupy shallow, nearshore nursery areas throughout the Gulf Islands region, although some fry are carried north into the Central Strait region or south into the San Juan Islands area. Barraclough and Phillips (1978) report capturing large numbers of juvenile pinks in the Fraser River Plume during April. I believe these fry were in transit to nearshore nursery areas in the Gulf Islands region. In the Nanaimo area, pink fry were caught in large numbers near shore from April to early June, but were rare in purse seine sets made over deep water in the same area until late May (Fig. 3). Juvenile pink salmon from central British Columbia river systems display the same behavior (LeBrasseur and Parker 1964, Healey 1967). For the first few weeks after entering the sea, pink fry stay close to shore, often in large schools in just a few centimeters of water. The fry disperse along the shore, but tend to accumulate in specific places. The characteristics that determine these areas of accumulation are unknown.

Table 3. Diets of Juvenile Salmon Captured By Two-Boat Trawl. Values are Estimated Percent of Stomach Contents By Volume: Sockeye Data are for the Fraser Plume in Late April and Early May; Pink, Chum, and Coho Data are for Saanich Inlet in Early June.

	Sockeye		Pink		Chum		Coho	
Diet Item	1966	1967	1966	1968	1966	1968	1966	1968
Copepoda	20	32	46	9	37	11	15	<1
Euphausiid	26	0	8	11	2	16	52	52
Amphipoda	<1	3	13	1	4	<1	30	8
Crab larvae	<1	<1	<1	2	<1	2	<1	11
Cladocera	0	10	0	<1	0	<1	0	0
Ostracoda	<1	7	0	0	0	0	<1	0
Insect (adult)	39	21	11	15	7	8	<1	26
Chaetognatha	0	1	10	0	16	<1	0	0
Oikopleura	6	4	11	17	34	40	0	0
Fish larva	9	21	<1	<1	<1	2	2	4
Euphausiid eggs	0	0	0	38	0	16	0	0
Polychaeta	0	0	0	6	0	4	0	0
Barnacle larvae	0	0	0	<1	0	0	0	0
Limacina	0	0	0	0	0	0	1	0

Table 4. Stomach Contents in Percent Volume of Major Taxa From Samples Taken Throughout Georgia Strait July to September 1975 and 1976.

	Coho		Chinook		Chum		Sockeye		Pink
	1975	1976	1975	1976	1975	1976	1975	1976	1976
Number of fish	482	52	166	37	465	14	26	11	22
Diet item									
Decapoda									
Crab zoea	1.8	0.2	1.4	0	1.6	0	6.1	0	1.5
megalops	26.2	10.9	19.3	16.0	7.5	14.1	9.3	1.4	9.4
juvenile	<0.1	0	0	0	0.3	0	0	0	0
Amphipoda	26.6	40.5	10.0	25.8	41.2	65.6	54.2	66.9	36.9
Caprellid	0.2	1.2	1.2	0	<0.1	0	0	0	0.5
Euphausiid	4.4	0.3	4.0	0.7	13.2	7.2	8.4	0	7.1
Copepoda	0.2	0	0.3	0	7.7	5.6	0.6	7.3	7.8
Shrimp larva	<0.1	1.8	0	0	0	0	0	0	0.2
Limacina	5.3	10.5	0.4	3.4	<0.1	0.8	0	<0.1	1.7
Polychaeta	<0.1	3.6	<0.1	1.2	5.6	3.7	0	0	0.2
Chaetognatha	0	0	0.1	0	2.8	0	<0.1	0	0.7
Ctenophora	0	0	0	0	0.3	0.8	0	13.7	0
Larvacea	0	0	0	0	2.3	0	0	0	0
Insect	0.4	1.6	2.3	24.2	4.3	0	10.9	10.5	32.0
Eggs	0	0	0	0	0.2	<0.1	0	0	0
Herring	13.4	3.6	16.1	8.3	3.1	0	0	0	0
Sandlance	1.3	0	0.2	0	0	0	0	0	0
Fish remains	19.9	25.3	46.8	20.2	9.2	2.1	10.5	0	1.7
All fish	34.6	28.9	63.1	28.5	12.3	2.1	10.5	0	1.7
Stomach contents as % of body weight	1.15	0.79	0.75	1.13	1.12	0.73			

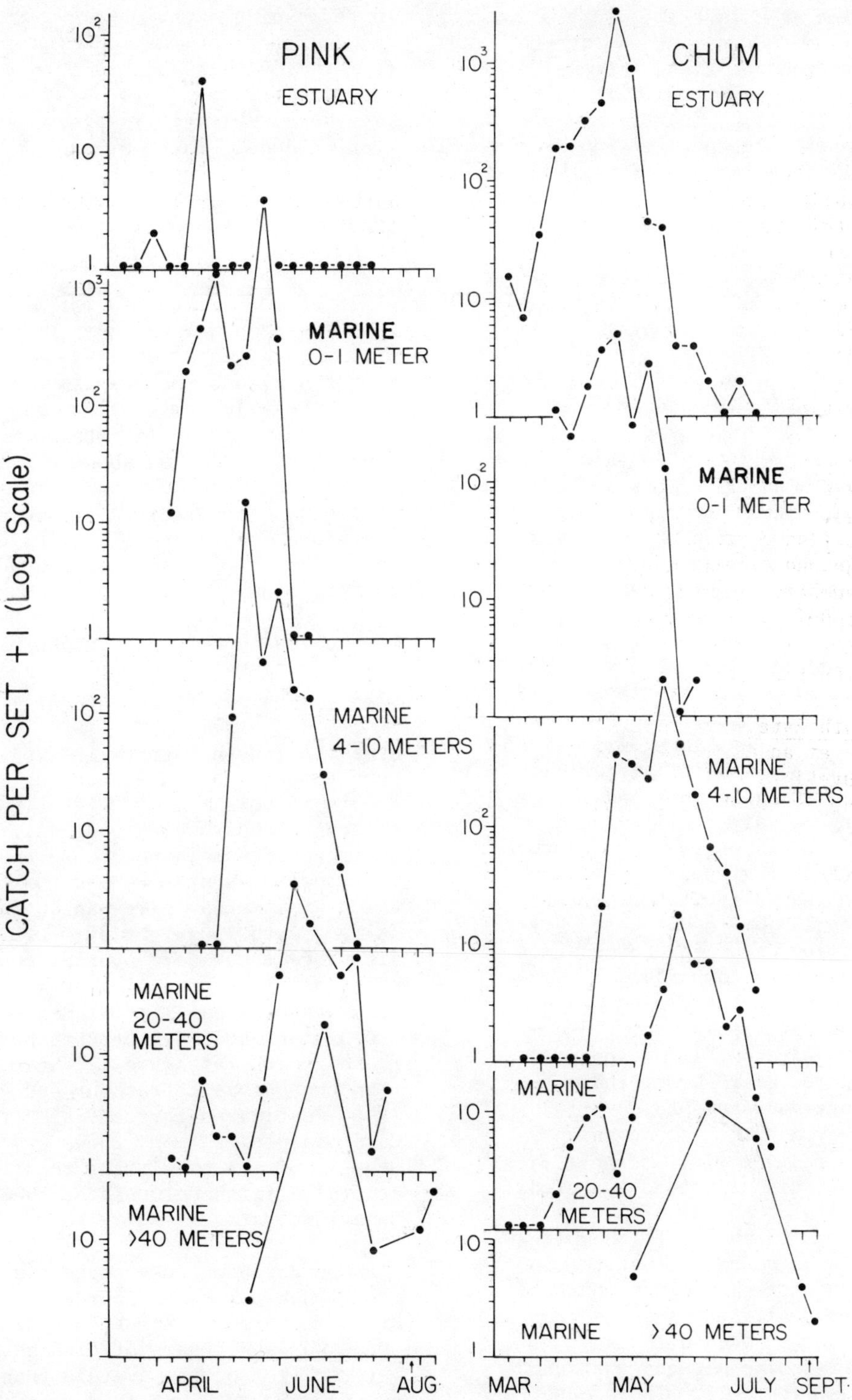

Figure 3. Weekly catch of chum and pink juveniles in different depth zones in the Nanaimo area compared with monthly catch in the Gulf Islands. The data for chum are the average of 1975 and 1976 catches. The data for pinks are from 1976 only. Catch-per-set is shown on a log scale.

In the Nanaimo area juvenile pink salmon move offshore from the beaches in late May and also become abundant along the east side of the outer chain of the Gulf Islands (Barraclough and Phillips 1978). By June, offshore purse seine catches peaked in both the Gulf Islands region and the Nanaimo area (Fig. 3), while pinks were rare in the Fraser Plume at this time of year (Barraclough and Phillips 1978). Catches in the Gulf Islands region declined dramatically in July indicating that most pinks had left Georgia Strait by this time (Table 1, Fig. 3).

Some juvenile pink salmon do persist in the strait long after the bulk of the population leaves. Why they do this is not clear, but these fish apparently leave the strait toward the end of September. In 1976, moderate numbers of pink juveniles persisted in the Gulf Islands region until September, yet few fish were found in other regions (Table 2).

Size and Growth Rate: Pink fry enter the Fraser Plume at an average length of about 35 mm. Juveniles captured by a two-boat trawl in the Fraser Plume during 1966 and 1968 averaged 34.6 mm in late April, 62 mm at the beginning of June and 100 mm at the beginning of July (Phillips and Barraclough 1978). Juvenile pinks captured in the Gulf Islands region from May to October 1976 increased from 36 mm fork length in May to 198 mm in October. Fry captured in shallow water in the Nanaimo area during April were comparable in size with those captured in the Fraser Plume. In late May, however, pinks captured over deep water averaged about 15 mm longer than those captured near shore, indicating that the offshore movement of juvenile pinks in late May may be size dependent, with larger juveniles moving offshore first. The late June or July departure of juvenile pinks from Georgia Strait also appears to involve the largest fish first. Sampling in the Gulf Islands revealed that the increase in length of fish between June and July was less than half the average increase between other months, even allowing for a slightly shorter time between the June and July samples (Table 1). These observations are consistent with those of LeBrasseur and Parker (1964) for central British Columbia pink salmon, except that pinks left Georgia Strait when 90-100 mm long, while in central British Columbia, pinks left Fitz Hugh Sound for the open Pacific when 60-80 mm long.

Size-dependent movement patterns complicate the interpretation of length increases in terms of growth. The average rate of increase for the May to October samples from the Gulf Islands in 1976 was 0.97 mm/day. This rate is probably an underestimate because large fish leave the population in June. The increase in length of pink juveniles in Fraser Plume samples during 1966 and 1968 was 0.85 mm/day, again probably an underestimate because large fish leave in late June.

If we break the data for the Gulf Islands into two stanzas, the May to June period prior to the major emigration of pinks from the Strait, and the July to October period after the major emigration, higher estimates of growth rate result. Between May and June pinks increased about 1.1 mm/day in length. Between July and October the increase was 1.0 mm/day. It seems likely, therefore, that juvenile pinks grow about 1.0 mm/day during their residence in Georgia Strait.

The length-weight relationship for juvenile pinks in the strait is $W_{(g)} = 1.62 \times 10^{-6} L_{(mm)}^{3.42}$. Growth at 1 mm/day over the length interval 35-100 mm, therefore, implies an instantaneous daily growth in weight of 0.0552 or about 5.7 percent of body weight per day. During their residence in nearshore nurseries the fry grow to about 50 mm in fork length, which at 1 mm/day gives an instantaneous daily growth in weight of 0.0737, or about 7.6 percent per day. During their pelagic existence they grow from 60-100 mm for an instantaneous daily growth in weight of 0.0437, or about 4.5 percent per day. These growth rates are slightly greater than those that LeBrasseur and Parker (1964) estimated for central British Columbia pink salmon. However, pinks in Georgia Strait are larger than those from central British Columbia at the same behavioral stage.

Some data on pinks suggest an inverse relationship between abundance and size. Pinks captured in the Fraser Plume in 1966 and 1968, years of high abundance, were smaller on the average than those captured in 1967 and 1969 when abundance was low (Phillips and Barraclough 1978). Pinks captured in the Gulf Islands in August 1974 averaged 122 mm fork length compared with 109 mm for the same time in 1976. Catch-per-set was only 2.5 in 1974 compared with 10.8 in 1976 (Beamish et al. 1976, Healey 1978). These fish demonstrated such a rapid growth rate during their first few months of life in the sea that it seems reasonable to suppose that growth might be density-dependent.

Diet: In order to grow at 7 percent of body weight per day, pink fry must feed actively much of the time. Food resources must be abundant. In the Fraser Plume during April and May, while in transit to nearshore nursery areas, young pinks feed most heavily on calanoid copepods. In nearshore nursery areas during April and May, however, pink juveniles feed mainly on epibenthic prey with harpacticoid copepods being the major food item (Kaczynski et al. 1973, Healey unpublished data).

In late May, when pinks move offshore, they again feed on zooplankton. Data for Saanich Inlet on southern Vancouver Island in early June 1966 and 1968 are fairly typical of pink salmon diets in Georgia Strait at this time. In 1966, copepods were by far the dominant item, with amphipods, insects, chaetognaths, and the larvacean *Oikopleura* of secondary importance. In 1968, however, euphausiid eggs dominated the diet, with *Oikopleura*, insects, euphausiids, and copepods of secondary importance (Table 3, data from Barraclough and Fulton 1968, Barraclough et al. 1968).

The diets of juvenile pink, chum and coho from the same samples taken in Saanich Inlet in 1966 and 1968 were compared. The relative contribution of the various food organisms changed in the same direction between years for all species (Table 3). It seems reasonable to conclude, therefore, that copepods were much less available in June 1968 than in 1966, and that in their absence pinks turned their attention to the next most available and preferred prey, which was euphausiid eggs. That the switch to euphausiid eggs was not simply a response to their overwhelming abundance, but involved a choice, is indicated by their relatively low contribution to the diet of chum and coho.

In late summer, pinks that remained in Georgia Strait fed mainly on amphipods and insects, with crab larvae, euphausiids and copepods being of secondary importance (Table 4).

As I did with sockeye, I interpret changes in the pink salmon diet between years, seasons and locations to reflect a balance between preference and availability. Small food organisms appear to be of particular importance to juvenile pink salmon in Georgia Strait, including harpacticoid copepods during their residence in nearshore nursery areas, and calanoid copepods and euphausiid eggs in late May and June before they leave the strait. The importance of amphipods and insects in late summer probably reflects the scarcity of copepods at this time as well as the large size of the few remaining pink juveniles in the strait.

Diet information, although valuable in providing insight into the food resource composition and feeding preferences, does not provide much insight into the foraging success of the fish. Available information on the amount of food in the pink salmon stomachs sampled from Georgia Strait, however, gives a better indication of the feeding conditions at different times of year.

In the Nanaimo area the stomach contents of pinks captured in shallow nearshore nurseries increased throughout their residence in these habitats (Fig. 4). In April, stomach contents averaged 1-2 percent of body weight, and by late May averaged 4-5 percent of body weight. Pink juveniles were common over depths of 4-10 m only during May and the first part of June. The stomach contents of fish captured over water of this depth ranged about 2-3 percent of body weight and showed a slight declining trend.

Stomach contents of juvenile pink captured over deep water ranged 2-3 percent of body weight for most of April and May, increased to 3.5 percent in early June, then dropped to less than 0.5 percent at the end of June. Stomach contents of pinks captured in July were slightly greater than those captured at the end of June, averaging a little more than 1 percent of body weight. I believe two points here are of particular interest. First, the fish left the nearshore nurseries when feeding conditions appeared to be improving there and entered a depth zone where their foraging success, as indicated by the amount in their stomachs, was lower. Second, the majority of pinks left Georgia Strait at a time when foraging success appeared to be very poor. Low stomach contents were also apparent for a brief period at the end of June 1968 in pinks captured in Saanich Inlet (Barraclough et al. 1968). It appears that the movement of pink juveniles away from the beaches in late May is not associated with poor feeding conditions in shallow water. The movement out of Georgia Strait in late June and early July may be a response to poor feeding conditions, however.

Chum Salmon

Timing of Migration and Areas of Concentration: The Fraser River produces

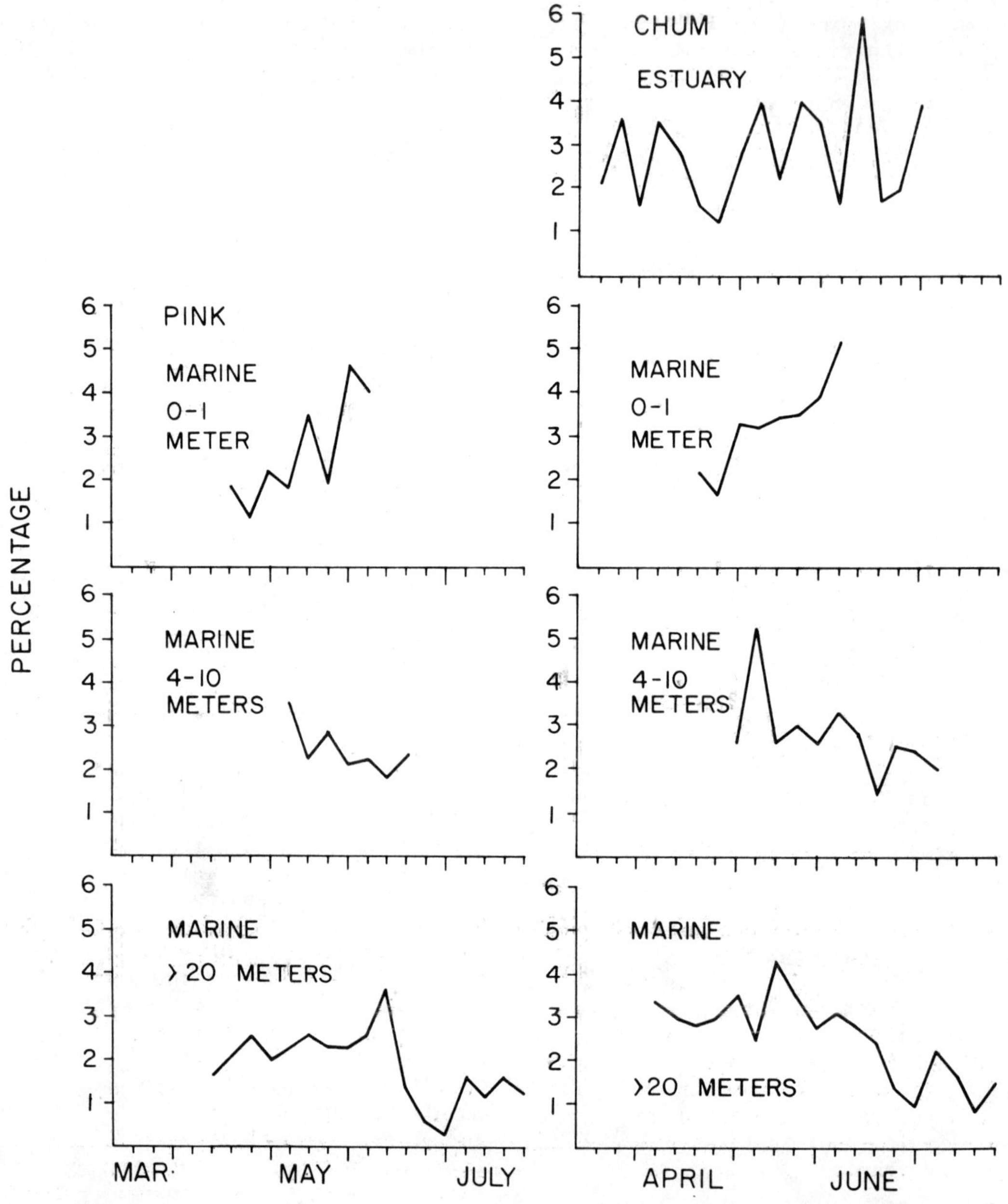

Figure 4. Weekly average weight of stomach contents for chum and pink juveniles in the Nanaimo area. Stomach contents are presented as a percentage of body weight. Data for chum is the average of 1975 and 1976 samples. Data for pinks are from 1976 only.

large numbers of chum. So do many other rivers that flow into Georgia Strait. Most chum fry migrate downstream in April and May. Those from the Fraser River disperse via the plume in much the same way as pinks, and occur together with pinks in nearshore nursery areas. Chum fry from smaller rivers disperse along shore after leaving the river mouths.

Unlike pink fry, significant numbers of chum rear in sloughs and side channels of the Fraser Delta and in other river estuaries for several weeks after migrating downstream (Dunford 1975; Healey et al. 1977*a*, Healey 1979 and unpublished data). Pink fry are seldom caught in these habitats, and those that are apparently spend no more than a few hours there. This seems inconsistent with the observation that pink fry in the Fraser Plume are more confined to low salinity water than chum fry (Barraclough and Phillips 1978). Pinks may be less flexible than chum, however, and after making the transition to seawater may be reluctant to re-enter estuaries.

Chum are not confined to low salinity nursery areas during the first few weeks of life in the sea. They occupy high salinity nursery areas as well, together with pink juveniles in even years. In the Nanaimo system, chum fry from the early part of the downstream run occupy the estuary nursery areas while fry from the late part of the run move to nearby marine nursery areas (Healey et al. 1977*a*, Healey 1979). As with pink juveniles, most chum occupy water of 1 m or less in depth until late May when they move away from the beaches into open water (Fig. 3). However, in the Nanaimo area during 1975 and 1976 there was the suggestion that part of the early run from the Nanaimo River dispersed as fry over deep water and reared there. The same may be true of pinks, and some fry of both species probably rear in open water rather than near shore in shallow water.

Most juvenile chum move offshore in May and June. Catches in Georgia Strait decline after June and presumably this is the time of their major exodus from the strait. Chum appear to leave the strait more slowly than the pinks, and catches in the Gulf Island region were still relatively high in July 1976 (Table 1). Due to their later timing of downstream migration, some chum may not be large enough to leave the strait in late June or early July. Some juvenile chum remain in the strait as late as October (Fig. 3, Table 1).

Many more chum remained in Georgia Strait after the major exodus in 1975 than in 1976, perhaps becuase fewer fry were present in the strait during 1976 (Table 2). Spawning escapement was lower in 1975 than 1974, and fewer chum were caught in the Nanaimo area during spring and early summer 1976 compared to 1975. In late summer 1975, chum were most abundant in the Gulf Islands region, less abundant in the Fraser Plume and North Strait, and least abundant in the Central Strait. In late summer 1976 chum were about equally abundant in the Gulf Islands region and North Strait, and less abundant in the Fraser Plume and Central Strait (Table 2). Specific areas of concentration included the Vancouver Island shore from Nanaimo to Saanich Inlet, and Swanson Channel in the Gulf Islands; along the outer chain of Gulf Islands and the outer reaches of Howe Sound in the Fraser Plume; and central Malaspina Strait and the west coast of Texada Island in the Central Strait.

<u>Size and Growth Rate</u>: Chum migrating downstream average about 36 mm in fork length. Juveniles captured in the Fraser Plume, in the estuaries and in nearshore nursery areas during March and April generally average less than 40 mm fork length. Chum captured in nearshore nursery areas remain small throughout their residence in Georgia Strait. By May, however, the larger juveniles begin moving offshore, and the average size of chum in catches away from the beaches steadily increased (Healey et al. 1977*b*). In the Fraser Plume, juvenile chum captured in mid-July averaged more than 100 mm fork length (Phillips and Barraclough 1978). Chum captured in the Gulf Islands region in 1976 increased from 72 mm in May to 182 mm in October (Table 1). During their major exodus from Georgia Strait in late June and early July, chum averaged 80-100 mm in fork length.

The sizes of chum captured throughout Georgia Strait in 1975 and 1976 suggest that juveniles in the northern part of the strait are larger. Since most chum adults enter Georgia Strait from the north, many juveniles probably leave the strait that way. The larger size of juveniles in the North Strait would be in keeping with a general northward movement in the strait. However, the length frequencies of juveniles captured in late summer 1975 also suggest that stocks may differ in their growth patterns, and this complicates the interpretation of average lengths (Fig. 5). Chum captured in the Gulf Islands in late July 1975 showed a single size mode at 100 mm fork length. In the Fraser Plume, Central Strait and North Strait, the length-frequency distribution was bimodal. In the Fraser Plume the modes were at 75 and 195 mm with most fish in the mode at the smaller size. In the Central Strait the modes were at 100 and 180 mm and most fish were in the mode at the larger size. In the North Strait, only a few fish represented the small sized mode, and most juvenile chum were around 175 mm in fork length.

The bimodal pattern was not repeated in samples taken at the same time in 1976. Chum captured in the Gulf Islands in late summer 1976 averaged 154 mm compared with 102 mm in 1975. In the Fraser Plume the few chum captured in 1976 averaged almost 180 mm, comparable to the larger fish captured in 1975. Chum were smallest in the Central Strait in 1976 averaging 150 mm, while those from the North Strait were similar in size to those captured in the Gulf Islands region. There was no bimodality in the length-frequency distributions

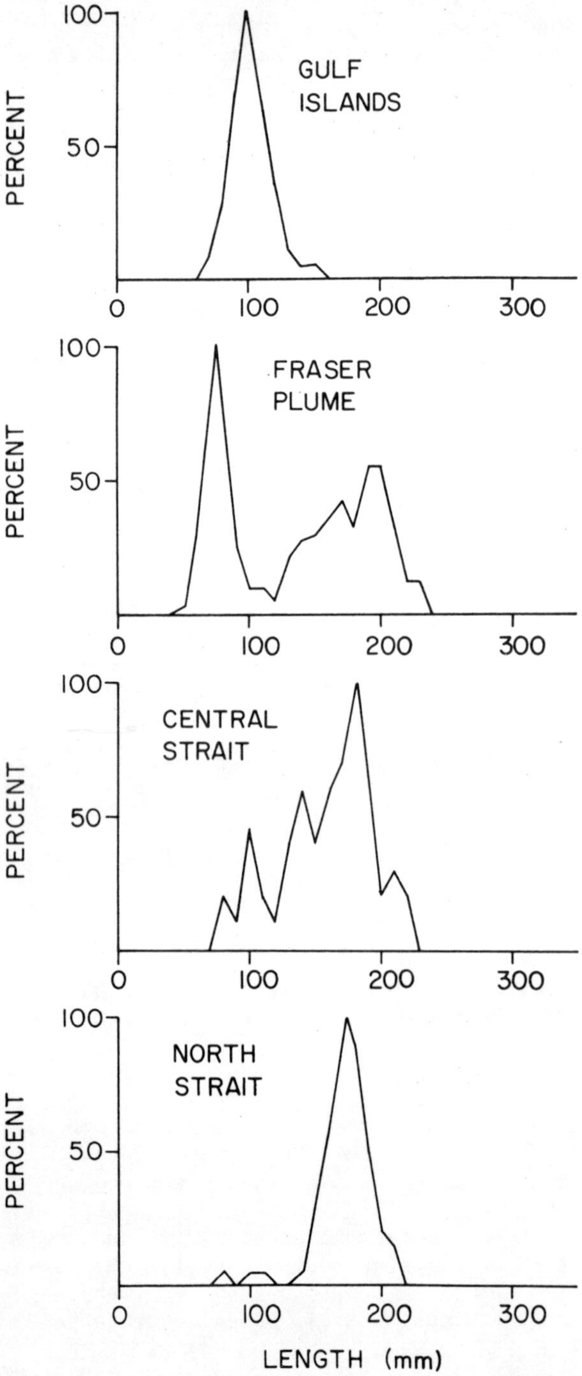

Figure 5. Length-frequency distributions for chum captured in different regions of Georgia Strait in late summer 1975. The number at each length is expressed as a percentage of the number at the modal length.

in 1976, but frequent misidentification of pink and chum in the field in 1976 complicates interpreting these data (Healey 1978).

The average increase in length of chum from samples taken in the Fraser Plume between 1966-1969 was 0.80 mm/day (Phillips and Barraclough 1978), and from samples taken monthly in the Gulf Islands in 1976 was 0.72 mm/day (Table 1). The length-weight relationship for chum captured in Georgia Strait is $W_{(g)} = 1.96 \times 10^{-6} L_{(mm)}^{3.38}$. Growth in length of about .75 mm/day, therefore, implies an instantaneous daily growth in weight of 0.0409, or about 4.2 percent per day over the length range 35-100 mm. These growth estimates are lower than comparable estimates for pink juveniles, and are likely underestimates for the same reasons as the estimates for pinks. In addition, the existence of various size groups in the chum population as observed in 1976 makes interpreting of simple average lengths in terms of growth even more tenuous.

Marked chum from the Nanaimo River provided a known population from which early marine growth could be estimated in 1975 and 1976. Differences in growth between years and between marked groups within a year were not significant. The average daily increase in length for all groups was about 1 mm/day, and the average instantaneous daily increase in weight was 0.059 or 6.1 percent per day (Healey et al. 1977*b*, Healey 1979). These estimates are also lower than those for pink juveniles during their first few weeks of ocean life. Chum fry may possibly grow more slowly than pinks. Chum fry are larger than pink fry at the time of downstream migration and, on the average, appear to leave Georgia Strait a little later than pinks. Their time of peak downstream migration is somewhat earlier than pinks, and consequently they reside in Georgia Strait longer than pinks. Since chum leave Georgia Strait at the same time as pinks, their growth rates may well differ as much as the above growth estimates indicate.

A constant growth rate for all chum stocks can not explain the bimodal length distribution observed in late summer 1976 (Fig. 5). It seems unlikely that the timing of downstream movement for different stocks differs sufficiently to account for the occurrence of two groups so different in size. If the juveniles in the smaller size mode in the Fraser Plume grew at 1 mm/day, they would have had to enter the sea at the end of June. Usually, chum fry complete their downstream movement before June. Alternatively, if these fry entered the sea in May, their growth rate could not have exceeded 0.5 mm/day. The larger juveniles in the Fraser Plume, on the other hand, must either have entered the sea very early (early March), or must have grown more rapidly (about 1.3 mm/day) than expected to reach 195 mm by early August. Since the North Strait was not sampled until September 1975, the juvenile chum captured there were of a size consistent with downstream migration in May followed by a growth rate of about 1 mm/day. The length frequency for the Central Strait, however, poses the same problems as that for the Fraser Plume. The average size of chum captured in 1976 was generally consistent with downstream migration in April and May followed by growth of 1 mm/day.

Diet: While they reside in estuaries and nearshore nursery areas, chum feed mainly on harpacticoid copepods (Healey 1979, Kaczynski et al. 1973), although Dunford (1975) found that chum in upper Fraser marshes feed mainly on chironomid larvae and cladocera.

In 1966 and 1968 in Saanich Inlet, chum that moved offshore fed mainly on *Oikopleura* (Table 3). *Oikopleura* was also a dominant food in the chum diet in the Nanaimo area during June 1975 (Healey, unpublished data), and Chatham Sound in northern British Columbia (Manzer 1969). Copepods and chaetognaths were important in the diet of chum from Saanich Inlet in 1966, but were replaced by euphausiids and euphausiid eggs in 1968 (Table 3). In late summer 1975 and 1976, the chum diet was mainly amphipods, with copepods and larval fish of secondary importance in 1975 and crab megalops in 1976. These differences in diet apparently reflect differences in the food item abundance between years as was described for pink juveniles. Preferences appear also to be involved and chum seem particularly fond of *Oikopleura*. Comparing changes in chum diet between years and locations with the diets of other salmon captured at the same time suggests that chum also prefer chaetognaths and polychaetes. Chum were obviously less attracted by euphausiid eggs in Saanich Inlet than pinks were, and in general, chum take larger prey than pinks.

Regional differences in the chum diet were apparent in late summer 1975 and provide further insight into diet preferences (Table 5). In the Gulf Islands region, fish larvae, crab megalops and copepods dominated the diet. Amphipods were the dominant item in all other regions and comprised 64 percent of the diet in the Central Strait. Fish larvae continued to be important in the Fraser Plume and Central Strait, but were unimportant in the North Strait. Neither crab megalops nor copepods were important diet items outside the Gulf Islands region. Euphausiids, however, were important in the Fraser Plume and increased to importance in the Central and North Strait. Polychaetes were an important diet item in the North Strait, and their occurrence in moderate numbers in chum stomachs while other species ignored them confirms that chum prefer them as a diet item. The regional comparison of diets further suggests that chum prefer euphausiids over various other diet items such as crab megalops.

In 1975 and 1976, I recorded stomach content weights of juvenile chum captured in different depths of water in the Nanaimo area. These data provide an index of foraging success compared to that for pinks. In the Nanaimo estuary average stomach contents varied from 1.2 to 5.9 percent of body weight between March and June (Fig. 4). There was no apparent seasonal trend except that stomach contents tended to be low when chum were most abundant in the estuary. However, more chum reared in the estuary in 1975 than 1976 and their stomach contents were greater on average in 1975, which suggests a relationship between food supply and carrying capacity of the estuary (Healey 1979).

During 1975 and 1976, average stomach contents of chum in nearshore marine nurseries increased continuously from around 2 percent of body weight in mid-April to 5.1 percent of body weight in early June (Fig. 4). In the 4-10 m depth zone, chum stomach contents were variable but tended to decline with time (Fig. 4). Stomach contents of chum captured over deep water ranged 2.5 to 3.5 percent of body weight for most of April, May and early June, with stomach contents in 1976 generally higher than 1975. In late June, however, stomach contents dropped dramatically to a little over 1 percent of body weight. The drop in 1976 was much greater than in 1975 with stomach contents falling to less than 0.5 percent in 1976 but remaining above 1.5 percent in 1975. Average stomach contents increased somewhat in July, but still were at a low level compared with April and May (Fig. 4). The seasonal changes in stomach contents of chum in different depth zones were similar to those for juvenile pinks, although chum had more in their stomachs on the average. In late summer 1975 and 1976 the stomach contents

Table 5. Chum Stomach Contents in Percent Contribution by Value of Major Taxa for Four Regions of Georgia Strait in Late Summer 1975.

	Gulf Islands	Plume	Central	North
Number of fish examined	213	187	49	16
Diet item				
Decapoda				
Crab zoea	5.8	0.3	<0.1	<0.1
megalops	24.3	5.0	0.3	0.4
juvenile	0.4	0	0	0
Amphipoda	11.8	43.6	64.2	45.1
Caprellid	<0.1	0	0	0
Euphausiid	2.2	10.8	14.2	25.4
Copepoda	18.7	5.2	4.0	2.8
Ostracoda	<0.1	0	0	0
Limacina	0	<0.1	0.1	0
Polychaeta	0	3.4	0.3	18.8
Chaetognatha	1.4	5.8	1.2	2.8
Ctenophora	0.7	0.4	0	0
Larvacea	5.2	3.9	0	0
Insect adult	3.8	7.8	2.3	3.2
Eggs	0.9	0.1	0	0
Herring	2.6	4.1	5.9	0
Fish remains	22.1	5.9	7.6	1.2
All fish	24.7	10.0	13.5	1.2
Stomach contents as % of body weight	1.25	1.07	1.05	1.12

of chum remaining in Georgia Strait averaged about 1 percent of body weight or less (Table 4).

Seasonal changes in stomach content weights suggest that the best feeding conditions for chum are in May. As with pinks, juvenile chum left nearshore nurseries when feeding conditions were improving there and entered the pelagic zone where their feeding success appeared lower. Their departure from Georgia Strait in late June was associated with the lowest recorded stomach contents. Chum also showed reduced stomach contents in Saanich Inlet in late June 1968 (Barraclough et al. 1968).

The abundance of chum was positively correlated with stomach contents in a number of instances. I have already noted that chum were less abundant in the Nanaimo River estuary in 1976 than in 1975 and had less food in their stomachs in 1976 (Healey 1979). The average stomach contents of chum in the Nanaimo area generally were lower in 1976 than in 1975, and chum were less abundant in 1976. In samples taken throughout Georgia Strait in late summer 1975, stomach contents were greatest in the Gulf Islands region where chum were most numerous, and least in the Central Strait where chum were least abundant (Table 5). Stomach contents were greater in late summer 1975 than 1976, and chum were more abundant in 1975 (Tables 2, 4) indicating better feeding conditions throughout the summer and autumn of that year. This was associated with a higher population density of chum early in the season, and a larger number of juveniles remaining in Georgia Strait after the main exodus in June and July. Within the strait, however, the fish also appeared to seek regions where feeding conditions were best. This was indicated by the relationship between stomach content and regional distribution in late summer 1975.

Coho Salmon

Timing of Migration and Areas of Concentration: Coho smolts enter Georgia Strait in May and June from virtually every tributary river and stream, and apparently disperse rather rapidly throughout the strait. Juvenile coho are rare in the open water of the strait prior to May, although a few were captured in the Fraser Plume in April (Barraclough and Phillips 1978). Yearling smolts become common in purse seine and trawl samples at the end of May and remain common throughout the summer. Although people often observe coho around docks and other shore structures in early summer, few were captured in nearshore locations in the Nanaimo area. The larger purse seines made the greatest catches over deep water, but the larger and faster swimming coho may have been able to escape the small purse seine and beach seine used to sample nearshore stations.

Tag returns from coho indicate that while many Georgia Strait coho rear in the open Pacific, many also rear to maturity in Georgia Strait. The proportion of the population remaining in the strait apparently varies from year to year, possibly according to oceanographic conditions (personal communication, 1976, W. P. Wickett, Fisheries and Oceans Canada, Nanaimo). I had expected that sampling throughout the summer would indicate the timing and possibly the extent of the outmigration of coho smolts as was shown for pink, chum and sockeye. Purse seine samples taken monthly in the Gulf Islands region from May to October 1976 and weekly in the Nanaimo area from April to August 1975 and 1976 did not reveal any decline in coho catch that would signal the timing of an out-migration (Healey et al. 1977*a*, Healey 1978, and unpublished data) (Table 1). Possibly the coho which leave Georgia Strait to rear in the open ocean do so very quickly so that no appreciable population builds in the strait, or they may not leave until late autumn.

In late summer 1975 and 1976 coho were most abundant in the Gulf Islands region, and relatively abundant in the North Strait (Table 2). An alternative explanation for the absence of a seasonal peak in the coho catch in the Gulf Islands region exists. Coho may prefer this region, and congregate in it from other parts of the strait at the same time as others leave for the open ocean.

Coho were more abundant in late summer 1975 than in 1976 (Table 2). This is at variance with potential smolt production, as the estimated number of spawners in 1973 was only about half that in 1974. The differences in catch may reflect variable egg to smolt survival or year-to-year differences in the number of coho choosing to reside within Georgia Strait.

In the late summer surveys, coho concentrated most along the Vancouver Island shore in all regions of the strait except for a small concentration off the northwest tip of Texada Island.

Size and Growth Rate: Coho average about 100-120 mm in fork length when they enter Georgia Strait. The mean length of smolts captured in the Gulf Islands region from May to October 1976 increased 1.1 mm/day. Their instantaneous daily increase in weight was 0.0188, or about 1.9 percent per day (Table 1). Mathews and Buckley (1976) reported a similar increase in weight of Puget Sound hatchery coho during their first six months at sea. Certain features of the available data on seasonal changes in length and weight are perplexing, however, and complicate the interpretation of these changes as representative of growth.

Samples captured by a two-boat trawl in the Fraser Plume from April to July 1966-1969 do not show a consistent increase in average length with time. In fact, coho captured in June were often smaller than those captured in May and July (Phillips and Barraclough 1978). Coho captured throughout Georgia Strait from July to September 1975 appeared to increase in length about 1.2 mm/day but those captured from August to September 1976 showed no increase with time (Healey 1978). The small average size of coho in the Fraser Plume in June may be due to recruitment of small smolts from the Fraser River during that same month. The rapidly increased size of coho in late summer 1975 occurred when sampling of Georgia Strait proceeded from south to north and no apparent increase in size occurred in 1976 when sampling proceeded from north to south. I interpret this to mean that coho in the northern part of the strait are larger than those in the south. Regional differences in the size of coho, coupled with an apparent northward movement of tagged coho within Georgia Strait make one suspicious of any simple interpretation of seasonal changes in average size within a region. Accepting these reservations, however, the best estimates of coho growth during their first six months in Georgia Strait is about 1 mm/day in length. The length-weight relationship for coho in Georgia Strait is $W_{(g)} = 1.62 \times 10^{-6} L_{(mm)}^{3.42}$. Growth of 1 mm/day over the length range 100-280 mm, therefore, implies an instantaneous daily increase in weight of 0.0196, or about 2 percent per day.

Diet: Euphausiids dominated the diet of coho captured in Saanich Inlet in June 1966 and 1968. Since the diets of pink and chum suggested that euphausiids were less abundant in 1966 than 1968, their predominance in the coho diet probably indicates that coho prefer them. In 1966, amphipods and copepods composed the rest of the diet while in 1968 it was insects and crab larvae (Table 3). The switch in important secondary items between the years parallels changes in pink and chum diets in the same area, and probably reflects changes in the availability of these diet items between years. Larval and juvenile fish made only a small contribution to the coho diet in the samples from Saanich Inlet (Table 3). By late summer, however, fish are an important part of the coho diet and made up 29 to 35 percent of their diet in 1975 and 1976 (Table 4). The rest of the diet in late summer consisted mostly of amphipods and crab megalops *(Cancer gracilis)*, and amphipods were more important than fish in 1976. The Pteropod mollusc *(Limacina)* was also important in the diet of coho in late summer (Table 4).

Like chum, coho showed considerable regional differences in diet in late summer 1975 (Table 6). Crab megalops (47 percent) and fish (45.6 percent) made up most of their diet in the Gulf Islands. Fish declined in importance through the Fraser Plume, Central and North Strait regions, and were only 24.6 percent of the diet in the North Strait. Crab megalops were also less important in the Fraser Plume and Central Strait and were only 5.9 percent of the diet in the North Strait. Amphipods, unimportant in the Gulf Islands region, made up a significant and increasing proportion of the coho diet as one moved north in the strait. *Limacina* were a significant diet item only in the North Strait, but here they made up 20 percent of the diet. Godfrey (1968) observed similar diets for coho juveniles in the Gulf Islands region, Fraser Plume and Central Strait regions in June 1965. Most of the differences between Godfrey's observations and those reported here can be

Table 6. Coho Stomach Contents in Percent Contribution by Volume of Major Taxa for Four Regions of Georgia Strait in Late Summer 1975.

	Gulf Islands	Plume	Central	North
Number of fish examined	164	124	98	96
Diet item				
Decapoda				
Crab zoea	5.5	<0.1	1.6	<0.1
megalops	47.0	22.0	30.0	5.9
juvenile	0	0	0	0.1
Shrimp larvae	0.1	0.1	0	0
Amphipoda	0.4	32.0	36.0	38.2
Caprellid	0.2	0.1	0	0.4
Euphausiid	0.2	6.3	0.9	10.1
Copepoda	0	0.3	0	0.3
Limacina	0.3	0.8	0.2	20.0
Polychaeta	0	0	0	0
Insect adult	0	1.5	<0.1	0.2
Herring	13.6	26.0	0.5	13.6
Sandlance	3.0	0	1.6	0.6
Fish remains	29.0	11.0	29.1	10.4
All fish	45.6	37.0	31.2	24.6
Stomach contents as % of body weight	1.38	1.1	0.58	1.55

ascribed to the fact that he sampled in June when coho were smaller and invertebrate larvae were at an earlier developmental stage.

Coho also showed a positive correlation between abundance and the amount of food in their stomachs in late summer 1975 and 1976. Comparing between regions in 1975, coho were most abundant in the Gulf Islands region and North Strait. In these two regions, coho stomach contents were 1.38 and 1.55 percent of body weight. In the Fraser Plume and Central Strait, stomach contents were only 1.1 and 0.58 percent of body weight (Table 6). Coho were more abundant on average in 1975 than in 1976, and in 1975 their stomach contents averaged 1.15 percent of body weight compared with 0.79 percent in 1976 (Table 4). Apparently coho, as chum, found better feeding conditions in Georgia Strait in 1975 and this may have influenced the numbers remaining in the strait during late summer.

Chinook Salmon

Timing of Migration and Areas of Concentration: Chinook are produced in most of the medium to large rivers tributary to Georgia Strait, but the Fraser produces the most by far. Chinook have, perhaps, the most complex early life history of all Pacific salmon species.

After emerging from the gravel, chinook fry may remain in freshwater or move directly to estuaries and rear in intermediate salinity water (Healey 1980). Those remaining in freshwater go to sea in their first year with many migrating about two months after emergence. Some stay in the river for a year or longer and then migrate in the spring as smolts. When chinook that rear in freshwater migrate, they may hold in the estuary for a few weeks before moving into high salinity water (Reimers 1971), or go directly to sea. Fry rearing in either an estuary or river grow at about the same rate although in some instances there appears to be a sufficient difference in their growth rates for estuary reared fish to be distinguished from river reared fish. Juveniles that rear in the estuary leave it about the time those reared in the river move downstream.

Chinook that go to sea during their first year are termed "ocean type". Those going to sea as yearling or older smolts are termed "stream type". In British Columbia, ocean-type fish predominate. The rarer stream-type fish occur mainly in the larger river systems such as the Fraser, but smaller systems such as the Nanaimo produce some.

Sampling in the Nanaimo area from 1975 to 1977 revealed few chinook in surface waters during the spring (March and April). Those caught were principally ocean-type fish with one ocean check on their scales. In late April, they disappeared and yearling stream-type fish replaced them. In late May, these stream-type fish disappeared and underyearling ocean-type fish that had recently migrated from the rivers and estuaries replaced them. More underyearling fish were caught than the yearling fish, and underyearling fish remained abundant through the summer and fall (Healey 1980). Monthly sampling in the Gulf Islands region from May to October 1976 showed that the pattern for the Nanaimo area was a general one, at least for the Gulf Islands region (Table 1).

Catch-per-set in the Gulf Islands from May to October 1976 was from two to three chinook (Table 1). In May, most of the catch was fish in their second year with a few fish in their third year, while from June to October, first year ocean-type fish dominated the catch.

As with coho, some Georgia Strait chinook rear to maturity inside the strait while others leave to rear in the ocean. And, as with coho, chinook catches in the Gulf Islands did not reveal any peak in abundance followed by a decline that would indicate a departure from Georgia Strait. The only noticeable decline in catch-per-set occurred during winter in the Nanaimo area, and in spring when new fish began entering the strait and the fish that had overwintered declined in abundance. Chinook that rear outside Georgia Strait must either leave quickly or delay their departure until the winter or spring following their migration to sea. The exception might be the stream-type fish, as they were present in catches for only a short time in the spring. It must be remembered, however, that the observations reported here refer to the surface waters of Georgia Strait accessible to purse seining. Some segments of the chinook population may have become unavailable to this gear in winter and spring merely by moving deeper in the water.

In late summer 1975 and 1976 chinook were most abundant in the Fraser Plume. In 1975 they were less abundant in the Gulf Islands and virtually absent from the Central and North Strait regions (Table 2). In 1976, chinook were only slightly less abundant in the Gulf Islands, Central and

North Strait compared with the Fraser Plume. Overall, chinook were more abundant in 1976 than in 1975.

Within the Gulf Islands region, chinook were most abundant along the Vancouver Island shore. Unlike coho, they did not extend into Swanson Channel. In the Fraser Plume, chinook were most abundant off the Fraser Delta and into Howe Sound. In the Central and North Strait, chinook were concentrated on the mainland shore, and especially in Malaspina Strait and around the north end of Texada Island. There was a tendency for chinook and coho to segregate somewhat within the strait, This was most obvious in late summer 1976 when chinook concentrated on the east side of Georgia Strait and coho on the west, and in winter 1974 when chinook were concentrated in the north of the Gulf Islands region and coho in the south (Healey 1978).

Size and Growth Rate: In keeping with their complex early life history, chinook also show a complex size distribution within Georgia Strait. Catches in the Gulf Islands during May to October 1976 showed a multimodal size frequency distribution (Fig. 6). Two modes predominated in May, one at 290 mm and the second at 450 mm fork length. These were mainly ocean-type fish with one or two ocean checks on their scales.

In June, three modes were evident. One was at 80 mm, one between 140-160 mm and the third between 320-360 mm fork length. These three modes, with some variation, persisted throughout the sampling. The mode at the large size, which declined continuously in importance throughout the June to October period, represents the continued presence of two-year-old fish. The two small-sized modes apparently represented two types of one-year-old ocean-type fish. It is tempting to hypothesize that these modes reflect a difference between stream and estuary reared fish; however, downstream migrant chinook in June were comparable in size with those that reared in estuaries.

The multimodal length frequency distribution pattern seen in the Gulf Islands was repeated in other years and in some of the other regions of the strait (Fig. 7). In 1975, the distribution for the Gulf Islands was bimodal with modes at 120-140 mm and 160 mm. In the Fraser Plume there was one major mode at 110 mm and a strong positive skew suggesting a second mode around 200 mm. The few fish captured in the Central and North Strait in 1975 were almost all larger than 170 mm. In 1976, the distribution in the Gulf Islands was bimodal with strong modes at 150 and 220 mm and the suggestion of a third mode around 400 mm. The Fraser Plume showed two definite modes at 130 and 190 mm in 1976. The Central and North Strait each showed one mode at 130 mm with a strong positive skew. The good catches in the Central and North Strait regions in 1976 were apparently due to an abundance of small chinook that were not present in these regions in 1975.

The multimodal character of the size distributions for chinook rules out any estimate of growth from changes in average size over time. However, the three modes observed in the Gulf Islands in 1976 increased regularly in size at about the same rate, which was 0.75-0.85 mm/day. The length-weight relationship for chinook in Georgia Strait is $W_{(g)} = 3.53 \times 10^{-6} L_{(mm)}^{3.282}$. Assuming growth in length of 0.8 mm/day for each size group, the instantaneous daily growth in weight was 0.0213 for the small young-of-the-year, 0.0133 for the large young-of-the-year and 0.0073 for the two-year-old fish. Juvenile chinook marked on the Nanaimo estuary and recaptured in the Nanaimo area grew 1.32 mm/day and their instantaneous daily growth in weight was 0.0566 over the size range 44-115 mm fork length (Healey 1980). Chinook, therefore, apparently grow more rapidly during their few weeks of life in the estuary than after entering the marine environment.

Diet: During their estuary residence, juvenile chinook feed on a wide variety of invertebrate fauna and appear much less dependent than chum on small copepods. Insect larvae and adults are an important component of their diet at this stage. So are amphipods, decapod larvae and calanoid copepods (Sibert and Kask 1978, Healey unpublished data).

After chinook leave the estuary, larval and juvenile fish become their dominant food with various invertebrate prey of secondary importance (Godfrey 1968, Beamish et al. 1976, Healey unpublished data). In late summer 1975 fish made up 63 percent of the chinook diet, but in 1976 fish comprised only 29 percent. The poor representation of fish in 1976 presumably reflects reduced availability as all other salmon species took less fish in 1976 as well (Table 4). Other important diet items in late summer 1975 were crab megalops and amphipods. These two items and insects made up most of the chinook diet in 1976.

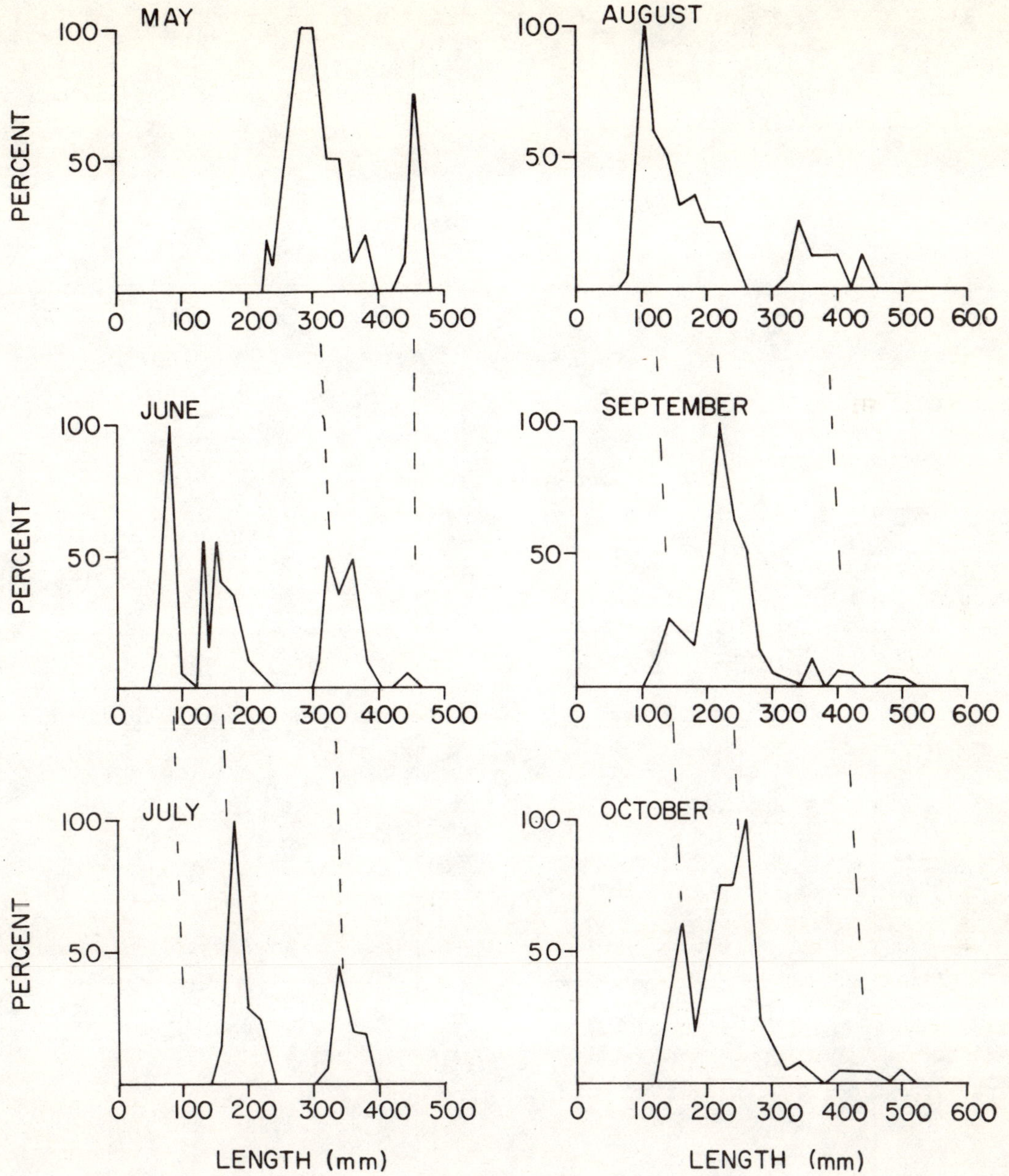

Figure 6. The length-frequency of chinook captured in the Gulf Islands region of Georgia Strait, May to October 1976. The number at each length is expressed as a percentage of the number at the modal length.

Regional differences in chinook diet were apparent in late summer 1975 as they were in chum and coho diets (Table 7). Juvenile fish, primarily small herring, made up 79 percent of chinook diet in the Gulf Islands. Crab larvae were the only other significant contributor. In the Fraser Plume and Central Strait, fish were only about 37 percent of their diet. In the plume, crab larvae, amphipods and euphausiids made up most of the invertebrate fraction of the diet, while in the Central Strait crab megalops were the single most important diet item at 46 percent of stomach volume. Too few fish from the North Strait were examined to give a detailed description of chinook diet in this region; however, fish were the dominant diet item in those stomachs examined.

Chinook apparently prefer fish, particularly herring, but can obviously capitalize on available invertebrate food resources. They seem to prefer larger invertebrates like amphipods, crab megalops and euphausiids over smaller ones and thus appear to have food preferences similar to coho.

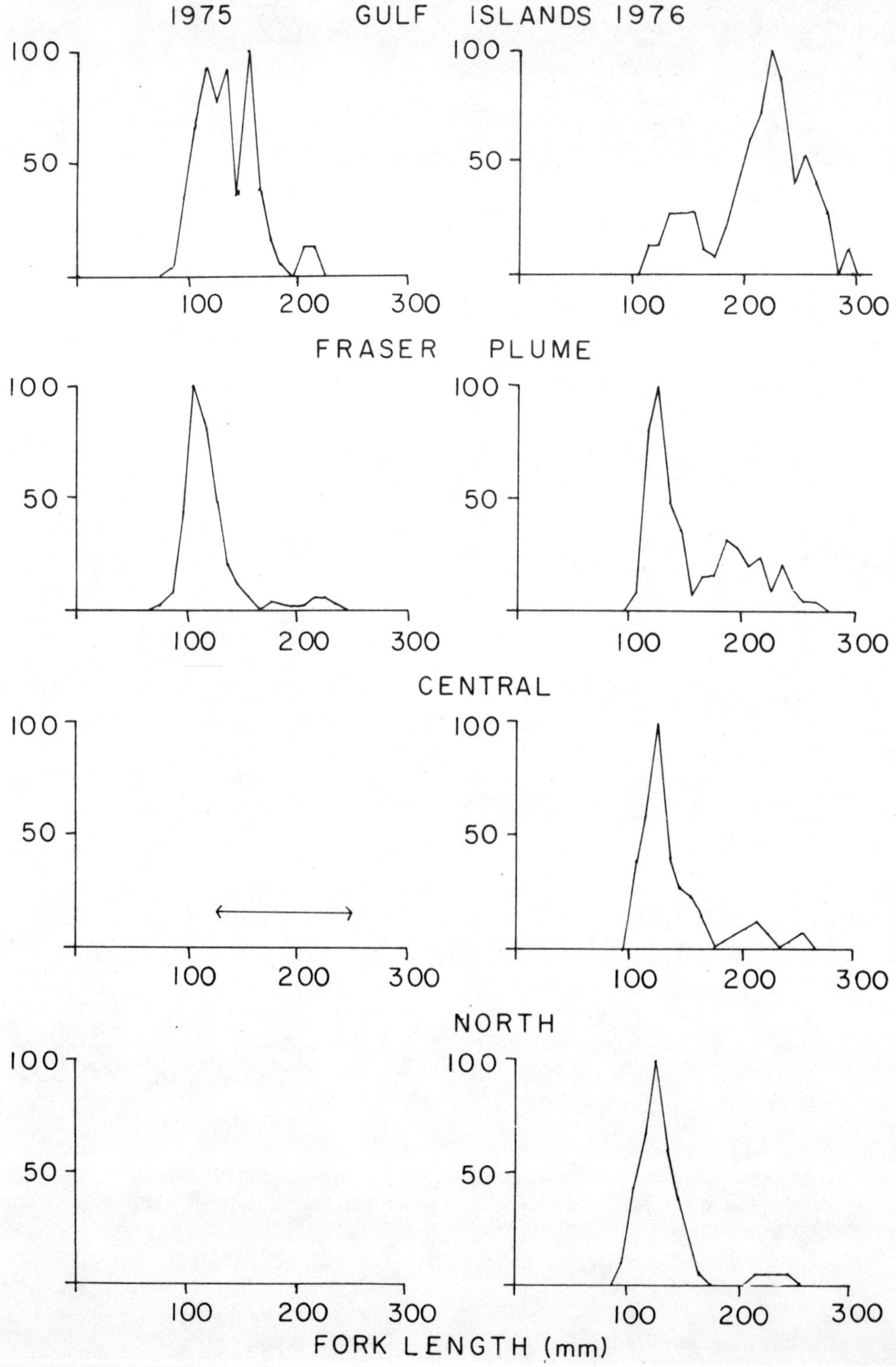

Figure 7. The length-frequency of chinook captured in different regions of Georgia Strait in late summer 1975 and 1976. The number at each length is expressed as a percentage of the number at the modal length. Too few chinook were captured in the Central and North Strait to permit a length-frequency distribution for these regions in 1975. The length range of those captured is shown in the panel for the Central Strait in 1975.

Table 7. Chinook Stomach Contents in Percent Contribution by Volume of Major Taxa for Three Regions of Georgia Strait in Late Summer 1975.

	Gulf Islands	Plume	Central
Number of fish examined	65	86	11
Diet item			
Decapoda			
Crab zoea	4.7	0.7	0
megalops	11.5	17.7	46.4
juvenile	0	0	0
Amphipoda	0.8	24.1	13.9
Caprellid	0	0	0
Euphausiid	1.7	11.9	2.4
Copepoda	0.8	0.4	0
Limacina	0	0	0
Polychaeta	<0.1	0	0
Chaetognatha	0	0.3	0
Insect adult	1.5	7.6	0
Herring	30.5	14.2	19.8
Sandlance	0	0.6	0
Fish remains	48.6	22.5	17.4
All fish	79.1	37.3	37.2
Stomach contents as % of body weight	1.11	1.51	0.40

A relationship between chinook abundance and feeding conditions, as was noted for coho and chum, apparently exists. Comparing between regions in 1975, stomach contents averaged 1.51 percent of body weight in the Fraser Plume where chinook were most abundant. The average was lower in the Gulf Islands region and very low in the Central Strait (0.4 percent) (Table 7). In comparing 1975 to 1976 when chinook were more abundant, particularly in the Central and North Strait regions, stomach contents averaged 1.13 percent of body weight compared with 0.75 percent in 1975 (Table 4). The fact that chinook were more abundant and had more food in their stomachs during a year when larval and juvenile fish appear to have been unavailable emphasizes the ability of chinook to capitalize on abundant invertebrate foods.

DISCUSSION AND CONCLUSIONS

Part of my objective is to consider whether the migratory behavior of juvenile salmon in Georgia Strait can be related to food limitation. For chum in the restricted area of the Nanaimo River estuary, where both the fish and their food could be studied in detail, this appears to be the case (Healey 1979, Sibert 1979). The disappearance of juvenile chum from the estuary coincided with the disappearance of their dominant food resource. For the entire strait, however, directly demonstrating food limitation becomes a formidable and perhaps impossible task. Such an analysis is further frustrated by the fact that plankton samples seldom reveal an abundance of the kinds of organisms that are found in the stomachs of young salmon (Barraclough 1967, Robinson et al. 1968, Barraclough et al. 1968, Barraclough and Fulton 1968). Do young salmon really feed on comparatively rare prey or do the plankton net samples simply not represent plankton in the way young fish see them? Despite the obvious difficulties in directly analyzing the relationship between young salmon and their food supply for Georgia Strait as a whole, certain features of their early life in the sea suggest that food resource limitation may be important at specific times.

The way young salmon use various nursery areas within Georgia Strait suggests resource partitioning. Pink, chum and chinook that migrate downstream as fry spend most of their first month in the ocean in very shallow water close to shore. Chum and chinook coexist in estuaries, and pink and chum in high salinity nurseries. An obvious question is why pinks do not utilize estuarine nursery areas in the same way chum do? If available food is limiting in estuary nurseries, as was suggested by studies in the Nanaimo estuary, perhaps pinks cannot compete with the larger chum and chinook fry. All three species appear to be opportunistic feeders and they often have the same dominant diet items when they coexist. Their diets do differ, however, in ways that I interpret to reflect important differences in both feeding preferences and feeding strategy.

When dominant food items are the same, their relative contribution is generally different and secondary food items are usually not the same. Chinook, in particular, feed on different things than chum during their estuarine residence. Harpacticoid copepods and shrimp larvae are dominant foods of chum throughout their estuarine residence. But chinook feed on harpacticoids for only a few days after migrating downstream, and thereafter concentrate on amphipods, insect larvae and adults, and mysids. Pinks and chum do feed on the same organisms in nearshore nurseries but in different proportions. In Puget Sound during spring 1970, both species took large numbers of harpacticoid copepods, but these prey formed a smaller proportion of the pink diet (Kaczynski et al. 1973). Pinks took almost as many barnacle nauplii and invertebrate eggs as they did harpacticoids, and tended to seek smaller prey than chum. In Georgia Strait during June, chum preyed heavily on *Oikopleura* while pinks were less attracted to larvaceans and continued to feed on small organisms. These diet differences occurred while the young fish lived together, often in mixed schools and presumably presented with the same range of foods. This suggests that the differences are a mechanism to reduce competition. For such a mechanism to have arisen, food resources must have been limiting in the past if not at present.

Despite the suggestion that competition for food may influence feeding preferences of pinks and chum in nearshore habitats, the offshore movement of both species in May appears unrelated to food shortage. In fact, the offshore movement is ill-timed to take advantage of pelagic plankton blooms in the open strait, which are often declining by May. The stomach contents of pink and chum caught in shallow water increases throughout April and May, yet these fish move offshore to begin exploiting a food resource that appears to be much less abundant and beginning to decline.

Pink, chum and sockeye, which actively forage over deep water in May and June, face decreasing food resources at a time when their food demands are increasing. The major concentration of pink and sockeye, and possibly chum at this time, occurs in the Gulf Islands region. Consequently, only a small proportion of the strait's food producing area supports the bulk of the juvenile salmon population. It is not surprising therefore, that differences in the diet of the three species appear most obvious at this time. Chum concentrate on *Oikopleura*; pinks on small copepods and invertebrate eggs; and sockeye on copepods, amphipods and insects. Weights of stomach contents of pink and chum in June suggest the fish suffer rather severe food shortages just before they leave Georgia Strait. Conditions are right for food shortage in June, and stomach contents indicate poor foraging success. This suggests that pink, chum and sockeye may leave the strait in response to poor feeding conditions. After most of the fish depart, feeding conditions improve a bit, and the young-of-the-year sockeye entering Georgia Strait shortly after the smolts leave feed well and grow rapidly at the food densities prevailing in July and August.

Coho and chinook present a somewhat different picture, as many of them remain in Georgia Strait throughout the summer and autumn. These species depend less on invertebrate food, however, and probably do not face the same decline in food resources after May that pink, chum and sockeye experience. Chinook stomach contents do not decline in late June as do pink and chum stomach contents. Both coho and chinook feed preferentially on larger invertebrates as well as fish. For example, while pink and chum in Saanich Inlet take copepods, euphausiid eggs and *Oikopleura*, coho feed on euphausiid adults and insects. In late summer both coho and chinook concentrate on amphipods, crab megalops and fish. Because of their piscivorous habits and their preference for large invertebrates, coho and chinook may not compete with pink, chum and sockeye. However, they do potentially compete with each other, which may explain their tendency to segregate spatially in Georgia Strait. Competition is further suggested by preferential differences in the invertebrate fraction of their diet. Perhaps the most obvious indication of preference is that of coho for *Limacina*.

Differences in the numbers of coho, chinook and chum remaining in Georgia Strait in late summer appears related to food supply. Fewer fish remain when feeding conditions are poor. The exodus of chum from Georgia Strait in 1976 appeared more complete than in 1975; and it is perhaps significant that the stomach contents of chum in late June 1976 dropped to less than 0.5 percent of body weight while in 1975 it remained above 1.5 percent throughout the period of exodus. The regional distribution of these three species in late summer is apparently also related to food availability. These species, and probably the others, apparently distribute themselves in relation to their food supply, at least over the geographic range that they can effectively sample during their daily movements.

The pieces of evidence that support an intimate relationship between the distribution, dispersal and residence of juvenile salmon in Georgia Strait and food resources are tantalizing rather than conclusive. Although I have assumed the possibility of competition among some or all of the species, and certain interpretations hinge on that assumption, stomach contents are a wholly inadequate index of competition for food. Differences in diet may simply reflect feeding preferences or feeding strategies that would reasonably be assumed to be different in different species. However, it is not necessary to invoke interspecific competition to demonstrate food limitation as intraspecific competition for preferred prey would be sufficient. If I reject the hypothesis that differences in diet among similar species in the same geographic location result from competitive interaction, then the hypothesis of food limitation rests on three observations: The suggestion of density-dependent growth in some species; the low stomach contents of pink and chum at the time of their major exodus; and the relationship between stomach contents and regional and annual abundance of coho, chinook and chum.

The offshore movement of pink and chum in May is the most perplexing observation contrary to the general hypothesis of food limitation. The fish move from a region of good feeding conditions to poorer feeding conditions. Although I can rationalize this observation as an anticipation of a future food shortage, or predator avoidance, it must remain an unresolved contradiction until more data become available.

The capacity of Georgia Strait to support juvenile salmon cannot be estimated from the observations presented here. But the data do imply that food shortages may occur in late summer even at present population densities of juvenile salmon. Salmon enhancement could only aggravate present shortages. Nearshore habitats, with the exception of estuaries, appear least likely to be overloaded. Stomach contents of fish captured in these areas were relatively high and their growth rates during nearshore residence very rapid. Problems appear most likely to rise with pink and chum when they forsake the beaches and begin cropping a declining pelagic food resource. Food shortages at this time could result in premature departure from Georgia Strait or a prolonged period of low rations in June. Sockeye, present at the same time and probably exploiting the same food resource, may experience the same difficulties. For coho and chinook no immediate prediction can be made about a critical period for food limitation. Reduced chinook stocks at present make it unreasonable to expect any obvious food supply problem as long as herring remain abundant. These two species, nevertheless, appear to congregate in regions of best feeding conditions. Therefore, problems with food supply, if coho and chinook numbers increase dramatically, should not be dismissed lightly.

LITERATURE CITED

Aro, K. V., P. L. Miller, and J. McDonald. 1977. Catches and escapements of Pacific salmon in British Columbia 1965-1975. Fish. Mar. Serv. Data Rep. 39. 67 pp.

Barraclough, W. E. 1967. Data record. Number, size and food of larval and juvenile fish caught with a two-boat surface trawl in the Strait of Georgia, 25-29 April 1966. Fish. Res. Bd. Can. MS Rep. 922. 54 pp.

Barraclough, W. E., and J. D. Fulton. 1968. Data record. Food of larval and juvenile fish caught with a surface trawl in Saanich Inlet during June and July 1966. Fish. Res. Bd. Can. MS Rep. 1003. 78 pp.

Barraclough, W. E., D. G. Robinson, and J. D. Fulton. 1968. Data record. Number, size composition, weight, and food of larval and juvenile fish caught with a two-boat surface trawl in Saanich Inlet, 23 April-21 July 1968. Fish. Res. Bd. Can. MS Rep. 1004. 305 pp.

Barraclough, W. E., and A. Phillips. 1978. Distribution and abundance of juvenile salmon in the southern Strait of Georgia during the period April to July from 1966 to 1969. Fish. Mar. Serv. Tech. Rep. No. 826. 47 pp.

Beamish, R. J., F. P. Jordan, J. R. Scarsbrook, and R. Page. 1976. An initial study of fishes inhabiting the surface waters of the Strait of Georgia, *M.V. CALIGUS*, July-August 1974. Fish. Res. Bd. Can. MS Rep. 1377. 37 pp.

Dunford, W. E. 1975. Space and food utilization by salmonids in marsh habitats of the Fraser River estuary. Unpublished M.S. Thesis. Univ. of British Columbia. 81 pp.

Giovando, L. F. 1973. The effluent outfall proposed for the Five Finger Island area, Nanaimo, B.C.: Oceanographic and related considerations. Environ. Can. Mar. Sci. Directorate, Pac. Mar. Sci. Rep. 73-12. 83 pp.

Giovando, L. F., and S. Tabata. 1970. Measurements of surface flow in the Strait of Georgia by means of free-floating current followers. Fish. Res. Bd. Can. Tech. Rep. No. 163. 69 pp.

Godfrey, H. 1968. Some observations on juvenile marine chum, chinook and coho salmon taken in waters adjacent to Georgia Strait in 1965. Fish. Res. Bd. Can. MS Rep. No. 955. 19 pp.

Godfrey, H., and E. A. R. Ball. 1974. Data record, with comments. Fall and winter fishing for juvenile salmon in Georgia Strait 1972-73. Fish. Res. Bd. Can. MS Rep. 1294. 59 pp.

Healey, M. C. 1967. Orientation of pink salmon *(Oncorhynchus gorbuscha)* during early marine migration from Bella Coola River system. J. Fish. Res. Bd. Can. 24(11):2321-2338.

Healey, M. C. 1978. The distribution, abundance and feeding habits of juvenile Pacific salmon in Georgia Strait, British Columbia. Fish. Mar. Serv. Tech. Rep. 788. 49 pp.

Healey, M. C. 1979. Detritus and juvenile salmon production in the Nanaimo estuary: 1. Production and feeding rates of juvenile chum salmon (*Oncorhynchus keta*). J. Fish. Res. Bd. Can. 36:488-496.

Healey, M. C. 1980. Utilization of the Nanaimo River estuary by juvenile chinook salmon, *Oncorhynchus tshawytscha*. Fish. Bull. 77:653-668.

Healey, M.C., R. V. Schmidt, F. P. Jordan, and R. M. Hungar. 1977*a*. Young salmon in the Nanaimo area 1975: 1. Distribution and abundance. Fish. Res. Bd. Can. MS 1369. 161 pp.

Healey, M. C., R. V. Schmidt, F. P. Jordan, and R. M. Hungar. 1977*b*. Juvenile salmon in the Nanaimo area, 1975: 1. Distribution and abundance. Fish. Res. Bd. Can. MS 1369. 161 pp.

Hutchinson, A. H., and C. C. Lucas. 1931. The epithalassa of the Strait of Georgia. Can. J. Res. 5:231-284.

Kaczynski, V. W., R. J. Feller, J. Clayton, and R. J. Gerke. 1973. Trophic analysis of juvenile pink and chum salmon *(Oncorhynchus gorbuscha* and *O. keta)* in Puget Sound. J. Fish. Res. Bd. Can. 30:1003-1008.

LeBrasseur, R. J., and R. R. Parker. 1964. Growth rate of central British Columbia pink salmon *(Oncorhynchus gorbuscha)*. J. Fish. Res. Bd. Can. 21(5):1101-1128.

Manzer, J. I. 1969. Stomach contents of juvenile Pacific salmon in Chatham Sound and adjacent waters. J. Fish. Res. Bd. Can. 26(8):2219-2223.

Mathews, S. B., and R. Buckley. 1976. Marine mortality of Puget Sound coho salmon *(Oncorhynchus kisutch)*. J. Fish. Res. Bd. Can. 33:1677-1684.

Parker, R. R. 1965. Estimation of sea mortality rates for the 1961 brood year pink salmon of the Bella Coola area, British Columbia. J. Fish. Res. Bd. Can. 22(6):1523-1554.

Parker, R. R. 1968. Marine mortality schedules of pink salmon of the Bella Coola area, British Columbia. J. Fish. Res. Bd. Can. 25(4):757-794.

Parsons, T. R., K. Stephens, and R. J. LeBrasseur. 1969*a*. Production studies in the Strait of Georgia. Part I. Primary production under the Fraser River Plume, February to May 1967. J. Exp. Mar. Biol. Ecol. 3(1):27-38.

Parsons, T. R., R. J. LeBrasseur, J. D. Fulton, and O. D. Kennedy. 1969*b*. Production studies in the Strait of

Georgia. Part II. Secondary production under the Fraser River Plume, February to May 1967. J. Exp. Mar. Biol. Ecol. 3(1):39-50.

Parsons, T. R., R. J. LeBrasseur, and W. E. Barraclough. 1970. Levels of production in the pelagic environment of the Strait of Georgia, British Columbia: a review. J. Fish. Res. Bd. Can. 27(7): 1251-1264.

Peterman, R. M., and M. Gatto. 1978. Estimation of functional responses of predators on juvenile salmon. J. Fish. Res. Bd. Can. 35:797-808.

Phillips, A. C., and W. E. Barraclough. 1978. Early marine growth of juvenile Pacific salmon in the Strait of Georgia and Saanich Inlet, British Columbia. Fish. Mar. Serv. Tech. Rep. No. 830. 19 pp.

Reimers, P. E. 1971. The length of residence of juvenile fall chinook salmon in Sixes River, Oregon. Ore. Fish. Comm. Res. Div. Research Brief. 99 pp.

Robinson, D. G., W. E. Barraclough, and J. D. Fulton. 1968. Data record. Number, size composition, weight and food of larval and juvenile fish caught with a two-boat surface trawl in the Strait of Georgia, 1-4 May 1967. Fish. Res. Bd. Can. MS Rep. No. 964. 105 pp.

Sibert, J. 1979. Detritus and juvenile salmon production in the Nanaimo estuary: 2. Meiofauna available as food for juvenile salmon. J. Fish. Res. Bd. Can. 36:497-503.

Sibert, J., and B. Kask. 1978. Do fish have diets? *In* Shepherd, B. G., and R. M. J. Ginetz (eds.), Proceedings of the 1977 Northeast Pacific chinook and coho salmon workshop. Fish. Mar. Serv. Can. Tech. Rep. 759:48-56.

Stephens, K., J. D. Fulton, and O. D. Kennedy. 1969. Summary of biological oceanographic observations in the Strait of Georgia, 1965-1968. Fish. Res. Bd. Can. Tech. Rep. No. 110. 97 pp.

Stockner, J. G., D. D. Cliff, and K. R. S. Shortreed. 1979. Phytoplankton ecology of the Strait of Georgia, British Columbia. J. Fish. Res. Bd. Can. 36: 657-666.

Tully, J. P., and A. J. Dodimead. 1957. Properties of the water in the Strait of Georgia, British Columbia, and influencing factors. J. Fish. Res. Bd. Can. 14(3):241-319.

Waldichuk, M. 1957. Physical oceanography of the Strait of Georgia, British Columbia. J. Fish. Res. Bd. Can. 14(3): 321-486.

Waldichuk, M. 1958. Drift bottle observations in the Strait of Georgia. J. Fish. Res. Bd. Can. 15(5):1065-1102.

Primary Productivity of the North Pacific Ocean—A Review

L. A. Hobson

INTRODUCTION

Temperate waters of the North Pacific Ocean are feeding grounds for salmon, and the primary productivity of these waters ultimately determines salmon productivity. Thus, knowledge of primary production and its control by nutrient distribution, irradiation and temperature is important to managing salmon fisheries on the high seas. The purpose of this paper is to review published data pertaining to primary production of the various domains (Dodimead et al. 1963) of the North Pacific (Fig. 1). Although certain reports were not available, I believe that sound, acceptable generalizations about the primary productivity of the North Pacific can be made based on the literature at hand.

METHODS

All primary productivity measurements reported in this paper were determined by the light-dark bottle method using carbon 14 to trace photosynthesis (Steemann-Nielsen 1952) in either *in situ* or simulated *in situ* experiments. This method is generally believed to measure carbon accumulation by cells (i.e. net photosynthesis minus excretion), but is subject to several uncertainties. It may overestimate carbon accumulation if only a small amount of carbon compared to cell carbon is fixed during experiments (Hobson et al. 1976). The magnitude of this error cannot be determined until we increase our understanding of respiration rates in the light. In addition, ignoring the influence of dark respiration on daily carbon accumulation may lead to overestimation of accumulation rates, but again our understanding of environmental influences on rates of dark respiration is limited. Marra (1978) has shown that some cells moving vertically in the mixed layer may have higher productivities than cells in bottles, and our estimates of organic production based on the light-dark bottle technique may sometimes be too low. The net effect of these potential errors on estimates of primary productivity is not known.

All measurements of phytoplankton biomass are based on chlorophyll *a* concentrations averaged over the euphotic zone. These pigment concentrations must be multiplied by a ratio that relates phytoplankton biomass to chlorophyll *a* concentration. Usually a carbon to chlorophyll *a* ratio is used. This ratio may vary from 25 in nutrient-rich water to 100-150 in nutrient-poor water (Strickland 1965), and may decrease when irradiation decreases, reaching a value of 10 in poorly illuminated water (Hobson and Ketcham 1974). The magnitude of this potential error in estimating biomass of phytoplankton is unknown.

DISCUSSION

Regulation of Primary Production

Production of unicellular marine algae equals the product obtained by multiplying the number of photosynthesizing cells times net photosynthetic rates. In turn, net photosynthetic rates equal the rates that CO_2 is reduced to new cell matter minus subsequent oxidation of this matter by respiration and photorespiration.

Rates of CO_2 reduction (gross photosynthesis) are primarily a function of irradiance, temperature and nutrient availability. In the absence of light between 400 and 700 nm wavelength, the gross photosynthetic rate is zero. Photo-

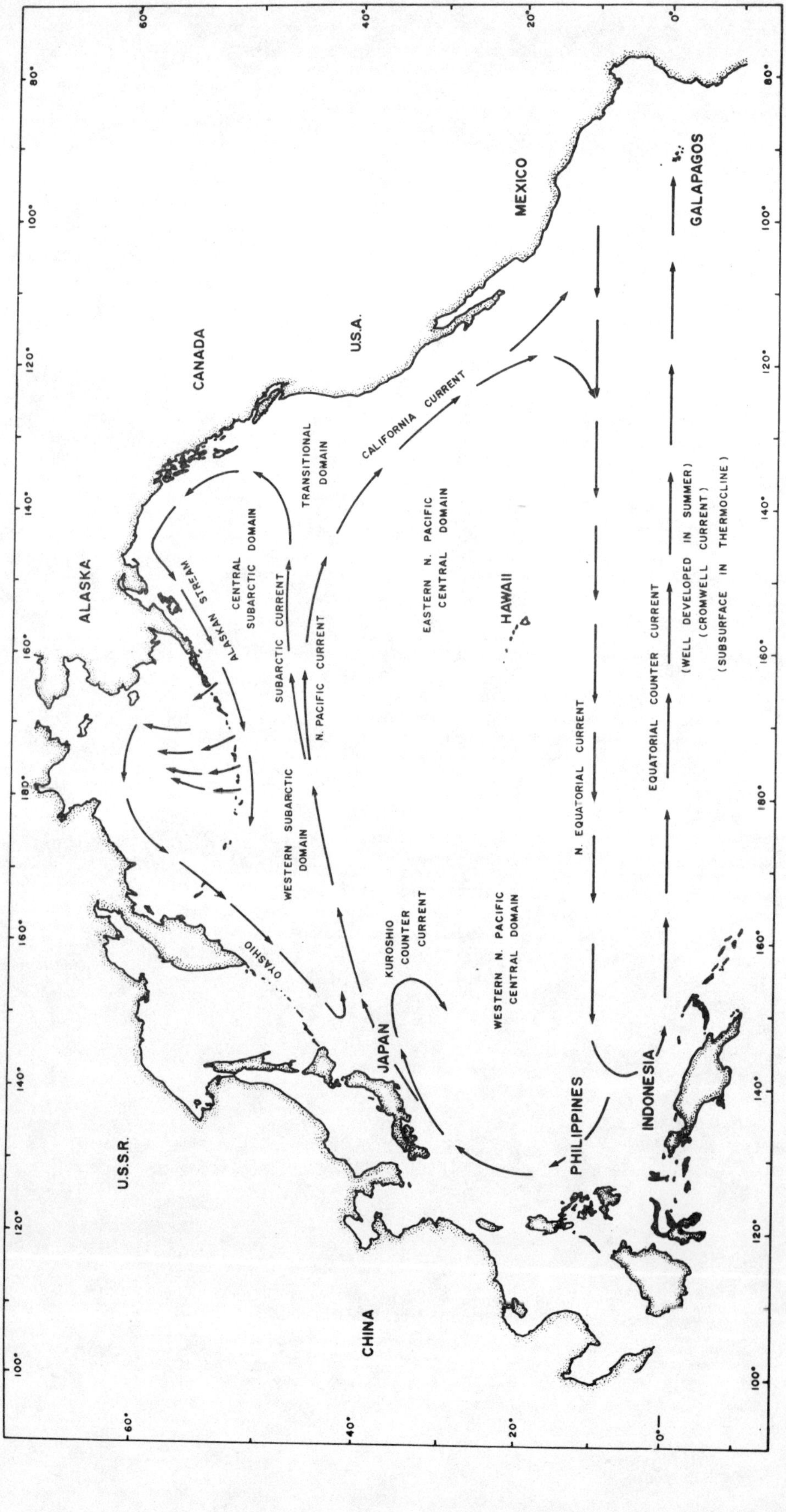

Figure 1. Generalized scheme of surface circulation and structure of the North Pacific Ocean.

synthesis rapidly increases as irradiance levels increase (light-limited), approaches an upper limit (saturated) and may decrease at very high levels (inhibited) (e.g. Platt et al. 1977). Available irradiance and the quantum efficiencies of cells determine the light-limited rates of photosynthesis. Theoretically, quantum efficiency should undergo only small variations caused by changes in pigment composition (Bannister 1974). This is supported by observations of similar values for alpha (proportional to quantum efficiency) among different species of phytoplankton given similar culture conditions (Dunstan 1973). However, other studies show that alpha decreases in nutrient-deficient cells (Jørgensen 1970) and quantum efficiency changes in cells given different photoperiods (Govindjee et al. 1968, Senger and Bishop 1967). In addition, Platt and Jassby (1976) and Taguchi (1970) have shown that alpha increases during summer in phytoplankton populations in the sea. To a certain extent, low alpha values seen in winter populations and deep in the water column (Yentsch and Lee 1966) may be correlated with increases in cellular chlorophyll *a* content (Preszelin 1976, Sargent 1940, Steemann-Nielsen and Jørgensen 1968) of individual photosynthetic units (Schmid and Gaffron 1966). This indicates a physiological adaptation to low irradiation.

Cellular activities of the carboxylating enzyme, ribulose diphosphate carboxylase (RuDP-Case), probably control light saturated rates of photosynthesis. This assumes that control mechanisms are the same in algae as in higher plants (Bjorkman 1968*a*, Collatz 1977, Ku and Edwards 1977, Neales et al. 1971, Treharne and Nelson 1975, Wareing et al. 1968). In higher plants, irradiation history (Bjorkman 1968*b*, 1973), temperature (Philips and McWilliam 1971, Treharne and Cooper 1969, Treharne and Nelson 1975), and presumably nutritional status regulate the activity of RuDP-Case.

Variables that affect respiration and photorespiration regulate net photosynthetic rates in addition to variables that affect gross photosynthesis. Temperature and perhaps irradiance regulate cellular respiration rates. In higher plants, light inhibits dark respiration via metabolites released by photosynthesizing chloroplasts (e.g. Mangat et al. 1974). This inhibition has also been observed in algae (Lloyd et al. 1977, Reid 1970), but Bidwell (1977) obtained evidence indicating that respiration proceeds at rates in the light nearly equal to those in the dark. Whether or not photorespiration reduces photosynthetic rates in marine algae is not yet known. Evidence exists that photorespiration occurs in algae on tropical reefs (Tolbert and Osmond 1976). In the open sea, however, it seems unlikely that oxygen and carbon concentrations become high and low enough respectively, to stimulate photorespiration (Beardall and Morris 1975).

To summarize, irradiation and photoperiod, temperature and nutrient availability primarily control net photosynthetic rates. Net photosynthetic rates, adjusted for losses due to excretion, result in organic matter accumulating as daughter cells in the phytoplankton. These cells may either continue to accumulate organic matter and increase primary production rates, sink below the euphotic zone, or be eaten by herbivores.

Primary Production

Few studies of seasonal variations in primary productivity of offshore waters have been initiated. Of these, data have been collected in the Central Subarctic Domain (Larrance 1971, McAllister 1962); Transitional Domain (Anderson 1964, Hobson 1966*a*, Larrance 1971); and in eastern equatorial waters (Owen and Zeitzschel 1970). Figure 2 presents data from these and other studies conducted over restricted time intervals. No data have been included for water shallower than 100 m to the sea floor or where upwelling was obviously progressing. Productivity values in these waters are usually much larger and not representative of those found in offshore waters.

Primary productivity values in water influenced by the North Equatorial Current and Equatorial Counter Current averaged about 0.20 and 0.24 gC m^{-2} day $^{-1}$ with little seasonal variation. Values appeared slightly higher during February and March than at other times.

Most productivity measurements over the large expanse of the North Pacific Central Domain have been made from June to February. These values average about 0.15 gC m^{-2} day^{-1} with little variation. Two values are available, one in March (0.61 gC m^{-2} day $^{-1}$) and the other in June (0.43 gC m^{-2} day $^{-1}$), that indicate higher values during the spring season than at other times. However, these two measurements were made near the northern boundary of the North Pacific Central Domain (Larrance 1971), and perhaps indicate greater seasonal change in this region than farther south. There is a clear need for a more complete survey of phyto-

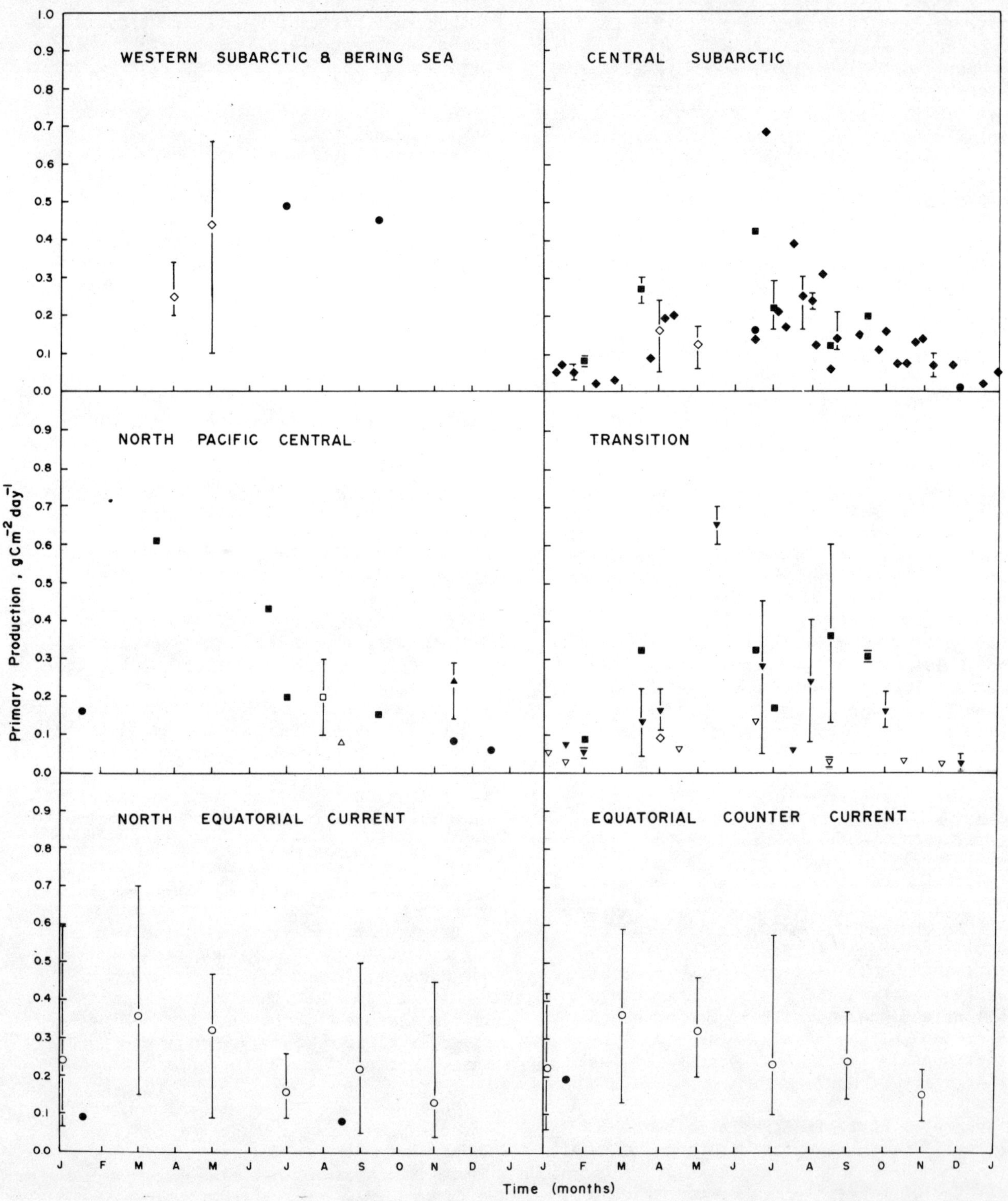

Figure 2. Seasonal variation of primary productivity in various domains of the North Pacific Ocean. Open circles (Owen and Zeitschel 1970), closed circles (Taniguchi 1972), open square (Aruga and Ichimura 1968), closed squares (Larrance 1971), open triangles (Gunderson et al. 1976), closed triangles (Eppley et al. 1973), open inverted triangles (Hobson 1966*a*), closed inverted triangles (Anderson 1964), open diamonds (Parsons and Anderson 1970), and closed diamonds (McAllister 1962).

plankton production in this area than has previously been carried out.

In the Transitional Domain and Central Subarctic Domain, primary production values are less than 0.1 gC m^{-2} day $^{-1}$ from December to March. Values in these domains increase and reach their maximum of about 0.6-0.7 gC m^{-2} day $^{-1}$ in May and June. A measurement of 1.1 gC m^{-2} day $^{-1}$ was made on 1 June in the Central Subarctic Domain (McAllister 1962) and was not included in Figure 2. In this domain, few measurements have been made in May. The seasonal cycle of production in the Transitional Domain differs from that in the Central Subarctic Domain. Values vary greatly in the Transitional Domain from July to October while they uniformly decrease in the subarctic domains. Few data from the Western Subarctic Domain and Bering Sea were found, but April-September values suggest that this area is more productive than the Central Subarctic Domain.

Annual production estimates for the North Pacific were based on data in Figure 2, and vary between 60 and 70 gC m^{-2} in the Central Subarctic Domain, Transitional Domain, and North Pacific Central Domain. The estimate for the latter domain was made assuming production values vary between 0.1 and 0.2 gC m^{-2} day $^{-1}$ throughout the year. This estimate would approach 100 gC m^{-2} year $^{-1}$ in the northern region if spring and summer increases in production occur as suggested by data that Larrance (1971) published. To the south of the North Pacific Central Domain, annual production varies between 85 and 90 gC m^{-2} in waters of the North Equatorial Current and Equatorial Counter Current. These estimates, based on data that Owen and Zeitzschel (1970) published, were collected in the eastern region. Measurements that Taniguchi (1972) presented suggest a lower annual production in equatorial waters to the west. The similarity of these annual production values to those further north is perhaps surprising, but similarities are probably caused by nutrient limitation in southern regions and light limitation in northern regions of the North Pacific.

Nutrient Distribution

Phytoplankton typically utilize nitrogen as nitrate, nitrite and ammonium ions. Other sources include organic nitrogen (principally as urea) and molecular nitrogen used by blue-green bacteria. Phosphate ions and organic phosphorus meet phytoplankton requirements for this element. In addition to nitrogen and phosphorus, Bacillariophytes and Chrysophytes require silicon as silicate ions. Phytoplankton may also require many trace metals and vitamins.

Of the major nutrients, the lack of nitrogen most often limits productivity (Ryther and Dunstan 1971), perhaps, at times, in conjunction with phosphorus (Perry 1972). Nitrogen can limit primary productivity if low concentrations of this element cause photosynthetic rates of cells to decrease, or if the species composition changes to species adapted to utilize low nitrogen concentrations with an accompanying decrease in biomass, or both. Dugdale (1976) recently reviewed in detail the dynamics of the nitrogen cycle in the sea. Phytoplankton remove this nutrient from water in the euphotic zone to produce proteins, nucleic acids and other nitrogen-containing organic compounds. These organic materials subsequently become distributed into animals that release ammonia and urea back into waters in and below the euphotic zone. A net transfer of nitrogenous compounds to waters below the euphotic zone takes place where the compounds eventually oxidize to nitrate. Advection and turbulent diffusion processes operating in the ocean replenish the euphotic zone with nitrate.

Two large gyres dominate the surface circulation of water in the North Pacific (Fig. 1). An anticyclonic movement is bounded by the North Equatorial, Kuroshio, North Pacific and California currents. The Subarctic, Alaskan Stream and Oyashio currents bound the cyclonic movement. Surface winds, which flow in response to pressure fields in the atmosphere, maintain these currents (e.g. Tabata 1975). The position and speeds of surface winds vary seasonally and cause changes in ocean current speed, volume transport and position. However, the anticyclonic movement lying between latitudes of about 10° N and 45° N generally causes convergence and downwelling relative to the cyclonic circulation between latitudes of about 45° N and 60° N where divergence and upwelling occurs. The Subarctic Boundary separates these two regions (Dodimead et al. 1963). The downwellings and upwellings in them produce a southerly flow of surface water toward the North Pacific Central Domain and a northerly flow of subsurface water rich in nutrients to the subarctic domains. Park (1967) discussed this transport and described an upwelling of nitrate-rich water into the euphotic zone north of the Subarctic Boundary (Fig. 3). This upwelling accounts for NO_3-N enrichments that are detectable throughout the year in surface waters of the subarctic domains and vary between 1.0 and 20 μM (Fig. 4).

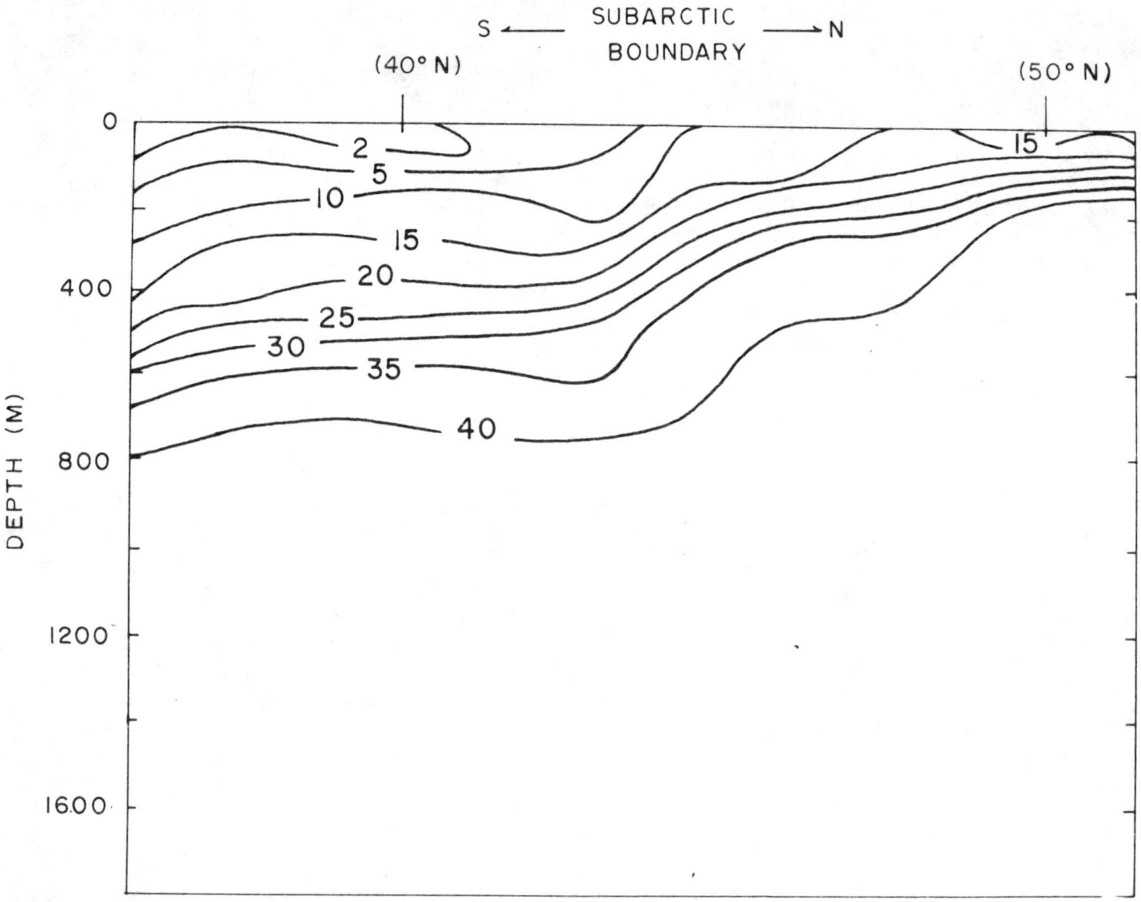

Figure 3. Spatial distributions of NO_3-N concentrations (μM) across the Subarctic Boundary of the North Pacific Ocean, after Park (1967).

To the south in the North Pacific Central Domain, NO_3-N concentrations vary between undetectable and about 0.01 μM during summer months, and undetectable and 0.50 μM during winter months (Fig. 4). In these waters, concentrations of NH_4-N and urea-N are usually greater than NO_3-N (Eppley et al. 1973, 1977), indicating that excretory wastes of animals provide most of the nitrogen available to phytoplankton. During summer months, ammonia may also disappear from surface waters because blue-green bacteria fix molecular nitrogen (Mague et al. 1977), a process that ammonia inhibits.

East of the areas mentioned above is the Transitional Domain. There, NO_3-N concentrations vary between 11 μM during winter and undetectable during summer. In summer, ammonia and urea are probably important to continued phytoplankton production. Also, in coastal waters of this domain, upwelling transports nutrients into the euphotic zone intermittently, starting off Central America during spring and progressing north off Washington and British Columbia during summer (Smith 1969, Wooster and Reid 1963).

South of the westward flowing North Equatorial Current is the Equatorial Counter Current which is best developed in summer (Knauss 1963), and the Cromwell Current. They transport water to the east. Upwelling occurs in waters near and north of the equator between the eastward and westward flowing currents. This transports nutrients into the euphotic zone (Fig. 5), and leads to increased nitrate concentrations in equatorial waters (e.g. Barber and Ryther 1969, Beers and Stewart 1971, Taniguchi 1972) (Fig. 4). Research remains to be done regarding the influences of advective processes, such as eddies produced by the Kuroshio Current (Dantzler 1976) and the Subarctic and Pacific Counter currents (Tabata 1976*a*), on nutrient concentrations in the euphotic zone.

Wind mixing of surface waters extends to depths between 80 and 100 m in subarctic domains, and 100 and 140 m in the North Pacific Central Domain during winter months (Tabata 1976*b*). The depths of this mixing are sufficient to transport nutrients into the euphotic zone from the nutricline

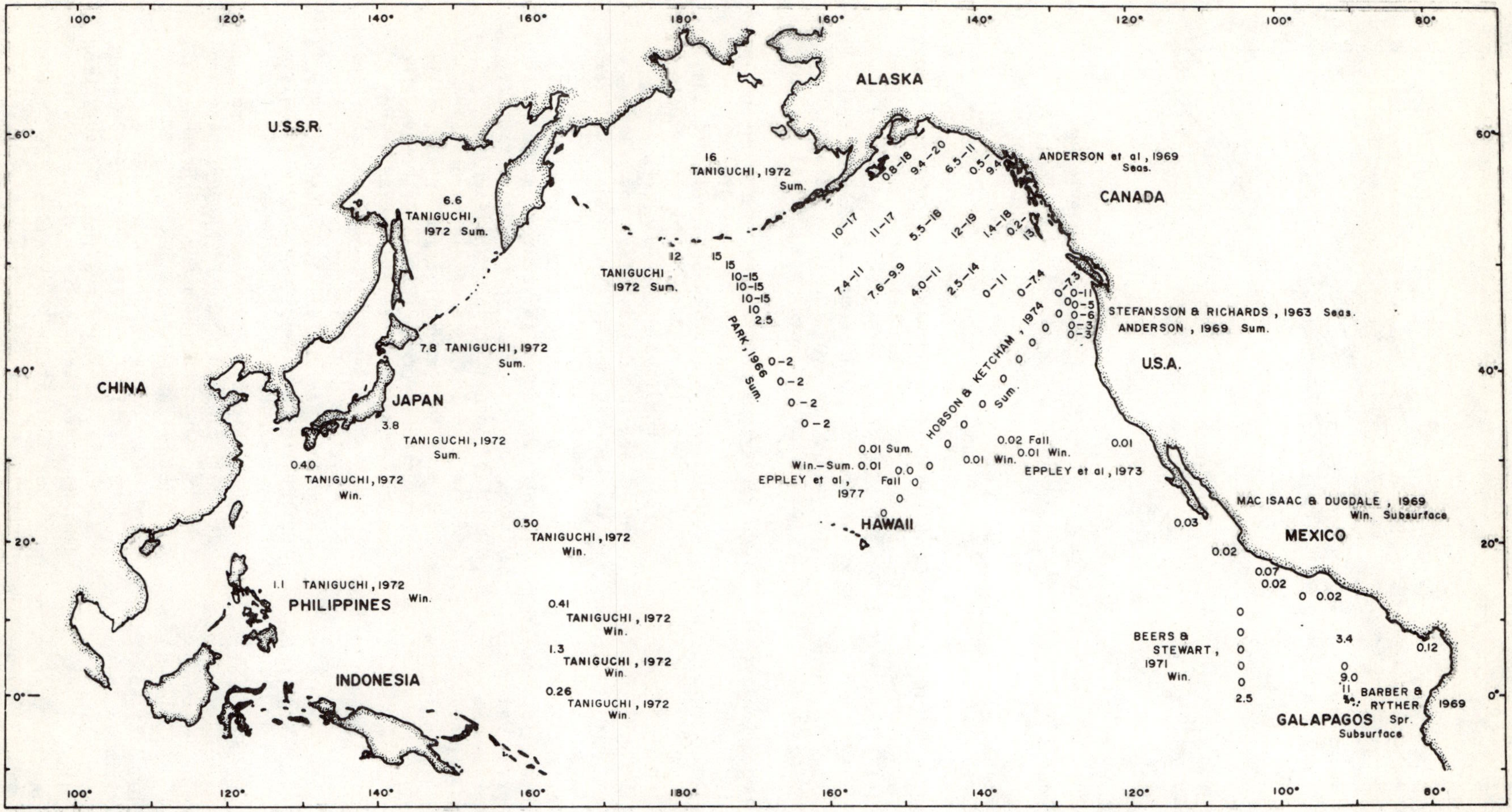

Figure 4. Spatial and seasonal distributions of NO_3-N concentrations (μM) in surface and near-surface water of the North Pacific Ocean.

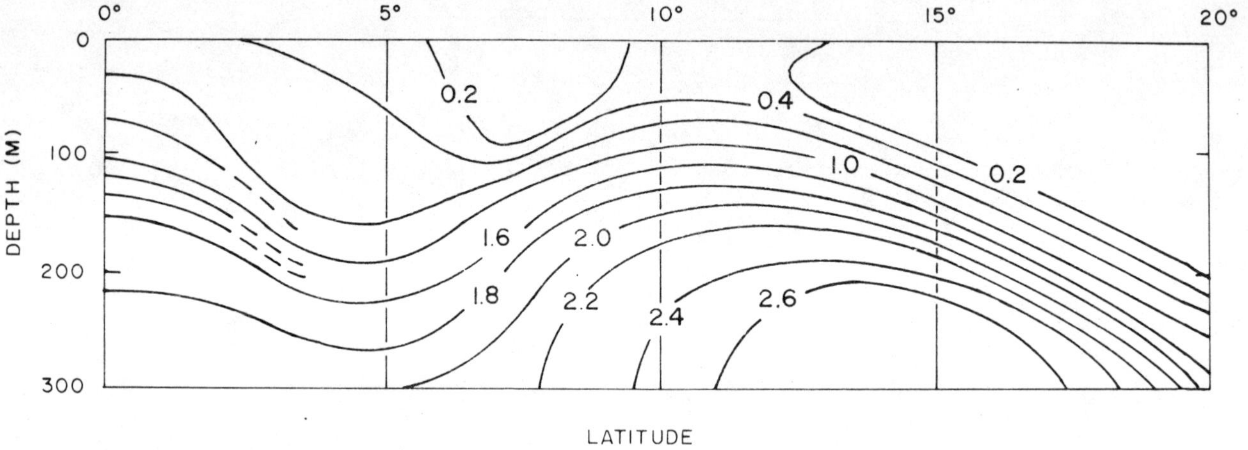

Figure 5. Spatial distribution of PO_4-P concentration (μM) north of the equator in the Pacific Ocean, after Sverdrup et al. (1942).

occuring at depths between 50 and 75 m in the Transitional Domain (Anderson 1969, Hobson and Ketcham 1974), but not from the nutricline found at depths below 130 m in the North Pacific Central Domain (Hobson and Ketcham 1974). During winter, mixed depths in eastern equatorial waters are between 30 and 35 m, shoaling toward the equator with a nutricline ranging from 50 m at 10° 38' N to 25 m at 0° 39' N (Beers and Stewart 1971). Theoretically, storm winds in equatorial waters could fertilize surface waters and lead to relatively large phytoplankton blooms.

Nutrient data for the North Pacific lead to the prediction that maximum phytoplankton production would take place in subarctic domains and minimum production in the North Pacific Central Domain. However, this prediction is limited to late spring and early summer (Fig. 2). During other months, environmental and biological variables, including irradiation, temperature and phytoplankton biomass, limit productivity in subarctic domains.

Irradiation

The influences of irradiation on primary production can be estimated. To do this, variations in depths where daily rates of net photosynthesis are zero must be examined. These compensation depths were calculated for different latitudes in the North Pacific using irradiation data from Sverdrup et al. (1942); transmission coefficients of seawater based on data presented by Eppley et al. (1973), Gunderson et al. (1976), Takahashi et al. (1972) and Taniguchi (1972); and a compensation irradiation of 2.9 joules cm^{-2} day^{-1} (Hobson and Ketcham 1974).

Seasonal variations in compensation depth at the equator range from 90 to 100 m. The compensation depth range reaches maximum values in the northern Gulf of Alaska and Bering Sea varying between 0 m in December and about 45 m during late spring and early summer (Fig. 6). Comparing these depths partially explains the relatively large productivity values in equatorial waters relative to subarctic domains because productivity per unit volume, which may be low in equatorial waters, is integrated over greater depths than in subarctic domains. The seasonally uniform productivity data reflects the lack of seasonal change in subsurface irradiation near the equator (Fig. 2). To the north, seasonal variation in compensation depths would suggest high primary productivity from at least March through September in subarctic domains and throughout the year in the Transitional Domain. Actual production values are relatively high in these domains from March to September, but are quite low in winter in the Transitional Domain (Fig. 2). Perhaps the assumed compensation irradiation is too low for winter phytoplankton and thus the real compensation depth during fall and winter is nearer the sea surface than shown in Figure 6. Preliminary studies I conducted in a temperate estuarine environment give a seasonal average for compensation irradiation at 6.8 joules cm^{-2} day^{-1} for net photosynthesis, but variance among measurements masked any seasonal change (Fig. 7). We need further information about winter levels of compensation irradiation. Another source of information indicates that phytoplankton in temperate latitudes may be adapted to seasonal variations in irradiation. The seasonal cycle of light-saturated rates of net photosynthesis normalized on chlorophyll *a* (P_{max}) gives maximum and minimum responses

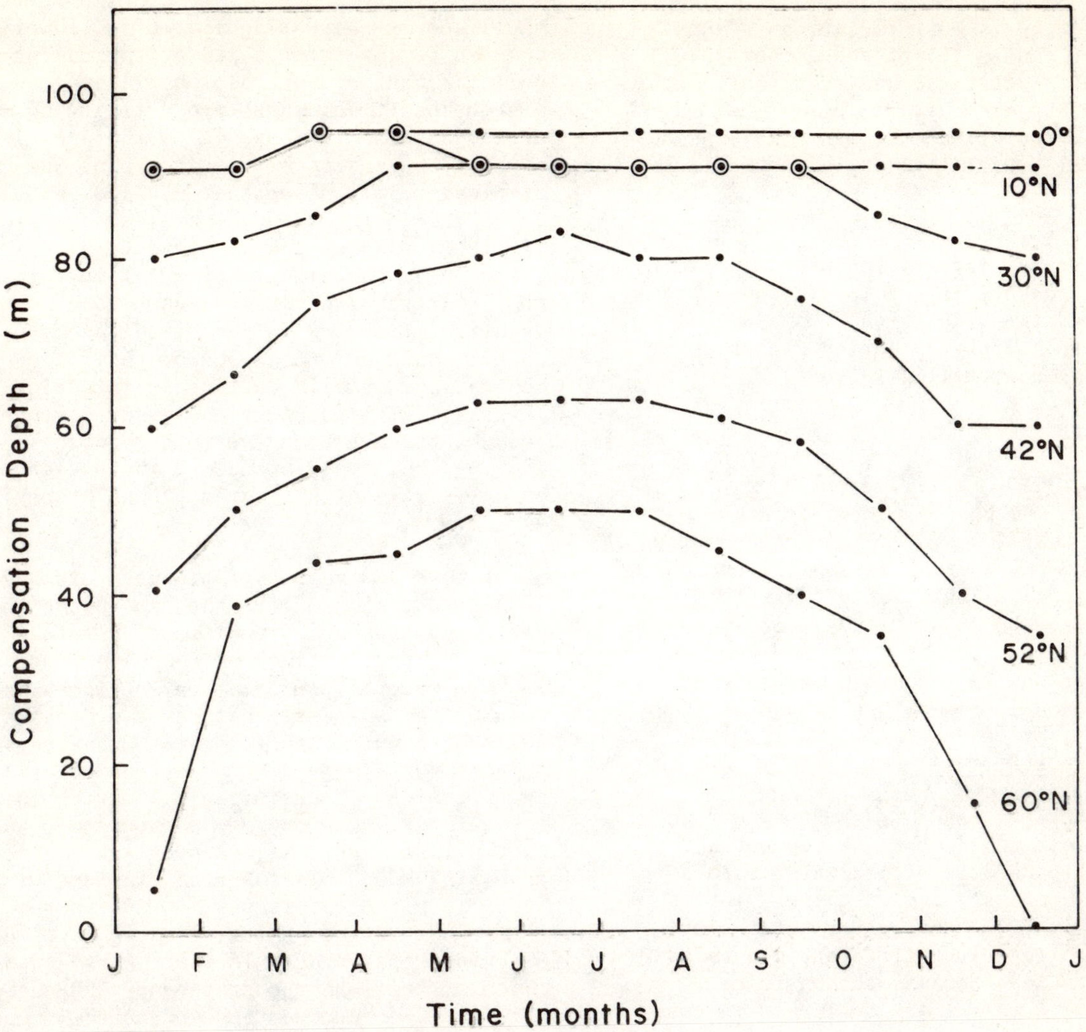

Figure 6. Seasonal variation in calculated compensation depths at various latitudes of the North Pacific Ocean. See text for details.

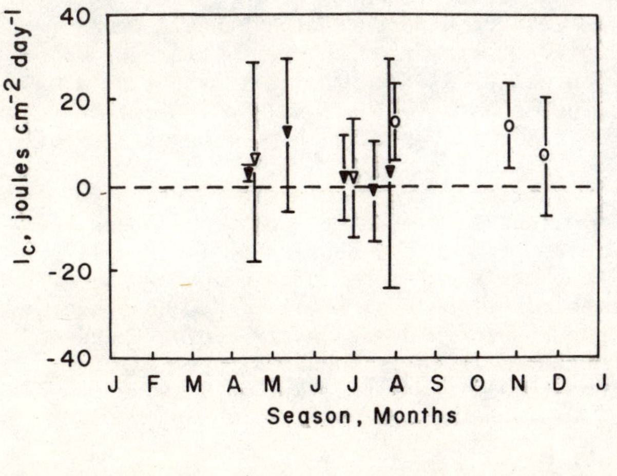

Figure 7. Seasonal variation in compensation irradiation for net photosynthesis of phytoplankton assemblages in Saanich Inlet, British Columbia ($48^{o}40'$ N lat., $123^{o}30'$ W long.). Values were derived by extrapolation of lines based on measurement of daily photosynthetic rates using ^{14}C as a tracer at various irradiations to net photosynthetic rates of zero. Vertical bars represent 95 percent confidence limits of mean values.

of photosynthesis to irradiation in summer and winter months (Platt and Jassby 1976). Results of statistical analyses that I have conducted suggest that day length, and thus seasonal irradiation (Steemann-Nielsen and Hansen 1961), participates in controlling P_{max} (Fig. 8).

In summary, irradiation probably limits seasonal productivity cycles in subarctic domains and the Transitional Domain, but temperature and phytoplankton biomass may also act as controlling factors.

Temperature

Eppley (1972) reviewed the literature on temperature and phytoplankton growth. He concluded that low winter temperatures may depress growth rates of phytoplankton, but few data are available to test his conclusion. Platt and Jassby (1976) concluded that seasonal cycles of P_{max} in phytoplankton populations in temperate waters were due to seasonal temperature changes.

Phytoplankton Biomass

Chlorophyll *a* concentrations appear to be directly related to nutrient concentrations in the North Pacific with highest values in subarctic domains and near the equator in January (Fig. 9). Whether or not these differences represent different biomasses is not yet known because carbon to chlorophyll *a* ratios may be larger in subtropical and tropical waters than in the subarctic domains.

The predicted presence of large phytoplankton biomasses in subarctic domains using chlorophyll *a* concentrations in winter months is surprising because ratios of critical (Sverdrup 1953) to mixed depths are usually less than 1.0 from December through February in these waters (Parsons et al. 1966). Thus, only a net decrease in biomass could occur in the mixed layer unless cells are unevenly distributed. If flagellated cells dominated the phytoplankton during winter, as suggested by results from temperate inshore waters (Fig. 10) and the Transitional Domain (Hobson 1966*b*), then they might maintain a distribution near the surface, at least during calm periods. The large chlorophyll *a* concentrations could then be explained if these flagellated cells were adapted to low irradiation by increasing cellular contents of chlorophyll *a*. As winter conditions pass, productivity of subarctic domains increases as irradiance increases, waters stabilize and warm, and critical to mixed depths reach 1.0 and greater in March through May (Parsons et al. 1966). These changes lead to a spring increase in the biomass of relatively immobile diatoms, which are known to be evenly distributed in the mixed layer during winter (Hobson 1966*b*). Chlorophyll *a* concentrations usually decline at this time (Fig. 9). This decline has generally been thought to have been due to herbivore grazing, but could be due to the presence of cells with higher carbon to chlorophyll *a* ratios.

Diatoms dominate phytoplankton biomasses in nutrient-rich waters such as in the subarctic domains and Transitional Domain during spring, regions of upwelling, and inshore waters. Information, beginning to accumulate, suggests that these organisms are adapted to nutrient-rich waters that can support very large concentrations of cells and colonies. These organisms may contain the carboxylating enzyme, phosphoenolpyruvate carboxylase (PEP-Case) (Beardall et al. 1976), requiring HCO_3^- as a substrate instead of CO_2. Thus, in bloom conditions, as CO_2 is removed from seawater and pH increases in daytime, CO_2 concentrations could limit photosynthetic rates of cells that contain only RuDP-Case. This could occur even though the cells contain carbonic anhydrase, an enzyme whose kinetics usually favor production of HCO_3^- from CO_2. Cells having PEP-Case could continue to use abundant supplies of HCO_3^-, even at a relatively high pH, to donate CO_2 to RuDP-Case or to produce biochemical substrates that the cell requires. Photosynthesis would continue, albeit at a lower efficiency than when using CO_2 from the environment as a carbon substrate (personal communication, W. J. Morris., Biology Dept., University of Victoria, Victoria, B.C.).

Nutrient-rich waters are relatively important in providing food to man because diatoms frequently consist of large cells. These cells may form large colonies that herbivores, such as small fish, can graze. Thus, photosynthetic energy can be transported efficiently to animal species of commercial importance (Ryther 1969).

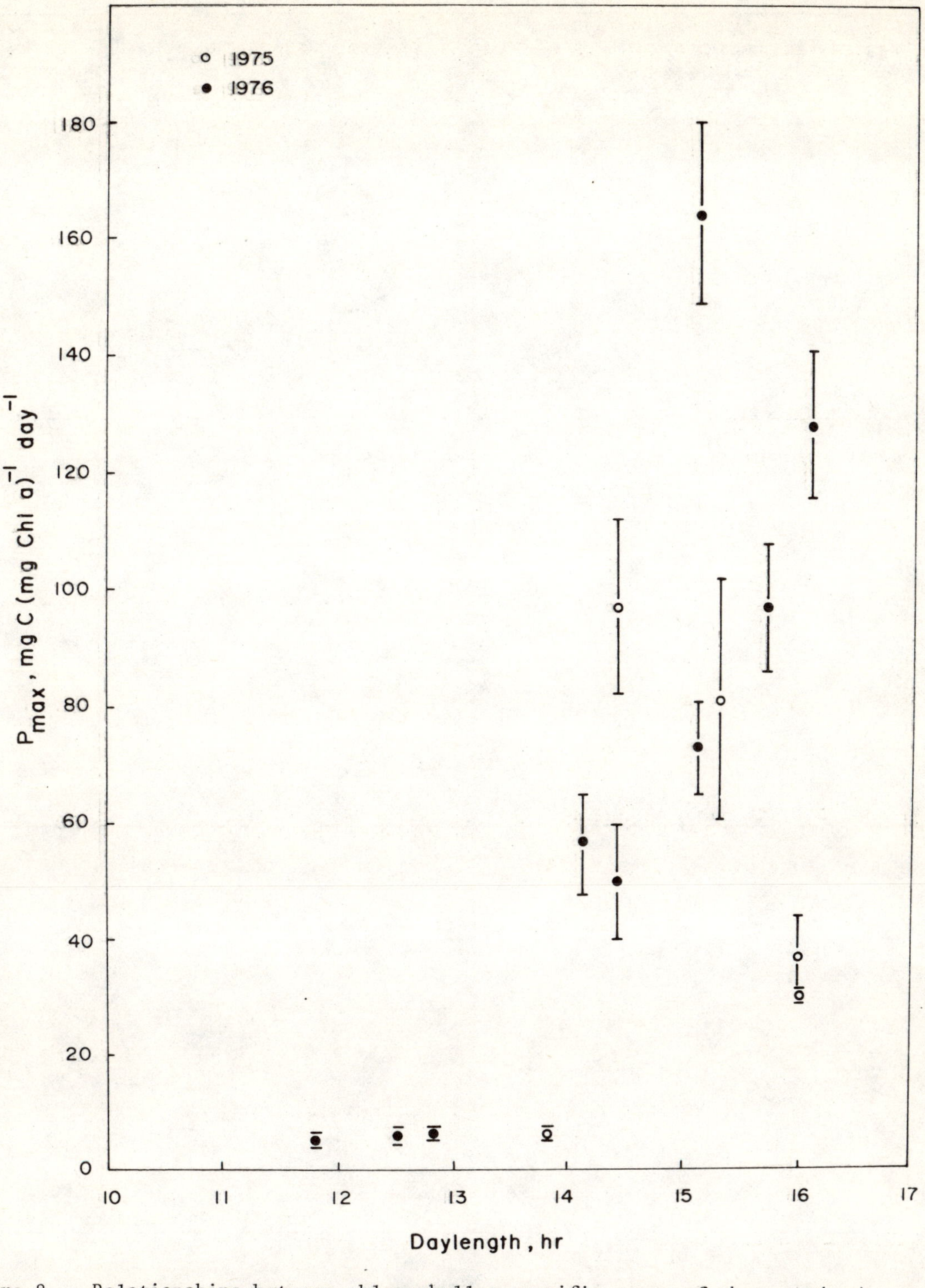

Figure 8. Relationships between chlorophyll *a* specific rates of photosynthesis and day length of phytoplankton assemblages in Saanich Inlet.

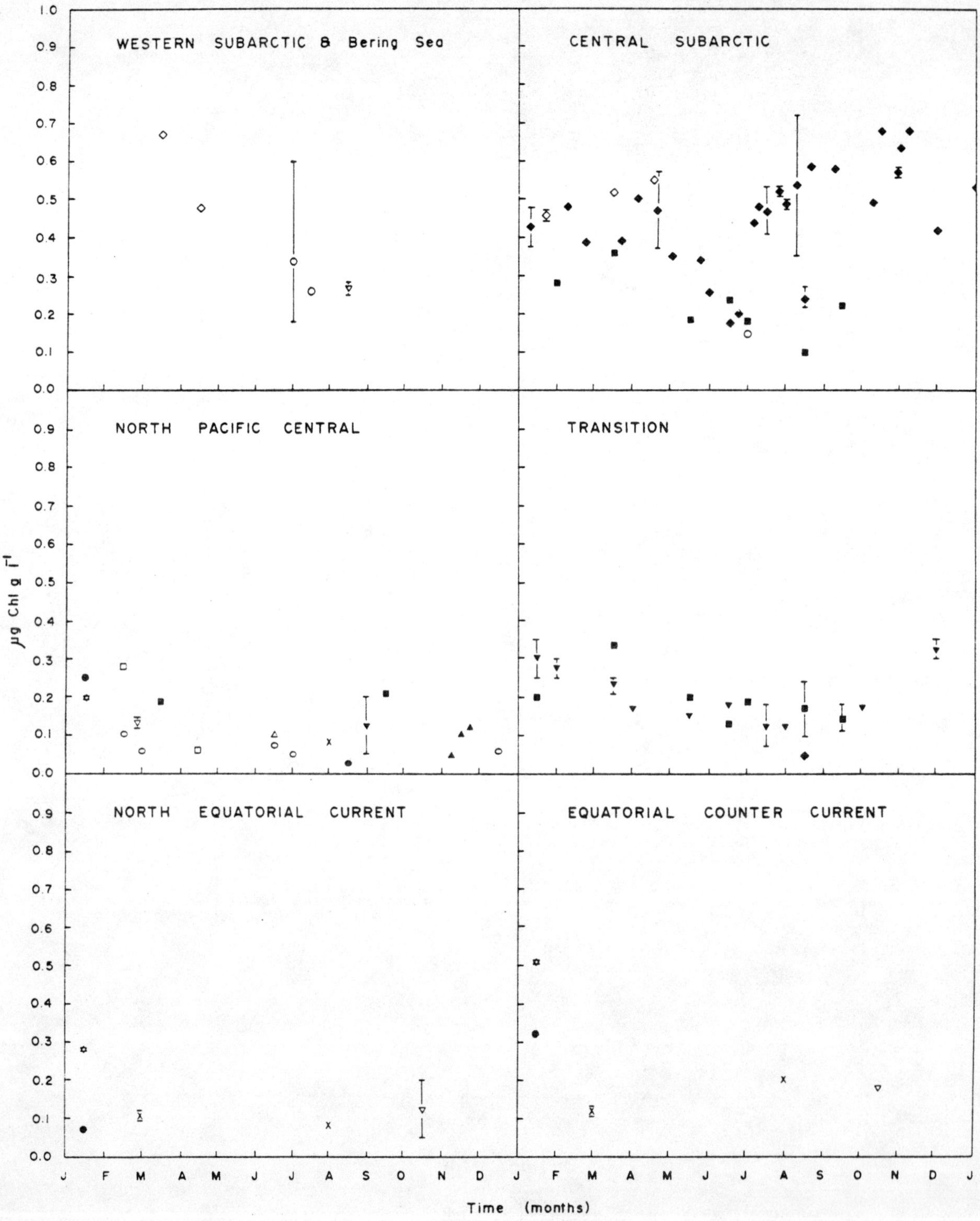

Figure 9. Seasonal variations of chlorophyll *a* concentrations in various domains of the North Pacific Ocean. Open circles (Taniguchi 1969), closed circles (Taniguchi 1972), open squares (Aruga and Ichimura 1969), closed squares (Larrance 1971), open triangles (Gunderson et al. 1976), closed triangles (Eppley et al. 1973), open inverted triangles (Takahashi et al. 1972), closed inverted triangles (Anderson 1964), open diamonds (Parsons and Anderson 1970), closed diamonds (McAllister 1962), crosses (Nishizawa et al. 1971), stars (Kawarada and Sano 1969), open hexagons (Eppley et al. 1977), closed hexagons (Hobson and Ketcham 1974), and crosses with bars (Beers and Stewart 1971).

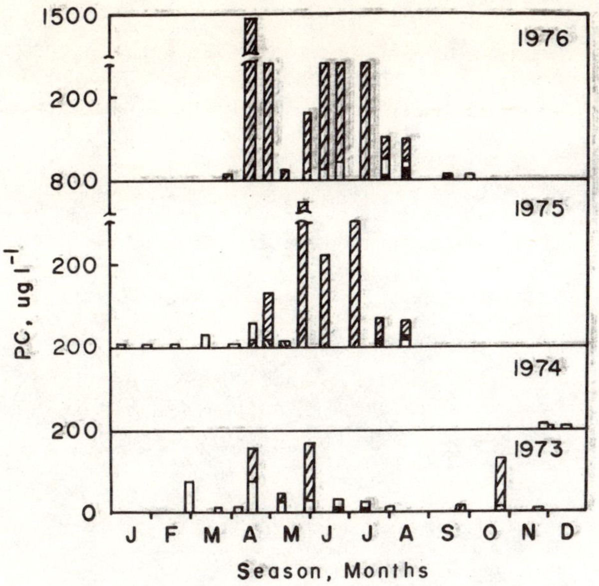

Figure 10. Seasonal variations in phytoplankton carbon and taxonomic composition in Saanich Inlet. μ-Flagellates (☐), diatoms (▨), dinoflagellates (■), and euglenophytes (▧). Phytoplankton carbon was determined by inverted microscopy (Hobson et al. 1976). No samples were taken during winter, spring, and summer seasons of 1974.

ACKNOWLEDGEMENTS

Preparation of this review and collection of original data were supported by grants from the National Research Council of Canada and the Faculty Research Fund of the University of Victoria. I thank Mrs. F. A. Hartley for drafting figures. Thanks also to Dr. J. McInerney for providing a useful critique of this manuscript.

LITERATURE CITED

Anderson, G. C. 1964. The seasonal and geographic distribution of primary productivity off the Washington and Oregon coasts. Limnol. Oceanogr. 9: 284-302.

Anderson, G. C. 1969. Subsurface chlorophyll maximum in the northeast Pacific Ocean. Limnol. Oceanogr. 14:386-391.

Anderson, G. C., T. R. Parsons, and K. Stephens. 1969. Nitrate distribution in the subarctic northeast Pacific Ocean. Deep Sea Res. 16:329-334.

Aruga, Y., and S. Ichimura. 1968. Characteristics of photosynthesis of phytoplankton and primary production in the Kuroshio. Bull. Misaki Mar. Biol. Inst., Hyoto Univ. 12:1-20.

Bannister, T. T. 1974. Production equations in terms of chlorophyll concentration, quantum yield, and upper limit to production. Limnol. Oceanogr. 19:1-12.

Barber, R. T., and J. H. Ryther. 1969. Organic chelators: factors affecting primary production in the Cromwell Current upwelling. J. Exp. Mar. Biol. Ecol. 3:191-199.

Beardall, J., and I. Morris. 1975. Effects of environmental factors on photosynthesis patterns in *Phaeodactylum tricornutum* (Bacillariophyceae). II. Effects of oxygen. J. Phycol. 11:430-434.

Beardall, J., D. Mukerji, H. E. Glover, and I. Morris. 1976. The path of carbon in photosynthesis by marine phytoplankton. J. Phycol. 12:409-417.

Beers, J. R., and G. L. Stewart. 1971. Micro-zooplankton in the plankton communities of the upper waters of the eastern tropical Pacific. Deep Sea Res. 18:861-883.

Bidwell, R. G. S. 1977. Photosynthesis and light and dark respiration in freshwater algae. Can. J. Bot. 55:809-818.

Björkman, O. 1968*a*. Carboxydismutase activity in shade-adapted and sun-adapted species of higher plants. Physiol. Plant. 21:1-10.

Björkman, O. 1968*b*. Further studies on differentiation of phytosynthetic properties in sun and shade ecotypes of *Solidago virgaurea*. Physiol. Plant. 21:84-99.

Björkman, O. 1973. Comparative studies on phytosynthesis in higher plants. Pages 1-63 *in* Giese, A. C. (ed.), Photophysiology, Vol. 8. Academic Press, N.Y.

Collatz, G. J. 1977. Influence of certain environmental factors on photosynthesis and photorespiration in *Simmondsia chinensis*. Planta. 134:127-132.

Dantzler, H. L., Jr. 1976. Geographic variation in intensity of the North Atlnatic and North Pacific eddy fields. Deep Sea Res. 23:783-794.

Dodimead, A. J., F. Favorite, and T. Hirano. 1963. Salmon of the North Pacific Ocean. II. Review of the oceanography of the subarctic Pacific region. Bull. Inter. North Pacific Fish. Comm. 13:1-195.

Dugdale, R. C. 1976. Nutrient cycles. Pages 141-172 *in* Cushing, D. H. and J. J. Walsh (eds.), The ecology of the seas. W. B. Saunders Co., Philadelphia.

Dunstan, W. M. 1973. A comparison of the photosynthesis-light intensity relationship in phylogenetically different marine microalgae. J. Exp. Mar. Biol. Ecol. 13:181-187.

Eppley, R. W. 1972. Temperature and phytoplankton growth in the sea. Fish. Bull. 70:1063-1085.

Eppley, R. W., E. H. Renger, E. L. Venrick, and M. M. Mullin. 1973. A study of plankton dynamics and nutrient cycling in the central gyre of the North Pacific Ocean. Limnol. Oceanogr. 18:534-551.

Eppley, R. W., J. H. Sharp, E. H. Renger, M. J. Perry, and W. G. Harrison. 1977. Nitrogen assimilation by phytoplankton and other microorganisms in the surface waters of the central North Pacific Ocean. Mar. Biol. 39:11-120.

Govindjee, R. E., Rabinowitch and Govindjee. 1968. Maximum quantum yield and action spectrum of photosynthesis and fluorescence in *Chlorella*. Bioc. Biophys. Acta. 162:539-544.

Gunderson, K. R., J. S. Corbin, C. L. Hanson, M. L. Hanson, R. B. Hanson, D. J. Russell, A. Stollar, and O. Yamada. 1976. Structure and biological dynamics of the ogliotrophic ocean photic zone off the Hawaiian Islands. Pac. Sci. 30:45-68.

Hobson, L. A. 1966*a*. The seasonal and vertical distribution of suspended particulate matter in an area of the northeast Pacific Ocean. Ph.D. Thesis. Univ. of Wash., Seattle. 107 pp.

Hobson, L. A. 1966*b*. Some influences of the Columbia River effluent on marine phytoplankton during January, 1961. Limnol. Oceanogr. 11:223-234.

Hobson, L. A., and D. E. Ketcham. 1974. Observations of subsurface distributions of chlorophyll *a* and phytoplankton carbon in the northeast Pacific Ocean. J. Fish. Res. Bd. Can. 31:1919-1925.

Hobson, L. A., W. J. Morris, and K. T. Pirquet. 1976. Theoretical and experimental analysis of the ^{14}carbon technique and its use in studies of primary production. J. Fish. Res. Bd. Can. 33:1715-1721.

Jørgensen, E. G. 1970. The adaptation of plankton algae. V. Variation in the photosynthetic characteristics of *Skeletonema costatum* cells grown at low light intensity. Physiol. Plant. 23:11-17.

Kawarade, Y., and A. Sano. 1969. Distribution of chlorophyll and phaeophytin in the western North Pacific. Oceanogr. Mag. 21:137-146.

Knauss, J. A. 1963. Equatorial current systems. Pages 235-252 *in* Hill, M. N. (ed.), The sea, Vol. 2. Interscience Publ., New York.

Ku, S., and G. E. Edwards. 1977. Oxygen inhibition of photosynthesis. Plant Physiol. 59:991-999.

Larrance, J. D. 1971. Primary production in the mid-subarctic Pacific region, 1966-68. Fish. Bull. 69:595-613.

Lloyd, N. D. H., D. T. Canvin, and D. A. Culber. 1977. Photosynthesis and photorespiration in algae. Plant Physiol. 59:936-940.

Lorenzen, C. J. 1966. A method for the continuous measurement of *in vivo* chlorophyll concentration. Deep Sea Res. 13:223-227.

MacIsaac, J. J., and R. C. Dugdale. 1969. The kinetics of nitrate and ammonia uptake by natural populations of marine phytoplankton. Deep Sea Res. 16:45-57.

Mague, T. H., F. C. Mague, and O. Holm-Hansen. 1977. Physiology and chemical composition of nitrogen-fixing phytoplankton in the central North Pacific Ocean. Mar. Biol. 41:213-227.

Mangat, B. S., W. B. Levin, and R. G. S. Bidwell. 1974. The extent of dark respiration in illuminated leaves and its control by ATP levels. Can. J. Bot. 52:673-681.

Marra, J. 1978. Phytoplankton photosynthetic response to vertical movement in a mixed layer. Mar. Biol. 46:203-208.

McAllister, C. D. 1962. Photosynthesis and chlorophyll *a* measurements at Ocean Weather Station "P", July 1959 to November 1961. Fish. Res. Bd. Can., Manuscr. Rep. Series 126. 14 pp.

Neales, T. F., K. J. Treharne, and P. F. Wareing. 1971. A relationship between net photosynthesis, diffusive resistance, and carboxylating enzyme activity in bean leaves. Pages 89-96 *in* Hatch, M. D., C. B. Osmond and R. O. Slatyer (eds.), Phytosynthesis and photorespiration. J. Wiley and Sons, Inc., New York.

Nishizawa, S., A. Taniguchi, and T. Ichikawa. 1971. An oceanographic section from Lat. 20° S to 20° N at 175° E in July to August 1968. Kaiyo Rep. 3. 16 pp.

Owen, R. W., and B. Zeitzschel. 1970. Phytoplankton production: seasonal change in the oceanic eastern tropical Pacific. Mar. Biol. 7:32-36.

Park, K. 1967. Chemical features of the Subarctic Boundary near 170° W. J. Fish. Res. Bd. Can. 24:899-908.

Parsons, T. R., and G. C. Anderson. 1970. Large scale studies of primary production in the North Pacific Ocean. Deep Sea Res. 17:765-776.

Parsons, T. R., L. F. Giovando, and R. J. LeBrasseur. 1966. The advent of the spring bloom in the eastern subarctic Pacific Ocean. J. Fish. Res. Bd. Can. 23:539-546.

Perry, M. J. 1972. Alkaline phosphatase activity in subtropical central North Pacific waters using a sensitive fluorometric method. Mar. Biol. 15:113-119.

Philips, P. J., and J. R. McWilliam. 1971. Thermal responses of the primary carboxylating enzymes from C_3 and C_4 plants adapted to contrasting temperature environments. Pages 97-104 *in* Hatch, M. D., C. B. Osmond and R. O. Slayter (eds.), Photosynthesis and photorespiration. J. Wiley and Sons, Inc., New York.

Platt, T., and A. D. Jassby. 1976. The relationship between photosynthesis and light for natural assemblages of coastal marine phytoplankton. J. Phycol. 12: 421-430.

Platt, T., K. L. Denman, and A. D. Jassby. 1977. Modeling the productivity of phytoplankton. Pages 807-856 *in* Goldberg, E. (ed.), The sea, Vol. 6.

Prézelin, B. B. 1976. The role of peridinin-chlorophyll *a*-proteins in the photosynthetic light adaptation of the marine dinoflagellate, *Glenodinium* sp. Planta. 130:225-233.

Reid, A. 1970. Energetic aspects of the interaction between photosynthesis and respiration. Pages 231-246 *in* Prediction and measurement of photosynthetic productivity. Proc. IBP/PP Tech. Meet. Wageningen.

Ryther, J. H. 1969. Photosynthesis and fish production in the sea. Sci. 166:72-76.

Ryther, J. H., and W. M. Dunstan. 1971. Nitrogen, phosphorus and eutrophication in the coastal marine environment. Sci. 171:1008-1013.

Sargent, M. C. 1940. Effect of light intensity on the development of the photosynthetic mechanism. Plant Physiol. 15:275-290.

Schmid, G. H., and H. Gaffron. 1966. Photosynthetic units. J. Gen. Physiol. 52:212-239.

Senger, H., and N. I. Bishop. 1967. Quantum yield of photosynthesis in synchronous *Scenedesmus* cultures. Nature 214: 140-142.

Smith, R. L. 1968. Upwelling. Pages 11-46 *in* Barnes, H. (ed.), Oceanography and marine biology, Vol. 6. George Allen and Unwin Ltd., London.

Steemann-Nielsen, E. 1952. The use of radioactive (C^{14}) for measuring organic production in the sea. J. Cons. Int. Explor. Mer. 18:117-140.

Steemann-Nielsen, E., and V. K. Hansen. 1961. Influence of surface illumination on plankton photosynthesis in Danish waters (56° N) throughout the year. Physiol. Plant. 14:595-613.

Steemann-Nielsen, E., and E. G. Jørgenson. 1968. The adaptations of plankton algae. I. General part. Physiol. Plant. 21: 401-413.

Stefánsson, U., and F. A. Richards. 1963. Processes contributing to the nutrient distributions off the Columbia River and Strait of Juan de Fuca. Limnol. Oceanogr. 8:394-410.

Strickland, J. D. H. 1965. Production of organic matter in the primary stages of the marine food chain. Pages 478-610 *in* Riley, J. P., and G. Skirrow (eds.), Chemical oceanography, Vol. 2. Academic Press, London.

Sverdrup, H. U. 1953. On conditions for the vernal blooming of phytoplankton. J. Cons. Int. Explor. Mer. 18:287-296.

Sverdrup, H. U., M. W. Johnson, and R. H. Fleming. 1942. The oceans. Prentice-Hall, Inc., Inglewood Cliffs, New Jersey.

Tabata, S. 1975. The general circulation of the Pacific Ocean and a brief account of the oceanographic structure of the North Pacific Ocean. Pt. I. Circulation and volume transports. Atmos. 13:133-168.

Tabata, S. 1976*a*. Evidence of a westward-flowing "Subarctic Countercurrent" in the North Pacific Ocean. J. Fish. Res. Bd. Can. 33:2168-2196.

Tabata, S. 1976*b*. The general circulation of the Pacific Ocean and a brief account of the oceanographic structure of the North Pacific Ocean. Pt. II. Thermal regime and influence on the climate. Atmos. 14:1-27.

Taguchi, S. 1970. Seasonal variations of photosynthetic behaviour of phytoplankton in Akkeshi Bay, Hokkaido, with special reference to low photosynthetic rate in summer associated with large percentage of dwarf cells. Bull. Plankton Soc. Japan. 17:65-77.

Takahashi, M., K. Satake, and N. Nakamoto. 1972. Chlorophyll distribution and photosynthetic activity in the North and Equatorial Pacific Ocean along 155° W. J. Oceanogr. Soc. Japan. 28:27-36.

Taniguchi, A. 1969. Regional variations of surface primary production in summer and the vertical stability of water affecting the production. Bull. Fac. Fish., Hokkaido Univ. 20:169-179.

Taniguchi, A. 1972. Geographical variation of primary production in the Western Pacific Ocean and adjacent seas with reference to the interrelations between various parameters of primary production. Mem. Fac. Fish., Hokkaido Univ. 19:1-34.

Tolbert, N. E., and C. P. Osmond, eds. 1976. Photorespiration in marine plants. CSIRO, Melbourne and University Park Press, Baltimore.

Treharne, K. J., and C. J. Nelson. 1975. Effect of growth temperature on photosynthetic and photorespiratory activity in tall fescue. Pages 61-69 *in* Marcelle, R. (ed.), Environmental and biological control of photosynthesis. Dr. W. Junk, The Hague.

Wareing, P. F., M. M. Khalifa, and K. J. Treharne. 1968. Rate-limiting processes in photosynthesis at saturating light intensities. Nature 220:453-457.

Wooster, W. S., and J. L. Reid, Jr. 1963. Eastern boundary currents. Pages 253-280 *in* Hill, M. N. (ed.), The sea. Vol. 2. Interscience Publ., New York.

Yentsch, C. S., and R. W. Lee. 1966. A study of photosynthetic light reactions, and a new interpretation of sun and shade phytoplankton. Jr. Mar. Res. 24:319-337.

Estuarine Influence of Salinity, Temperature, and Food on the Behavior, Growth, and Dynamics of Bristol Bay Sockeye Salmon

Richard R. Straty and Herbert W. Jaenicke

INTRODUCTION

Fishery scientists generally agree that a major part of total salmon mortality in the ocean occurs during their brief residence as juveniles in estuarine and coastal waters (Parker 1962). Pink salmon *(Oncorhynchus gorbuscha)* in British Columbia experienced a much greater natural mortality during the initial period of life (40 days) in the coastal marine waters than during the remaining period of life (410 days) (Parker 1965).

Experience shows that mortality of salmon in the ocean is quite variable and often unpredictable. The imprecision of predicting year class success stems mostly from our lack of knowledge about how the estuarine environment affects juvenile salmon. Most would agree that conditions in the estuarine and nearshore environment affect the distribution, abundance, movements and survival of salmon; however, the complexities of physical and biological processes involved make it difficult to define cause and effect relations. Much research has been conducted on the responses of salmon to variations in experimental environments, but the results have only limited application to salmon behavior in natural environments. This report summarizes information from the Bristol Bay region of western Alaska (Fig. 1) on variations in the estuarine and marine environment and resulting responses of seaward-migrating sockeye salmon.

The National Marine Fisheries Service investigated the early sea life of sockeye salmon, *Oncorhynchus nerka*, in Bristol Bay in 1966-67 and 1969-72. The overall objective was to determine what portion of marine mortality of sockeye salmon occurs as juveniles in the estuary. The results provide information on responses of juvenile sockeye to environmental conditions and insight into some probable causes of marine mortality.

In this paper, we use results of our investigations in Bristol Bay and published research from other areas to describe and hypothesize how the estuarine environment affects the behavior, growth and survival of sockeye salmon. Information on the behavior of juvenile sockeye salmon may apply to some extent to other salmonid species.

DISCUSSION

Marine Life of Bristol Bay Sockeye

The marine life of the sockeye salmon can be conveniently divided into three phases: seaward migration through estuarine and nearshore waters; ocean life in the offshore water; and spawning migration from the offshore water through the estuary to the river of origin.

Sockeye smolts begin their seaward migration by about 15 May from the lake and river systems draining into inner Bristol Bay (Fig. 1). They usually complete this migration by 15 July (Straty 1974). Several days after the smolts leave the lake outlets, they enter the bay and begin their journey through the bay and eastern Bering Sea enroute to feeding grounds in the North Pacific Ocean. The period of seaward migration to the North Pacific lasts longer than six months (Straty 1974). During this period young sockeye attain the bulk of their first year of ocean growth.

Life in the ocean begins when immature sockeye enter the North Pacific. They reside there for one to four years, growing to sexual maturity before returning to spawn in their home river systems that drain into Bristol Bay.

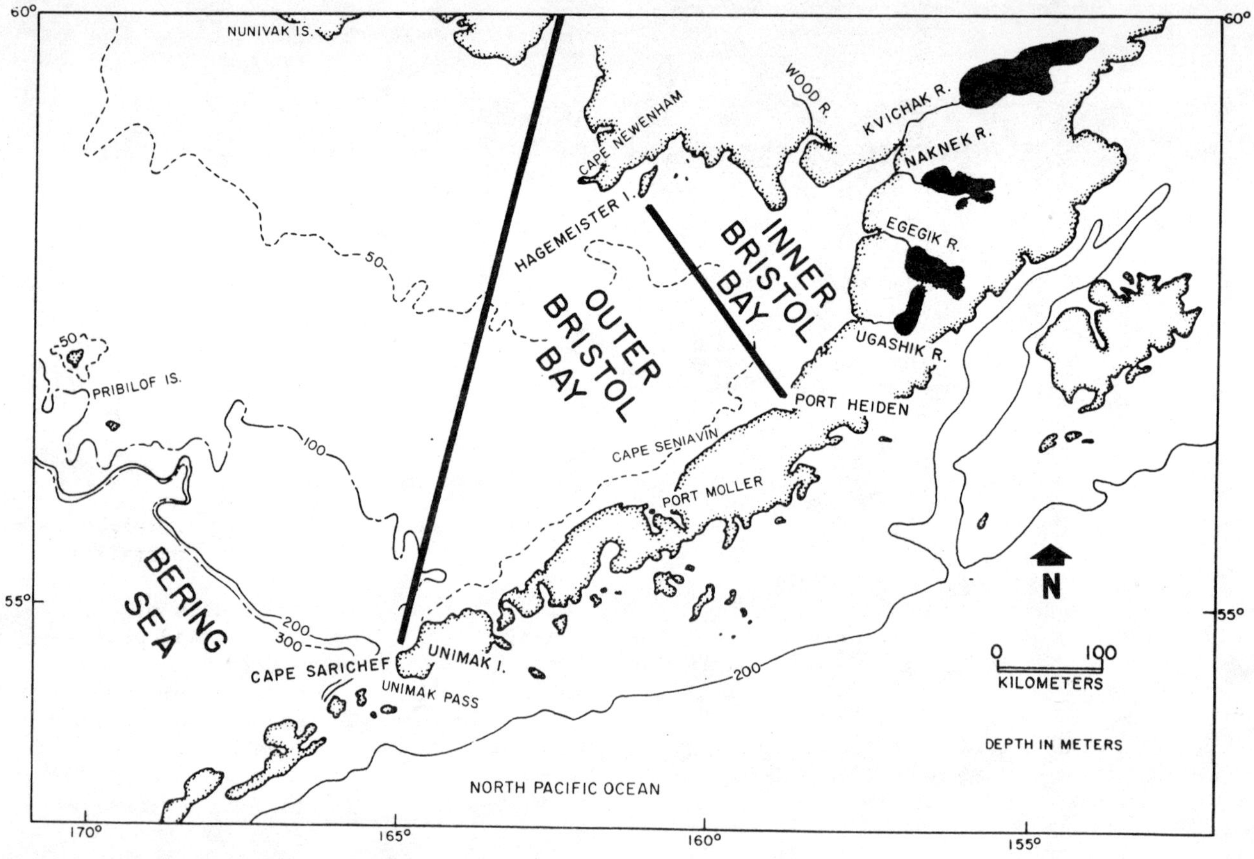

Figure 1. Bristol Bay, showing inner and outer bays.

The spawning migration from the North Pacific begins in early May (Straty 1975). Salmon move directly toward Aleutian Island passes and then eastward through the Bering Sea into Bristol Bay (Hartt 1962). Mature sockeye first appear in the bay during mid-June. Their greatest abundance usually occurs between 1-10 July. By late July or early August, most adults have entered their home river systems to spawn.

Both seaward-migrating and mature sockeye are present in Bristol Bay during June and July, but their migration paths do not appear to overlap significantly except in the vicinity of some adjacent small bays and river mouths. Seaward migration is largely near the coast (Fig. 1) as far seaward as Port Moller (Straty 1974). Mature sockeye returning to spawn remain, for the most part, in the offshore waters of the bay until they reach the vicinity of their home river systems (Straty 1975).

Estuarine Environment of Bristol Bay

Salinity, sea temperature and food abundance are the principal environmental factors that have been investigated with respect to the early marine life of sockeye salmon. Salinity gradients may serve to orient migrating juvenile (McInerney 1964) and adult salmon (Straty 1974) in the turbid waters of inner Bristol Bay and, therefore, influence their distribution. Temperature profoundly affects the metabolism of fish, regulating processes such as activity, feeding, growth and sexual maturation. In addition, the sensitivity of fish to temperature change (Murray 1971) provides them with a mechanism for selecting thermal ranges most conducive to their survival and, therefore, influences their distribution. Finally, the abundance and distribution of food when young salmon enter the sea influences their period of residence in an area, the rate and amount of marine growth and survival.

The information presented here on the variations of salinity, temperature and food ab undance (zooplankton) is taken from the results of investigations conducted primarily in inner Bristol Bay and that portion of outer Bristol Bay as far seaward as Port Moller (Dodimead et al. 1963; Straty 1974, 1975, and 1976).

Salinity: The major sockeye producing river systems of Bristol Bay discharge into the inner bay. Waters of the inner bay and lower portions of most of these systems are well mixed and extremely turbid due to the stirring action caused by winds and tidal currents with velocities up to 6 knots (Straty 1977). The water circulates in a counterclockwise pattern from spring through fall (i.e. the direction of the current is northeastward along the Alaska Peninsula toward the inner bay and northwestward from the inner bay past Cape Newenham) (Fig. 1). The northwesterly current carries most of the freshwater from the major river systems discharging into the inner bay. As a result, pronounced horizontal salinity gradients characterize the inner bay (Fig. 2). Salinity is higher and more uniform in the outer bay (Fig. 3).

Sea Temperature: Juvenile sockeye salmon are most abundant in the upper 10 meters of the water column during seaward migration (Straty 1974). Water temperatures to 10 m were generally lower offshore than nearshore from May to September. Sea temperatures in Bristol Bay showed large annual variations during the period of peak seaward migration of sockeye (Fig. 4). Sea temperature data available for the same periods in outer Bristol Bay and in the eastern Bering Sea before 1967 indicated that temperatures as low as those that occurred in 1971 were not unusual but were less common than those that prevailed in 1967, 1969 and 1970.

Food: In Bristol Bay, zooplankton and small fish were the major prey of the young salmon during their first year at sea (Dell 1963, Straty and Jaenicke 1971, Straty 1974, Carlson 1976). Zooplankton were sampled throughout Bristol Bay in the last spring and through the summer of 1969-71 (Straty 1974) and 1972. Copepods were the most abundant zooplankter; euphausiids and chaetognaths made up most of the remainder.

Along the seaward migration route of the juvenile sockeye (Fig. 5), zooplankton were far more abundant in outer Bristol Bay than within the inner bay (Fig. 6). Average zooplankton densities in the outer bay exceeded the inner bay averages on all occasions (Table 1). A Fisher sign test ($p < .001$) (Hollander and Wolfe 1973) rejects the null hypothesis that there is no difference between average densities of food organisms between the inner and outer bay. A three-way analysis of variance (month vs. year vs. location) of the average densities also indicates that the location effect (inner vs. outer) is highly significant. Thus, the sooner the juvenile sockeye salmon entered the outer bay, the sooner they encountered more zooplankton. Coastal waters (shallower than 37 m) contained a significantly lower abundance of zooplankton than offshore waters (deeper than 37 m) in the early summer of 1969 and 1972 as shown by t-tests on the plankton counts. Coastal waters are more turbid, more mixed and less saline than offshore waters, which likely decreases primary production and, ultimately, secondary production (Gran and Braarud 1935, Jermolajev 1958). We believe that these factors were, in part, responsible for the differences in the standing crop of zooplankton between the inner and outer bays (Fig. 7). In addition, Straty (1974) speculated that large numbers of smolts leaving Bristol Bay could also measurably reduce the available zooplankton, particularly when the fish were concentrated in a narrow band.

Finally, examination of the organisms in the zooplankton and in stomachs of juvenile sockeye revealed that larger zooplankton and nekton (euphausiids, larval and young Pacific sand lance, *Ammodytes hexapterus*) were more abundant beyond the muddy waters of the inner bay. On occasion an important food item of young sockeye was very scarce in Bristol Bay. For example, euphausiids were virtually absent in the plankton in the inner bay in June 1972, yet were much more abundant there in June of 1969-71.

In summary, zooplankton abundance in Bristol Bay increased seaward and offshore; larger zooplankters were also more abundant seaward.

Responses of Juvenile Sockeye to Environmental Factors

Salinity Gradients: On the basis of laboratory experiments with five species of Pacific salmon including sockeye, McInerney (1964) proposed that juvenile salmon can use salinity gradients as a directive cue during seaward migration. He showed that juvenile salmon display a temporal progression of salinity preference changes, which parallel the salinity gradients typical of river outflows. The final salinity preference shown by young salmon was for the open ocean.

The general seaward migration route of juvenile sockeye from major Bristol Bay river systems is from the river mouths to the southeast side of inner bay (Straty 1974). This movement is through the region of the most pronounced salinity gradients (Fig. 8) and supports McInerney's experimental conclusions. The direction that juvenile sockeye from the Wood River system

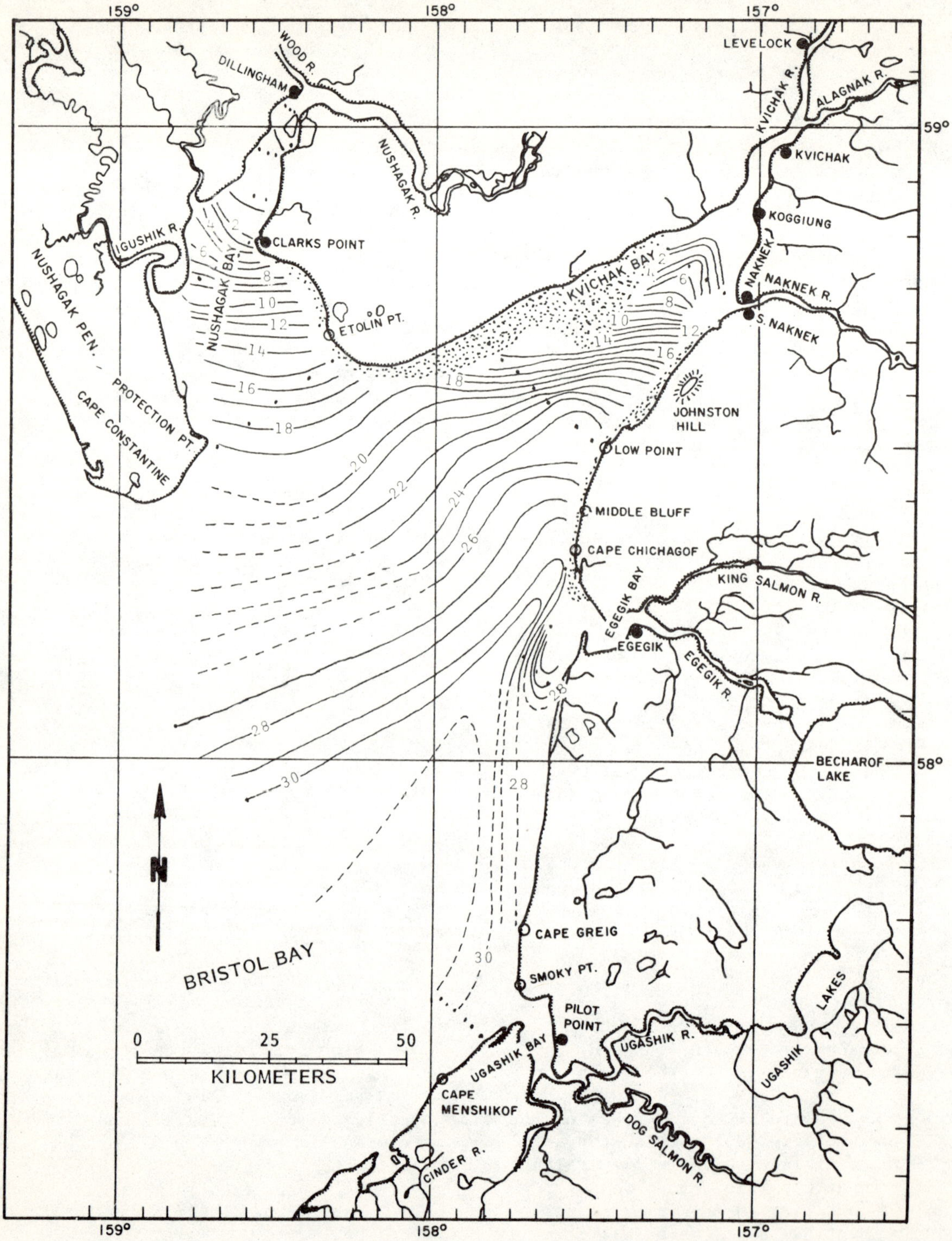

Figure 2. Surface salinity (o/oo) at low tide, inner Bristol Bay, July and August 1966 (from Straty 1977).

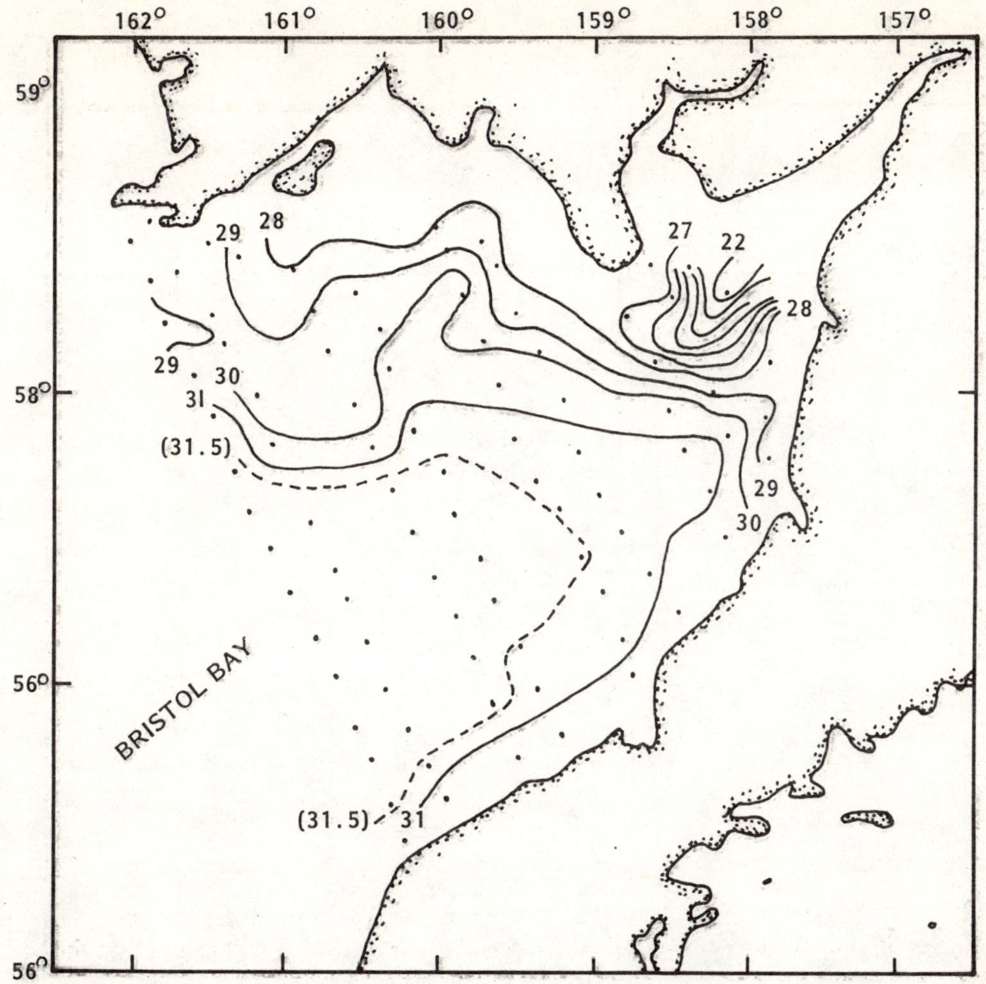

Figure 3. Surface salinity (o/oo) of Bristol Bay, August 1939 (from Dodimead, et al. 1963).

moved best exemplifies this apparent response to salinity gradients in inner Bristol Bay. These fish left their river of origin on the north side of the inner bay and migrated to the southeast side of the bay. This is the most direct route through turbid waters of the inner bay to clearer, high-salinity water of the outer bay. Salinity gradients become much less pronounced in the outer bay, and juvenile sockeye may use other environmental cues for continued seaward migration.

Temperature: Fish can perceive temperature gradients as small as 0.03°C (Murray 1971). The aggregation of fish within a limited preferred temperature range has been observed in nature and in the laboratory (Brett 1952, Ferguson 1958, Garside and Tait 1958, Jarvaid and Anderson 1967). The selection of a preferred temperature is related to the available temperatures which offer fish an opportunity for maximum expression of activity without causing undue stress. Included within the scope of this activity are feeding and swimming, which must be sufficient to permit an animal's existence in the face of competition and predation.

Temperature limits an animal's metabolic rate and, therefore, its feeding activity, swimming speed, growth and development. Metabolic processes increase with rising temperature until an optimum is reached. Growth is most efficient where water temperature is sufficiently warm to stimulate appetite but sufficiently low to avoid high maintenance requirements and stress. Figure 9 illustrates the influence of temperature on food conversion (the ratio of fish weight change to food weight). Growth increased with temperature to a plateau followed by a rapid decline with further increase in temperature. When young sockeye salmon were placed in water and water temperature was

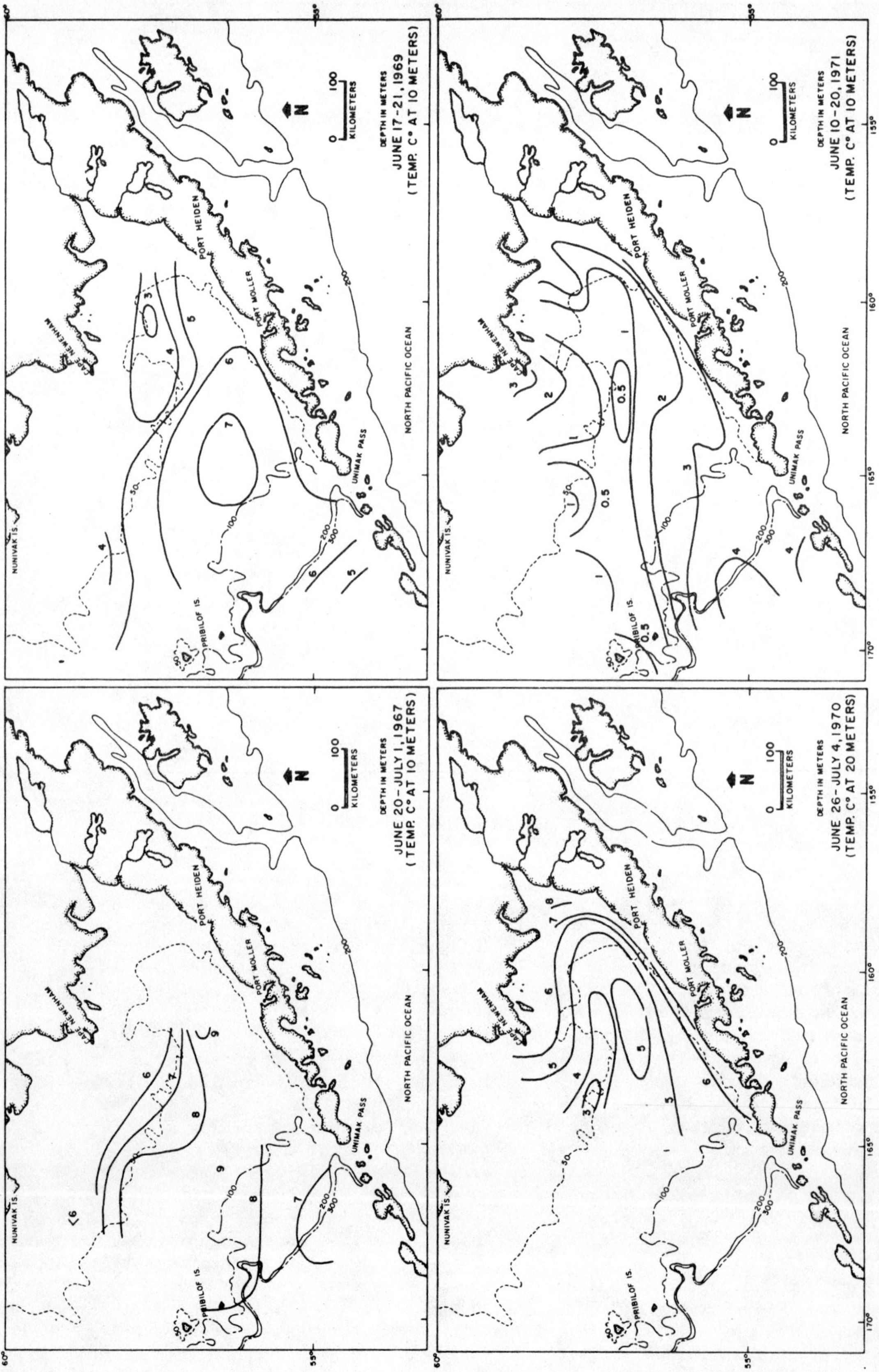

Figure 4. Sea temperatures in Bristol Bay and southeastern Bering Sea in mid-June and early July 1967, 1969, 1970 and 1971 (from Straty 1974)..

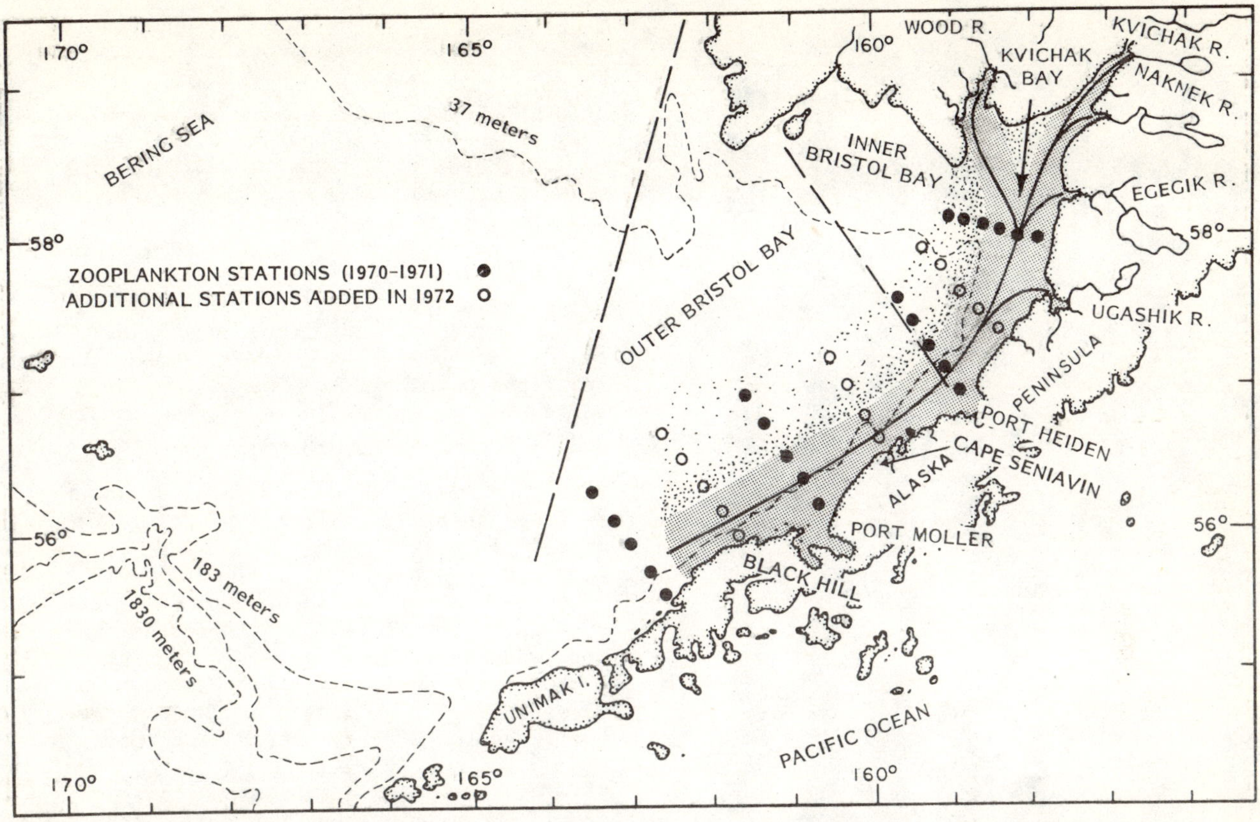

Figure 5. Migration route of juvenile sockeye salmon through Bristol Bay, 1969 and 1970. Also shown is the location of stations sampled for zooplankton with bongo nets in 1970-72. Different stations within the migration route were used in 1969.

gradually raised from 17.2°C to 25.6°C over a period of a week, the fish refused to eat and their respiration rate became excessive (Donaldson and Foster 1940). When the temperature was lowered to 21.1°C, the fish began to eat and increased slightly in weight. At approximately 10°C, growth and food utilization were highest for the temperatures studied. Fish did not grow at temperatures below 4°C. The chief factor limiting consumption and growth at low temperatures appeared to be the slow rate of digestion (Brett and Higgs 1970).

Environmental temperature also increases or decreases the swimming speed of fish. In laboratory tests the swimming speed of under-yearling sockeye and coho salmon increased with increasing temperature and then decreased as the upper lethal temperature was approached (Fig. 10) (Brett et al. 1958).

Since surface temperatures of the North Pacific Ocean, Bering Sea and Bristol Bay may vary more than 15°C each year, a wide range of food consumption, swimming and growth rates can be expected throughout the year. A wide range in these responses can also be expected during the short period that sockeye migrate seaward through Bristol Bay due to variations in temperature (Fig. 4).

A small change in water temperature between 4°C and 10°C may have a significant effect on growth (Fig. 9) and swimming speed (Fig. 10). Year-to-year differences in water temperature encountered during seaward migration in Bristol Bay range to 6°C (Fig. 4). Such annual differences in temperature could influence the migration pattern, distribution and survival of juvenile sockeye.

The survival of sockeye during their seaward migration and ocean residence partially depends on the size of the smolts at migration time (Foerster 1954). Foerster found that survival increased with increased size of smolts. Conditions that favor a rapid increase in the size of sockeye during seaward migration, such as warmer sea temperatures, might also be expected to increase their survival. As mentioned

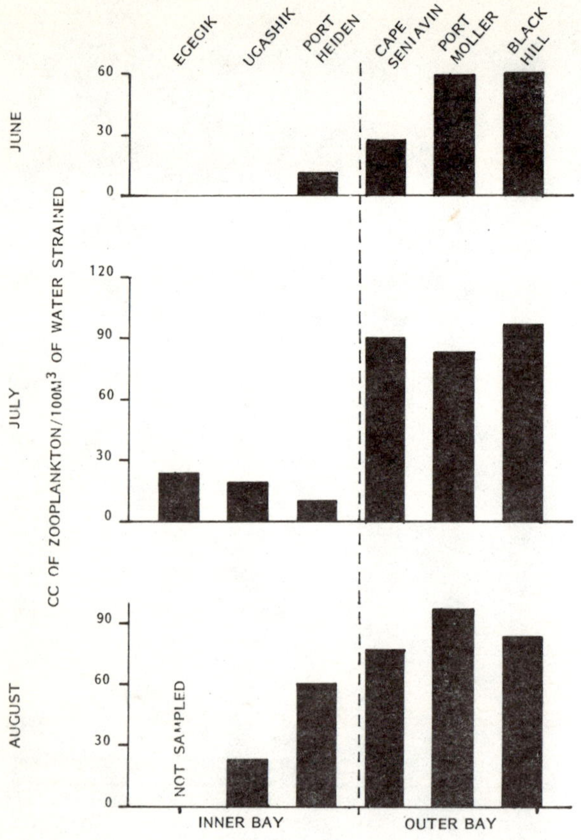

Figure 6. Zooplankton abundance along migration route of juvenile sockeye salmon in inner and outer Bristol Bay, summer 1972.

earlier, the seaward migration of juvenile sockeye through inner Bristol Bay is near the coast rather than offshore, and it is this area that has the highest sea temperatures (Fig. 4). In 1971, when extremely low sea temperatures prevailed during seaward migration, juvenile sockeye were distributed very close to the coast in shallow water where the warmest water occurred at the time. Preference for warmer temperatures is one possible factor contributing to the concentration of juvenile sockeye in nearshore waters of Bristol Bay.

Studies of juvenile sockeye distribution in Bristol Bay (Straty 1974) also indicated considerable annual variation in the time that the fish arrived and length of residence in the inner and outer bays. This variation may be due, in part, to annual differences in sea temperature. For example, juvenile sockeye were present in the Port Moller vicinity of outer Bristol Bay (Fig. 1) by June 1969 and 1970. However, in 1971, the coldest summer of the study, they had not arrived at Port Moller by early July.

The presence of only a few seaward-migrating sockeye in the outer bay in July 1971 can be attributed to: (1) a slower rate of seaward migration due to a reduction in swimming speed as a result of the unusually low sea temperatures (Fig. 4) and (2) later than usual out-migration of smolts from some major river systems. The ultimate effect of the low temperature would be to increase the migration time of juvenile sockeye in inner and outer Bristol Bay and in the Bering Sea resulting in later than usual entrance into the North Pacific.

Straty (1974) examined scales of juvenile sockeye caught in Bristol Bay during seaward migration in 1966, 1967, 1969 and 1970, and compared the lengths of these fish with those of individual stocks sampled on the river systems at the time of out-migration. He concluded that very little, if any, growth had occurred during migration through the inner bay. Marine growth had been delayed until the sockeye had entered the outer bay.

Annual variation in the amount of estuarine and early marine growth may be in large part due to the nature of the early marine growth period. Based on laboratory experiments by Donaldson and Foster (1940), growth is assumed to occur only when temperature exceeds 4°C. We have selected seawater temperatures from two extreme years, 1967 and 1971, to illustrate the extent of variations in the period of growth during seaward migration of sockeye in Bristol Bay (Fig. 11). The monthly sea temperature values represent mean sea surface temperatures at 57° lat., 163° long., 100 miles northwest of Port Moller.

The period of growth is represented in Figure 11 by the area above the 4°C line extending from the time of seaward migration (vertical line). In 1967, seaward migrating sockeye experienced a much longer growing period than migrants in 1971. The area under the curve representing the 1967 temperatures is 2.5 times greater than the area under the curve for 1971.

The scales of adult sockeye, which migrated seaward as smolts in 1967 and 1971 were examined to determine if scale growth and growth patterns differed significantly between the last freshwater annulus and the first marine annulus. The scales of over 700 adult fish from three river systems

Table 1. Average Zooplankton Volumes (cc/100 m³ Water Strained by Bongo Net) By Month In Inner and Outer Bristol Bay, 1969-72.

	1969		1970		1971		1972	
	Inner Bay	Outer Bay	Inner Bay	Outer Bay	Inner Bay	Outer Bay	Inner Bay	Outer Bay
June	15	74[1]	23	36	22	48	4	50[1]
July	18	61[1]	--	--	--	--	18	103[1]
August	43	59	28	29	14	53[1]	41	77
September	--	--	16	33[1]	24	57[1]	--	--

[1] T-tests indicate these values were significantly different (0.05 level) from those in the inner bay that month.

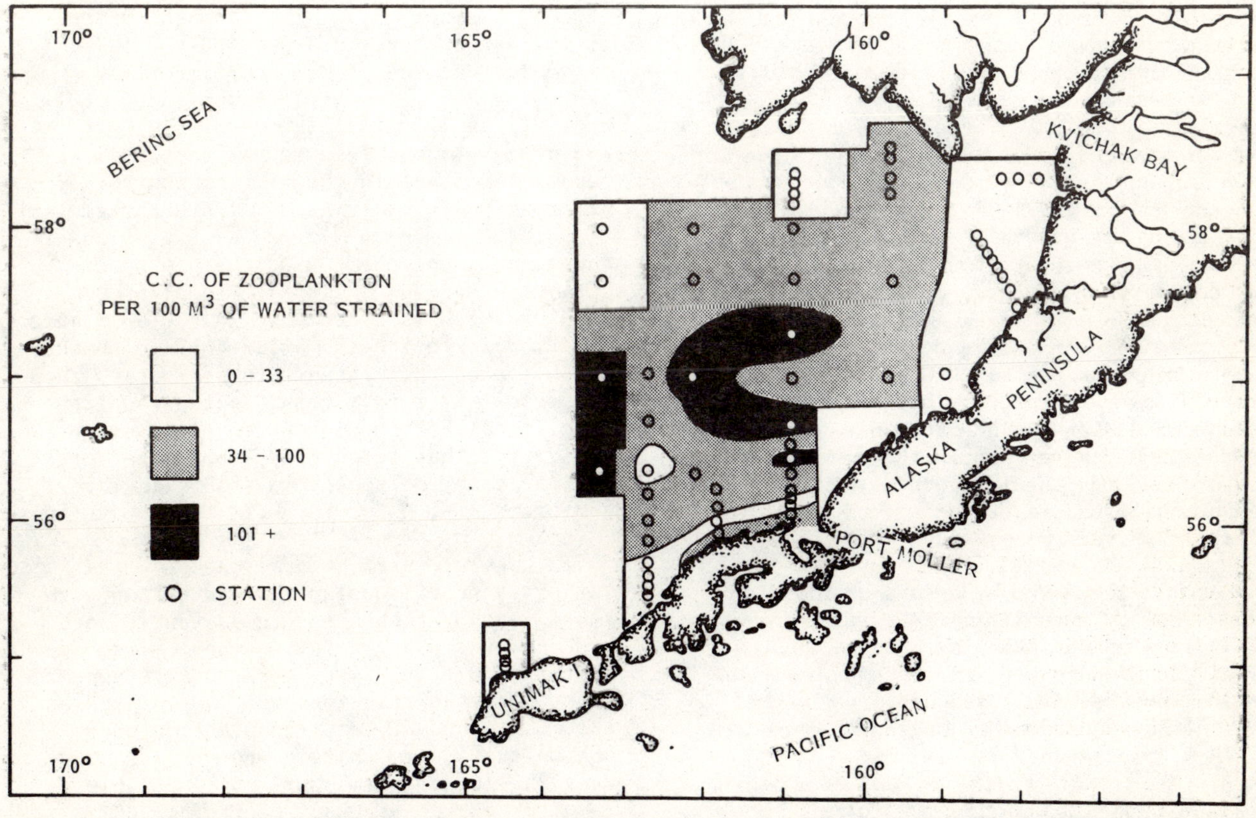

Figure 7. Variations in zooplankton abundance in Bristol Bay, July 1969.

in Bristol Bay were examined and the following parameters recorded for the first marine growth zone:

1. Number of circuli
2. Width of zone
3. Width of first 10 circuli
4. Width of second 10 circuli
5. Width of five widest consecutive circuli
6. Position of five widest consecutive circuli

Data were analyzed by analysis of variance. For each stock examined, all parameters were significantly different between the two years, indicating that scale growth and pattern were reflecting effects of high and low temperatures the first marine year. Indirectly, from these same scales, we hope to calculate the size of the 1967 and 1971 juvenile sockeye at the end of their first marine year to (1) relate first year marine growth to sea surface temperatures and (2) determine if sockeye growing rapidly in the first marine year return to spawn a year earlier than sockeye that grow more slowly.

An answer to this latter question could explain why the 1971 seaward migrants, which experienced the two coldest years of marine life on record (1971-72), remained an extra year at sea in unusually high numbers. Seventy-nine percent of the juvenile sockeye salmon that entered the sea in 1971 returned as 3-ocean fish, which is a two to four times higher incidence than any other year during the period 1955-1974. If extremely low temperatures that sockeye encounter during ocean life cause more fish to remain an extra year in the ocean, then weather records should prove helpful in tempering forecasts of the age of returning Bristol Bay sockeye salmon.

Zooplankton: The kinds and quantity of food eaten by juvenile sockeye salmon vary among areas of Bristol Bay (Straty 1974). Results of food habit studies are discussed by area progressively seaward from upper Bristol Bay and refer to migrants sampled in 1969 unless otherwise indicated.

Kvichak Bay: In the turbid, shallow estuarine waters of Kvichak Bay (Fig. 5), copepods made up 93 percent or more of the food volumes in stomachs of juvenile sockeye sampled in June and July (Fig. 12). Insects and cladocerans formed nearly all the remaining food. All these organisms are similar to those kinds that sockeye eat in freshwater (Jaenicke and Kirchhofer 1976, Hoffman 1978, 1979).

Ugashik Bay: In this area, 90 km seaward of Kvichak Bay, the Pacific sand lance (*Ammodytes hexapterus*) was the major food (86 percent by stomach volume) of smolts sampled in July. Nekton, instead of small zooplankton or insects, became important prey for the sockeye as they moved seaward into progressively deeper, less turbid and more saline waters.

Port Heiden: At this outer boundary of the inner bay, large zooplankton or nekton again made up the bulk of smolt stomach contents sampled in June, July and August (Fig. 12). Principal foods of the migrants off Port Heiden were euphausiids (61 percent by stomach volume in June) or *Ammodytes* (76 percent in July, 90 percent in August).

Port Moller: At Port Moller, 315 km seaward of Kvichak Bay, large organisms were the main prey of juvenile sockeye in June and July followed by small organisms in August and September. Euphausiids accounted for over 72 percent of the stomach contents from young sockeye in June 1969 and 1970. *Ammodytes* was the second most important food item. By July, *Ammodytes* composed 90 percent of the juvenile sockeye diet. Late in the summer, two rarely sought prey items became important in the diet of the sockeye off Port Moller: In August 1969, the small cladoceran *Podon leuckarti* (G. Sars); and in August and September 1970, the pteropod *Limacina helicina* (Phipps). This change in diet suggests a decrease in the abundance of *Ammodytes* off Port Moller in late summer, which forces the migrants to seek other food. *Ammodytes* was rarely found in the zooplankton samples off Port Moller after early summer indicating that these prey items were not available to the sockeye migrants in late summer.

In general, area comparisons show that larger forms of zooplankton and nekton were found in the stomachs of the juvenile sockeye caught seaward. During their first four months in Bristol Bay, young sockeye appeared to be opportunistic feeders. They preyed on the most abundant large zooplankton and nekton available, changing to smaller prey when the larger forms were not available. In other marine waters, Foerster (1968) found that the food spectrum of young sockeye was also broad. If favored prey were absent, the fish consumed substitute organisms.

This opportunistic feeding is consistent with the optimal foraging hypothesis that sockeye ignore small prey if larger prey are available (Doble and Eggers 1978).

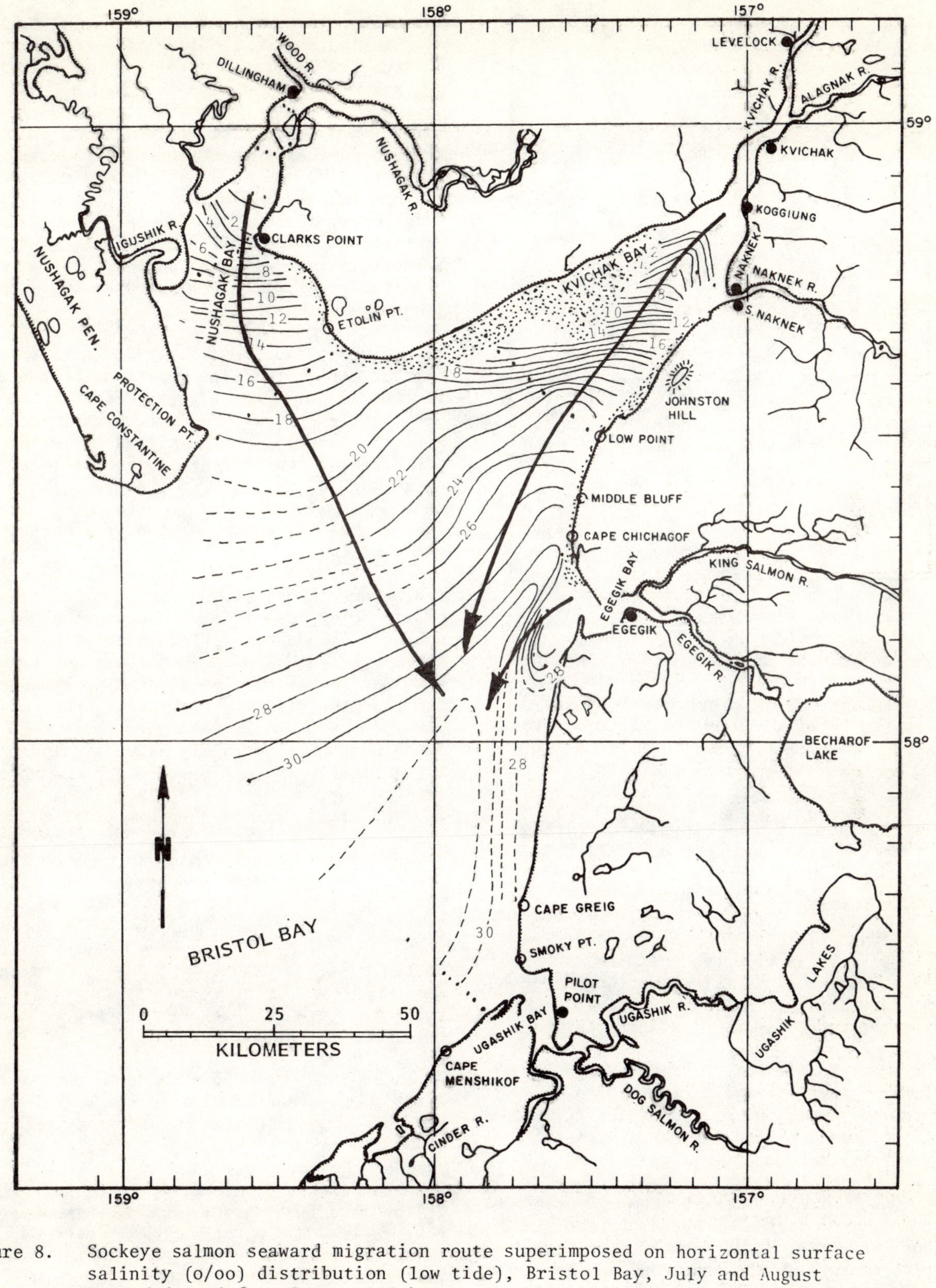

Figure 8. Sockeye salmon seaward migration route superimposed on horizontal surface salinity (o/oo) distribution (low tide), Bristol Bay, July and August 1966 (adapted from Straty 1974).

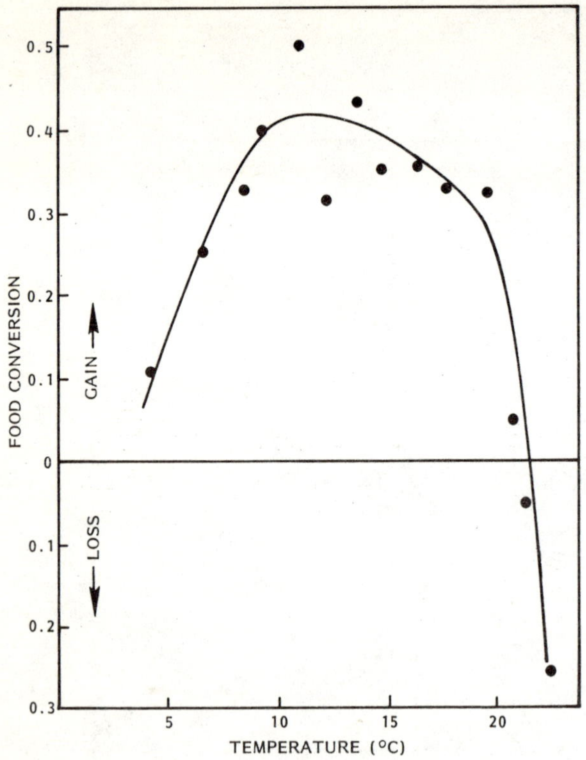

Figure 9. Growth of young sockeye salmon in 2-week periods as shown by food conversion at different water temperatures (adapted from Rounsefell 1975; data from Donaldson and Foster 1940).

Eggers (1977) cited many examples of size-selective predation by fish; and Carlson (1974) and Doble and Eggers (1978) showed evidence of young sockeye selectively feeding on larger organisms in freshwater. This was also true for Bristol Bay sockeye. For example, while copepods became more abundant seaward along the migration route, the sockeye rarely fed upon them after leaving the muddy waters of Kvichak Bay.

The stomachs of sockeye sampled in inner and outer Bristol Bay were examined for stomach fullness (Fig. 12). Results reflect the distribution of zooplankton described earlier (Fig. 6).

In 1969, juvenile sockeye generally consumed more food farther seaward, farther offshore and later in the summer. In June, juvenile sockeye in the inner bay and inshore waters of the outer bay had little or no food in their stomachs (Fig. 13). In July, migrants sampled everywhere had fuller stomachs, especially in the offshore waters of the outer bay. By August, sampling indicated that sockeye had left the inner bay and inshore waters of the outer bay. Sockeye sampled off Port Heiden (the boundary between the inner and outer bay) in August contained more food than they had all season (Fig. 13). In contrast, off Port Moller in August, a severe reduction in the stomach fullness was evident, which was thought to be due to a change in diet. Sockeye sampled in July contained mostly large prey, while those sampled in August contained mostly small prey. This change from large to small prey could result in a significant difference in stomach fullness (LeBrasseur 1966).

In August and September 1970, sockeye sampled off Port Moller revealed the trend for the fish to ingest greater quantities of food later in the summer, which coincides with the movement of the migrants farther offshore into water containing more zooplankton.

The smaller amounts of food that sockeye consumed in the inner bay and in the inshore waters of the outer bay earlier in the summer are thought to be due to a comparatively lower abundance of zooplankton and nekton. Abundance of these food items was usually less than 33 cc per 100 m^3 of water strained in these areas (Fig. 7). Although the young sockeye along the same transect generally fed on the same major prey both inshore and offshore, the migrants offshore, where plankton volumes ranged from 34 to 101+ cc/100 m^3, often contained significantly greater quantities of the same prey.

The response of juvenile sockeye to this greater abundance of food seaward was evident. Generally, fish in the outer bay were larger than those in inner bay. The appearance of widely spaced marine-type circuli on the scales of these fish also showed that marine growth did not occur or accelerate until the juvenile sockeye had entered the outer bay (Straty 1974).

In addition to food abundance, several other factors contributed to the faster growth of juvenile sockeye once they entered the outer bay. By the time juvenile sockeye reached the outer bay in July and August, zooplankton (Table 1), sea temperatures (Fig. 11) and the feeding activity of salmon were near their seasonal maxima. Rising sea temperatures during the summer would accelerate feeding activities, digestion rates and growth potential. After entering the outer bay, where food abundance was greater, juvenile sockeye apparently moved seaward at a more leisurely pace than they had through the inner bay (Straty 1974). This apparent slower rate of seaward migration

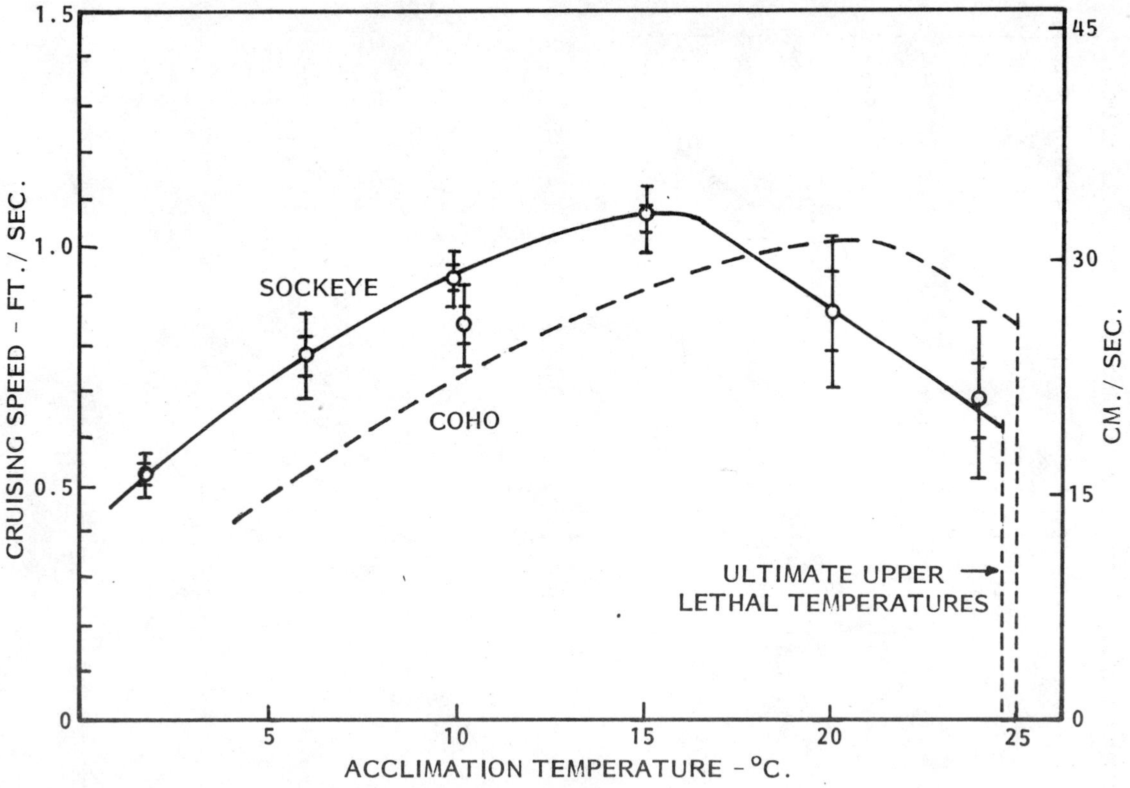

Figure 10. Variations in cruising speed for temperature-acclimated underyearling sockeye and coho adjusted in each case to common mean lengths of 6.9 cm and 5.4 cm, respectively. The samples were cultured under similar conditions and are of comparable age, 4 to 6 months from hatching (from Brett et al. 1958).

may be due to sockeye remaining for a prolonged period of time to feed where excellent food conditions exist (Birman and Konovalov 1968, Mitans 1970).

Not only was food more abundant in the outer bay, but larger forms of food items occurred there. Apparently the type of food present, especially with respect to particle size, has a major influence on growth (Paloheimo and Dickie 1966). Only when sockeye fed on larger prey, such as *Ammodytes* in July, did they have fuller stomachs and begin to grow at a faster rate. Nikolsky (1963) noted that "usually the period of most intense feeding occurs at the time of appearance of some mass form of food which forms the basis for its food". It is probable that the distribution and abundance of *Ammodytes* (the first large prey organism encountered in abundance by juvenile sockeye over a large area of Bristol Bay) may be a significant factor determining the time of rapid initial marine growth.

One additional factor contributing to better feeding conditions for juvenile sockeye in the outer bay and offshore is the increased clarity of the water here compared with the more turbid inner bay and nearshore waters. Juvenile sockeye feeding in clearer waters would benefit from increased visibility, as visibility may govern their prey selection and capture efficiency (Doble and Eggers 1978).

The amount of energy derived during the foraging process is a function of the prey selectivity by the migrants. Fish that feed on larger organisms, such as euphausiids or *Ammodytes*, would derive a higher caloric intake per energy expenditure than when feeding on smaller organisms such as copepods. Therefore, the growth potential of juvenile salmon should vary with the size and abundance of available prey.

When the abundance of food is low, as in early summer and in the inner bay and nearshore waters of the outer bay, the larger juvenile sockeye probably prey more successfully than smaller juveniles on what is available. Competition among juvenile sockeye of different sizes would be particularly intense in some years when large numbers (estimated by Rogers (1977) as up

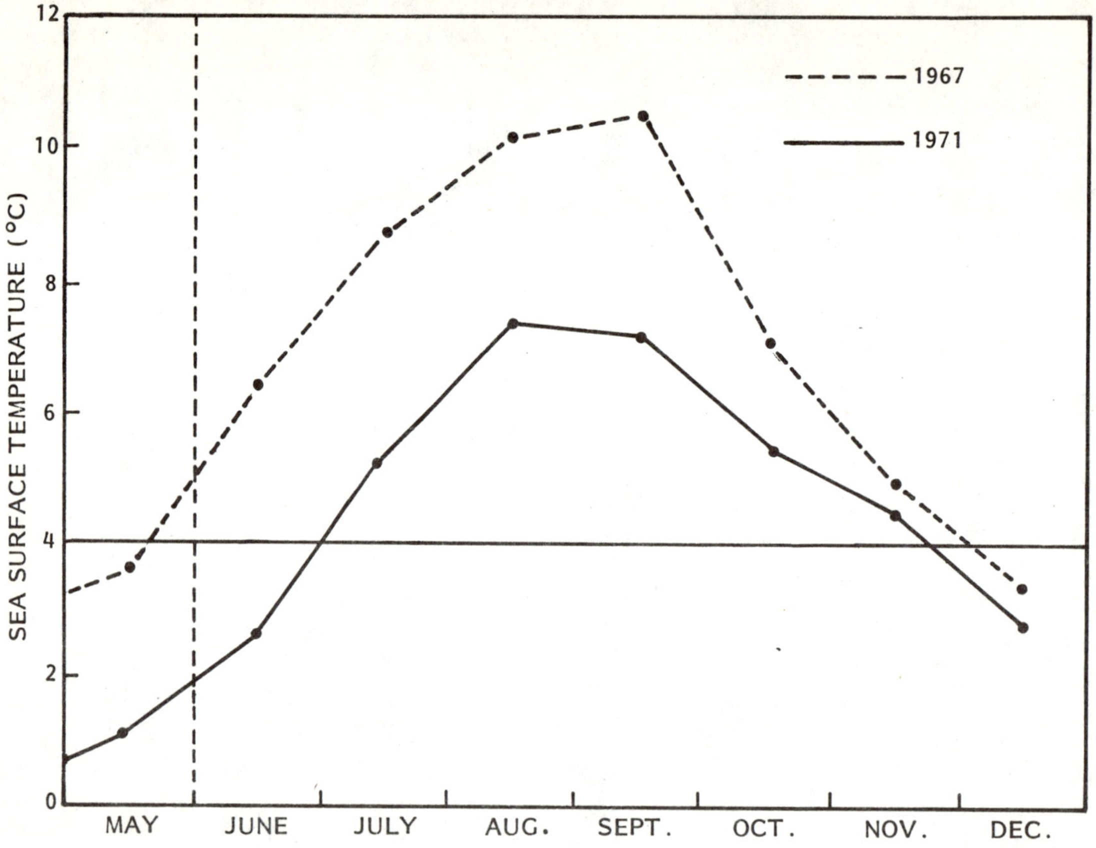

Figure 11. Average monthly sea surface temperatures in Bristol Bay in an area 100 miles northwest of Port Moller during May-December 1967 and 1971. Vertical line represents time of peak seaward migration of sockeye salmon.

to 650 million) are present in the bay, or during cold years when the migration path was narrower and closer to shore. Under these conditions, the Age II smolts, which are larger and leave the lake systems first (Hartman et al. 1967) may have an advantage with respect to initial feeding compared with Age I smolts that are smaller and migrate later. (An Age II smolt has spent two winters in freshwater; an Age I smolt has spent one winter. These two age groups make up nearly all the sockeye seaward migrants entering Bristol Bay.)

The Age II-Age I order of out-migration from freshwater was maintained in the outer bay with Age II preceding Age I sockeye by almost a month off Port Moller. Thus, the Age II sockeye would continue to have a growth advantage over the Age I migrants by haveing first chance to feed on available prey. The larger sockeye may also be more able to use larger prey (Doble and Eggers 1978) such as *Ammodytes*. The probable growth advantage of Age II over Age I juveniles should result in a higher survival for the Age II fish (e.g. Age II migrants would outgrow predators at a faster rate than the smaller Age I migrants). An advantage of large over small size may explain, in part, the higher ocean survival of Age II versus Age I sockeye observed in Bristol Bay stocks (Pella and Jaenicke 1978); the high ocean survival of larger juvenile sockeye within single stocks (Foerster 1954, Manzer 1972); and among stocks around the Pacific rim (Ricker 1962).

SUMMARY AND CONCLUSIONS

The information presented provides some insight into the natural variability of the Bristol Bay estuarine environment and the responses of seaward-migrating sockeye to these variations. Because of limited data, we are restricted to hypothesizing how variations in the estuarine environment may affect an increase or decrease in early marine survival. We speculate that higher sea temperatures in Bristol Bay will result in more rapid growth and a faster rate of seaward migration. Presumably, rapid early marine growth is conducive to high marine

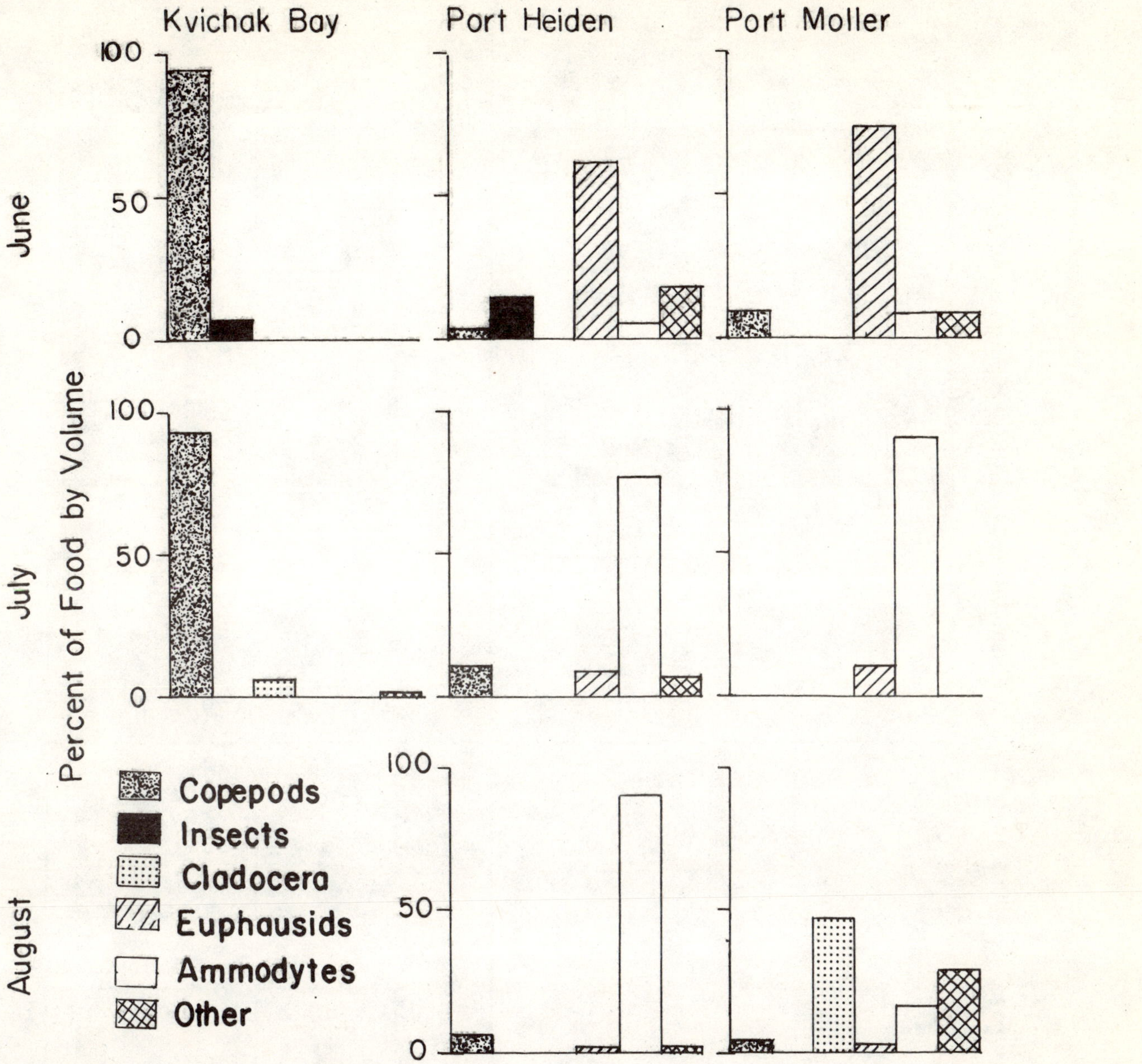

Figure 12. Stomach contents of 100 juvenile sockeye salmon from each of three areas of Bristol Bay, summer 1969.

survival. A faster seaward migration would result in earlier entrance into the North Pacific Ocean. Low sea temperatures would have the reverse effect. We do not know if early entrance into the North Pacific and rapid growth (therefore, larger size) before the onset of the first winter at sea are conducive to a higher survival rate for sockeye than would occur if the situation were reversed. By reducing metabolic demand and thus inhibiting feeding activity (Gilhousen 1962), low sea temperatures could be a survival asset if food was scarce during some period of seaward migration. Low sea temperatures, which inhibit rapid growth and activity, could prolong the period during which juvenile sockeye would be subject to size-dependent predation by other fish and warm-blooded animals, such as marine birds and mammals. The distribution of fish-eating marine birds, which number in the millions in Bristol Bay, is very similar to the distribution of juvenile sockeye salmon (Straty and Haight *in press*). Predation on juvenile sockeye salmon by marine mammals in Bristol Bay is also documented (Anonymous 1957).

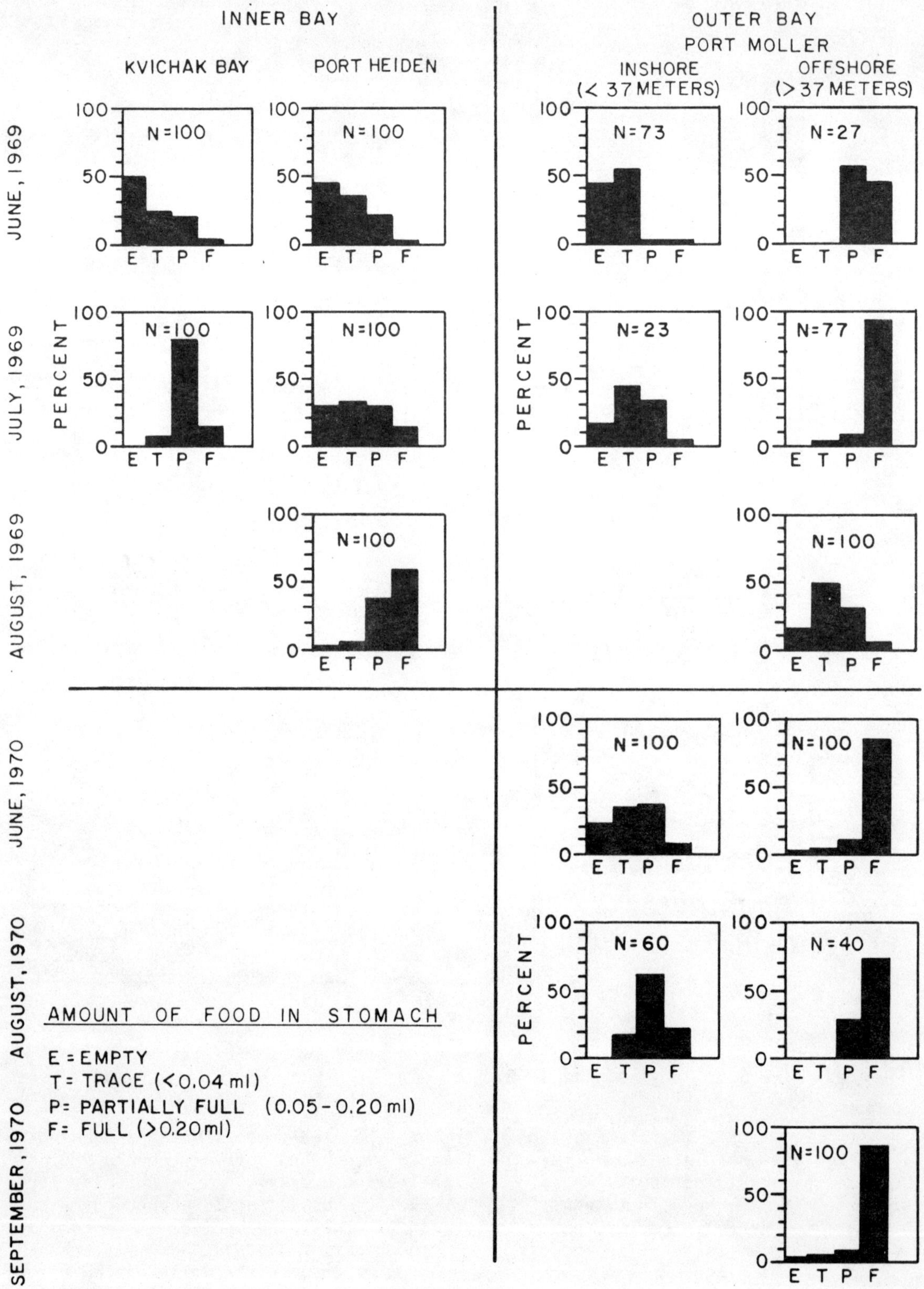

Figure 13. Fullness of stomachs of seaward-migrating sockeye salmon by month in inner and outer Bristol Bay, 1969 and 1970 (from Straty 1974).

Information is still required on the impact of both salmonid and non-salmonid competition for the food available to the juvenile sockeye. We need more information on the effect of competition between Age II and Age I sockeye smolts. In addition, when a single stock dominates the total biomass of juvenile sockeye in Bristol Bay, as has occurred in some years (Pella and Jaenicke 1978), we want to know that stock's impact on the food supply and growth of the other stocks.

We also do not know how the passage of the entire adult sockeye salmon run through the offshore waters of Bristol Bay impacts the abundance of zooplankton. The run, which in some years has exceeded 50 million fish, migrates through Bristol Bay enroute to home river systems to spawn from early June to late July. Sexually mature sockeye are known to feed until shortly before they enter their home river systems. Thus, the adult run, which may vary in size annually by an order of magnitude, represents a potential and variable source of competition with juvenile sockeye. This competition has not been evaluated. Seaward-migrating sockeye salmon, and probably the juveniles of other salmon species, move offshore seaward of Port Moller after the adult run has passed. At this time the juveniles are larger and capable of eating larger food.

The abundance, distribution and types of zooplankton available inshore and offshore influenced the growth and distribution of young sockeye during their first few months in Bristol Bay. We found that the sooner the juvenile sockeye left the inner bay, the sooner they encountered more abundant food, consumed more food and increased their growth rate. As Andrievskaya (1970) noted, commensurate with growth and the increased requirement for food, young pink salmon in Russian waters began to prey on larger oceanic forms of plankton and nekton, a trend which was linked with their migration to the open sea. From our work, causes for rapid growth of young sockeye in July and later in outer Bristol Bay are thought to be due to warming sea temperatures, more time for intensive feeding, greater availability of larger prey seaward and offshore, and increased ability of the sockeye to consume that prey.

We have barely begun to understand how fluctuations in estuarine and coastal marine conditions affect the distribution, movement and abundance of the sockeye salmon. The complexity, variability and interaction of physical, chemical and biological processes operating during the early sea life make it difficult to predict many cause-and-effect relations or to describe the niche of juvenile salmon in the marine environment.

During the last decade, we have become more aware of the importance of acquiring and integrating oceanographic and fisheries information. This increased awareness has resulted, in part, from the increasing ability to produce and release large numbers of viable juvenile salmon into the ocean. We are interested in forecasting the success of such releases in terms of returns and in the impact of these releases on the other salmonid and non-salmonid species.

The importance of environmental and resource relationship studies in the management of salmonid and non-salmonid fisheries is obvious. The difficulty, we believe, lies in justifying the required 20- to 40-year studies on a cost-effective basis so that funding will be made available to conduct them without interruption. The solution would appear to be coordinated studies conducted jointly by all governmental and private agencies sharing an interest in the ocean salmon fisheries.

ACKNOWLEDGMENTS

We are particularly indebted to Lin Sonnenberg for her editorial suggestions in preparation of this paper, initial editing and typing of several revisions of the manuscript; to Ann Kramer for her analysis of the zooplankton data and help in preparing that section; and to Malin Babcock for her analysis of the food habit data.

We also owe our thanks to Robert Ellis and Jack Bailey for their critical reviews of the paper and their valuable suggestions, and to Elmer Landingham who prepared the figures.

LITERATURE CITED

Andrievskaya, L. D. 1970. Feeding of Pacific salmon fry in the sea. Fish. Res. Bd. Can. Transl. Ser. No. 1423. 16 pp.

Anonymous. 1957. Study plan II Beluga. Alaska Fish. Bd. and Alaska Dep. Fish. Ann. Rep. No. 8:54-56.

Birman, I. B., and S. M. Konovalov. 1968. Distribution and migration at sea of the local stock from Kurile Lake of red salmon, *Oncorhynchus nerka* (Walbaum). Am. Fish. Soc. Prob. Ichthyol. 8:581-587.

Brett, J. R. 1952. Temperature tolerance in young Pacific salmon genus *Oncorhynchus*. J. Fish. Res. Bd. Can. 9:265-323.

Brett, J. R., M. Holands, and D. F. Alderice. 1958. The effect of temperature on the cruising speed of young sockeye and coho salmon. J. Fish. Res. Bd. Can. 15:587-605.

Brett, J. R., and D. A. Higgs. 1970. Effect of temperature on the rate of gastric digestion in fingerling sockeye salmon, *Oncorhynchus nerka*. J. Fish. Res. Bd. Can. 27:1967-1779.

Carlson, H. R. 1976. Foods of juvenile sockeye salmon, *Oncorhynchus nerka*, in the inshore coastal waters of Bristol Bay, Alaska, 1966-67. Natl. Mar. Fish. Serv., Fish. Bull. 74:458-462.

Carlson, T. J. 1974. The life history and population dynamics of *Bosmina coregoni* in Pedro Bay, Iliamna Lake, Alaska. M.S. Thesis. Univ. of Wash., Seattle. 142 pp.

Dell, M. B. 1963. Oceanic feeding habits of the sockeye salmon, *Oncorhynchus nerka* (Walbaum), in Aleutian waters. M.S. Thesis. Univ. of Mich., Ann Arbor. 40 pp.

Doble, B. D., and D. M. Eggers. 1978. Diel feeding chronology, rate of gastric evacuation, daily ration and prey selectivity in Lake Washington juvenile sockeye salmon *(Oncorhynchus nerka)*. Trans. Am. Fish. Soc. 107:36-45.

Dodimead, A. J., F. Favorite, and T. Hirano. 1963. Salmon of the North Pacific Ocean. Part II. Review of oceanography of the subarctic Pacific region. Int. North Pac. Fish. Comm. Bull. 13. 195 pp.

Donaldson, L. R., and F. J. Foster. 1940. Experimental study of the effect of various water temperatures on the growth, food utilization and mortality rates of fingerling sockeye salmon. Trans. Am. Fish. Soc. 70:339-346.

Eggers, D. M. 1977. The nature of prey selection by planktivorous fish. Ecol. 58:46-59.

Ferguson, R. G. 1958. The preferred temperature of fish and their midsummer distribution in lakes and streams. J. Fish. Res. Bd. Can. 15:607-624.

Foerster, R. E. 1954. On the relation of adult sockeye salmon *(Oncorhynchus nerka)* returns to known smolt seaward migrations. J. Fish. Res. Bd. Can. 11:339-350.

Foerster, R. E. 1968. The sockeye salmon, *Oncorhynchus nerka*. Bull. Fish. Res. Bd. Can. 162. 422 pp.

Garside, E. T., and J. S. Tait. 1958. Preferred temperature of rainbow trout *(Salmo gairdneri)* and its unusual relationship to acclimation temperature. Can. J. Zool. 36:563-567.

Gilhousen, P. 1962. Marine factors affecting survival of Fraser River pink salmon. Pages 105-111 *in* Wilimovsky, W. J. (ed.), Symposium on pink salmon, H. R. MacMillan lecture in fisheries. Univ. British Columbia, Vancouver, B. C.

Gran, H. H., and T. Braarud. 1935. A quantitative study of the phytoplankton in the Bay of Fundy and the Gulf of Maine (including observations on hydrography, chemistry and turbidity). J. Biol. Bd. Can. 1:279-467.

Hartman, W. L., W. R. Heard, and B. Drucker. 1967. Migratory behavior of sockeye salmon fry and smolts. J. Fish. Res. Bd. Can. 24:2069-2099.

Hartt, A. C. 1962. Movement of salmon in the North Pacific Ocean and Bering Sea as determined by tagging, 1956-58. Int. North Pac. Fish. Comm. Bull. 6. 157 pp.

Hoffman, M. A. 1978. Diet of sockeye salmon *(Oncorhynchus nerka)* and threespine stickleback *(Gasterosteus aculeatus)* from the limnetic zone of Lake Nanavaugaluk, Alaska, summer 1975. Northwest and Alaska Fish. Center, Auke Bay Lab., Natl. Mar. Fish. Serv., Auke Bay, Alaska, 99821. MR-F No. 147. Unpubl.

Hoffman, M. A. 1979. Food habits of sockeye salmon fry and threespine sticklebacks in the littoral of Lake Nunavaugaluk, Alaska. M.S. Thesis. Univ. of Alaska, Juneau.

Hollander, M., and D. A. Wolfe. 1973. Nonparametric statistical methods. John Wiley and Sons, New York, NY. 503 pp.

Jaenicke, H. W., and D. A. Kirchhofer. 1976. Food of sockeye salmon, *Oncorhynchus nerka*, and associated fishes in Lake Nunavaugaluk, Alaska. Northwest and Alaska Fish. Center, Auke Bay Lab., Natl. Mar. Fish. Serv., Auke Bay, Alaska, 99821. MR-F No. 137. Unpubl.

Jarvaid, M. Y., and J. M. Anderson. 1967. Thermal acclimation and temperature selection in Atlantic salmon, *Salmo salar*, and rainbow trout, *S. gairdneri*. J. Fish. Res. Bd. Can. 24:1507-1513.

Jermolajev, E. G. 1958. Zooplankton of the inner Bay of Fundy. J. Fish. Res. Bd. Can. 15:1219-1228.

LeBrasseur, R. J. 1966. Stomach contents of salmon and steelhead trout in the northeastern Pacific Ocean. J. Fish. Res. Bd. Can. 23:85-100.

Manzer, J. I. 1972. On the relation between lacustrine growth and production of age 1.2 sockeye *(Oncorhynchus nerka)* in Great Central Lake prior to experimental nutrient enrichment. Fish. Res. Bd. Can. Tech. Rep. 324. 31 pp.

McInerney, J. E. 1964. Salinity preference: an orientation mechanism in salmon migration. J. Fish. Res. Bd. Can. 21:995-1018.

Mitans, A. R. 1970. The feeding of Baltic salmon smolts in the river and in the sea. Am. Fish. Soc. Prob. Ichthyol. 10:89-94.

Murray, R. W. 1971. Temperature receptors. Pages 121-133 *in* Hoar, W. S., and D. J. Randall (eds.), Fish physiology. Vol. V. Academic Press, Inc., New York, N.Y.

Nikolsky, G. V. 1963. The ecology of fishes. (Transl. by L. Birkett). Academic Press, Inc., New York, N.Y. 352 pp.

Paloheimo, J. E., and L. M. Dickie. 1966. Food and growth of fishes. III. Relations among food, body size and growth efficiency. J. Fish. Res. Bd. Can. 23:1209-1248.

Parker, R. R. 1962. Estimations of ocean mortality rates for Pacific salmon *(Oncorhynchus)*. J. Fish. Res. Bd. Can. 19:561-589.

Parker, R. R. 1965. Estimation of sea mortality rates for the 1961 brood-year pink salmon of the Bella Coola area, British Columbia. J. Fish. Res. Bd. Can. 22:1523-1554.

Pella, J. J., and H. W. Jaenicke. 1978. Some observations on the biology and variations of populations of sockeye salmon of the Naknek and Ugashik systems of Bristol Bay, Alaska. Northwest and Alaska Fish. Center, Auke Bay Lab., Natl. Mar. Fish. Serv., Auke Bay, Alaska, 99821. Manuscr.

Ricker, W. E. 1962. Comparison of ocean growth and mortality of sockeye salmon during their last two years. J. Fish. Res. Bd. Can. 19:531-560.

Rogers, D. E. 1977. Determination and description of knowledge of the distribution, abundance and timing of salmonids in the Gulf of Alaska and Bering Sea. A supplement to the final report. Fish. Res. Inst., Univ. Wash., Seattle. Unpubl. manuscr.

Rounsefell, G. A. 1975. Ecology, utilization and management of marine fisheries. C. V. Mosby Co., St. Louis. 516 pp.

Straty, R. R. 1974. Ecology and behavior of juvenile sockeye salmon *(Oncorhynchus nerka)* in Bristol Bay and the eastern Bering Sea. Pages 285-320 *in* Hood, D. W., and E. J. Kelley (eds.), Oceanography of the Bering Sea with emphasis on renewable resources. Proceedings of the International Symposium on Bering Sea Study, Univ. Alaska. Inst. Mar. Sci. Occas. Publ. 2.

Straty, R. R. 1975. Migratory routes of adult sockeye salmon, *Oncorhynchus nerka*, in the eastern Bering Sea. U.S. Dep. Comm. Natl. Ocean. and Atmos. Admin. Tech. Rep., Natl. Mar. Fish. Serv. SSRF-690. 32 pp.

Straty, R. R. 1977. Current patterns and distribution of river waters in inner Bristol Bay, Alaska. U.S. Dep. Comm. Natl. Ocean. and Atmos. Admin. Tech. Rep., Natl. Mar. Fish. Serv. SSRF-713. 13 pp.

Straty, R. R., and R. E. Haight. *In press*. Interaction of marine birds and commercial fish in the eastern Bering Sea. U.S. Fish Wildl. Serv. Wildl. Res. Rep.

Straty, R. R., and H. W. Jaenicke. 1971. Studies of the estuarine and early marine life of sockeye salmon in Bristol Bay, 1965-67. Northwest and Alaska Fish. Center, Auke Bay Lab., Natl. Mar. Fish. Serv., Auke Bay, Alaska, 99821. Unpubl. manuscr.

Density-Dependent Growth of Bristol Bay Sockeye Salmon

Donald E. Rogers

INTRODUCTION

The growth of sockeye salmon *(Oncorhynchus nerka)* can significantly affect their subsequent survival and production. A primary objective of research on sockeye salmon has been to determine what factors affect growth and survival and how these factors operate (Foerster 1968). Hereditary differences in growth probably contribute to observed variations in size of fish from different populations. Environmental differences also contribute to annual variation in size within populations as well as differences among populations (Ricker 1972). Growth is particularly affected by water temperature, abundance of food, and fish density.

Relationships between abundance and size of adult Pacific salmon have been examined by previous workers. Kamchatkan pink salmon are smaller in odd-numbered years when abundance is high than in the even-numbered years when abundance is low. Sockeye salmon returning to Lake Kurile in Kamchatka also tend to be larger in even-numbered years when pinks are scarce, than odd-numbered years when pinks are abundant. Nevertheless, there has not been a significant correlation demonstrated between annual abundance and mean weight of adult salmon returning to Kamchatka. Differences in size between odd and even years may be caused largely by hereditary factors (Krogius 1960, Foerster 1968, Ricker 1972). This conclusion is supported by observations in the Fraser River where annual variation in the weight of sockeye salmon was not related to abundance, but there was evidence for hereditary effects on growth (Killick and Clemens 1963).

Contribution No. 531, College of Fisheries, University of Washington, Seattle, Washington.

The annual migrations of sockeye salmon to Bristol Bay are the largest for this species in the world. Density-dependent growth of juveniles in two Bristol Bay lakes has been described by Mathisen (1969) and Rogers (1973). Immature Bristol Bay sockeye salmon have been sampled annually on the high seas since 1956 in an area south of Adak Island (Fig. 1). However, density-dependent growth of immature fish has not been observed in saltwater (Rogers 1976).

I will use data on the age and length of Bristol Bay sockeye gathered since 1957 by the Bureau of Commercial Fisheries, Fisheries Research Institute and Alaska Department of Fish and Game for the following purposes:

1) Review evidence on density-dependent growth of juveniles in Bristol Bay lakes;
2) Analyze statistics on immature sockeye at sea with respect to annual variations in size and abundance;
3) Calculate annual mean lengths and weights of adult sockeye in Bristol Bay; and
4) Evaluate density-dependent growth at sea.

METHODS

Annual mean lengths and weights of sockeye smolts in Bristol Bay are presented in Alaska Department of Fish and Game (ADF&G) informational leaflets and in various unpublished reports by the Bureau of Commercial Fisheries (BCF) and Fisheries Research Institute (FRI) at the University of Washington. In all cases anesthetized fish were measured live. Length was determined to the nearest millimeter from tip of snout to fork and weight to the nearest 0.1 gram. Annual estimates of the abundance of smolts were often inaccurate. Number of parent spawners in escapements were usually fairly accurate and provided a useful measure of relative abundance.

Immature sockeye salmon captured by purse seine south of Adak were determined to be primarily of Bristol Bay origin, based on tag returns (Hartt 1966, French et al. 1976). The annual age composition and relative abundance of immature sockeye generally corresponds to the age composition and abundance of the run returning to Bristol Bay in the following year. Immature salmon caught at sea with purse seines were measured live from tip of snout to fork to the nearest 5 mm. Regressions of sample mean lengths on date were used to estimate the annual mean lengths on 21 July for each age group sampled. A common weight-length regression was used to estimate mean weights (Rogers 1976).

Age and length measurements from Bristol Bay catches and escapements over the period 1957-1962 were provided by Dr. C. DiCostanzo (BCF) and were obtained from FRI data files. Age, length and weight measurements for the period 1963-1973 were provided by Mr. R. Paulus (ADF&G). Weight measurements from escapements were available only for the years 1964, 1965 and 1971. Sample sizes for age and length of adult fish caught in the commercial fishery usually included 2,000 to 3,000 fish. Sample sizes from escapements to each river usually included 1,000 to 2,000 fish. Approximately two adults out of each 1,000 returning to Bristol Bay were sampled for age, sex and length. Weight measurements were obtained from random samples of about 400 fish from individual fishing districts.

Adult length was measured to the nearest millimeter from mid-eye to tail fork. Adult weight was measured to the nearest tenth of a pound. Adults in the escapement were measured live, and those from the catch were measured dead. Average mean lengths and weights of live fish in commercial catches were estimated by subtracting 2 mm in length and adding 11 g in weight to carcasses. These corrections were determined from measurements on 101 fish while live and after death.

The standard errors for the mean lengths of the more abundant age groups were typically 1-2 mm. Mean weights were sometimes calculated from mean lengths and weight-length regressions. When no weight measurements were available in a given year, mean weight was estimated from regressions of sample mean weights on mean lengths from other years. The standard errors of mean weights ranged from 91 g to 136 g in the various fishing districts.

Annual numbers of fish (by age and sex) in Bristol Bay districts were obtained from ADF&G informational leaflets. Historical catch statistics for Alaskan and Asian fisheries were obtained from International North Pacific Fisheries Commission (INPFC 1974).

Selectivity of gillnets affects sizes of fish in the catch and escapement. Mesh size was 5 ½ inches from 1928 to 1960 and has been 5 3/8 inches since. During 1957-1973, the overall rates of exploitation varied from 0.09 to 0.80 among years and districts. However, the rate of exploitation has remained highest for age .3 females, lowest for age .2 females and intermediate for males. Gillnets tend to catch larger age .2 males and females and the smaller age .3 males. There is little selectivity for size on age .3 females. The mean length or weight of fish could only be calculated for an entire district and not for discrete populations.

The average air temperature during the winter and spring at King Salmon (Naknek-Kvichak District), date of ice breakup in Lake Aleknagik (Nushagak District), average water temperature in three Bristol Bay rivers during 1-15 June and the average temperature 1-30 miles south of Adak during July were used as annual estimates of environmental conditions. Fish size (average mean weight) was related to these parameters and to annual estimates of the number of fish in the run by linear correlation analysis and by stepwise multiple regression to determine affect of abundance on size.

RESULTS

I assumed that intraspecific competition generally had a much greater affect on growth of sockeye than interspecific competition. Sockeye salmon are the most abundant species in most Bristol Bay lakes. They also experience high annual variation in abundance. Abundance of sockeye affects the growth of competitor species as well as their own growth (Rogers 1973).

Pink salmon are numerically the most abundant salmonid in the North Pacific Ocean. However, pink salmon are generally concentrated to the southwest of the main summer concentration of Bristol Bay sockeye salmon in the North Pacific Ocean. Chum salmon, mostly from Asia, are also probably more abundant than Bristol Bay sockeye salmon in most years. Since their oceanic distribution greatly overlaps the distribution of Bristol Bay sockeye salmon (Neave et al. 1976),

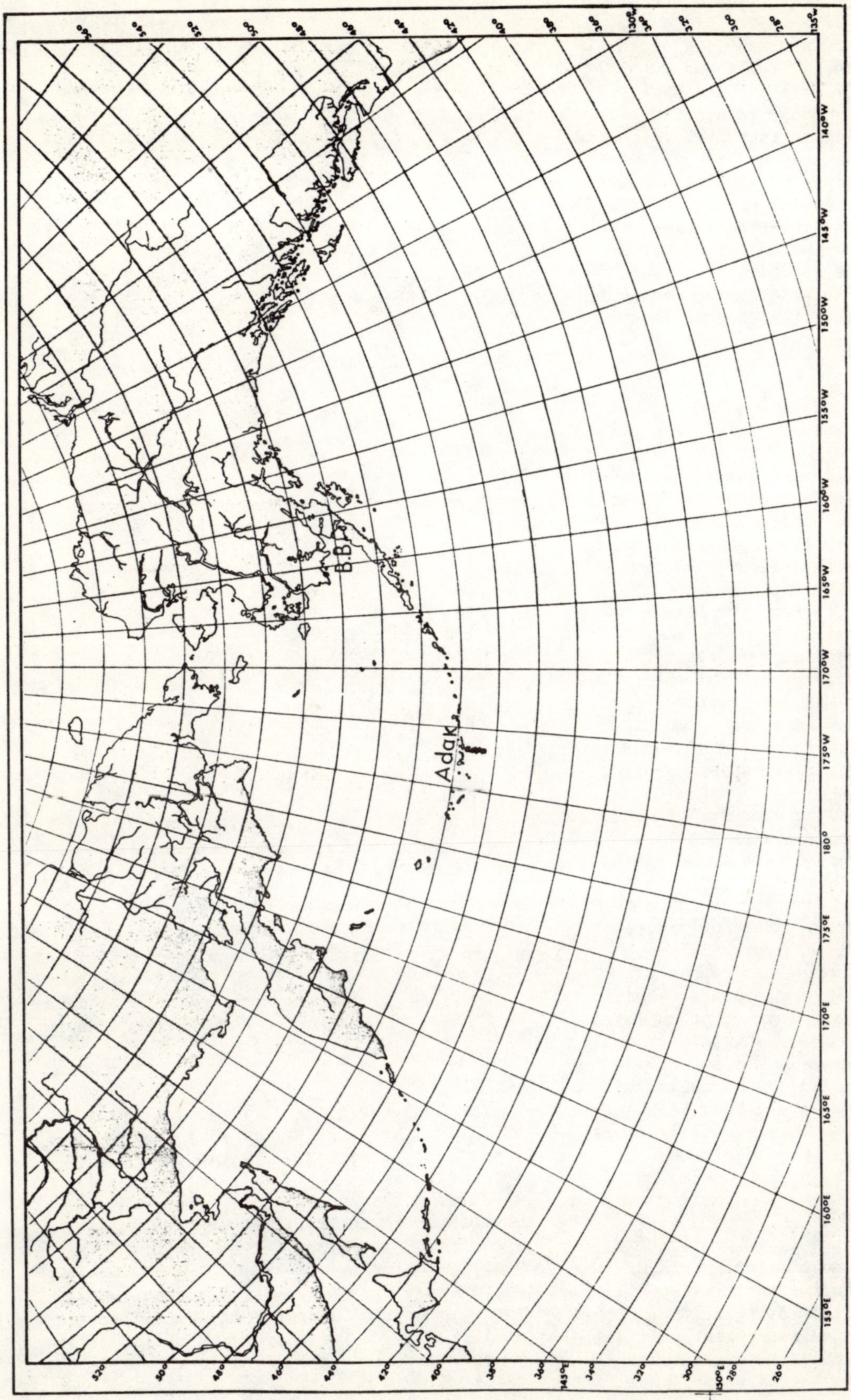

Figure 1. Bristol Bay (B. B.) and the purse seine sampling areas south of Adak Island.

their abundance could affect the ocean growth of Bristol Bay sockeye.

Since the early 1950s, the Asian runs of sockeye salmon have remained quite small relative to the Bristol Bay runs. The Asian and central Alaskan runs were estimated by doubling the annual catch which assumes a 50 percent annual rate of exploitation. The majority of sockeye caught on the high seas was of Asian origin except in 1957 and 1961 according to Fredin and Worland (1974). It is evident from catch statistics that Bristol Bay fish largely determine annual variation in abundance of sockeye salmon in the North Pacific Ocean.

Smolts

The annual migration of smolts from Bristol Bay rivers usually begins in mid-May and continues to the end of July. The period of migration is shortest in the Kvichak River (about two weeks) and longest in the Wood River (about two months). The beginning of the annual migration is directly related to the date of ice breakup (Burgner 1962). The earliest migrations are from the Egegik and Kvichak rivers and the latest from the Nushagak River. Smolts migrate through Bristol Bay over a relatively long period of time, extending from May into August (Straty 1972). As a consequence, there is considerable variability in their period of ocean residence during their first summer at sea.

Average size of smolts from any given lake system is largely determined by density of spawners. Mean weights of age 1 and 2 smolts from the Wood and Kvichak rivers were inversely related to the density of parent spawners (Fig. 2). Wood River smolts are mostly age 1 fish and are typically the smallest smolts in Bristol Bay (Table 1). Population densities of sockeye are usually higher in the Wood River Lakes than in lakes of the Naknek and Ugashik systems, where density-dependent growth is not so obvious (Fig. 3). The lowest density of spawners has typically occurred in the Egegik (Lake Becharof), and the few years of observations indicate that this system produces the largest smolts.

Immatures off Adak

Information on oceanic distribution of Bristol Bay sockeye during their first summer and winter at sea is limited. In the spring following the first winter at sea, Bristol Bay sockeye appear to concentrate in the western Gulf of Alaska. They have been observed to be abundant south of the Aleutians in July of their second summer

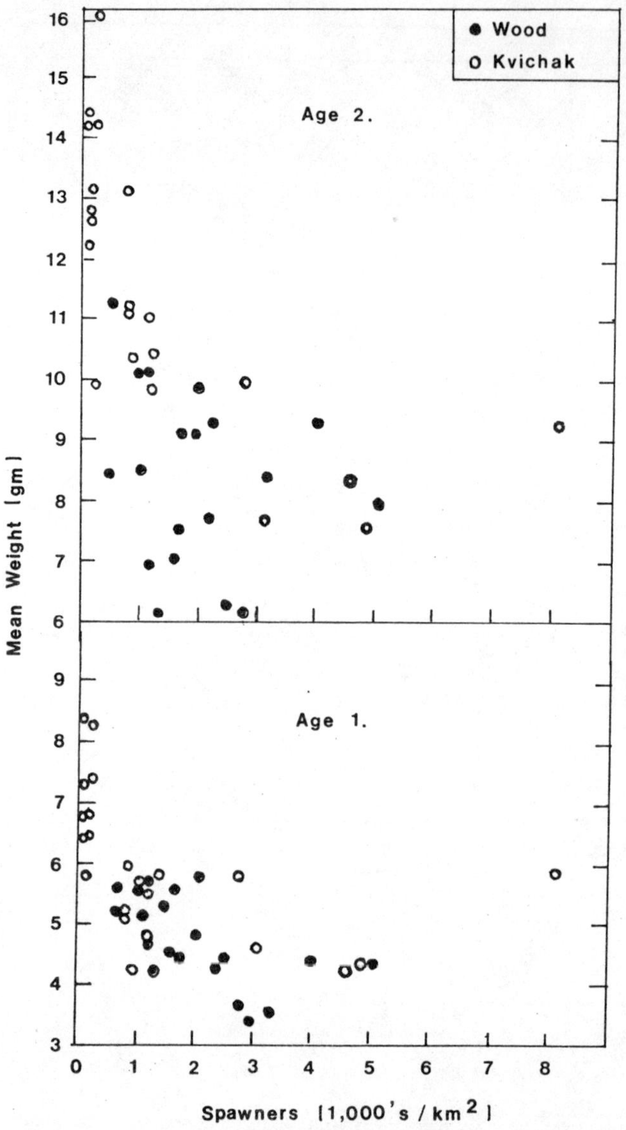

Figure 2. Plots of the mean weight of smolts on the relative density of parent spawners (number per km^2 of lake area) in the Wood River and Kvichak lake systems.

at sea and to migrate in a predominantly westerly direction (French et al 1976). Few age .1 sockeye salmon were observed in seine catches south of Adak before the end of June. Presumably, their oceanic distribution during their first year at sea is influenced by weather in the spring of the year they migrated to sea. Climatic observations since 1956 show cold spring temperatures in Bristol Bay on 1964, 1965, 1971 and 1972 (Fig. 4). Earliest times of ice breakup were observed in 1958, 1970 and 1974.

The annual mean lengths of immature chum salmon caught south of Adak are shown in

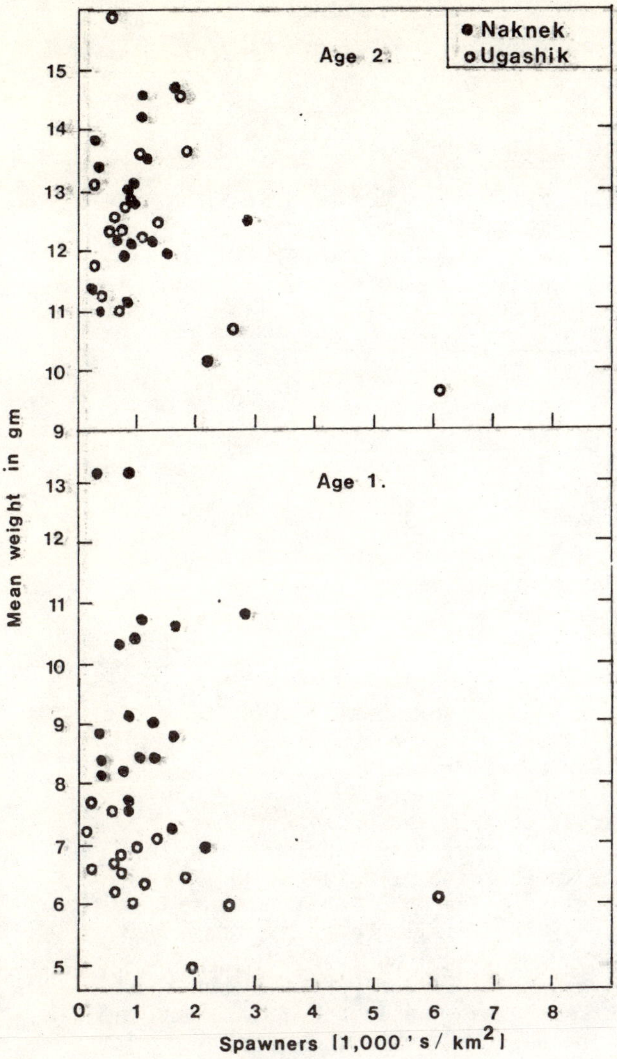

Figure 3. Plots of the mean weight of smolts on the relative density of parent spawners (number per km^2 of lake area) in the Naknek and Ugashik lake systems.

Figure 5. Observations for sockeye are the unweighted mean lengths of freshwater age. Age 1.1 sockeye salmon are about the same size as age 0.1 chum salmon. With the exceptions of 1964 and 1972, when age .1 fish were small, and 1971, when age .1 fish were large, there was relatively little annual variation in the size of age .1 sockeye.

Annual mean weights of age .1 sockeye were correlated with the average water temperatures in Bristol Bay rivers in the preceding year (i.e., when they migrated to sea). The correlations were .55 for age 1.1 and .64 for age 2.1. The deviations about the regression line for age 2.1 were related somewhat to the Bristol Bay run in the following year (partial correlation of .45). Therefore, 40 percent of the annual variation in size of age 2.1 fish was explained by temperature at the time they migrated to sea and 20 percent by abundance.

Size of age .2 chum and sockeye sampled off Adak trended upward over the period 1958-77 (Fig. 5). The year of sampling when combined with the mean air temperature recorded at the town of King Salmon (Fig. 5) accounted for 67 percent of the annual variation in the mean weight of age 1.2 sockeye and 66 percent of the variation in mean weight of age 2.2 sockeye. No measure of abundance was found to be correlated to mean weights of age .2 sockeye. It is possible that the yearly trend in the mean weights of age .2 fish may have been partially caused by a change in the distribution of sampling effort off Adak. After 1968, sampling effort was shifted southward where fish tended to be larger. The average catches of sockeye and chum were generally lower after 1970 than before (Fig. 6). This was partly caused by the seaward shift of sampling effort, at least for chum, which tend to concentrate nearer to Adak than sockeye.

Mean length of age .2 chum salmon was correlated with average catch (Fig. 7). Observations since 1968 tended to fall above the regression line and were not as closely correlated with the catch as the earlier observations. Age .2 chum salmon were usually more abundant than age .2 sockeye, but no correlation was found to occur between their annual relative abundance and the annual size of sockeye.

Adults in Bristol Bay

The cylical abundance of Bristol Bay sockeye is largely determined by size of the Kvichak runs. The other systems in Bristol Bay do not have pronounced cycles. However, runs to Egegik and Ugashik are typically above average in a peak cycle year for Kvichak.

The large peak-cycle runs are characterized by a high proportion of age .2 fish. Sockeye returning to Bristol Bay were smaller than average when the runs were very large due to the predominance of age .2 fish. In the year following a peak run, the fish were predominantly age .3, thus they were larger. Very few fish return at age .4. Age .2 fish are usually the faster growing individuals, since the slower growing individuals tend to remain at sea an extra year. Males are typically faster growing

Table 1. Means and Ranges of the Annual Mean Lengths and Weights of Sockeye Salmon Smolts From Bristol Bay Rivers 1954-1976.

	Location	Number of years	Length (mm) Mean	Length (mm) Range	Weight (g) Mean	Weight (g) Range
Age I	Wood River	19	83.1	77*-89	4.8	3.5*-5.8
	Kvichak	22	88.9	80-98	5.9	4.2-8.4
	Naknek	20	100.8	91-113	9.3	6.9-13.1
	Ugashik	16	91.6	81-96	6.6	5.0-7.7
	Egegik	5	102.2	99-106	--	--
Age II	Wood River	19	100.5	90-114	8.3	6.1-11.3
	Kvichak	22	110.7	97-122	11.4	7.6-16.4
	Naknek	20	112.8	106-120	12.6	10.1-14.7
	Ugashik	16	114.4	104-125	12.4	9.6-15.9
	Egegik	5	117.2	115-120	--	--

*In 1977 the age I smolts from Wood River averaged 70 mm in length and 3.4 g in weight (Alaska Department of Fish and Game, P. Krasnowski).

than females and tend to mature at an earlier age. A higher proportion of age .2 fish tend to return at age 2.2 than age .1 fish at age 1.2. There was considerable annual variation in age of return both among and within lake systems. Part of this is probably due to variation in abundance. Annual variation in age composition was lower in the Wood River Lakes than in other lake systems.

There were some significant differences in size of fish between lake systems within a fishing district. For the Naknek-Kvichak District, fish of a given age and sex returning to the Kvichak River were usually larger than fish returning to the Naknek River. Naknek age .2 fish were the smallest in Bristol Bay and females were occasionally larger than males, which was not observed in other systems.

In the Nushagak District, fish returning to the Igushik River were usually larger than those returning to the Wood River. Within the Wood River system, Roberson (1966) demonstrated statistically significant differences in the mean lengths of sockeye spawners from different spawning grounds.

The largest sockeye were observed in the Togiak District, which has the smallest run of the Bristol Bay fishing districts. The smallest sockeye were in the Naknek-Kvichak District (Tables 2 and 3). Sockeye in the Togiak and Nushagak districts were heavier for a given length than sockeye in the other districts (Figs. 8 and 9). Fish of age 1. tended to be heavier at a given length than fish of age 2.

The annual mean weights of fish of any given age and sex usually significantly correlated with the abundance of sockeye returning to Bristol Bay and to the individual districts in the case of Naknek-Kvichak, Egegik and Ugashik (Table 2). Correlations were not consistently significant in the case of Nushagak and Togiak districts. The annual mean weights tended to vary in a similar pattern among the districts. The annual mean weights tended to vary in a similar pattern among the districts, especially for age 2.2 and 2.3 in the Naknek-Kvichak and Egegik districts and for ages 1.2 and 1.3 in the Naknek-Kvichak and Nushagak districts (Fig. 10). There were generally low correlations between abundance of fish and the mean weight of age 1.2 fish. These fish were the smallest in 1960 when they were the most abundant age group (Table 4). In 1965, there were relatively few age 1.2 fish in the samples, and the estimates of their mean weights were not as reliable as for other ages.

The correlations between weight and the number of fish in the Bristol Bay run were generally higher for females than males; whereas, the slopes of the regressions were generally steeper for males than females (Fig. 11). The regressions for each age group (sexes combined) are shown in Figure 12. Observations from 1957 were omitted from this analysis because samples were small and the Japanese high seas fishery caught an exceptionally high proportion of the run in 1957.

Spring air temperature at King Salmon was correlated significantly with the deviations

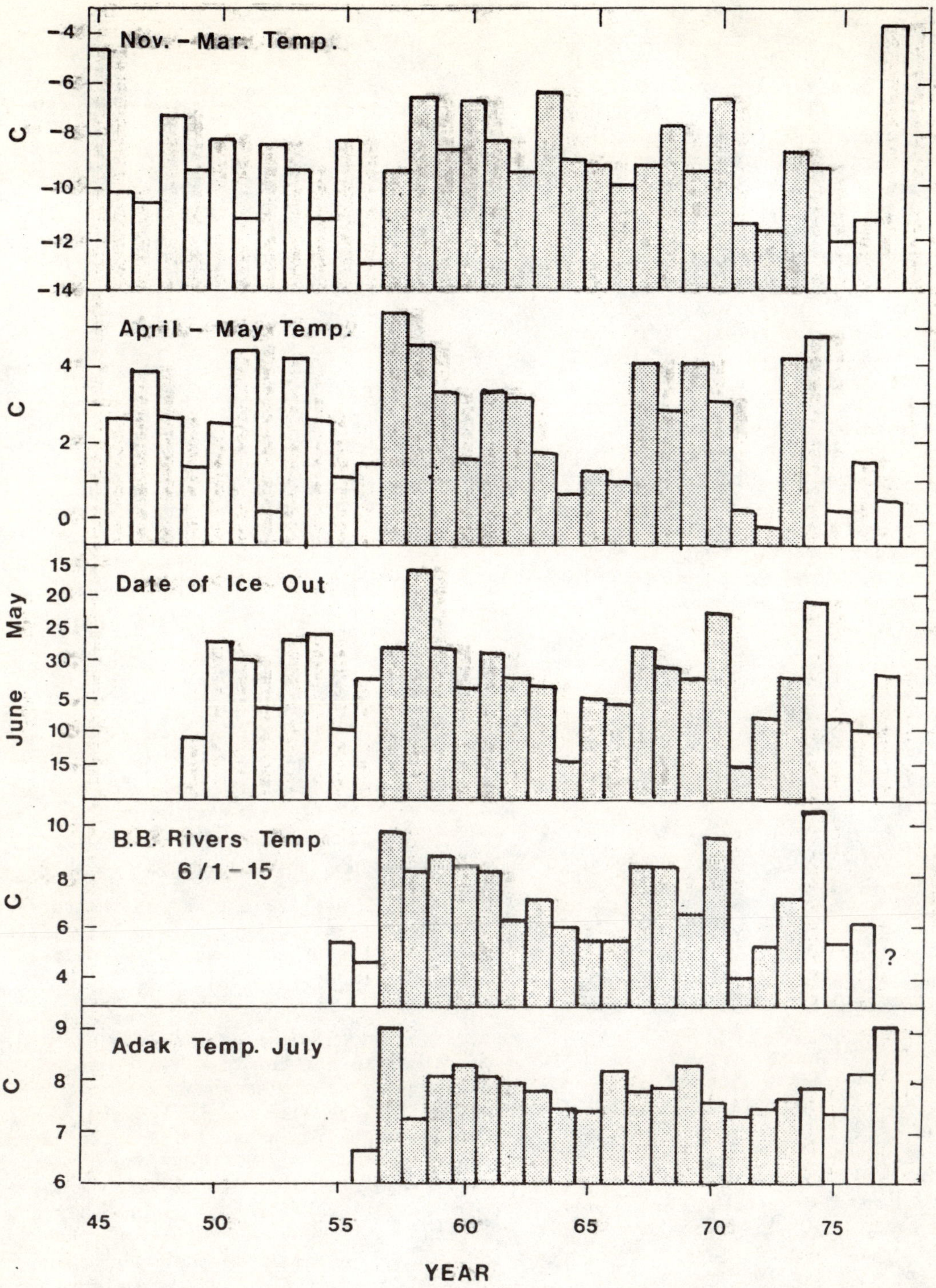

Figure 4. Annual climatological and water temperature data. Mean air temperature November-March at King Salmon (Bristol Bay); mean air temperature April-May at King Salmon; date of ice breakup in Lake Aleknagic (Wood River); mean water temperature in the Kvichak, Naknek and Wood River, 1-15 June; and the mean surface temperature 1-20 miles south of Adak.

about the regressions of mean weight on run size for age groups 2.2 and 2.3 (partial correlations of .69 and .52). Some correlation was also evident for age group 1.3 (partial correlation of .40). Air temperature and the abundance of the run explained 88 percent, 72 percent and 69 percent of the annual variation in the mean weights of ages 2.2, 1.3 and 2.3 (Fig. 13).

There could be some hereditary effect on the sizes of age 2.2 and 1.3 fish because they mature five years after spawning and their mean weight fluctuated in a five-year cycle. However, the mean weights of the age 2.3 fish also varied in a five-year cycle, although they mature six years after spawning. Therefore, environmental factors (abundance and temperature) probably affected size to a much greater extent than hereditary factors.

Density-dependent growth at sea appears to occur during the final year and probably during late spring, early summer when sockeye are concentrated on their homeward migration to Bristol Bay. They pass through Bristol Bay as adults in about a two-week period in late June and early July. Their seaward migration as smolts requires a much longer period. During their ocean residence they are distributed over a large area so that competition for food is probably not intense until they concentrate north of the Alaska Peninsula on their migration back to freshwater. The mean weights of sockeye in all of the Bristol Bay stocks tend to be correlated with the size of the total run which is largely determined by the Kvichak run. Therefore, the abundance of Kvichak fish affects production in the other stocks.

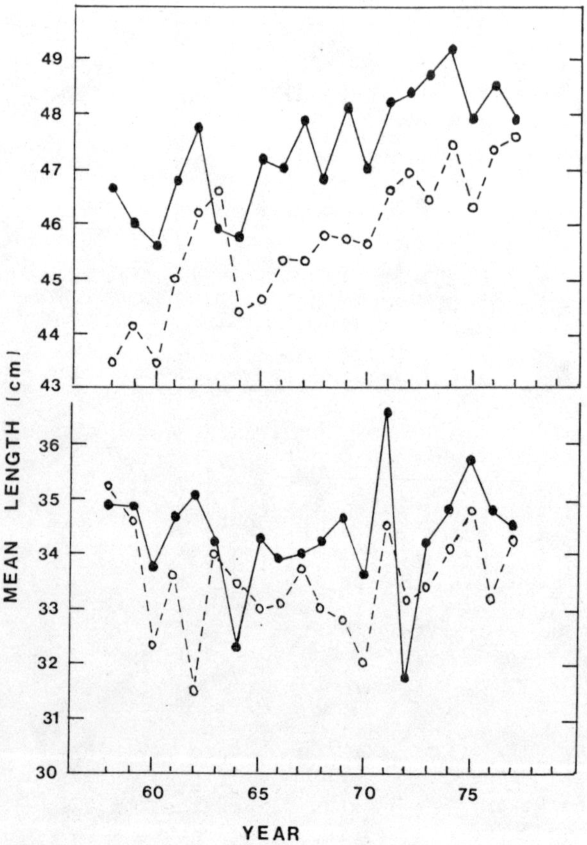

Figure 5. Annual mean lengths of age .2 (top) and age .1 (bottom) immature sockeye (solid points) and chum (open points) south of Adak on 21 July.

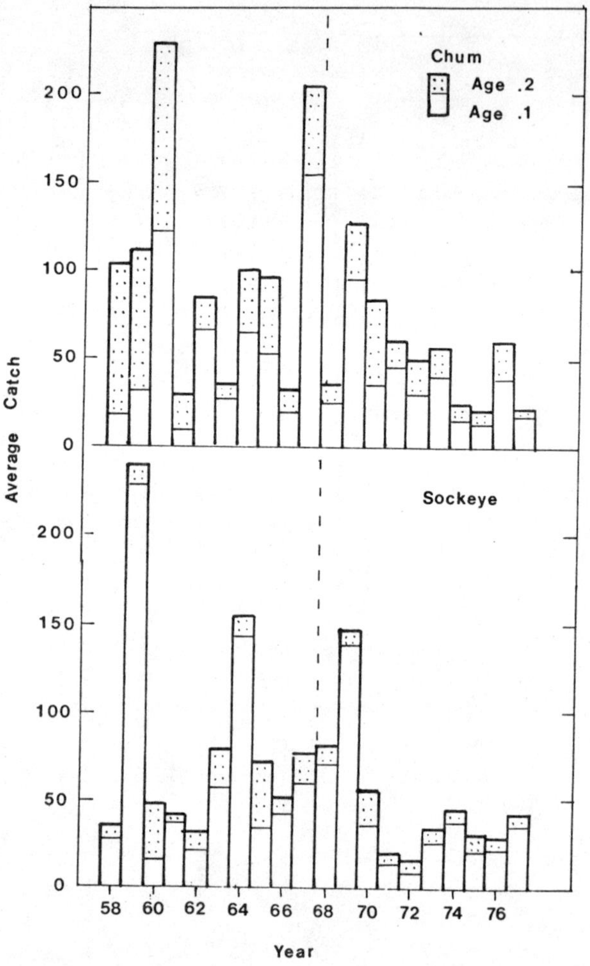

Figure 6. Average catches of immature chum and sockeye salmon 1-50 miles south of Adak from 1-July-10 August, 1958-1977.

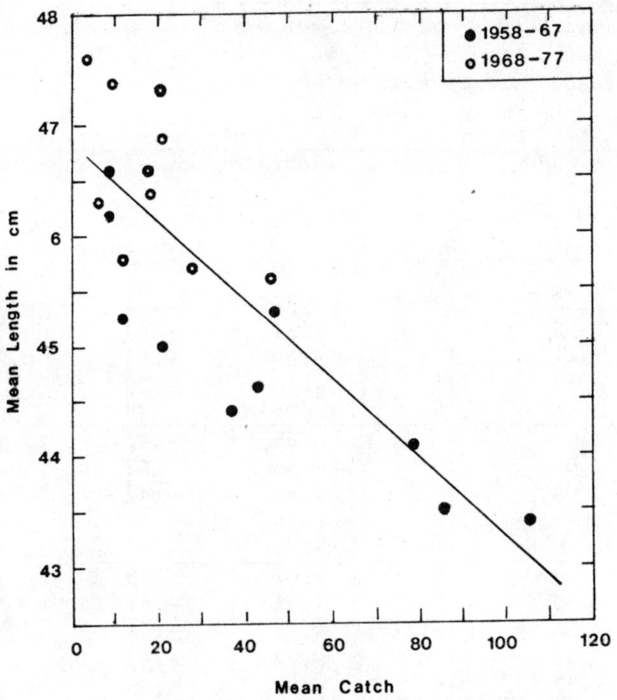

Figure 7. Mean length of age .2 chum salmon south of Adak on 21 July as a function of the mean catch during 1 July-10 August. y = 46.85 - 0.036X, r = -.83.

SUMMARY

Bristol Bay produces more sockeye salmon than any other geographic area in North America or Asia. Growth of juvenile sockeye in Bristol Bay lake systems is affected by the density of parent spawners and the length of the growing season (ice-free period). Smaller sockeye smolts emigrate from lakes with high density and short growing seasons (e.g. Wood River) than from lakes with low density and long growing seasons (e.g. Egegik). Average size of all smolts in an annual migration from Bristol Bay is largely determined by the distribution of parent spawners among the lake systems. Smolts migrate through Bristol Bay from about mid-May to the end of August. Timing of the annual migrations is affected by spring weather conditions and dates of ice breakup. There are differences in the timing of migrations from the lake systems and among years. Juveniles spend a variable length of time at sea in their first summer.

The annual mean weights of immature age .1 sockeye off Adak were correlated with the spring weather (water temperature) in the previous year when they migrated to sea. The abundance of sockeye had only a small effect on the size of age 2.1 immatures. There was a significant trend in the annual mean lengths of age .2 immature sockeye and chum salmon, possibly related to changes in sampling location. The average size of age .2 sockeye on 21 July was affected by the spring weather in the same year, but no effect of abundance on size was detected. The annual mean lengths of age .2 chum salmon (primarily Asian stocks) were correlated with their average catch, and there was evidence that their growth was density-dependent.

There were significant differences in the mean weights of adult sockeye in the runs within districts of Bristol Bay and among the districts. Annual mean weights of all sockeye of a given age were correlated with the number of fish in the run. The correlations were highest for ages 2.2 and 1.3 which are five years old at maturity. The abundance of the annual runs from 1957 to 1973 followed a five-year cycle which was caused by the cyclic abundance of the Kvichak run. There may have been some hereditary influences on the annual variation in the weights of these age groups. However, the annual mean weights of age 2.3 fish (six-year-olds) were also correlated with the abundance of the run and the mean weights of age 1.2 sockeye were below average in the years of highest abundance. Therefore, density-dependent growth was probably the major cause of the annual variation in the mean weights of ages 2.2, 2.3 and 1.3. Spring weather conditions also had a significant but small effect on annual mean weights. The density-dependent growth probably occurred when maturing sockeye were concentrated in waters north of the Alaska Peninsula on their return migration to Bristol Bay. Adult fish return to Bristol Bay over a relatively short period of about two weeks in late June-early July.

Table 2. Means and Standard Deviations of the Annual Mean Lengths and Weights of Sockeye Salmon in Bristol Bay Runs and the Linear Correlation Coefficients Between Mean Weights and the Numbers of Fish in the District and Total Bristol Bay Runs.

			Length (mm)		Weight (kg)		Correlation coefficient (weight)	
Age	Sex	District	Mean	S.D.	Mean	S.D.	District run	Bristol Bay run
1.2	M	Naknek-Kvichak	508	8	2.20	.12	-.49	-.51
		Egegik	521	8	2.36	.16	-.51	-.64
		Ugashik	522	8	2.41	.12	-.49	-.61
		Nushagak	515	9	2.34	.17	-.10	-.20
		Togiak	534	6	2.68	.15	.04	-.43
	F	Naknek-Kvichak	498	10	1.94	.13	-.38	-.41
		Egegik	509	10	2.08	.15	-.36	-.62
		Ugashik	509	11	2.08	.13	-.44	-.42
		Nushagak	494	8	1.93	.13	-.26	-.38
		Togiak	512	5	2.20	.11	.26	-.05
2.2	M	Naknek-Kvichak	522	11	2.40	.15	-.67	-.70
		Egegik	531	9	2.50	.17	-.61	-.81
		Ugashik	534	11	2.56	.18	-.39	-.65
		Nushagak	528	14	2.51	.27	-.22	-.42
		Togiak	540	8	2.77	.16	-.03	-.55
	F	Naknek-Kvichak	511	10	2.10	.12	-.83	-.85
		Egegik	515	10	2.11	.14	-.59	-.78
		Ugashik	513	11	2.11	.13	-.58	-.74
		Nushagak	507	10	2.10	.18	-.45	-.51
		Togiak	523	8	2.36	.15	.03	-.26
1.3	M	Naknek-Kvichak	579	8	3.26	.22	-.77	-.77
		Egegik	587	8	3.45	.20	-.47	-.61
		Ugashik	591	11	3.52	.21	-.72	-.78
		Nushagak	583	8	3.53	.22	-.28	-.59
		Togiak	595	10	3.83	.31	-.29	-.56
	F	Naknek-Kvichak	560	8	2.79	.15	-.75	-.77
		Egegik	568	10	2.88	.19	-.40	-.56
		Ugashik	571	9	2.95	.16	-.60	-.77
		Nushagak	555	8	2.82	.14	-.12	-.63
		Togiak	564	8	2.99	.15	-.36	-.62
2.3	M	Naknek-Kvichak	585	9	3.38	.17	-.62	-.64
		Egegik	591	8	3.45	.18	-.56	-.65
		Ugashik	594	12	3.53	.23	-.48	-.46
		Nushagak	585	10	3.53	.26	-.09	-.34
		Togiak	596	11	3.81	.34	-.32	-.65
	F	Naknek-Kvichak	566	7	2.86	.13	-.60	-.62
		Egegik	572	8	2.92	.15	-.55	-.72
		Ugashik	572	9	2.94	.17	-.37	-.56
		Nushagak	557	9	2.82	.14	-.37	-.48
		Togiak	569	7	3.09	.13	-.24	-.32

Table 3. Means and Standard Deviations of the Number of Fish (Millions) in the District Runs and the Total Bristol Bay Runs for Years that Were Used to Calculate the Statistics in Table 2.

District	Years	District run Mean	District run S.D.	Bristol Bay run Mean	Bristol Bay run S.D.	Correlation (r) district-Bristol Bay
Naknek-Kvichak	1957-73	11.4	12.1	16.7	13.7	.99
Egegik	1957-73	2.0	1.0	16.7	13.7	.84
Ugashik	1958-73	.9	.8	17.0	14.1	.71
Nushagak	1959-73	2.3	1.0	17.8	14.3	.29
Togiak	1963-73	.3	.1	17.2	15.6	.48

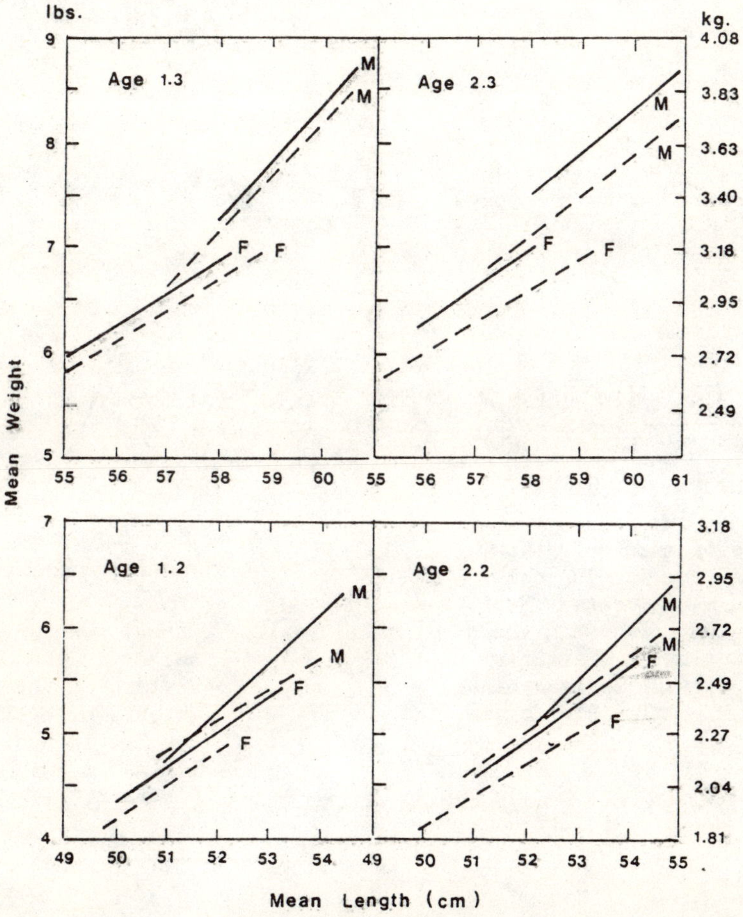

Figure 8. Regressions of mean weight on mean length from samples collected from the commercial fishery catches. Solid lines represent fish from the Nushagak and Togiak districts and dashed lines fish from other districts; M = male, F = female.

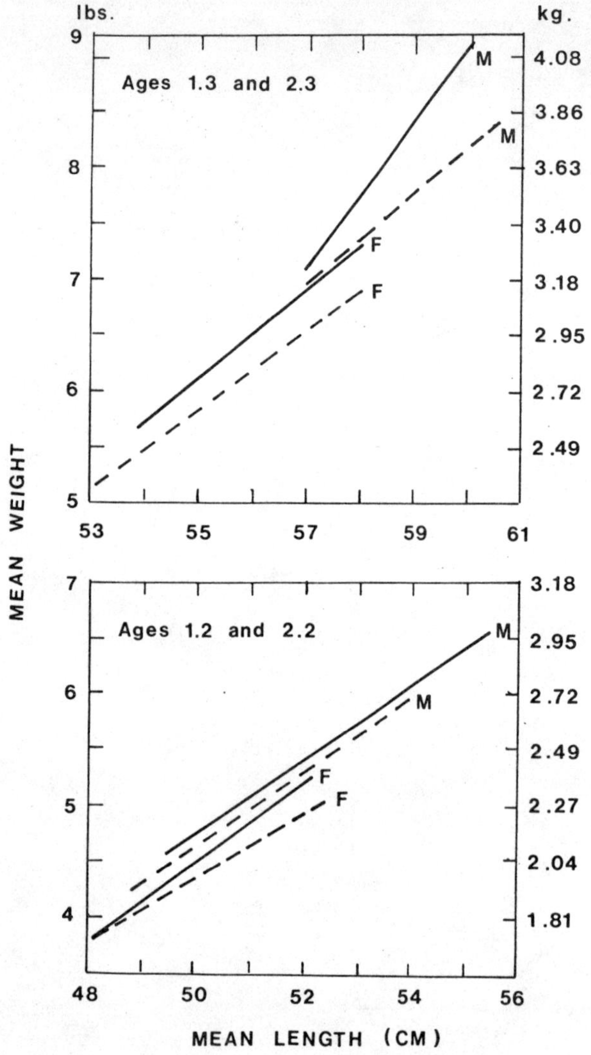

Figure 9. Regressions of mean weight on mean length from samples collected from escapements. Solid lines represent fish from the Nushagak and Togiak districts and dashed lines fish from other districts; M = male; F = female.

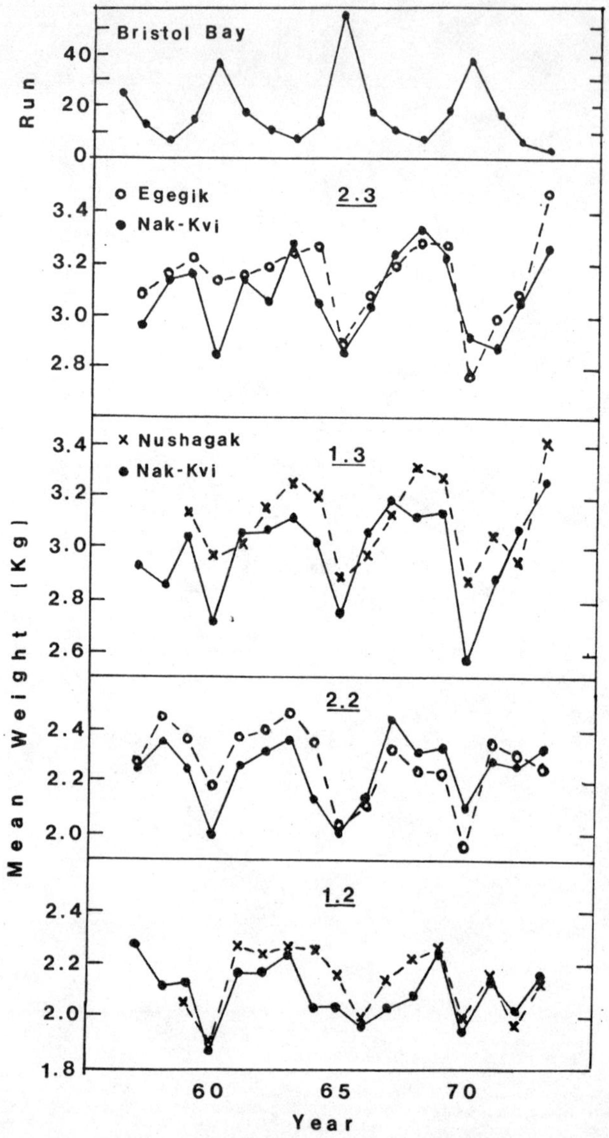

Figure 10. Annual mean weights of sockeye salmon in the Naknek-Kvichak District by age, and the mean weights of ages 2.2 and 2.3 in the Egegik District and ages 1.2 and 1.3 in the Nushagak District.

Table 4. The Sockeye Salmon Runs (In Millions) By District and Age for the Years 1960, 1965 and 1970; and the Average Runs for the Other Years During 1958-1973.

Year	District	Age 1.2	1.3	2.2	2.3
1960	Naknek-Kvichak	24.11	1.03	.75	.65
	Egegik	1.83	.14	.61	.57
	Ugashik	2.86	.03	.11	.05
	Nushagak	1.06	1.45	.48	.16
	Togiak	.12	.20	.01	.02
1965	Naknek-Kvichak	.59	1.03	42.12	.60
	Egegik	.07	.25	3.94	.35
	Ugashik	.21	.22	1.38	.10
	Nushagak	.30	1.33	.16	.07
	Togiak	.08	.22	.02	.01
1970	Naknek-Kvichak	1.34	.77	29.83	.66
	Egegik	.22	.06	1.78	.19
	Ugashik	.64	.03	.22	.01
	Nushagak	1.05	1.67	.31	.09
	Togiak	.18	.15	.03	.00
Average runs in other years, 1958-1973					
	Naknek-Kvichak	1.52	1.15	2.32	1.21
	Egegik	.07	.30	.70	.62
	Ugashik	.08	.22	.20	.09
	Nushagak	1.01	.78	.22	.09
	Togiak	.08	.14	.02	.02

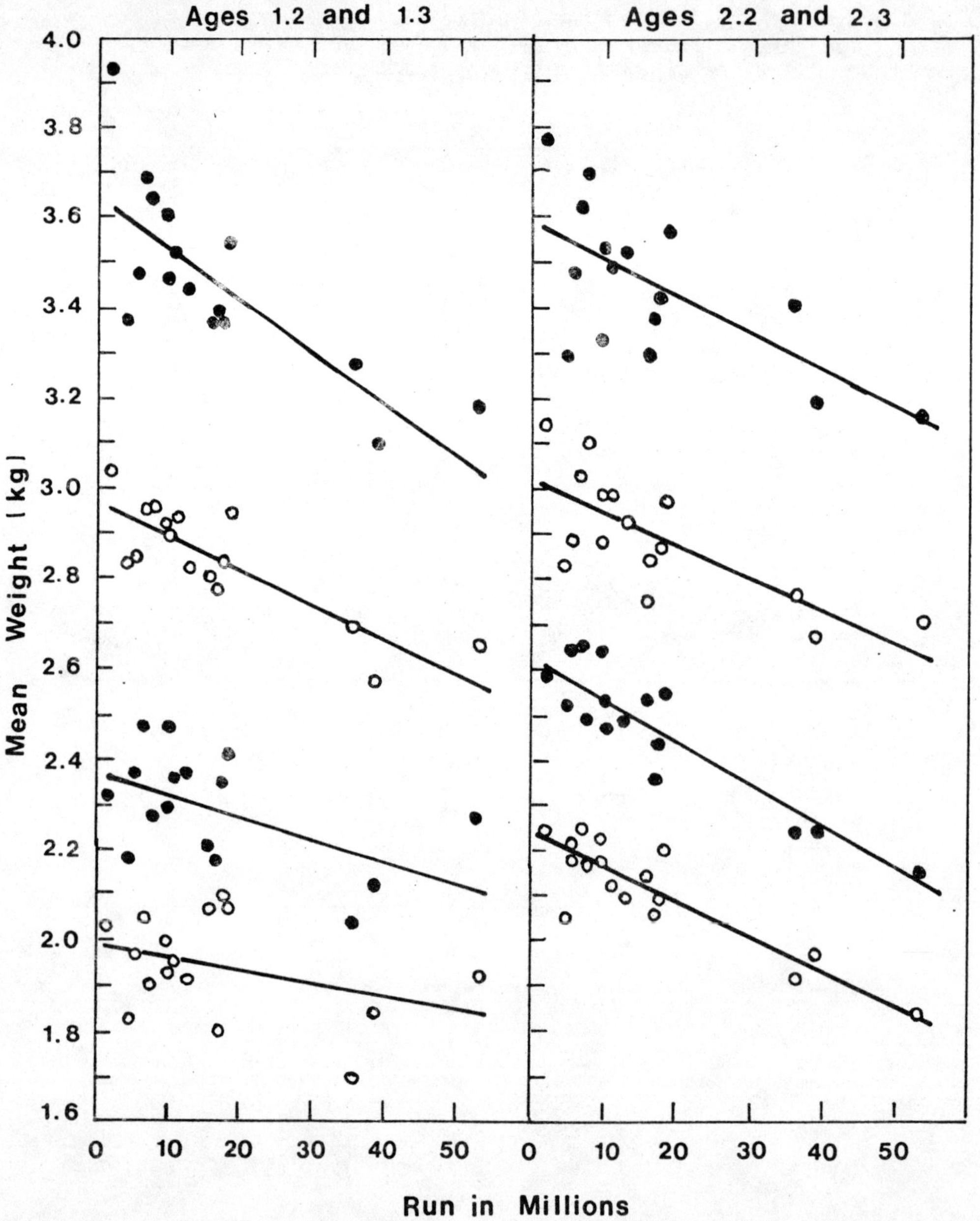

Figure 11. Mean weights of sockeye salmon in the Bristol Bay runs as functions of the number of fish in the run.

Age 1.3 males	y = 3.64 - .0109X,	r = -.76
1.3 females	y = 2.96 - .0074X,	r = -.84
1.2 males	y = 2.36 - .0044X,	r = -.49
1.2 females	y = 1.99 - .0029X,	r = -.36
Age 2.3 males	y = 3.58 - .0081X,	r = -.66
2.3 females	y = 3.01 - .0071X,	r = -.73
2.2 males	y = 2.61 - .0090X,	r = -.86
2.2 females	y = 2.23 - .0075X,	r = -.87

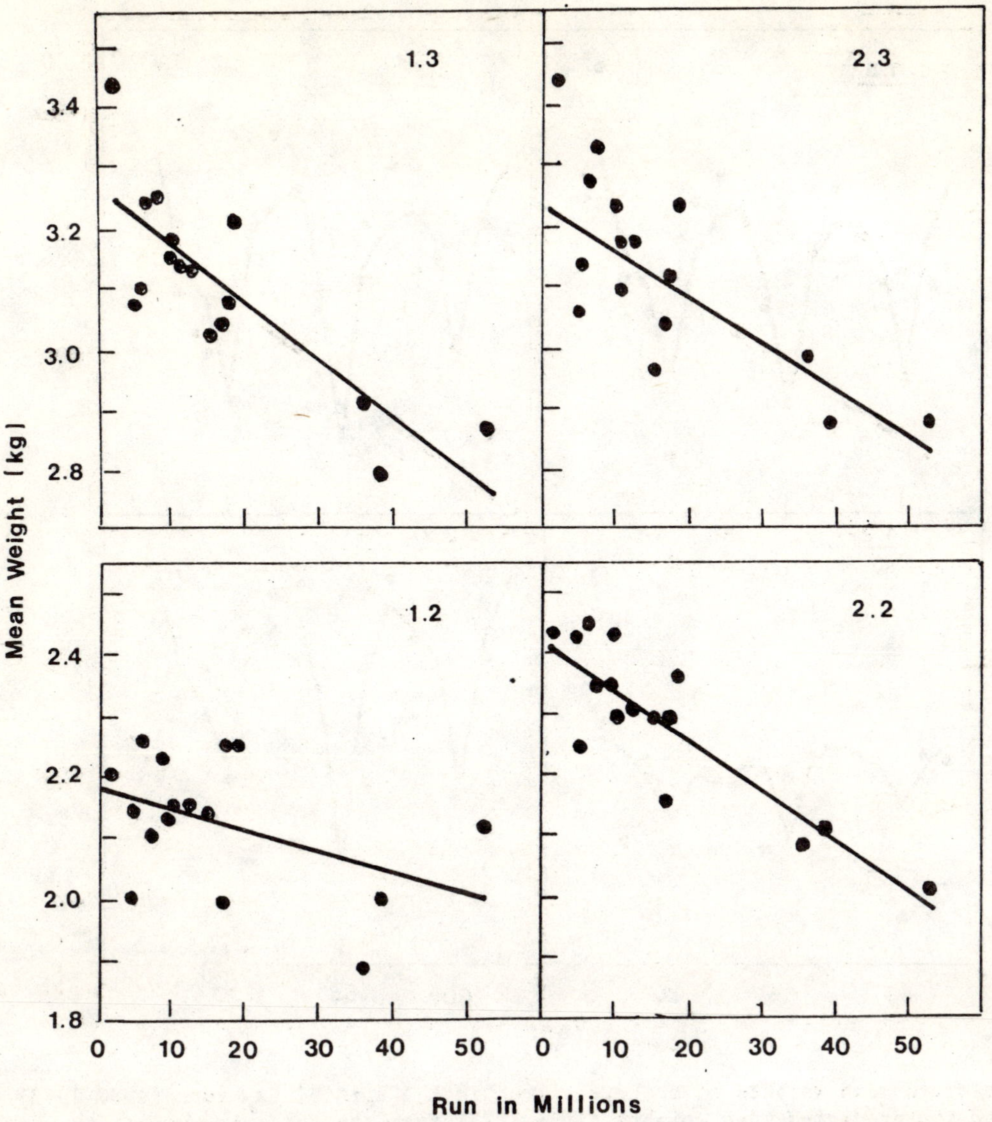

Figure 12. Mean weights of sockeye salmon in the Bristol Bay runs as functions of the number of fish in the run (1958-1973).

Age 1.2 y = 2.179 - .0033X, r = -.43
2.2 y = 2.421 - .0083X, r = -.87
1.3 y = 3.258 - .0092X, r = -.82
2.3 y = 3.261 - .0084X, r = -.75

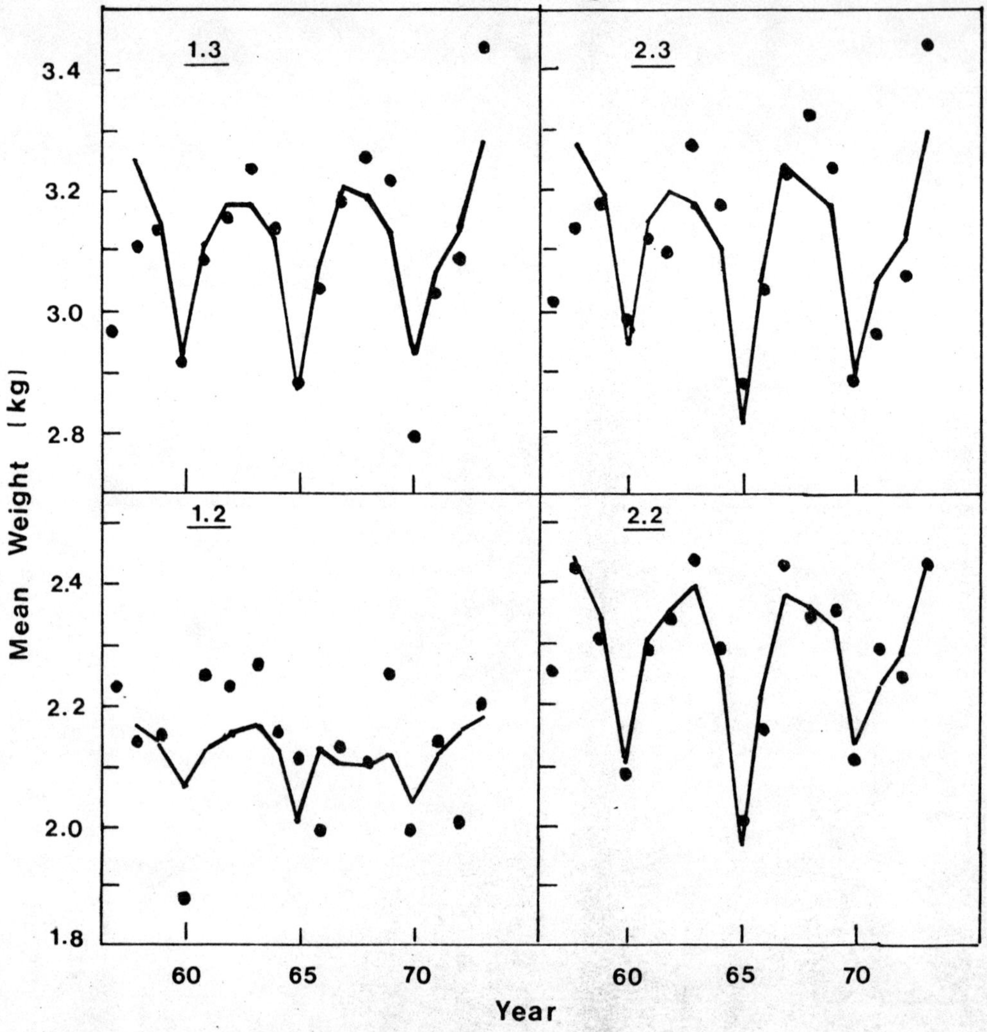

Figure 13. Observed mean weights of sockeye salmon in the Bristol Bay runs (1957-1973) and the predicted mean weights (1958-1973) from the following equations where X_1 = Bristol Bay run (millions) and X_2 = April-May air temperature (C):

Age 1.2 $y = 2.18 - .0033X_1$, $F = 3.09$, $r = -.43$
1.3 $y = 3.26 - .0086X_1 + .025X_2$, $F = 16.81$, $R = .85$
2.2 $y = 2.34 - .0076X_1 + .030X_2$, $F = 45.39$, $R = .94$
2.3 $y = 3.16 - .0076X_1 + .036X_2$, $F = 14.08$, $R = .83$

LITERATURE CITED

Burgner, R. L. 1962. Studies of red salmon smolts from the Wood River Lakes, Alaska. Univ. Wash. Publ. Fish. New Series 1:315-348.

Fredin, R. A., and D. D. Worland. 1974. Catches of sockeye salmon of Bristol Bay origin by the Japanese mothership salmon fishery, 1956-1970. Int. N. Pac. Fish. Comm. Bull 30.

French, R., H. Bilton, M. Osako, and A. Hartt. 1976. Distribution and origin of sockeye salmon *(Oncorhynchus nerka)* in offshore waters of the North Pacific Ocean. Int. N. Pac. Fish. Comm. Bull. 34. 113 pp.

Foerster, R. E. 1968. The sockeye salmon, *Oncorhynchus nerka*. Fish. Res. Bd. Can. Bull. 162. 422 pp.

Hartt, A. C. 1966. Migrations of salmon in the North Pacific Ocean and Bering Sea as determined by seining and tagging, 1959-1960. Int. N. Pac. Fish. Comm. Bull. 19. 141 pp.

International North Pacific Fisheries Commission. 1974. Historical catch statistics for salmon of the North Pacific Ocean. Manuscript version (second draft) of compilation proposed for publication in the INPFC Bulletin.

Killick, S. R., and W. A. Clemens. 1963. The age, sex ratio and size of Fraser River sockeye salmon, 1915 to 1960. Int. Pac. Sal. Fish. Comm. Bull. 14. 140 pp.

Krogius, F. V. 1960. [The rate of growth and age groupings of sockeye salmon (*Oncorhynchus nerka*) (Walbaum) in the sea.] Vopr. Ikhtiol., 16:67-88 [FRB Translation No. 413.]

Mathisen, O. A. 1969. Growth of sockeye salmon in relation to abundance in the Kvichak District, Bristol Bay, Alaska. Fisk, Dir. Skr. Ser. HavUnders. 15:172-185.

Neave, F., T. Yonemori, and R. G. Bakkala. 1976. Distribution and origin of chum salmon in offshore waters of the North Pacific Ocean. Int. N. Pac. Fish. Comm. Bull. 35. 79 pp.

Ricker, W. E. 1972. Hereditary and environmental factors affecting certain salmonid populations. Pages 19-160 *in* Simon, R. C., and P. A. Larkin (eds.), The stock concept in Pacific salmon. Univ. British Columbia.

Roberson, K. 1966. Report on age and size of sockeye salmon sampled from spawning grounds in the Wood River Lakes, 1965. Univ. Wash., Coll. Fish., Fish. Res. Inst. Circ. 66-68. 20 pp.

Rogers, D. E. 1973. Abundance and size of juvenile sockeye salmon, *Oncorhynchus nerka*, and associated species in Lake Aleknagik, Alaska, in relation to their environment. Natl. Ocean. Atmos. Admin. Fish. Bull. 71(4):1061-1075.

Rogers, D. E. 1976. Forecast of the sockeye salmon run to Bristol Bay in 1976. Univ. Wash., Coll. Fish., Fish. Res. Inst. Circ. 76-1. 49 pp.

Straty, R. B. 1972. Ecology and behavior of juvenile sockeye salmon *(Oncorhynchus nerka)* in Bristol Bay and the eastern Bering Sea. Oceanography of the Bering Sea - Occasional Publ. 2, Univ. Alaska Inst. Mar. Sci.

Population Structures of Indigenous Salmonid Species of the Pacific Northwest

Fred M. Utter, Donald Campton, Stewart Grant, George Milner,
James Seeb, and Lisa Wishard

INTRODUCTION

Studies at the National Marine Fisheries Service Northwest and Alaska Fisheries Center (NWAFC) (Hodgins 1972) to define the genetic structure of salmonid populations had a biochemical basis by the mid-1960s. These early studies (Utter et al. 1974) became the foundation of more recent applications of electrophoretic techniques to evaluate genetic variations of salmonid proteins. Genetic variations defined by electrophoretic studies have been reviewed by Utter et al. (1973), Utter et al. (1974), Utter et al. (1976), and Allendorf and Utter (1979).

The more recent genetic research at NWAFC has centered on populations of *Salmo* and *Oncorhynchus* sp. native to the Pacific Northwest. Although our understanding of the population structures of salmon and trout remains incomplete, our capability to examine specific loci for genetic differences is increasing steadily. This paper summarizes current knowledge on genetic variations in natural populations of sockeye salmon *(Oncorhynchus nerka)*, pink salmon *(O. gorbuscha)*, chum salmon *(O. keta)*, chinook salmon *(O. tshawytscha)*, coho salmon *(O. kisutch)*, rainbow (steelhead) trout *(Salmo gairdneri)* and coastal cutthroat trout *(S. clarki clarki)*.

METHODS

Geographic patterns of genetic variations in protein systems among salmon and trout populations can be used to identify different populations. Protein systems used to evaluate differences among populations are chosen on the basis of their simple inheritance, and genetic differences among populations are evaluated statistically by comparing the frequencies of genotypes or alleles.

Electrophoretic techniques provide an investigator with the capability to collect large amounts of data with relative ease. No other known method is as convenient for making genetic comparisons among congeneric or conspecific populations. Utter et al. (1974) and Allendorf and Utter (1979) describe methods for collecting and interpreting genetic data by electrophoresis. Useful references on histochemical staining include Siciliano and Shaw (1975) and Harris and Hopkinson (1976).

Statistical methods for analyzing data on gene variations among loci, individual fish, populations and species come mostly from procedures developed in studying *Drosophila* and man. The Hardy-Weinberg principle (Stern 1943) is the foundation of many of the analytical processes. It states that the expected genotypic proportions at a polymorphic locus are the square of the allelic frequencies in large, random mating populations where no selective difference exists among genotypes. Significant deviations from expected Hardy-Weinberg proportions may be the result of factors such as drawing the sample from a mixture of populations having different allelic frequencies; a small number of parents giving rise to an inbred population; or selection.

Statistical tests for significant differences between two groups of individuals are made either from genotypic or allelic frequencies. Tests include measuring deviations from expected Hardy-Weinberg proportions, chi-square tests for independence and measuring differences of allelic frequencies from normal approximations of binomial data. Significant differences occurring at a single locus or at several loci can be used to differentiate populations. Summing the differences of allelic frequencies at several loci provides a basis to calculate the degree of genetic similarity or difference between two populations. Rogers (1972) and Nei (1972) describe two methods for doing this. Calculated values range from 0 (complete genetic difference of loci examined) to 1 (complete genetic identity of loci examined). Dendrograms are sometimes constructed from matrices of such pairwise comparisons among populations (Sneath and Sokal 1973).

Allelic frequencies must remain stable within a population over time to use electrophoretic data to define genetic differences among populations and to estimate contributions of separate populations to mixed fisheries. Estimates taken at different times may be combined to increase the precision of statistical estimators where frequencies remain stable. Unstable frequencies require annual sampling of spawning populations and render the electrophoretic data virtually useless for managing fisheries on mixed stocks.

Allelic frequencies are assumed to remain stable in large populations of salmonids over succeeding generations and among year classes. This assumption is based largely on data where adult and juvenile fish from the same stream tend to express similar allelic frequencies at polymorphic loci, even though the samples include fish from different year classes. Allelic frequency comparisons supporting this assumption are presented in this report and describe different ages, generations and year classes of salmonid populations for which relatively extensive data have been obtained. Enzymes used in these studies are listed in Table 1.

RESULTS AND DISCUSSION

Comparisons of Allelic Frequencies Among Year Classes and Generations

Changes in allelic frequencies of pink salmon have been observed at two locations in Washington. Table 2 presents data for loci (*AGP-1* and *MDH-3*) from five consecutive generations of an early run of pink salmon at the Dungeness River (Aspinwall 1974, Seeb and Grant 1976). These data, along with data for two other loci (*PGM-1* and *AAT-3*) obtained from the Dungeness River and Hoodsport Hatchery populations, provide an opportunity to study changes in allelic frequencies.

There were no observed changes in allelic frequencies for the common alleles of *AGP-1*

Table 1. Protein Enzyme Systems Used in Electrophoretic Comparisons of Genotypes

Enzyme	Abbreviation
Albumin	Alb
Alcohol dehydrogenase	ADH
Alpha glycerophosphate dehydrogenase	AGP
Aspartate amino transferase.	AAT
Creatine kinase	Ck
N-acetyl-B-D-Galactosaminidase	GAL
Glutamic-pyruvic transaminase	GPT
Isocitrate dehydrogenase	IDH
Lactate dehydrogenase	LDH
Malate dehydrogenase	MDH
Malic enzyme	ME
Peptidase	PEP
6 phosphogluconate dehydrogenase	6PGD
Phosphoglucose isomerase	PGI
Phosphoglucomutase	PGM
Phosphomannose isomerase	PMI
Sorbitol dehydrogenase	SDH
Tetrazolium oxidase	TO
Transferrin	Tfn
Glycyl leucine peptidase	GL
Leucyl glycylglycine peptidase	LGG
Phenylalanylproline peptidase	PhAP

Table 2. Allelic Frequencies with (in Parentheses) 95 Percent Confidence Intervals Over Sequential Generations for Four Loci in Pink Salmon Collected from Two Areas of Washington.

Location	Year	N	*AGP-1*	*PGM-1*	*AAT-3*	*MDH-3*
Hoodsport	1973[2]	109	.959(.027)	.853(.048)	.719(.061)	.954(.028)
Hoodsport	1975[2]	135	.956(.025)	.828(.046)	.673(.057)	.978(.018)
Dungeness	1969[1]	39	.987(.067)	...	...	.936(.055)
Dungeness	1971[1]	93	.957(.030)	...	...	.962(.028)
Dungeness	1973[2]	35	.943(.055)	.971(.040)	...	.914(.067)
Dungeness	1975[2]	44	.932(.054)	.977(.032)	.549(.106)	.932(.054)
Dungeness	1977	225	.923(.025)	.916(.026)	.708(.043)	.976(.014)

[1] Data from Aspinwall 1974.

[2] Data from Seeb and Grant 1976.

and *MDH-3*. Significant changes occurred both within and between populations at the *PGM-1* and *AAT-3* loci where frequencies in the Dungeness population approached frequencies in the Hoodsport population at both loci. These results were puzzling at first. However, an examination of Washington Department of Fisheries records indicated that fry from the Hoodsport Hatchery had been released from the Dungeness Hatchery in 1976 (1975 brood year). Thus, the run returning to the Dungeness Hatchery in 1977 included survivors from the release of fry originating from Hoodsport plus progeny from natural spawners (Foster et al. 1977). Results are readily explained by: 1) similar frequencies of *AGP-1* and *MDH-3* alleles in the Dungeness and the Hoodsport pink populations, which should remain stable over generations; 2) different frequencies of *PGM-1* and *AAT-3* alleles between the two populations; and 3) an alteration of the *PGM-1* and *AAT-3* frequencies in the 1977 brood year for pink salmon returning to the Dungeness Hatchery as a direct result of large plantings of Hoodsport pink salmon into the Dungeness River.

A second set of data on changes in allelic frequencies pertains to steelhead. Summer-run steelhead from the Skamania Hatchery (Washington State Department of Game) on the Washougal River near Vancouver, Washington have been sampled for five years to study the genetic effect of transplanting stocks. Figure 1 shows that most frequencies of the common allele of a given locus were not significantly different ($P < .05$) between collections. However, the frequencies of the common allele of *MDH-3* from three collections and of the common allele of *TO* from one collection lie outside of the confidence intervals of two or more samples for each of these biochemical systems. These four exceptional frequencies (out of 37 data points) are somewhat higher than the one out of 20 such observations that would be expected from chance alone at the 95 percent confidence level in a stable, random mating population. But the fish returning to the Skamania Hatchery do not represent such a population. The hatchery run was derived from native fish of the Washougal River and also included fish from the Klickitat River approximately 50 miles upstream (personal communication, James Morrow, Washington State Department of Game). Each of the founding year classes comprised a different mixture of progenitor stocks. Some differences among year classes of such a heterogeneous stock would therefore be expected despite factors tending to reduce these differences such as the overlap among year classes and the capability of steelhead for multiple-year spawning. Under these conditions significantly different frequencies between some collections are therefore anticipated. Evidence of allelic frequency stability among cohorts over time is seen in the *MDH-3* frequencies of the juveniles sampled in 1973 and 1974 and in the adults of 1976 and 1977, considering the predominant three-year span from juvenile to adult. These data suggest that the initial inequalities of allelic frequencies expected among year classes may be approaching a point of equilibrium.

Finally, a third set of data exists concerning allelic frequencies from the four loci in juvenile and adult winter-run steelhead trout from the Washington Department of Game Chambers Creek Hatchery (Tacoma, Washington) (Fig. 2). Chambers Creek Hatchery stock was primarily derived from a natural run of Chambers Creek during

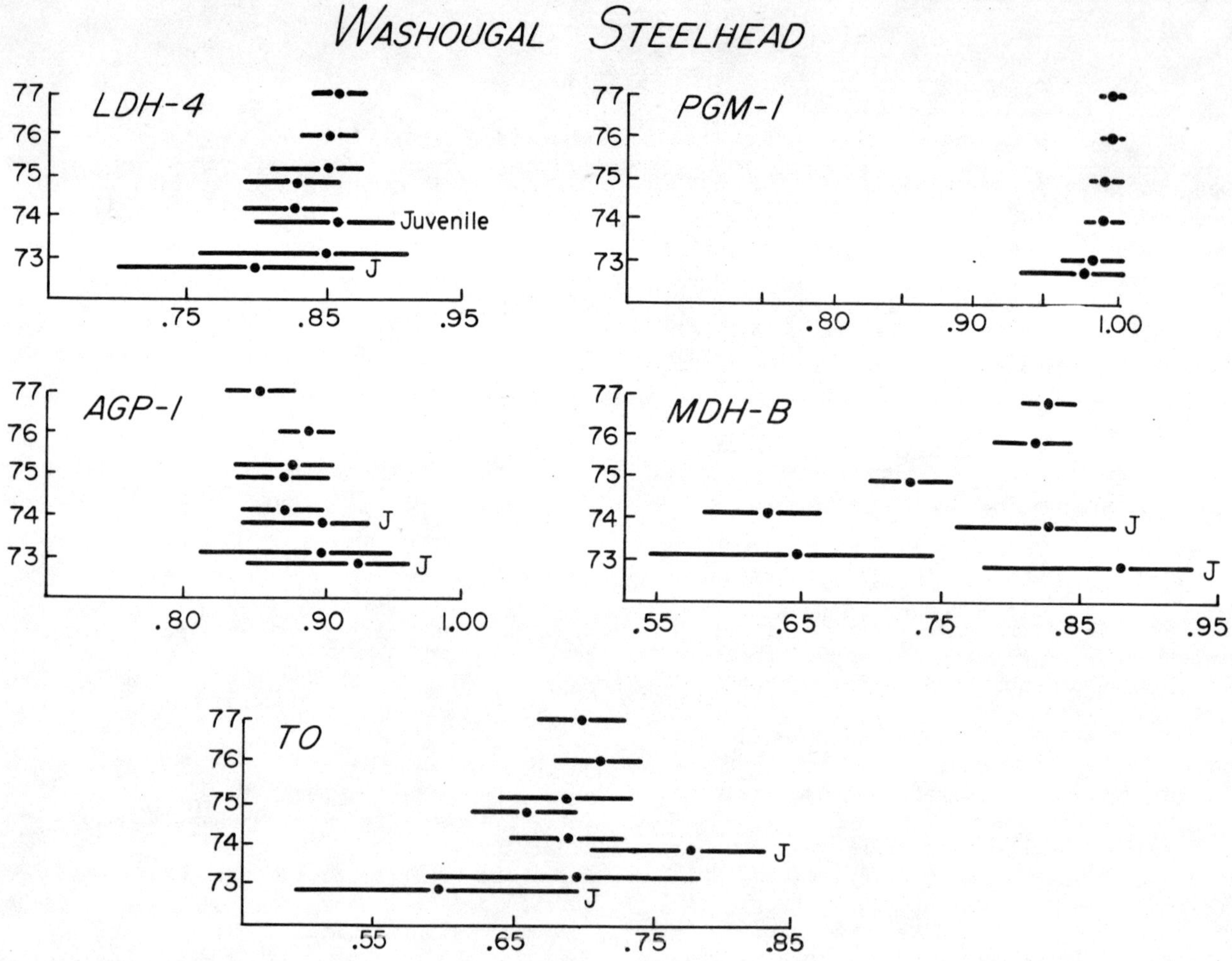

Figure 1. Allelic frequencies and 95 percent confidence intervals at five loci for juvenile and adult Washougal steelhead trout from the Skamania Hatchery over a five year interval.

the late 1940s (personal communication, Art Westrope, Washington State Department of Game). Only a single data point (*MDH-3*) was found to lie outside of the 95 percent confidence intervals. (A single aberrant observation out of 30 lies well within the expected range for this level of confidence). The data indicate that the Chambers Creek steelhead stock presently appears to be close to an equilibrium state with regard to allelic frequencies.

The three sets of data support the assumption of consistencies in allelic frequencies between generations and among (overlapping) year classes. It is important that allelic frequency records over time be maintained, wherever possible, for salmonid populations. Such data are necessary to document the level of genetic consistency within populations, and can provide valuable and unique insights into the causes of changes in populations as they occur.

Genetic Structures of Populations Within Species

This section describes the better known aspects of the genetic variation among salmonids native to western North America. Five salmon species and two trout species are considered.

The number of currently known polymorphic systems in these species (Appendix Table 1) is approximately double the number of variant systems known in rainbow trout five years ago (Utter et al. 1974). This substantial increase in our knowledge of genetic variation, and the likelihood of additional variants being revealed, suggests that a large reservoir of genetic variation exists, which is potentially available for examining genetic structures of salmonids.

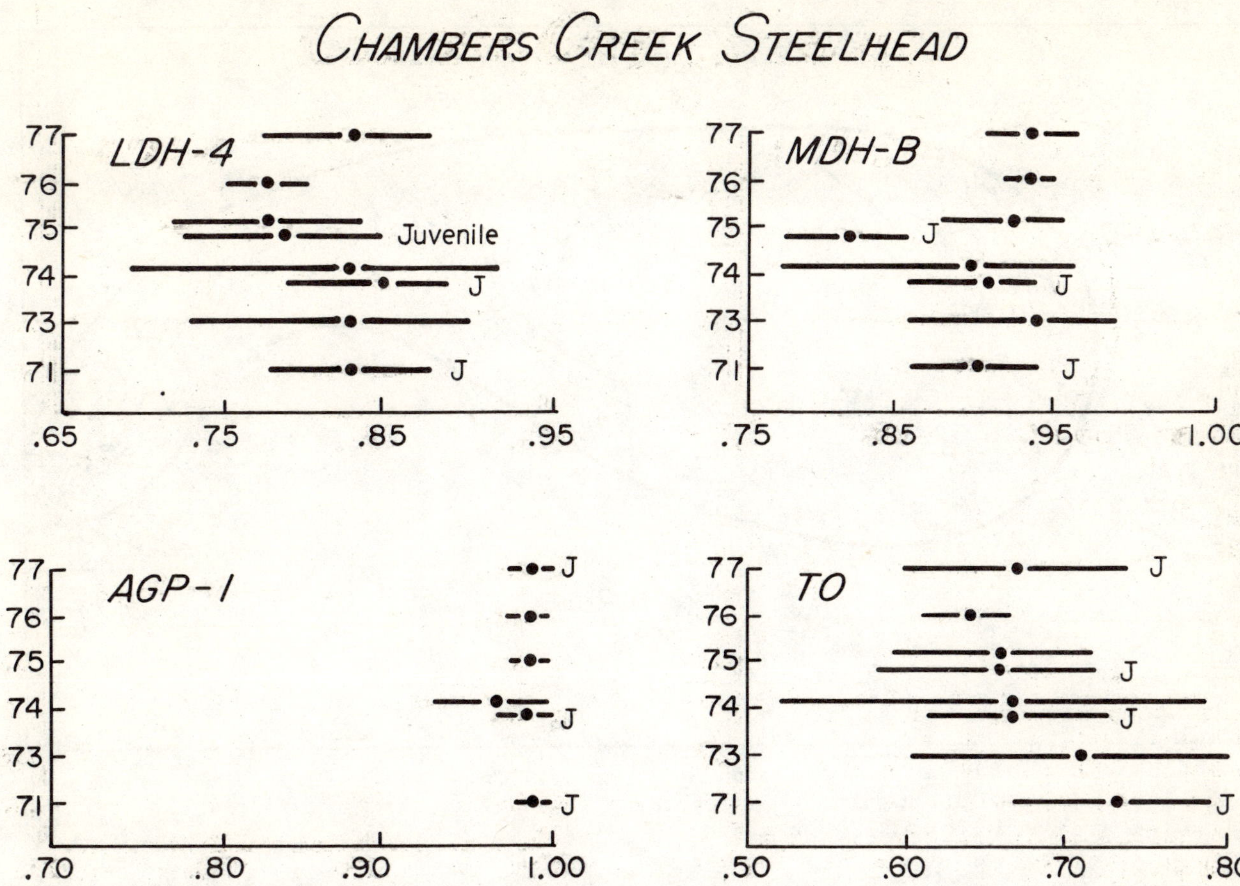

Figure 2. Allelic frequencies and 95 percent confidence intervals at four loci for juvenile and adult steelhead trout from the Chambers Creek Hatchery over a seven year interval.

Sockeye Salmon: Because more precise knowledge was needed about the origins of sockeye salmon harvested in the Japanese high seas fishery, this species was the initial primary target of immunological and electrophoretic studies.

Early electrophoretic data on sockeye were not very useful because alleles of two highly polymorphic systems (*LDH-4, PGM)* occurred at similar frequencies in both American and the Asiatic stocks (Hodgins et al. 1969, Utter and Hodgins 1970, Hodgins and Utter 1971, Altukhov 1975). However, genetic differences among sockeye populations have subsequently been observed to increase southeastwardly from the Copper River drainage in Alaska into regions where stocks do not intermingle significantly with Asiatic stocks. Sampling of sockeye salmon populations from the Skeena River (northern British Columbia) southward to Puget Sound and the Quinault River (Washington coast) indicated higher frequencies of the common *LDH-4* and *PGM* alleles than in populations from western Alaska and Asia (Utter et al. 1974). These differences were also found in non-anadromous populations of sockeye (kokanee). It is now assumed that high frequencies of these two alleles are typical of sockeye populations in the southern part of this species' North American range. Subsequent samplings, however, have proven that this assumption is an oversimplification (Fig. 3). Data from populations sampled southward through the Fraser River drainage into Washington are consistent with the concept of a single major population group (Utter and Hodgins 1970, Hodgins and Utter 1971). However, two collections of anadromous sockeye from Columbia River tributaries (May and Utter 1974) and collections of kokanee from Issaquah Creek near Seattle (personal communication, James Seeb, Washington Department of Fisheries) have allelic frequencies that differ from any other known sockeye population. Fish from these latter locations possess high frequencies of *PGM* variants typically found in populations from western Alaska and Asia. The Issaquah Creek

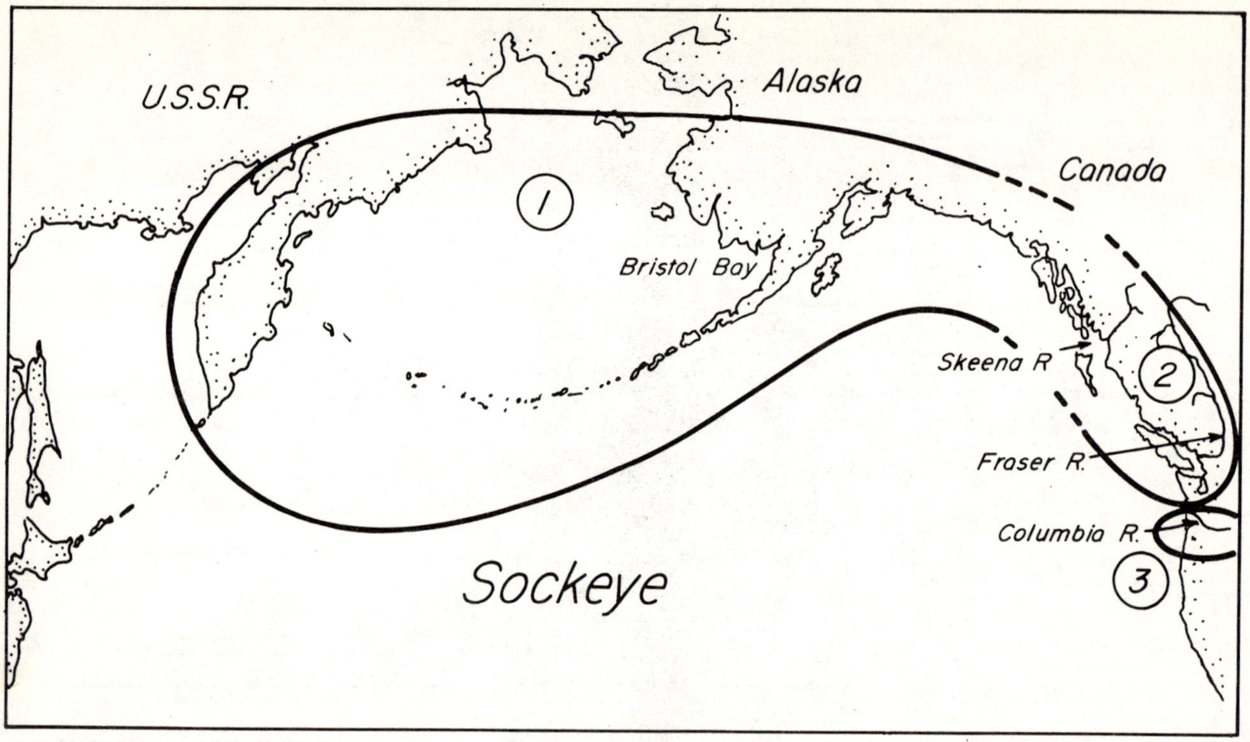

Figure 3. Major population units of sockeye salmon identified through different frequencies of allelic proteins.

kokanee population is also more polymorphic for the same *LDH-4* allele that occurs in the more western populations. Columbia River and Issaquah sockeye also possess variant alleles that have not been observed elsewhere in this species. *MDH-3* and *TO* variants were found in Columbia River sockeye, and a very high frequency of an *LDH-l* variant was found in Issaquah Creek kokanee.

The occurrence of previously undetected alleles in the southeastern extreme of this species' natural range is puzzling. One possible explanation is a diphyletic origin of Washington's sockeye populations following the last glacial era (about 10,000 years ago). One group may have originated from northern populations with the second group descending from southern stocks that may have existed during the last glacial period.

The similarity of allelic frequencies observed within a particular region coupled with low average heterozygosity observed in sockeye when compared to other species (Utter et al. 1973, Allendorf and Utter 1979) caused us to expect little variation in allelic frequencies among different spawning populations within a given drainage. Recent surveys of sockeye populations in the Cook Inlet region of Alaska (Grant et al. *in press*), the Port Alberni region of British Columbia (Allendorf and Mitchell 1977), and the Lake Washington drainage near Seattle (personal communication, James Seeb, Washington Department of Fisheries) have shown considerably greater complexity of allelic frequencies within drainages than expected. Heterogeneity among three drainages supporting sockeye in Cook Inlet was lowest in the river system containing a single lake (Kasilof River) and highest in two river systems containing a complex of lakes (Kenai and Susitna Rivers). The Port Alberni investigations were limited to three lakes in a single drainage where differences in the frequencies of *PGM*, *MDH-3* and *SDH* alleles were sufficient to distinguish fish collected from each of the lakes. The most detailed data were obtained from the Lake Washington drainage (personal communication, James Seeb, Washington Department of Fisheries), which supports both anadromous and non-anadromous populations including Issaquah Creek kokanee. There are large genetic differences among populations of Lake Washington sockeye. Samples from two consecutive years suggest that simulataneous spawning of anadromous and kokanee populations in Issaquah Creek occurs without significant gene flow between groups.

Pink Salmon: Pink salmon populations are characterized by a substantial degree of randomness of allelic frequencies over large geographic areas. Some genetic differences have been observed among odd-year populations in Alaska and the Pacific Northwest, including the Fraser River. An *MDH-B* variant is virtually absent from even-year populations in Alaska, but occurs at varying frequencies in odd-year populations (Aspinwall 1974, Johnson et al. 1978). Frequencies of *AGP* variants appear to be significantly lower in Fraser River and Puget Sound populations than in either even- or odd-year populations returning to Alaska.

For a particular year class, differences observed between even and odd years within a particular stream are generally greater than differences that occur between streams. This peculiar distinction has been recorded by three independent investigations (Aspinwall 1974, Seeb and Wishard 1977*a*, Johnson et al. 1978) and appears to be a direct reflection of the rigid two-year life cycle of pink salmon. Differences contrasting odd-year with even-year fish include generally higher frequencies in odd-year fish for variant alleles of the *AAT-3*, *MDH-3* and *PGM* loci and lower frequencies of *MDH-l* variants.

Attributes of pink salmon populations that distinguish them from other salmonid species include: 1) relatively high amounts of genetic variation within populations (Allendorf and Utter 1979); 2) reasonably large fluctuations of allelic frequencies among breeding groups within limited geographic areas; and 3) minimal patterns of similarity that are useful for defining broad geographic population units for even or odd-year populations. It is possible to interpret these characteristics within the context of the species' life history. Pink salmon spawn largely in the lower reaches and intertidal areas of small coastal streams that are periodically subjected to drastic fluctuations in temperature, and quality and quantity of water. Some streams are also occasionally modified through shifts of spawning strata resulting from earthquakes such as that centered in the Prince William Sound area of Alaska in 1964 (Thorsteinson et al. 1971). Population sizes vary drastically under such diverse conditions (Royce 1962). Allelic frequencies of populations subjected to periodic "bottlenecks" are expected to fluctuate randomly as observed in pink salmon (Crow and Kimura 1970). Thus, the relatively large degree of randomness observed for allelic frequency distributions among pink salmon populations comes as no great surprise. What is surprising are the high levels of heterozygosity in pink salmon populations because periodic "bottlenecks" in the absence of gene influx from other populations are expected to reduce levels of heterozygosity (Crow and Kimura 1970). It is possible that sufficient straying occurs among pink salmon populations to maintain reasonably high levels of heterozygosity, but that this straying is insufficient to overcome the periodic alteration of allelic frequencies brought about by the "bottlenecks."

Chum Salmon: Genetic studies of chum salmon populations were initiated at the same time as sockeye studies because the Japanese fish both species extensively on the high seas in the same areas. Chum salmon have played a subordinate role to sockeye in genetic studies because they are not as important to American fishermen. However, some genetic data on Asiatic chum salmon populations have been published (Numachi et al. 1972, Altukhov 1975).

Three major groups of chum salmon populations are evident from the allelic distribution of two polymorphic loci that both Asiatic and American workers examined. Variants for *LDH-l* are found in virtually all populations examined from southeastern Alaska through Japan (Utter et al. 1973, Altukhov 1975, Seeb and Wishard 1977*b*). *MDH-3* variants are found at polymorphic frequencies (frequencies greater than .01) in Asiatic collections (Numachi et al. 1972, Altukhov 1975, Seeb and Wishard 1977*b*). Neither *LDH-l* nor *MDH-3* variants have been observed from the Fraser River southward despite intensive sampling of Puget Sound populations recently. Thus, three major groups -- tentatively defined as Asiatic, Alaskan and American -- are identified on the basis of variants at two loci (Fig. 4). Discontinuity of sampling precludes any precise definition of boundaries at this time.

Other variants have been useful for examining genetic structures of chum salmon populations in greater detail. Studies by Utter et al. (1973) indicated a lack of genetic variation in chum salmon from Washington based on observations of 20 loci. However, more recent studies have revealed at least six polymorphic loci in Washington populations (*AAT-l*, *2*, and *3*; *IDH*; *PMI*; *6PGD*). There appears to be distinct populations of Puget Sound chum salmon (Seeb and Wishard 1977*b*). Different frequencies of *IDH* alleles tend to separate stocks from north and south Puget Sound, and an important late run of fish entering the Nisqually River is characterized by exceptionally high frequen-

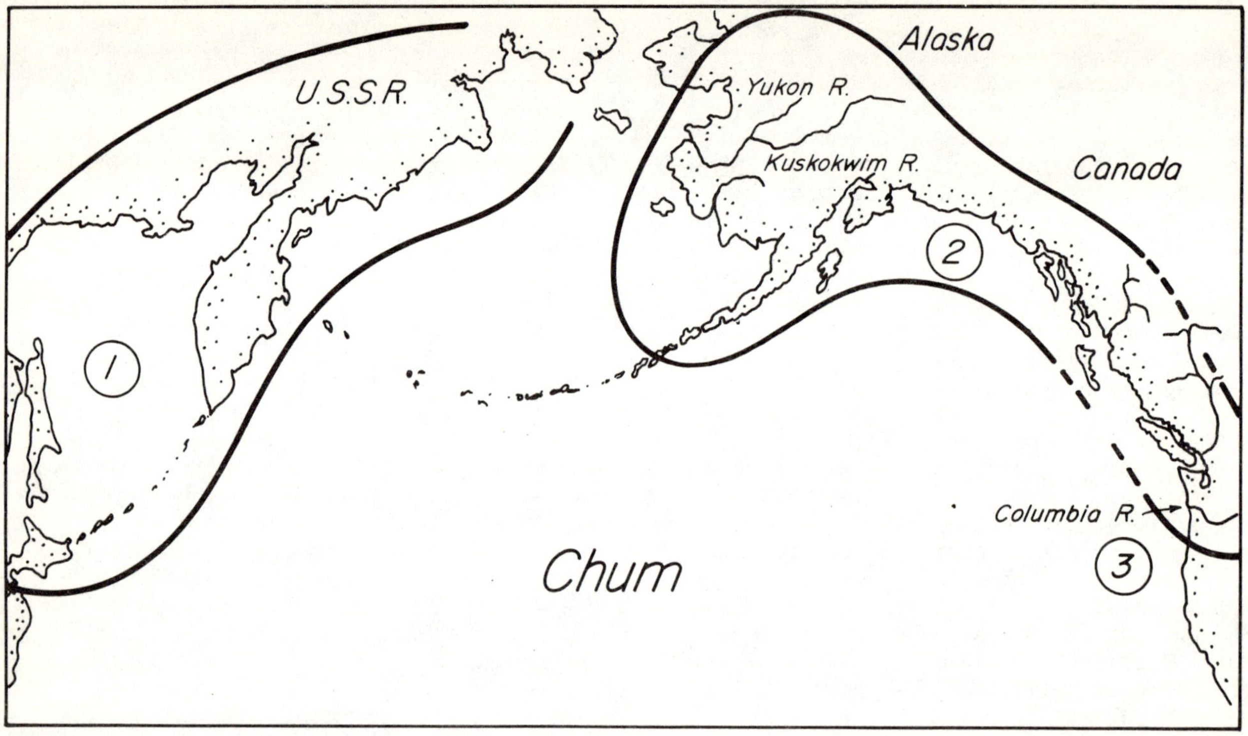

Figure 4. Major population units of chum salmon identified through different frequencies of allelic proteins.

cies of *6PDG* variants. A recent study of chum salmon populations of the Yukon and Kuskokwim drainages (Utter 1978) has indicated an apparent randomness of allelic frequencies within these two drainages except for a slight and consistently different frequency of *PMI* variants between the drainages.

Chinook Salmon: Chinook salmon populations were examined intermittently through collection efforts that focused on sockeye and chum salmon. This work, along with reports from other workers, resulted in a relatively early awareness of two polymorphic systems in chinook salmon, *TO* and *MDH-3* (Bailey et al. 1970, Utter 1971). A stronger emphasis has subsequently been directed toward studying chinook salmon populations, particularly those of the Pacific Northwest. These efforts considerably expanded the number of identified polymorphic systems and a more comprehensive understanding of the genetic structure of chinook salmon populations (May 1975, Kristiansson and MacIntyre 1976, Utter et al. 1976, Milner 1978, Utter 1978).

Chinook salmon populations that extend at least from the Mad River drainage in California through the Quillayute drainage in Washington share similar genetic characteristics. These populations are characterized by uniformly high frequencies of *PGI-2* variation and usually by the presence of *PGM-1* variation (Utter et al. 1976). Only low frequencies of *PGI-2* variants and no *PGM* variants have been observed in the rather extensive sampling of Columbia River and Puget Sound chinook populations. Preliminary data from the east and west coasts of Vancouver Island suggests that these characteristics do not extend into British Columbia populations.

Chinook salmon runs from the Columbia River have been examined for several years. Unlike coastal runs, no clearly identifiable groups of populations have been determined. Fall chinook salmon from hatcheries on the lower Columbia River share the common alleles of the *TO* and *PMI* loci. This may be more a reflection of the indiscriminant exchange of eggs among hatcheries than a close affinity of ancestral stocks (Simon 1972). Spring chinook from the lower Columbia River and its major tributaries tend to be quite distinct from fall chinook returning to this region, principally through considerably higher frequencies of the common *TO* allele.

Chinook populations of the upper Columbia and Snake rivers are tentatively identifiable either through distinctive frequencies of

PMI and *TO* alleles or the presence of low frequency variants at other loci that have not been seen in other populations. These differences must be regarded as provisional because they are generally based on single observations separated by large distances.

A few chinook salmon populations have also been examined in the Yukon and Kuskokwim rivers in Alaska. There appears to be genetic similarity characterized by high frequencies of the common allele of polymorphic loci among the populations from tributaries of both rivers. These populations were dissimilar from all other chinook populations examined except for populations in the upper Snake River drainage.

It is interesting to compare genetic variation within and among chinook salmon populations relative to the overall distribution of the species. The Columbia River is in the center of distribution with major populations both northward and southward. The Yukon and Kuskokwim rivers, on the other hand, are near the northern extremity of the range of chinook salmon in North America (Atkinson et al. 1967). It has long been postulated that populations near the center of the range should have higher levels of genetic variation than populations near the periphery because of the greater potential gene flow (Mayr 1970). Genetic data from chinook salmon support such a hypothesis. The Kuskokwim River and Yukon River populations are generally distinguished from those of the Columbia River by lower amounts of genetic variation among individuals and populations. Only those populations at the inland extremities of the Columbia River drainage (which themselves are peripheral populations within a large river system) resemble the Alaskan populations with regard to reduced levels of genetic variation.

Coho Salmon: The coho salmon has the lowest average heterozygosity value of the five Pacific salmon species in North America (Allendorf and Utter 1979) with only a single highly polymorphic locus - transferrin - among 24 loci examined in a broad survey of populations from California through Alaska. Populations from the Columbia and Fraser rivers can be distinguished from other populations on the basis of a single allele occurring in high frequencies in the former group contrasted with varying, but somewhat equal, frequencies of three alleles in the latter group (Utter et al. 1970, Utter et al. 1973, May and Utter 1974, May 1975, Seidel 1976, Utter et al. 1976, Suzumoto et al. 1977). Experimental data suggest that the distribution of transferrin alleles may be more a reflection of complex bacteriostatic properties of different transferrin alleles than of ancestral relationships (Suzumoto et al. 1977, Pratschner 1978).

Most other loci screened from collections of coho salmon taken throughout the Pacific Northwest are monomorphic. A variant allele of *LDH-4* occurs at moderate frequencies (up to .10) in streams in south Puget Sound and Hood Canal, but is virtually absent in collections from other Pacific Northwest streams (May 1975). Two unusual variants at the *PGM* and *LDH-1* loci occurred at polymorphic frequencies (greater than .05) at the Feather River Hatchery, California (Utter *unpublished data*, May 1975). Both of these variants may occur widely in the southern range of coho. At least some coho populations from the Fraser River and Puget Sound have polymorphic frequencies of variant alleles for *AGPD-3*, *AGPD-4* and two peptidase loci (Seeb *unpublished data*, Utter *unpublished data*). Present data are insufficient to define any geographic patterns for these variants. Coho salmon persist as the species with the lowest incidence of detected polymorphism in spite of these additional polymorphic loci. Much of the developmental work on methods to detect polymorphism has been directed at this species because of its apparent high incidence of monomorphism.

The low levels of genetic variation observed among coho populations can be readily explained if this species had a discontinuous distribution and restricted habitats. These factors would contribute to reduced gene flow among populations as in sockeye salmon, which has a similar low average heterozygosity (Allendorf and Utter 1978). However, coho salmon, more than any of its congeneric species, occupies a continuum of populations in diverse habitats over a broad geographic range (Aro and Shepard 1967). The species has readily adapted to transplantation and hatchery conditions (personal communication, R. Pressey, National Marine Fisheries Service) and is presently more abundant than historically over much of its range due to hatchery propagation. It appears anomalous that such an ubiquitous and adaptable species as the coho should express little generic variation in relation to other salmon species. Two opposing possibilities are suggested: 1) the level of genetic variation currently indicated by protein loci is not a valid reflection of genetic variation over the remainder of the coho salmon genome; and 2) the coho salmon has evolved a genome possessing little genetic variation but a highly adaptable phenotype.

Rainbow Trout: The genetic constitution of rainbow trout has been studied more intensively than any other salmonid species. Considerably greater genetic variation of proteins has been reported for rainbow trout than for Pacific salmon (Utter and Hodgins 1972, Utter et al. 1973). More recent studies of anadromous populations (steelhead) in Washington involving more than 30 loci (Allendorf 1975) have supported this conclusion. Continuing studies of rainbow trout place emphasis on more precise definition of Columbia River populations (Milner 1977, 1979). A more extended geographic survey of the gross population structures of steelhead populations based on data from two polymorphic loci (*LDH-4* and *TO*) has recently been published (Utter and Allendorf 1977).

The current data for frequencies of *LDH-4* and *TO* variants define at least two major geographic units of rainbow trout populations (Fig. 5). A coastal group tentatively extends at least from Kodiak Island southward through the Mad River drainage of northern California. Moderate to low frequencies of *LDH-4* variants and moderate frequencies of *TO* variants typify this group. An inland group, found exclusively in the Fraser and Columbia river drainages east of the Cascade range is identified through very high frequencies of *LDH-4* variants and low frequencies of *TO* variants. This major division presumably reflects two distinct lines dating into the last glacial era (Allendorf 1975). The inland group is postulated to have descended from fish migrating into a large freshwater impoundment resulting from the glacial diversion of the upper drainages of the Columbia and Fraser rivers. The coastal group presumably descended from Asiatic or American stocks that existed outside the glacial mass.

Neither the tendency toward anadromy nor the timing of upstream migration appear to be characteristics of distinct evolutionary lines. Allelic frequencies of both resident and migratory populations of a particular region were invariably similar (e.g. steel-

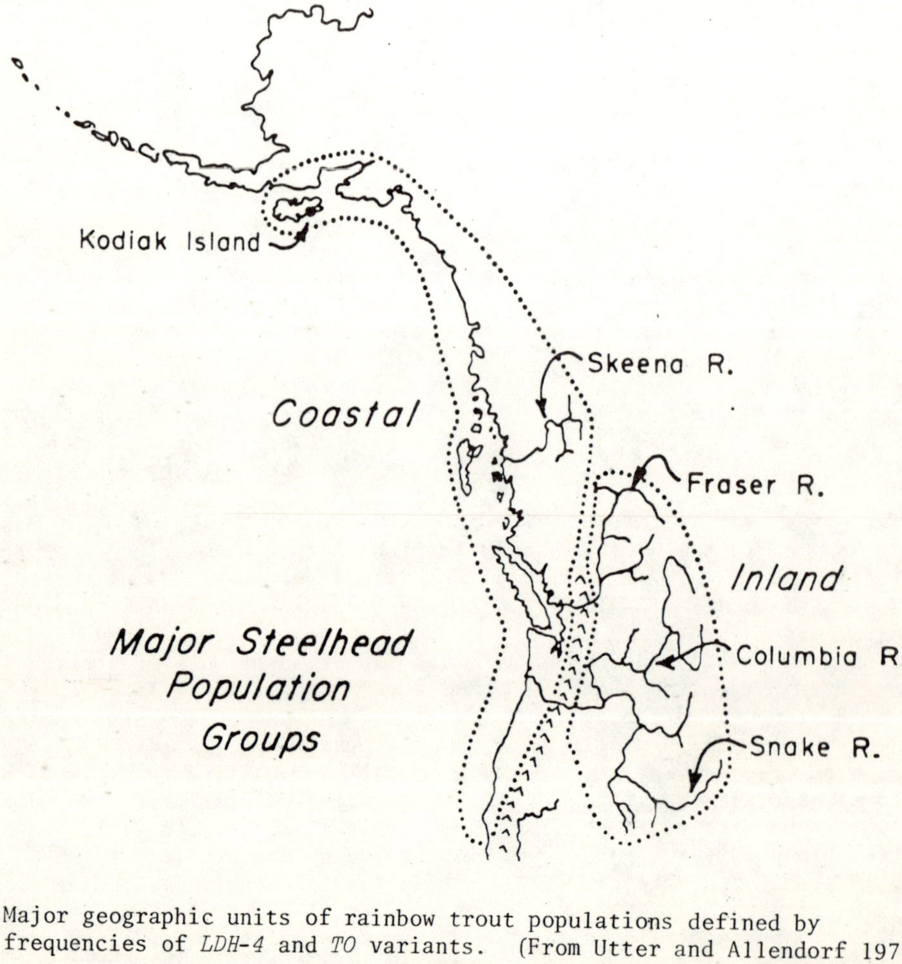

Figure 5. Major geographic units of rainbow trout populations defined by frequencies of *LDH-4* and *TO* variants. (From Utter and Allendorf 1977).

head of the upper Fraser River and "Kamloops" rainbow trout), while summer and winter-run steelhead of a particular drainage tended to resemble one another more than they resembled populations of adjacent drainages (Allendorf 1974; Thorgaard 1977*a*, 1977*b*. Thus, the terms "steelhead," "summer-run," and "winter-run" are useful for the description of life history patterns of a particular population, but are not an indication of close evolutionary relationships among populations of different areas.

The high frequency of *LDH-4* variants in the inland rainbow trout group separates it from the golden trout *(Salmo aquabonita)* and red-band trout. Present evidence indicates that these inland trout groups are closely related to rainbow trout (Gold 1977), but the absence of *LDH-4* variation (Utter and Allendorf 1977) indicates a divergence from the inland rainbow trout populations that predates at least the last period of glaciation.

The boundaries of the coastal group are uncertain because of the absence of sampling at the northern and southern extremes of distribution. The aberrant chromosome counts Thorgaard (1977*b*) observed in steelhead from the Mad River in California indicates a division at this point that was not reflected in the protein data. Non-anadromous hatchery strains Allendorf (1975) examined showed differences between inland and coastal groups. These populations presumably descended from fish taken from the McCloud River drainage of California during the last century (MacCrimmon 1971) and appear to reflect a third major group of rainbow trout populations that may converge with southern populations of the coastal group.

The major population groups of rainbow trout described to this point are virtually certain to have a considerable degree of internal structuring that will become apparent as more populations and loci are examined, and as complementary data are collected by other methods. Such structuring has already been indicated in the coastal group through the cytogenetic studies of Thorgaard (1977*a*, 1977*b*), and the presence of apparently unique protein variants in certain drainages (Allendorf 1975). Direct evidence for a substructure of the inland group was recently found in a comparison of three collections from widely separated areas of the upper Columbia River drainage through differing frequencies of peptidase variants (Milner 1977). The pattern of variation suggests a clear separation of populations from the Columbia River above its confluence with the Snake River and some populations of the Snake River drainage.

Cutthroat Trout: Data have been collected sporadically over a number of years from coastal cutthroat trout, primarily in conjunction with ongoing studies of rainbow trout. Most of these observations were directed toward determining biochemical differences between cutthroat trout and rainbow trout (Utter et al. 1973, Utter et al. 1976, Campton 1980). Cutthroat and rainbow are very closely related, occur sympatrically in many waters and are morphologically indistinguishable in early stages of life history. The ability to electrophoretically identify individuals of the two species and their hybrids has proven to be a valuable asset in stream surveys where both species occur sympatrically, and is routinely used by the Washington State Department of Game (WSDG) for this purpose (Burns et al. 1977). Although we have studied coastal cutthroat trout at the species level for a number of years, virtually nothing was known about the genetic structure of populations of this species prior to 1976 when a formal investigation was initiated to examine this question in conjunction with the establishment of different hatchery lines. These studies have been described by Campton (1980), and have now reached the point where some interesting facts have emerged concerning the population structure of coastal cutthroat trout.

Although cutthroat trout occur from northern California through Prince William Sound, Alaska, the present data are restricted to two proximal regions of western Washington-Hood Canal and northern Puget Sound. These two regions have been examined in considerable detail. Some collections included tributary streams within minor drainages and adjacent areas in a given stream. Coastal cutthroat trout have proven to be the most polymorphic salmonid species studied to date. Over 50 percent of more than 30 loci examined were polymorphic in one or more of the populations examined. Average heterozygosity values in this group of populations are well in excess of .10.

This detailed sampling and testing, coupled with known unique aspects of the life history relative to other indigenous salmonids, has led to new knowledge of this species' biology that has direct implications for its management. A major finding from the protein data was a general unity of populations within each region contrasted with a difference between regions of a magnitude similar to that separating the coastal and inland groups of rainbow trout. The linear distance separating the two regions is small, but the coastal distance is large because of the hundreds of miles of coastline separating the Stillaguamish River drainage at the southern extreme of the north Puget Sound region and Hood Canal. Tagging studies show that coastal cutthroat trout avoid open, deep water areas such as those separating Hood Canal and north Puget Sound (Johnston and Mercer 1976). This characteristic is consistent with the genetic difference separating populations of the two regions.

The apparent absence of gene flow between Hood Canal and north Puget Sound populations is sufficient evidence to preclude the use of a single hatchery stock derived from either region to enhance fishing in both regions. Interbreeding of native and strongly diverged exogenous fish carries the risk of disrupting highly adapated gene pools, giving rise to suboptimally adapted progeny, and an artificial dependence on continued hatchery supplementation to support the fishery (see Reisenbichler and McIntyre 1976).

One collection of yearling fish from Howe Creek of the Hood Canal region did not conform to the above picture of much greater genetic similarity within a region than between regions. In fact, the collection formed a separate cluster from the remainder of the Hood Canal populations and the cluster of north Puget Sound populations (Campton 1980). This population was also characterized by a low average heterozygosity value relative to all other anadromous coastal cutthroat trout populations examined. These peculiarities were consistent with a hypothesis of one or more population "bottlenecks" affecting the allelic frequencies and levels of genetic variation of this population coupled with an absence of gene flow from adjacent populations. Such a situation would be expected in non-anadromous (resident) populations in a small stream possessing a small population and experiencing little or no outside gene flow into the populations. Campton (1980) postulated that cutthroat residing in Howe Creek represented such an isolated population. This hypothesis was verified in a group of older fish from Howe Creek, that had the same genetic characteristics as young of the year.

A question concerning the systematic status of cutthroat trout native to different areas of the western United States has arisen as a result of cytogenetic and recent biochemical genetic examinations of these fishes (Gold 1977, Campton 1980). *Salmo clarki* presently includes a rather broad assortment of population units that are bound together through morphological similarities but appear to be rather widely diverged on the basis of cytological, biochemical and geographical evidence. Although it is beyond the scope of this paper to review this situation in detail, it is important to emphasize that the coastal cutthroat trout is a distinct taxonomic unit from all other groups of *Salmo* indigenous to western North America, including other groups of *Salmo clarki*. The diploid chromosome number (2N = 68) (Gold 1977) is four higher than any other cutthroat trout group. Coastal cutthroat are restricted to coastal drainages from Alaska through California and are largely sympatric with coastal steelhead trout populations. A comparison of the genetic similarities (measured by about 30 protein loci) among coastal cutthroat trout, "west-slope" cutthroat trout (from Pacific drainages of the Columbia River east of the Cascade Crest), "Yellowstone" cutthroat trout and coastal steelhead trout is particularly interesting. Each of these four groups are about equally diverged from one another, indicating a similar time interval since common ancestry (Campton, 1980). It appears, then, that at least some cutthroat trout lineages have diverged to the species level, and that a thorough multidisciplinary reexamination of the systematics of Pacific species of *Salmo* is appropriate at this time.

GENERAL CONSIDERATIONS REGARDING DATA AND BREEDING STRUCTURE

The present data amplify earlier conclusions from a less extensive survey of three species (coho and chinook salmon and rainbow trout) where a uniqueness of population structuring within a species was noted (Allendorf and Utter 1979). It has become increasingly evident that genetic data derived for one species are insufficient for drawing inferences regarding structures of closely related salmonid species. Differences in genetic structure become more evident from the data presented for seven species in this report, which also discusses possible causes and practical implications of genetic differences. In this section we will discuss four questions related to general aspects of genetic differences.

<u>What are the principal causes of these different patterns of variation in genetic structure?</u> This question lies at the center of one of the fundamental controversies of modern biology, i.e., is such genetic varia-

tion predominantly a reflection of processes of natural selection that directly affect the protein loci examined, or is most of this genetic variation more an echo of random events such as genetic drift or migration? The controversy is presently unresolved, in spite of the intense effort and thought that has been directed toward a solution for more than a decade, mainly because of the difficulties in obtaining reliable estimates of sizes of natural populations (Lewontin 1974). In large, random breeding populations, the effects of natural selection are much more apparent for loci where different coefficients of selection exist for a set of genotypes, while random factors tend to predominate in the distribution of genotypes in small populations (Nei 1975). The limited fecundity and reasonably small population sizes of salmonids places this group of organisms among those where genotypic distributions of natural populations would be principally a reflection of random factors. The above argument does not resolve the issue of selection vs. neutrality of polymorphic protein loci, but does provide a functional framework for interpreting most of the allelic frequencies observed in natural populations of salmonids. Thus, the distribution of protein variants described in this report for seven salmonid species (with the possible exception of transferrin variants of coho salmon) has been interpreted to result predominantly from random processes. This interpretation is supported by the following observations: 1) the major patterns of distribution apparently reflect geographic rather than environmental variables; 2) distinct allelic frequencies are observed in even and odd-year pink salmon from the same stream; 3) consistent allelic frequencies are observed among age groups, year classes and generations in "closed" populations; and 4) successfully transplanted fish obviously influence gene frequencies of the next generation in "open" populations (see Table 2).

It is apparent that allelic frequency data are highly useful indicators of the genetic structures of salmonid species and that the structure of each species is unique. This distinct structuring precludes generalizing from one species to another. Based on protein data collected from different populations of American eels *(Anguilla rostrata)*, Williams et al. (1973) stated "...the recognition of separate Mendelian populations on the basis of significantly different allelic frequencies in a number of other commercially important species is highly questionable..." Our findings invalidate this broad generalization for salmonids and suggest that it may be generally inappropriate.

<u>What are the ultimate limits of population definition based on genetic variations of proteins</u>? This question relates directly to the preceding question. The major factors influencing the frequencies of single gene variants in a population include 1) the number of breeding individuals in a particular generation; 2) the degree of genetic isolation from other populations; 3) the amount of time elapsed since gene exchange occurred from other populations; and 4) the differential influence of natural selection on variants of a particular protein.

We have already explained that the first three factors probably account for most of the variation observed among salmonid populations. Each factor is interrelated and varies from species to species and, to some degree, among populations within species.

Pink salmon populations are a good example of the apparent complexity of those genetic interactions. Pinks may be influenced sufficiently by periodic restrictions of population size within local breeding units to override the potentially stabilizing effects of gene flow between units from possible straying. This combination of genetic dynamics could explain the relatively high amount of genetic variation observed in pink salmon populations and the minimal degree of geographic structuring. The effects of time since gene exchange are apparent through the larger differences seen between year classes than are generally observed within them. The consistency over generations of gene frequencies of Washington pink salmon populations (excepting changes brought about by transplantation) occurred within a period when none of these populations were affected by drastic reductions of breeding individuals. The potential instability of Alaskan pink salmon populations minimizes the value of gene frequency data for estimating ancestral relationships among populations or defining broad geographic population units. However, stable local differences among populations, regardless of cause, are still potentially useful for measuring interactions among these populations (see Table 2).

Anadromous populations of coastal cutthroat trout, in contrast to pink salmon, tend to form distinct population units on a regional basis that appear to reflect ancestral relationships. This contrast is surprising at first glance because of life history similarities of the two species. Both species spawn in smaller coastal streams and, in a given generation, are potentially affected by drastic reductions in the number of spawning individuals. However, a number of differences exist between the life histories of the two species that appear to influence their genetic structures. Coastal cutthroat trout are capable of breeding more than once and fish of different ages participate in the breeding activities of a given year. Thus, an apparent "bottleneck" for coastal cutthroat trout is not nearly as drastic an event from the point of view of gene frequency stability as it is for pink salmon. A second major difference is that coastal cutthroat trout never venture into deeper saltwater and the potential for gene flow over large distances is therefore much more restricted. Both of these life history differences are consistent with the gene frequency patterns observed in coastal cutthroat trout.

These two examples demonstrate that knowledge of the ultimate resolving powers of gene frequency data in natural populations is dependent upon knowledge of the species' life history in question. This dependency does not reduce the value of gene frequency data for defining population structures. Rather, the dependency works in both directions. As more is known about the genetic structure of a species, more can be deduced about its life history patterns and vice versa.

How much data are required before a pattern of variation among populations of a species becomes evident? Again, there is no single, simple answer to this question any more than to the more general question of the number of points required to resolve trends from graphic projections of any set of data. A logical approach is to initially collect samples over as broad a geographic range as possible and to subsequently focus with more intense sampling on more restricted areas. Single data points are insufficient to define major population groups because of the possibility that differences observed from a particular group of fish may be localized aberrations from the effects of genetic drift. Two collections from proximal populations of a species having similar frequencies at one or more loci that differ significantly from frequencies of these loci in other populations are therefore regarded as the absolute minimum for positive identification of a distinct group of populations.

Will protein data replace traditional methods for studying population structures? The unique properties of genetic variants of proteins (i.e. simple inheritance coupled with ease of detection) permit genetic definitions at the individual, population and species level that were previously impossible to obtain. These capabilities have opened up a new dimension in the study and understanding of natural populations. These expanded capabilities are generally complementary to existing methodologies. External marks and tags remain indispensable tools because of their visibility. Coded wire tags combine relative ease of application and detection with a high potential information content. Intrinsic differences such as those observed in scale characters (Anas and Murai 1969) and trace element composition (Calaprice 1971) are primarily indicators of environmental differences and, as such, are completely complementary to genetic variants of proteins. Genetic variants of proteins are therefore, just one of the valuable tools available for obtaining information about structures and movements of fish populations. It is important that both the capabilities and limitations of a particular tool be properly understood so that the appropriate set of methods can be implemented in a particular study. Nevertheless, genetic data based on protein variation remain singularly superior to any other existing method for defining the genetic structures of populations.

SUMMARY

General consistencies of allelic and phenotypic frequencies at many electrophoretically detected loci have been observed within closed salmonid populations 1) throughout life cycles; 2) over successive generations; and 3) among year classes where overlap occurs. These data support the assumption that patterns of genetic variation observed among populations of these species tend to reflect life history rather than environmental variables. This assumption is the basis for interpreting known patterns of variation in seven salmonid species indigenous to the Pacific Northwest (sockeye, chum, pink, chinook and coho salmon, and rainbow and coastal cutthroat trout). Each species has a characteristic and unique pattern of variation and it is therefore inappropriate to generalize with regard to population structure between even such closely related forms as the two trout species. A major factor affecting the patterns and amounts of genetic variation is the effective population size. This parameter is, in turn, affected by a number of variables such as discrete or variable year classes, single or multiple spawnings, preferred spawning habitat, degree of anadromy, degree of immigration and emigration, population stability, location of population relative to conspecific populations and time since divergence of two populations. The tendency to form distinct populations in a particular drainage appears to be strongest in the coastal cutthroat and weakest in the pink salmon.

ACKNOWLEDGEMENTS

The outstanding technical assistance of Paul Aebersold and David Teel is gratefully acknowledged. The value of their efforts extended well beyond the collection and analysis of data. They have been jointly responsible for procedures that produce reliable identification of previously undetected polymorphic loci and an overall increase in the efficiency of data collection.

LITERATURE CITED

Allendorf, F. W. 1975. Genetic variability in a species possessing extensive gene duplication: Genetic interpretation of duplicate loci and examination of genetic variation in populations of rainbow trout. Ph.D. Thesis. Univ. of Wash., Seattle.

Allendorf, F. W. and N. Mitchell. 1977. Genetics of Alberni Inlet sockeye salmon populations. Report to Pacific Biological Station, Nanaimo, British Columbia.

Allendorf, F. W., and F. M. Utter. 1973. Gene duplication within the family Salmonidae: Disomic inheritance of two loci reported to be tetrasomic in rainbow trout. Genetics 74:647-654.

Allendorf, F. W., and F. M. Utter. 1976. Gene duplication in the family Salmonidae. III. Linkage between two duplicated loci coding for aspartate aminotransferase in the cutthroat trout *(Salmo clarki)*. Hereditas 82:19-24.

Allendorf, F. W., and F. M. Utter. 1979. Population genetics. Pages 407-454 *in* Hoar, W. S. and D. J. Randall (eds.), Fish physiology. Vol. VII. Academic Press, N.Y.

Allendorf, F. W., F. M. Utter, and B. P. May. 1975. Gene duplication within the family Salmonidae. II. Detection and determination of the genetic control of duplicate loci through inheritance studies and the examination of populations. Pages 415-432 *in* Markert, C. L. (ed.), Isozymes, Volume 4: Genetics and evolution. Academic Press, N.Y.

Altukhov, Y. P. 1975. Population genetics of fishes. Transl. Fish. Mar. Serv. Transl. Ser. No. 3548.

Anas, R. E., and S. Murai. 1969. Use of scale characters and a discriminant function for classifying sockeye salmon *(Oncorhynchus nerka)* by continent of origin. Int. North Pac. Fish Comm. Bull. 26:157-192.

Aro, K. V., and M. P. Shepard. 1967. Pacific salmon in Canada. Int. North Pac. Fish Comm. Bull. 23:225-327.

Aspinwall, N. 1974. Genetic analysis of North American populations of the pink salmon, *Oncorhynchus gorbuscha*, possible evidence for the neutral mutation-random drift hypothesis. Evol. 28:295-305.

Atkinson, C. E., H. J. Rose, and T. O. Duncan. 1967. Pacific salmon in the United States. Int. North Pac. Fish Comm. Bull. 23:43-224.

Bailey, G. S., A. C. Wilson, J. E. Halver, and C. L. Johnson. 1970. Multiple forms of supernatant malate dehydrogenase in salmonid fishes. J. Biol. Chem. 245: 5927-5940.

Burns, D., W. Young, and D. Hanson. 1977. Progress report for the Fish Management Division, Wash. State Dept. Game, Sept. 30.

Calaprice, J. R. 1971. X-ray spectrometric and multivariate analysis of sockeye salmon *(Oncorhynchus nerka)* from different geographic regions. J. Fish. Res. Bd. Can. 28:369-377.

Campton, D. E. 1980. Genetic structure of sea-run cutthroat trout *(Salmo clarki clarki)* populations in the Puget Sound region. M.S. Thesis. Univ. of Wash., Seattle.

Crow, J. F., and M. Kimura. 1970. An introduction to population genetics theory. Harper, N.Y.

Engel, W., J. Op't Hof, and U. Wolf. 1970. Geduplikation durch polyploide Evolution: die Isoenzyme der Sorbitdehydrogenase bei herings-und lachsartigen Fischen (Isospondyli). Humangenetik 9:157-163.

Foster, B., V. Fletcher, and B. Kiser. 1977. 1976 hatcheries statistical report of production and plantings. Wash. Dep. Fish. Prog. Rep. No. 30.

Gold, J. R. 1977. Systematics of western North American trout *(Salmo)*, with notes on the redband trout of Sheepheaven Creek, California. Can. J. Zoo.

Grant, W. S., G. B. Milner, P. Krasnowski, and F. M. Utter. *In press*. Biochemical genetic variation in sockeye salmon, *(Oncorhynchus nerka)*: the use of biochemical markers for stock identification in Cook Inlet, Alaska. J. Fish. Res. Bd. Canada.

Harris, H., and D. A. Hopkinson. 1976. Handbook of enzyme electrophoresis in human genetics. American Elsevier, N.Y.

Hodgins, H. O. 1972. Serological and biochemical studies in racial identification of fishes. Pages 199-208 *in* Simon, R. C., and P. A. Larkin (eds.), The stock concept in Pacific salmon, H. R. MacMillan lectures in fisheries. Univ. British Columbia.

Hodgins, H. O., W. E. Ames, and F. M. Utter. 1969. Variants of lactate dehydrogenase isozymes in sera of sockeye salmon *(Oncorhynchus nerka)*. J. Fish. Res. Bd. Can. 26:15-19.

Hodgins, H. O., and F. M. Utter. 1971. Lactate dehydrogenase polymorphism of sockeye salmon *(Oncorhynchus nerka)*. Cons. Int. Explor. Mer., Rapports et proces verbaux; 161:100-101.

Johnson, K. R., R. F. Donelly, W. K. Hershberger, and D. E. Bevan. 1978. Identification of Kodiak Island pink salmon populations based on biochemical genetic variation. Univ. Wash. Coll. Fish. Rep. FRI-UW-7801.

Johnston, J. M., and S. P. Mercer. 1976. Sea-run cutthroat in saltwater pens: Broodstock development and extended juvenile rearing (with a life history compendium). Wash. State Game Dep. Fish. Rep. Project AFS-57-1.

Kristiansson, A. C., and J. D. McIntyre. 1976. Genetic variation in chinook salmon *(Oncorhynchus tshawytscha)* from the Columbia River and three Oregon coastal rivers. Trans. Am. Fish. Soc. 105:620-623.

Lewontin, R. C. 1974. The genetic basis of evolutionary change. Columbia Univ. Press, N.Y.

MacCrimmon, H. R. 1971. World distribution of rainbow trout *(Salmo gairdneri)*. J. Fish. Res. Bd. Can. 28:663-704.

May, B. 1975. Electrophoretic variation in the genus *Oncorhynchus*: The methodology, genetic basis, and practical applications to fisheries research and management. M.S. Thesis. Univ. of Wash., Seattle.

May, B., and F. M. Utter. 1974. Biochemical genetic variation of the genus *Oncorhynchus* in Pacific Northwest populations: A progress report and program summary. Interim report of contractual work to Wash. Dep. Fish.

May, B., F. M. Utter, and F. W. Allendorf. 1975. Biochemical genetic variation in pink and chum salmon: Inheritance of intraspecies variation and apparent absence of interspecies introgression following massive hybridization of hatchery stocks. J. Hered. 66:227-232.

Mayr, E. 1970. Populations, species and evolution. Belknap Press, Cambridge, Mass.

Milner, G. B. 1977. Biochemical genetic variation: Its use in mixed fishery analysis. Pages 10-15 *in* Hassler, T. J., and R. R. VanKirk (eds.), Genetic implications of steelhead management. Spec. Rep. 77-1.

Milner, G. B. 1978. Final report, Columbia River stock identification study. Submitted to U.S. Fish and Widl. Serv.

Milner, G. B. 1979. Biochemical genetic variation in Columbia River steelhead trout *(Salmo gairdneri* richardson*)* populations and its use in mixed fishery analysis. Ph.D. Thesis. Univ. of Wash., Seattle.

Nei, M. 1972. Genetic distance between populations. Amer. Natur. 106:283-292.

Nei, M. 1975. Molecular population genetics and evolution. Nort-Holland, Amsterdam.

Numachi, K., Y. Matsumiya, and R. Sato. 1972. Duplicate genetic loci and variant forms of malate dehydrogenase in chum salmon and rainbow trout. Bull. Jap. Soc. Sci. Fish. 38:699-706.

Perriard, J., A. Scholl, and H. Eppenberger. 1972. Comparative studies on creatine kinase isozymes from skeletal-muscle and stomach of trout. J. Exp. Zoo. 182: 119-126.

Pratschner. G. A. 1977. Relative resistance of six transferrin phenotypes of coho salmon to cytophagosis, furunculosis and vibriosis. M.S. Thesis. Univ. of Wash., Seattle.

Royce, W. F. 1962. Pink salmon fluctuations in Alaska. Pages 15-34 *in* Wilimovsky, N. J. (ed.), H. R. MacMillan lectures in fisheries, Symposium on pink salmon.

Rogers, J. S. 1972. Measures of genetic similarity and genetic distance. Univ. Texas Publ. 7213:145-153.

Reinsenbichler, R. R., and J. D. McIntyre. 1977. Genetic differences in growth and survival of juvenile hatchery and wild steelhead trout, *Salmo gairdneri*. J. Fish. Res. Bd. Can. 34:123-128.

Reinitz, G. L. 1977. Inheritance of muscle and liver types of supernatant NADP-dependent isocitrate dehydrogenase in rainbow trout *(Salmo gairdneri)*. Biochem. Genetics 15:445-454.

Seeb, J., and W. S. Grant. 1976. Biochemical genetic variation in coho, chinook, chum and pink salmon: The use of electrophoretic markers in stock identification. Wash. Dep. Fish. Final rep. service contract 711.

Seeb, J. and L. N. Wishard. 1977*a*. Genetic characterization of Prince William Sound pink salmon populations. Report to Alaska Dep. Fish Game.

Seeb, J., and L. N. Wishard. 1977*b*. The use of biochemical genetics in the management of Pacific salmon stocks: Genetic marking and mixed fishery analysis. Wash. Dep. Fish. Final rep. service contract 792.

Seidel, P. R. 1977. Early life history study of three Washington Department of Fisheries coho salmon *(Oncorhynchus kisutch)* populations by interracial hybridization. M.S. Thesis. Univ. of Wash., Seattle.

Siciliano, M. J., and C. R. Shaw. 1976. Separation and visualization of enzymes on gels. Pages 185-209 *in* Smith, I. (ed.), Chromatographic and electrophoretic techniques, Vol. 2, Zone electrophoresis. Heinemann, London.

Simon, R. C. 1972. Gene frequency and the stock problem. Pages 161-196 *in* Simon, R. C., and P. A. Larkin (eds.), The stock concept in Pacific salmon. Inst. Fish. Univ. British Columbia, Vancouver.

Sneath, P. H., and R. R. Sokal. 1973. Numerical taxonomy. Freeman, San Francisco.

Stern, C. 1943. The Hardy-Weinberg law. Sci. 97:137-138.

Suzumoto, B. K., C. B. Schreck, and J. D. McIntyre. 1977. Relative resistances of three transferrin genotypes of coho salmon *(Oncorhynchus kisutch)* and their gematological responses to bacterial kidney disease. J. Fish. Res. Bd. Can. 34:1-8.

Thorgaard, G. H. 1977*a*. Chromosome studies of steelhead. Pages 20-25 *in* Hassler, T. J., and R. R. VanKirk (eds.), Genetic implications of steelhead management. Calif. Coop. Fish. Res. Unit. Spec. Rep. 77-1.

Thorgaard, G. H. 1977*b*. Chromosome rearrangements and sex chromosomes in the rainbow trout and sockeye salmon. Ph.D. Thesis. Univ. of Wash., Seattle.

Thorsteinson, F. V., J. H. Helle, and D. G. Birkholz. 1971. Salmon survival in intertidal zones of the Prince William Sound streams in uplifted and subsided areas. Pages 194-219 *in* The great Alaska earthquake of 1964: Biology. Nat. Acad. Sci. SIBN 0-309-01604-5.

Utter, F. M. 1971. Tetrazolium oxidase phenotypes of rainbow trout *(Salmo gairdneri)* and Pacific salmon *(Oncorhynchus* sp.) Comp. Biochem. Physiol. 39B:891-895.

Utter, F. M. 1978. Genetic variants of proteins in chum and chinook salmon from the Bering Sea and the Yukon and Kuskokwim rivers: Report of analyses of 1976 collections. Rep. to R.E.F.M. Div., Northwest and Alaska Fish. Center and the Alaska Dep. Fish and Game.

Utter, F. M., and F. W. Allendorf. 1977. Determination of the breeding structure of steelhead populations through gene frequency analysis. Pages 10-15 *in* Hassler, T. J. and R. R. VanKirk (eds.), Genetic implications of steelhead management. Calif. Coop. Fish. Res. Unit. Spec. Rep. 77-1.

Utter, F. M., F. W. Allendorf, and H. O. Hodgins. 1973. Genetic variability and relationships in Pacific salmon and related trout based on protein variations. Syst. Zoo. 22:257-270.

Utter, F. M., F. W. Allendorf, and B. May. 1976. The use of protein variation in the management of salmonid populations. Pages 373-384 *in* Taber, R. C., and F. Bunnell (eds.), Proceedings of the 41st North American Wildlife Conference.

Utter, F. M., F. W. Allendorf, and B. May. *In preparation*. The genetic basis of creatine kinase isozymes·in skeletal muscle of salmonid fishes.

Utter, F. M., W. E. Ames, and H. O. Hodgins. 1970. Transferrin polymorphism in coho salmon *(Oncorhynchus kisutch)*. J. Fish. Res. Bd. Can. 27:2371-2373.

Utter, F. M., and H. O. Hodgins. 1970. Phosphoglucomutase polymorphism in sockeye salmon. Comp. Biochem. Physiol. 36:195-199.

Utter, F. M., and H. O. Hodgins. 1972. Biochemical genetic variation at six loci in four stocks of rainbow trout. Trans. Am. Fish Soc. 101:494-502.

Utter, F. M., H. O. Hodgins, and F. W. Allendorf. 1974. Biochemical genetic studies of fishes: Potentialities and limitations. Pages 213-237 *in* Biochemical and biophysical perspectives in marine biology, Vol. 1. Academic Press, N.Y.

Utter, F. M., H. O. Hodgins, F. W. Allendorf, A. G. Johnson, and J. L. Mighell. 1973. Biochemical variants in Pacific salmon and rainbow trout: Their inheritance and application in population studies. Pages 329-339 *in* Schroder, J. H. (ed.), Genetics and mutagenesis of fish. Sprinter-Verlag, Berlin.

Utter, F. M., H. O. Hodgins, and A. G. Johnson. 1972. Biochemical of genetic differences among species and stocks of fish. Pages 98-101 *in* Annual Report for 1970. Int. North Pac. Fish. Comm.

Williams, G. C., R. K. Koehn, and J. B. Mitton. 1973. Genetic differentiation without isolation in the American eel, *Anguilla rostrata*. Evol. 27:192-204.

Wright, J. E., J. R. Heckman, and L. M. Atherton. 1975. Genetic and developmental analyses of LDH isozymes in trout. Pages 375-401 *in* Isozymes, Vol. III. Academic Press, N.Y.

APPENDIX

Symbols Used in Appendix Table 1

A -no variants seen
B -infrequent variants seen
C -polymorphic above one percent allelic frequency
.... -no data available
E -eye
G -gut
H -heart
L -liver
M -skeletal muscle (*MDH-M* is presumed to be mitochondrial locus in skeletal muscle)
S -serum

Abbreviated protein systems include:

Alb -albumin
ADH -alcohol dehydrogenase
AGP -alpha glycerophosphate dehydrogenase
AAT -aspartate amino transferase
CK -creatine kinase
GAL -M-acetyl-B-D-Galactosaminidase
GPT -glutamic-pyruvic transaminase
IDH -isocitrate dehydrogenase
LDH -lactate dehydrogenase
MDH -malate dehydrogenase
ME -malic enzyme
PEP -peptidase*
6PGD -6 phosphogluconate dehydrogenase
PGI -phosphoglucose isomerase
PGM -phosphoglucomutase
PMI -phosphomannose isomerase
SDH -sorbitol dehydrogenase
TO -tetrasolium oxidase (also commonly called superoxide dismutase-*SOD*)
Tfn -transferrin

*different peptidase for *PEP* include glycyl leucine (*GL*), leucyl glycyiglycine (*LGG*), and phenylalanylproline (*PhAP*). This summary excludes esterase and hemoglobin variants, which we do not regularly examine because of problems involving repeatability and instability.

Appendix Table 1. Summary of Variant Protein Systems in Seven Salmonid Species.

Locus	Tissue Distribution	Species Sock-eye	Pink	Chum	Coho	Chi-nook	Rain-bow	Cut-throat	Reference Sources
Alb	S,E	...	...	C	A	...	C	...	Altukhov 1975, *unpublished data* of G. A. E. Gall and B. Bentley.
ADH	L	A	A	A	B	C	C	C	Allendorf et al. 1976, Milne 1978, Campton 1980.
AGP-1,2	M	B	C	B	B	B	C	B	Seeb and Wishard 1977*b*, Utter and Hodgins 1972, Aspinwall 1974, May et al. 1975, Utter 1978, *unpublished data* of Seeb, Aebersold and Wishard.
-3,4	M,H	...	C	C	C	...	C	...	
AAT-1,2	M,H	A	A	C	B	C	C	c	May 1975, Allendorf and Utter 1976, May et al. 1975, Allendorf 1975, Seeb and Wishard 1977*a*.
-3	E	A	C	C	A	C	A	A	
CK-A1,2	M	A	A	A	B	A	C	A	Utter et al. *in preparation*, Perriard 1972.
-C1,2	G	...	...	...	...	...	B	...	
GAL	L	...	...	...	C	...	...	...	Aebersold and Wishard *unpublished data*.
GPT-2	M	C	A	A	A	A	A	C	Grant et al. *in preparation*, Campton 1980.
IDH-1,2	L	A	C	C	A	C	C	C	Allendorf and Utter 1973, Seeb and Wishard 1977*a*, Campton 1980, May 1975, Milner 1978, Reinitz 1977.
-3	M	A	A	A	A	A	C	A	
LDH-1	M	A	C	C	C	A	A	A	Utter and Hodgins 1972, Utter et al. 1973, Wright et al. 1975, May 1975, Campton 1980, Johnson et al. 1978, Milner 1978, Seeb and Wishard 1977*b*.
-3	E,M	A	A	...	B	A	A	A	
-4	L,E,M	C	B	C	C	B	C	B	
-5	E	B	A	A	A	C	C	C	
MDH-1,2	L,E,M	A	C	A	A	B	C	C	Bailey et al. 1970, Utter and Hodgins 1972, Numachi et al. 1972, Aspinwall 1974, Altukhov 1975, May 1975, Seeb and Wishard 1977*a*, Campton 1980, Milner 1978, Teel and Milner *unpublished data*.
-3,4	M	B	C	C	B	C	C	C	
-M	M	...	...	...	...	C	...	...	

Appendix Table 1. Summary of Variant Protein Systems in Seven Salmonid Species-continued.

Locus	Tissue Distribution	Species: Sockeye	Pink	Chum	Coho	Chinook	Rainbow	Cutthroat	Reference Sources
ME-I	M	A	C	...	A	A	A	A	Seeb and Wishard 1977*a*, Johnson et al. 1978, Campton 1980, Milner 1978, Gall and Bentley *unpublished data*.
-2	M	A	A	...	A	C	C	B	
-3	L	...	...	...	...	...	C	...	
PEP-1(GL)	M,L,E	A	A	A	C	C	C	C	Milner 1977, Campton 1980, *unpublished data* of all authors of this paper.
-2(GL)	E	A	A	A	C	A	A	A	
-3(LGG)	M,L,E	A	A	C	B	C	...	...	
-4(PhAP)	M,L.E	...	...	...	B	...	...	...	
6PGD	M,L,E	A	C	C	A	A	A	C	May 1975, Seeb and Wishard 1977*a*, 1977*b*; Campton 1980.
PGI-1,2	M	A	C	A	B	C	B	C	May 1975, Allendorf 1975, Utter et al. 1976, Seeb and Wishard 1977*a*, Campton 1980.
-3	M,E,L	A	A	A	B	A	C	B	
PGM-1	M	C	C	A	B	C	C	C	Utter and Hodgins 1970, 1972; May 1975, Seeb and Grant 1976, Kristiansson and McIntyre 1976, Campton 1980, *unpublished data* of all authors of this paper.
-2	L,E	C	A	...	B	...	C	C	
PMI	M,L,E	A	C	C	B	C	B	C	May 1975, Seeb and Grant 1976, Milner 1978, Campton 1980.
SDH-1,2	L	A	A	A	B	C	B	C	Engel et al. 1970.
TO-1	L,M	B	A	A	A	C	C	C	Utter 1971, May and Utter 1974, Campton 1980.
-2	H	...	...	...	...	C	...	...	Utter 1978.
Tfn	S,E	A	A	A	C	A	C	C	Utter et al. 1970, Utter and Hodgins 1972, Campton 1980.

Selection to Increase Yield of Coho Salmon, *Oncorhynchus kisutch*

John D. McIntyre

INTRODUCTION

A series of studies concerning the influence of genetic factors on production traits in coho salmon *(Oncorhynchus kisutch)* has been in progress since 1967 at Big Creek Hatchery near Knappa, Oregon. The Oregon Department of Fish and Wildlife manages the facility. Simon (1972) made a preliminary report of the initial studies. In it, he described heritability estimates which tentatively suggested ". . . that return of adult salmon to the fishery and to the hatchery can be substantially enhanced by selective breeding."

Because of the possible economic benefits, we initiated a program in 1971 to evaluate the potential for increasing returns of adult coho salmon via selective breeding. This report describes the results after two generations of selection.

Results are expressed by "yield." Yield is defined as the number of adults estimated to have been caught in the fisheries plus the number returning to the hatchery per thousand smolts released. Yield values are calculated for individual family groups representing two brood years. Brood year refers to the year of spawning. For coho salmon, fish produced in the 1968 brood year, for example, were released from the hatchery as smolts in April 1970. A relatively small proportion of these fish, predominantly males, matured and returned to the hatchery in November 1970 after remaining at sea for one summer. These precocious males (two-year-olds) are commonly called "jacks." All other fish (three-year olds) remained at sea where they were harvested in large numbers during the summer and fall of 1971. Those that escaped from the fisheries matured and returned to Big Creek Hatchery in 1971.

METHODS

In the summer and fall of 1971, coded-wire tags were recovered from samples of the 1968 brood of coho salmon from Big Creek that the fisheries harvested, and from fish that returned to Big Creek Hatchery. These fish originated from November 1968 matings of adult coho salmon according to the following scheme:

	Female A	Female B
Male 1	Family A1	Family B1
Male 2	Family A2	Family B2

The procedures included three steps:

1. Eggs of each female were divided volumetrically into two lots of approximately equal numbers;
2. Male 1 was used to fertilize one egg lot from each female; and,
3. Male 2 was used to fertilize the remaining eggs from each female.

This process was repeated 15 times producing 60 families.

Each family was reared as a juvenile group until the mean length of juveniles was about 7.5 cm (attained at about six months postfertilization). At this time, a coded-wire tag was injected into the snout of each fish. Codes are distinct for each family. After injecting the tag, all families were combined in an outdoor raceway and fed for about 12 additional months. At the time of tagging, the adipose fin was removed from 3 percent of the fish. These fish were sacrificed prior to releasing the remainder in order to estimate the numbers of tagged fish remaining in each family at release. Prior to release, all experimental fish were marked by removing the adipose fin.

Adults from the 1968 brood returned to Big Creek in November 1971. During sorting operations, adipose-marked fish were separated from unmarked production fish. Gametes obtained from each adult were placed in a separate container, which was identified by a number corresponding to a numbered disc-tag applied to the donor carcass. Fertilization was delayed (gametes were stored at 4°C) until family identity and a recapture record was established. The wire tag from each carcass was removed and decoded. Adults returning to the hatchery from each family were then added to those recaptured by the marine and Columbia River commercial and sport fisheries. Data on tag recoveries in the fisheries were obtained from the Washington Department of Fisheries, Oregon Department of Fish and Wildlife and California Department of Fish and Game.

Thirty pairs of fish from the 19 families with the highest yield values for three year olds were mated, one male per female. A control group was produced by mating 15 pairs of non-select fish that had returned to the hatchery. These procedures resulted in a total of 45 separate family groups that were reared separately until marked by methods previously described. The fish were then mixed in a common raceway. Approximately 1500 smolts were produced in each group from which 3 percent were removed to estimate the number of tagged fish released. Coded-wire tags were recovered from samples of fish from the fisheries in the summer and fall of 1974 to determine the yield of the select and control groups.

Yield values of three-year-olds in the 1971 brood families (1974 adults) were then used as the basis for a second generation of selection (1974 brood year). Matings were made to produce 29 select and 12 control families. Coded-wire tag recoveries in the summer and fall of 1977 again provided data to calculate yield values.

RESULTS

The yield of three year old fish produced by each family was used as the unit of selection for both the 1971 and 1974 brood years (Table 1). In 1971, we attempted to use fish from families with a high yield of three year olds and low yield of two year olds. In 1974, yield values for two years olds were ignored.

Tagged-fish returns from the 1971 brood were obtained from two year olds at the hatchery in the fall of 1973, and from the three year olds in the summer and fall fisheries and the hatchery in 1974 (Table 2). Returns from the 1974 brood similarly were obtained in the fall of 1976 and summer and fall of 1977 (Table 2).

The control group produced a higher average yield (5.2) of two year olds than the select group (4.2) in the 1971 brood. The total yield (two and three year olds) in this brood was the same in select and control groups. This result was not supported, however, in 1977 (1974 brood) from the second generation of selection where yield of two year olds was 1.3 in select groups and 0.3 in control groups.

Table 1. Ranges and Averages of Yield Values for the 1971 and 1974 Brood Families, and for Families From Which Parents of the 1971 and 1974 Brood Select Families Were Obtained.

	Range	Average
1971 brood families		
two year olds	0-30.2	7.8
three year olds	1.5-21.2	7.9
1971 brood select families		
two year olds	0-18.6	4.7
three year olds	11.0-21.2	14.5
1974 brood families		
two year olds	0-16.9	4.4
three year olds	4.4-28.4	13.2
1974 brood select families		
two year olds	0-10.9	2.6
three year olds	10.0-28.4	17.5

Table 2. Smolts Released and Number of Coded-Wire Tagged 1971 and 1974 Brood Big Creek Coho Salmon in Samples From Fisheries and From Mature Fish at Big Creek Hatchery.

	1971 Brood		1974 Brood	
	Select	Control	Select	Control
Smolts released	40,520	15,488	37,355	20,617
Three year olds recovered:				
Marine fishery	409	156	54	3
Columbia River gillnet	50	12	3	2
Sport fishery	1	1	0	0
Big Creek Hatchery	94	27	61	7
Two year olds recovered	170	80	49	7
Totals	724	276	167	19

Table 3. Yield of Three Year Olds in Select and Control Groups of Big Creek Coho Salmon.

Brood year	Yield Select	Yield Control	Probability that select=control (Sokal and Rohlf 1969, p. 607)
1971	13.7	12.7	0.34
1974	3.2	0.6	0.00

The apparent effect of selection has been to increase the yield of three year olds in the select group (Table 3). Higher overall yield values in the 1971 brood year may reflect a more favorable marine environment. The harvest and escapement of coho salmon along the Washington, Oregon and California coasts in 1977 (1974 brood year) were among the lowest recorded in recent years, suggesting low marine survival.

DISCUSSION

Theoretically, natural selection would be expected to maximize the frequency of characteristics associated with a survival advantage and thus preclude any response to artificial selection for greater survival. This consideration raises some interesting questions concerning the reasons why there was a response to the selection at Big Creek.

One possible explanation is that the selection response may only reflect an acceleration of the adaptive process that no doubt began when Big Creek coho salmon were originally introduced into an artificial environment for completing the freshwater phase of their life-cycle. The environment in the hatchery would provide new demands on the genetic system that must be met to maximize fitness. The artificial selection program then serves to facilitate this otherwise conservative process.

An alternative explanation is related to the fact that fish cultural practices often result in non-random breeding. For instance, selection of the largest, "best-looking" fish as brood fish is a widespread practice. Any such practice restricts genetic variability--the basic requirement for successful adaptation. Because survival (yield) is the only criterion of the selection practiced at Big Creek, it can be hypothesized that any previous non-random breeding has been overcome permitting "natural" adaptation to proceed. Again, in this instance, the artificial selection presumably accelerates adaptation.

Under either hypothesis, yield in control and select lines would be expected to converge in the long-term. They would converge because natural adaptation is expected to continue in the control line. It could be argued that yield of the control line may surpass that of the select in the long-term because of the artificial selection criterion being used. In addition, the select line may experience greater fluctuations in numerical strength than might be expected if natural adaptive processes were to prevail.

LITERATURE CITED

Simon, R. C. 1972. Genetic components of survival in coho salmon at Big Creek Hatchery (preliminary report). Pages 579-581 *in* 52nd Annual Conference of Western Association State Game and Fish Commissions.

Sokal, R. R., and F. J. Rohlf. 1969. Biometry. W. H. Freeman and Co., San Francisco. 776 pp.

Adaptive and Behavioral Responses to Artificial Propagation in a Stock of Chum Salmon, *Oncorhynchus keta*

James E. Lannan

INTRODUCTION

Ricker (1972) presented an articulate, historical perspective regarding the increasing awareness among fishery biologists of the heritable adaptations in geographical stocks of Pacific salmon (*Oncorhynchus* sp.). Subsequent contributions strengthened the inference of these adaptations. Fishery biologists now view stocks as populations that not only share a common environment, but also a common gene pool.

Biochemical techniques readily demonstrate the genetic diversity among stocks (Utter 1978); however, while this evidence has contributed to our present concept of stocks, it cannot be readily translated into performance implications.

Although we only partially understand at present the specific adaptation mechanisms, heritable behavioral modifications apparently provide one mechanism by which stocks adapt to specific reproductive environments. The heritable migratory patterns in juvenile sockeye salmon (*O. nerka*) provide one documented example of this (McCart 1967, Raleigh 1967, Brannon 1967, 1972).

In recent years, management-oriented biologists have exchanged a lively dialogue regarding the genetic implications of artificial propagation as a strategy for managing natural stocks of Pacific salmon. This makes the observation of a stock's adaptive responses to artificial propagation, as well as in nature, a matter of scientific and practical interest. The questions that arise regard the adaptive responses that may occur during the development of hatchery brood stocks.

The artificial propagation system may cause genotypic divergence from the naturally-reproducing stock in three ways. Most hatcheries use a randomized mating scheme because the basis for selecting a mate in nature is poorly understood. Thus, the combinations of genotypes resulting from hatchery production probably represent nearly all combinations of genetic units previously existing in the stock. In addition, approximately ten times more hatchery fish survive than naturally spawned fish. This greatly increases the diversity of genotypes recruited into estuarine and marine stocks. Finally, inherent properties of the hatchery production system that modify the selection regimen of hatchery fish could exist relative to those that occur in nature.

In Oregon, a pilot production program with chum salmon (*O. keta*) presented an opportunity to look for such adaptive responses. In this program, gametes were withdrawn from a vestigial, naturally-reproducing stock in Whiskey Creek, a tributary of Netarts Bay, Oregon. The fish have now been artificially propagated for two generations. This report describes selective processes and adaptive responses contributing to the evolution of this hatchery brood stock.

METHODS AND RESULTS

Whiskey Creek, the major tributary of Netarts Bay, Oregon, once supported a chum salmon population. From 1960 through 1969, the average peak count of chum salmon spawning in Whiskey Creek was 277 fish

---

Oregon Agricultural Experiment Station Technical Paper No. 4892.

(Berry 1976). The highest count occurring during that interval was 440 fish in 1962.

Beginning in 1969, gametes were taken from the Whiskey Creek stock and used in research to investigate artificial propagation technology for chum salmon. Juveniles produced during these investigations were released either as unfed fry or fingerlings into Netarts Bay. A pilot production project at Whiskey Creek began in 1971. Again, gametes were taken from the Whiskey Creek stock and placed in gravel incubators. The hatchery facilities and methods employed have been described elsewhere (Lannan 1975).

From 1971 through the 1974 brood year, a volitional release schedule was used. Unfed fry were allowed to migrate out of the hatchery into Netarts Bay by their own volition. From 1975 to the present, a proportion of total annual production was fed and released at an average size of approximately one gram per fish.

Starting with the 1971 brood year, and continuing to the present, hatchery records have been maintained on each of the following characters: fecundity; age at return; length and post-spawning weight of adults; time of entry into Whiskey Creek; time to hatching and emergence; survival to hatching and to swim-up; and time of emergence from gravel incubators. In addition, a fry trap was placed in Whiskey Creek. Observations on the time and size at emergence of naturally produced fry were recorded for the 1971 and 1972 broods.

In his studies that led to the development of the Netarts gravel incubation system, Poon (1977) observed that fry tended to emerge from the incubators earlier than naturally produced fry. He also presented evidence from which he inferred that fry left the hatchery at a slightly earlier stage of development. Timing and size of fry emerging from the hatchery were compared to fry emerging from the stream during the 1972 brood year. Figure 1 graphically presents results of this comparison. The major peaks of out-migration from the hatchery preceeded the corresponding peaks of fry emerging in the stream by several days. Although the positions of the respective fry traps imposed sampling bias, the trend likely related to inherent properties in the production system as well. The important point here is that hatchery fry enter the estuary somewhat earlier than naturally produced fry. This may be considered a form of artificial selection because it deviates from the adaptive out-migration timing existing in the Whiskey Creek stock as a consequence of natural selection.

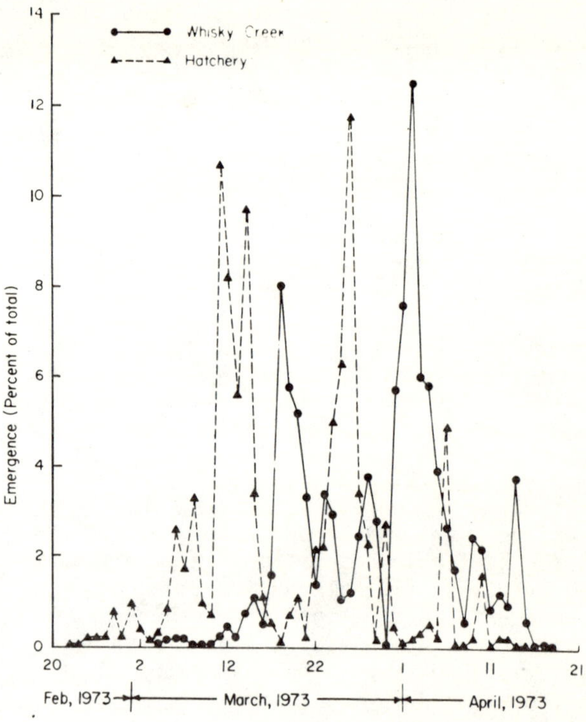

Figure 1. Comparison of out-migrations of chum salmon fry from the hatchery and the stream (1972 brood year).

The relative sizes of fry emerging from the hatchery and the stream were also compared during the 1972 brood year. Hatchery fry had an average fork length of 35.67 mm with a variance of 1.62 mm. The average fork length of the stream fry was 35.98 mm with a variance of .73 mm. The difference in average size was not statistically significant ($P > .05$). The hatchery fry were more variable in size than the stream fry, the difference being statistically significant ($P < .01$). The increased variance in size of hatchery fry likely related to the fact that approximately ten-fold more hatchery fish survived than those in the stream.

In summary, hatchery records reveal at least two ways that hatchery production caused deviations from the stock's inherent adaptations. First, for whatever reasons, hatchery-produced fry entered the estuary somewhat earlier than fry produced in the natural habitat. Second, a large proportion of hatchery-produced fish, which would not have survived incubation in the natural habitat, were recruited to the estuarine stock. The increased variation in size of out-migrant fry from the hatchery indicated qualitative as well as quantitative change in the stock's structure.

Hatchery records on time of entry, age at return, population sex ratio, fecundity, length and weight of returning adults were reviewed for apparent trends that might indicate a population response to selective processes occurring in the hatchery production system. Additionally, the time to hatching and emergence of the subsequent generation was observed. The results were divided into two catagories on the basis of whether population responses were or were not observed. Trends indicating that artificial propagation had influenced the average length, weight, fecundity, age at return or population sex ratio in the stock were not apparent. This does not mean that these characters were not modified, but rather that the year-to-year variation observed was no greater than that occurring in nature. Consequently any trends that might be present were statistically obscured.

Hatchery operations may have influenced the time interval during which adult fish enter Whiskey Creek. The date when the first fish of the run has been observed in the trap has shown remarkable consistency for six consecutive years, occurring either on 26 or 27 October. However, the total interval during which fish enter Whiskey Creek appeared to have been reduced. During 1972 and 1973 brood years, for example, chum salmon continued to arrive as late as the fourth week of November. In 1974 and subsequent years, migrations into Whiskey Creek were essentially complete by 20 November. I am of the opinion that this shift resulted not from inherent properties in the hatchery, but rather because of arbitrary egg dispersement policies. The Whiskey Creek hatchery had a capacity of 1.2 million eggs from 1971 through 1976. In 1977 the capacity was doubled. In years when surplus eggs are available, they are distributed by the state for use in other enhancement activities. Prior to 1977, the policy was to completely stock the Whiskey Creek hatchery before surplus eggs were diverted to other sources. Surplus eggs were dispersed during the 1972, 1973 and 1974 brood years. If heritable factors influence the date of entry into the natal stream, it follows that dispersing eggs from the late portion of the run would reduce the duration of the annual spawning migration.

Figure 2 shows thermal units accrued at the time of hatching and emergence as a function of brood years for the 1972-1977 broods. These data revealed a trend toward increasing thermal unit accrual to each developmental stage in succeeding years. The trend line, fitted by linear regression, infers that the rate increased at an average of 22.43°F days/year to hatching and 91.74°F days/year to emergence. The use of linear regression here does not mean that the increasing trend is necessarily linear. Rather, it tests the statistical hypothesis showing that the slope of the trend line does not equal zero. The data fit the linear model ($r^2 = .83$) suggesting that chance cannot explain the trend toward decreasing development rates.

I interpret the trend toward increased incubation time to emergence to be causally related to the early out-migration of hatchery fry. The Whiskey Creek stock is adapted to enter the estuary when estuarine conditions are favorable. Prematurely releasing hatchery fry into the estuary deviates from this adaptation and, as such, is a form of artificial selection. Consistent with the principal of genetic homeostasis (Lerner 1945), natural selection has resisted this trend and selected for genotypes with slower developmental rates. Thus, the hatchery stock enters the estuary when conditions are typically favorable. Stated another way, the hatchery stock is evolving towards the adaptation which existed in the ancestral, naturally reproducing stock.

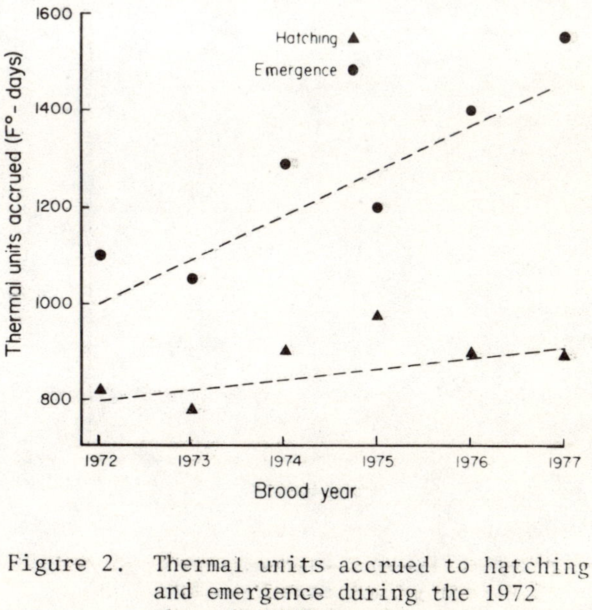

Figure 2. Thermal units accrued to hatching and emergence during the 1972 through 1977 brood years.

ACKNOWLEDGMENTS

This work was authorized under an agreement between the Oregon Agricultural Experiment Station and the Oregon Department of Fish and Wildlife, and is a result of research sponsored by the Oregon State University Sea Grant College Program, supported by the Natl. Oceanic and Atmos. Administration Office of Sea Grant, Department of Commerce, under Grants #2-37187, #04-3-158-4, #04-3-158-4, #04-158-2, #04-6-158-44004, and #04-6-158-44094. The U.S. Government is authorized to produce and distribute reprints for governmental purposes notwithstanding any copyright notation that may appear hereon.

LITERATURE CITED

Berry, R. L. 1976. Status of chum salmon runs in selected tributaries of Tillamook and Nestucca Bays, 1975. Ore. Dep. Fish and Wildl. Internal Rep.

Brannon, E. L. 1967. Genetic control of migrating behavior of newly emerged sockeye salmon fry. Int. Pac. Salmon Fish. Comm. Progr. Rep. 16. 31 pp.

Brannon, E. L. 1972. Mechanisms controlling migration of sockeye salmon fry. Int. Pac. Salmon Fish. Comm. Bull. XXI. 86 pp.

Lannan, J. E. 1975. Netarts Bay Chum Salmon Hatchery. Ore. State Univ. Sea Grant Coll. Prog. Publ. No. ORESU-H-75-001. 28 pp.

Lerner, I. M. 1954. Genetic homeostasis. Oliver and Boyd: Edinburgh. 134 pp.

McCart, P. 1967. Behavior and ecology of sockeye salmon fry in the Babine River. J. Fish. Res. Bd. Can. 24(2):375-428.

Poon, D. C. 1977. Quality of salmon fry from gravel incubators. Ph.D. Thesis. Ore. State Univ., Corvallis. 253 pp.

Raleigh, R. F. 1967. Genetic control of lakeward migrations of sockeye salmon, *Oncorhynchus nerka*, fry. J. Fish. Res. Bd. Can. 24(12):2613-2622.

Ricker, W. E. 1972. Hereditary and environmental factors affecting certain salmonid populations. Pages 27-160 *in* H. R. MacMillan lectures in fisheries, 1972. Univ. British Columbia.

Utter, F. M. 1978. See contribution in this volume.

Possible Role of the Pineal Gland in Migratory Behavior of Salmonids

Lavern J. Weber and John R. Smith

INTRODUCTION

Role of the Pineal in Hormone Production

The pineal gland in anadromous salmonids may play a role in the migratory behavior of these fish. Most pineal research on lower vertebrates has concentrated on the relationship between light, the production of hormones in the gland and hormonal influences on body coloration. A good deal of evidence has been collected suggesting that the pineal gland, upon receiving light information, directly or indirectly, produces hormones, which control many body functions.

Locy (1893), Hill (1894) and Holmgren (1918) described the pineal gland as a photoreceptor or "third eye" in lower vertebrates. Von Frisch (1911*a*, 1911*b*) showed that blinded fish alter their body color in response to light, possibly via sensory cells in the pineal.

It was McCord and Allen (1917) who determined that the pineal gland chemically controls body coloration in lower vertebrates, and Lerner et al. (1958) who found that the hormone melatonin, isolated from bovine pineal extracts, blanches the skin of amphibians. Later, several researchers determined that the pineal gland in lower vertebrates produces melatonin *in vitro* (Quay 1965; Fenwick 1970*a*; Hafeez and Quay 1970*a*, 1970*b*; Smith and Weber 1973, 1974). As more evidence accumulated, it became apparent that the teleost pineal gland acts as a photochemical transducer. Melatonin, the hormone localized in the pineal of lower vertebrates, probably serves as the chemical link.

Scientists studying mammals pursued a different line of inquiry. The convincing volume of data they accumulated indicates that the mammalian pineal is an endocrine gland. It controls the synthesis and release of melatonin and other hormones into the bloodstream (Kelly 1962, Kappers and Schade 1965).

Scientists who studied mammals also investigated the relationship between light and melatonin synthesis in the pineal gland. The activity of an enzyme involved in this synthesis, hydroxyindole-o-methyltransferase (HIOMT), is positively correlated with environmental lighting. Rats held in continuous darkness have a higher HIOMT activity level than rats held in continuous light (Wurtman et al. 1963, Wurtman et al. 1964, Axelrod et al. 1965). Furthermore, rats exposed to a 12-hour light to 12-hour dark photoregimen (12L:12D) display a diurnal rhythm in HIOMT activity. This activity dramatically increases in the initial hours of darkness; however, constant light or constant dark conditions, or an additional five hours of light, eliminates this rise.

More thorough treatments on the evolution of these and other aspects of pineal research can be found in Young (1935); Kitay and Altschutze (1954); Bagnara (1960, 1965); Kelly (1962); Kappers and Schade (1965); Wiener (1968); Wurtman et al. (1968); Fenwick (1970*a*, 1970*b*); Wolstenholme and Knight (1971); Rieter (1973); Relkin (1976, 1977); Wurtman and Moskowitz (1977).

DISCUSSION

Photoperiod and Pineal Activity in Salmonids

The endocrinological approach has hardly been used at all in research on the salmonid pineal gland. Only one previous effort attempted to correlate HIOMT activity with illumination (Hafeez and Quay 1970*b*). So we set out to determine if a relationship exists between photoperiod and HIOMT enzyme activity in the pineal gland of *Salmo gairdneri*, a migratory species of trout commonly referred to as steelhead. Because our data and results have been published (Smith and Weber 1974, 1976*a*, 1976*b*), we will discuss our results in a general way, then speculate on how the pineal may influence migration in salmonids.

We found that HIOMT activity in steelhead is photodependent and displays a diurnal rhythm similar to that found in rats. As with rats, HIOMT activity among the fish increased at night.

We studied the characteristics of this rhythm in steelhead further by using tanks set up indoors. Lighting, water temperature, feeding and other parameters were controlled. In three photoregimens (8L:16D, 12L:12D, 16L:8D), HIOMT activity rose in the first four hours of darkness. In the 8L:16D and 12L:12D photoregimens, the high HIOMT activity was clearly associated with the dark portion of the photoperiod. In the 16L:8D regime, activity was not as great and gradually decreased in the light phase. Our results oppose those that Hafeez and Quay (1970*a*, 1970*b*) reported. They used rainbow trout and concluded that constant light and dark conditions did not influence HIOMT activity. However, differences may be due to different methodologies or possibly species variations.

We next concluded that the eyes convey the light information necessary for photoperiodic enzyme regulation in the steelhead pineal gland. Fish were blinded by bilateral enucleation and then exposed to light. HIOMT activity was monitored. Results showed that blinding abolished the diurnal HIOMT rhythm, suggesting that the eyes convey light to the pineal gland. Further support for this came from the results of another experiment. We held intact, or non-blinded, fish in constant darkness. Darkness abolished the diurnal HIOMT rhythm in these fish.

Another aspect of blinding needed investigating to strengthen our conclusion that the eyes convey light to the pineal gland. Surgically blinding salmonids causes a generalized darkening of the skin (Neill 1940). This could possibly negate the contribution to HIOMT activity of direct stimulation of the pineal through the skull. To answer this question, we capped, or covered, the pineal region of intact fish to prevent direct illumination. These fish were held in an appropriate photoregimen. Capping did not alter the characteristic increase of activity during the first four hours of darkness indicating again that the eyes convey light to the pineal gland. To further corroborate this finding we conducted a second experiment where fish were blinded as before and held in a diurnal photoregimen. This time, however, the darkened skin above the pineal gland was removed to insure direct illumination of the area. Removing the skin did not reestablish the diurnal rhythm in blind fish.

Scientists who studied the mammalian pineal gland also concluded that the eyes convey light to the pineal gland. In rats, bilateral enucleation or superior cervical ganglionectomy abolished the diurnal HIOMT rhythm (Wurtman et al. 1964). The characteristic rise of HIOMT activity with the onset of darkness under a diurnal photoregimen was also abolished by these surgical procedures (Axelrod et al. 1965).

We should add that an age factor may have been the reason some observations indicate a non-neuronal pathway to the pineal. Several researchers have attributed a more functional role to the direct illumination of the pineal gland in the early lifestages of amphibians and mammals prior to complete development of the eyes (McCord and Allen 1917, Wyman 1924, Young 1935, Joss 1973, Zweig et al. 1966).

The actual relationship between HIOMT activity and melatonin production in trout is not entirely clear. We have only studied the activity of one enzyme in the biosynthetic pathway of melatonin. However, Gern et al. (1978) clearly show that photoperiod influences plasma melatonin concentrations in trout with a pattern very similar to HIOMT activity. Interestingly, however, this rhythm persisted in blinded trout, which suggests that HIOMT may not be the rate limiting enzyme in melatonin production in this species.

Possible Relationship of the Pineal to Migratory Behavior

We can speculate on the possible relationship between the pineal gland, melatonin production and the migratory behavior of anadromous salmonids. We want to stress that we intend to point out areas where existing evidence from our studies with steelhead, and other studies on lower vertebrates and mammals, suggests pineal involvement. Admittedly, species differences may exist, caution must be exercised in applying these generalizations on species other than salmonids.

The thyroid has been suspected of playing a prominent role in creating the migratory disposition in salmon (Hoar et al. 1952; Hoar et al. 1955; Hoar 1963; Baggerman 1959, 1960*a*, 1960*b*, 1962, 1963). Baggerman (1963) showed a correlation between thyroid activity and salinity preference in anadromous salmonids. Using thyroid inhibiting agents, she determined that low thyroid activity is associated with freshwater preference and high thyroid activity with saltwater preference.

Hoar (1963) reported another possible role for the thyroid in migration. He suggested that thyroxine and other hormones may play a permissive role. That is, these hormones increase metabolism and alter responsiveness of the fish to external stimuli. This would initiate appetitive behavior and result in an animal more susceptible to direct migratory influences. Interestingly, thyroxine does increase both locomotor activity of fish and their sensitivity to electrical stimuli (Hoar et al. 1952, Hoar et al. 1955, Baggerman 1959, 1962). Fenwick (1970*a*, 1970*b*) demonstrated that pinealectomized goldfish increase their responsiveness to light and electrical stimuli, thus supporting Hoar's conclusion that increased alertness accompanies an increase in thyroid activity.

Photoperiod regulates both thyroid activity and the pineal gland. It would appear that thyroid regulation facilitates seasonal synchrony with migration (Baggerman 1959, 1960*a*, 1960*b*). Interestingly, as the photoperiod becomes longer in spring, thyroid activity increases while at the same time melatonin synthesis in the pineal gland decreases. It could be that melatonin inhibits thyroid activity, and the increased photoperiod in spring releases the thyroid from melatonin inhibition. The increased thyroid activity may induce saltwater preference and appetitive behavior. In mammals, melatonin inhibits thyroid activity (Bashieri et al. 1963, De Prospo et al. 1968, Panda and Turner 1968). Melatonin, or dark conditions, also depresses locomotor activity in young salmon (Hoar et al. 1955, Byrne 1970).

There is another aspect of migration that involves the pineal gland. Smolts must physiologically alter their osmoregulatory mechanisms to tolerate salinity changes. Teleost prolactin, produced in the pituitary gland, may play an important role (Parry 1966, Bern 1967).

Prolactin has been implicated in the ability of saltwater-adapted stickleback (*Gasterosteus aleuticus*) to tolerate abrupt transfer to freshwater (Pickford and Phillips 1969, Schriebman and Kallam 1966, Lam and Hoar 1967, Lam 1968, Lam and Leatherland 1969). Pineal involvement in the osmoregulation of anadromous salmonids is suggested from certain mammalian studies. In rats, constant dark conditions and administering melatonin increases plasma prolactin levels (Kamberi et al. 1971, Relkin 1972). Removing the pineal gland abolished the effects of constant darkness (Relkin 1972, 1973). However, the role of the pineal in prolactin regulation is far from simple since blinding, and Na^+ deprivation inhibit the effects of pinealectomy (Donofrio and Reiter 1972, Relkin 1973, Shiino et al. 1974). Removing the pineal gland also dampens the normal diurnal rhythm of the pituitary prolactin, and because of this some have postulated that the pineal may regulate or mediate the normal prolactin rhythm (Vodicnik 1976).

We can speculate on the relationship between prolactin, melatonin, photoperiod and migration. Perhaps the longer photoperiod in spring decreases melatonin synthesis. Decreased melatonin levels decrease plasma prolactin levels. This impairs the smolt's ability to osmoregulate in freshwater and thus stimulates seaward movement.

The parr-smolt metamorphosis in juvenile salmonids is another aspect of migration to consider. At this stage of development, the growth rate accelerates (Baggerman 1960*b*). The pineal gland may be involved. Addair and Chidester (1928), expanding on the work of Dana and Berkeley (1913) and McCord (1914), found that pineal extracts increase the rate of metamorphosis in frogs. Huxley and Hogben (1922) concluded that the pineal gland plays an active role in the metamorphosis of amphibians. However, those studying mammals have shown that constant darkness and administering melatonin decreases growth hormone (GH) levels in the plasma (Relkin 1972, Smythe and Lazarus 1974*a*, 1974*b*). These results disagree with those of earlier scientists who felt that the pineal gland stimulates growth.

The observations of Smythe and Lazarus may resolve this disagreement. They found that melatonin initially elevated but then decreased plasma GH levels. This pattern was also exhibited by antiserotoninergic agents. Since melatonin also inhibits the action of serotonin (Quastel and Rahamimoff 1965, Rahamimoff and Bruderman 1965, Rahamimoff et al. 1965), Smythe and Lazarus felt that melatonin acts as a partial agonist, which explains the initial stimulation followed by a blockade. Serotoninergic stimulation in the hypothalamus may be responsible for GH responses, and the pineal may alter this control by secreting melatonin.

To speculate once more, perhaps the longer daylight periods in spring, which are associated with the parr-smolt metamorphosis in anadromous salmonids, decrease melatonin production. This may facilitate rapid growth, characteristic of this developmental stage, by freeing serotoninergic nerves in the hypothalamus from melatonin inhibition.

LITERATURE CITED

Addair, J. and F. E. Chidester. 1928. Pineal and metamorphosis: the influence of pineal feeding upon the rate of metamorphosis in frogs. Endocrin. 12:791-796.

Axelrod, J. 1970. The pineal gland. Endeavor 29:144-148.

Axelrod, J. 1971. Neural control of indoleamine metabolism in the pineal. Pages 35-47 *in* Wolstenholme, G. E., and J. Knight (eds.), The pineal gland. Edinburgh and London, Churchill Livingstone.

Axelrod, J. 1974. The pineal gland: a neurochemical transducer. Sci. 184:1341-1348.

Axelrod, J., R. J. Wurtman, and S. H. Snyder. 1965. Control of hydroxyindole-o-methyltransferase activity in the rat pineal gland by environmental lighting. J. Biol. Chem. 240:949-954.

Axelrod, J., R. J. Wurtman, and C. M. Winget. 1964. Melatonin synthesis in the hen pineal gland and its control by light. Nature 201:1134.

Baggerman, B. 1959. The role of external factors and hormones in migration of sticklebacks and juvenile salmon. Pages 24-37 *in* Gorbman, A. (ed.), Comp. endocrin.

Baggerman, B. 1960*a*. Factors in diadromous migrations of fish. *In* Hormones in fish. Symp. Zool. Soc. London 1:33-60.

Baggerman, B. 1960*b*. Salinity preference, thyroid activity, and the seaward migration of four species of Pacific salmon (*Oncorhynchus*). J. Fish. Res. Bd. Can. 17:295-322.

Baggerman, B. 1962. Some endocrine aspects of fish migration. Gen. Comp. Endocrin., suppl. 1:188-205.

Baggerman, B. 1963. The effect of TSH and anti-thyroid substnaces on salinity preference and thyroid activity in juvenile Pacific salmon. Can. Journ. Zool. 41:307-319.

Bagnara, J. R. 1960. Pineal regulation of the body lightening reaction in amphibian larvae. Sci. 132:1481-1483.

Bagnara, J. R. 1965. Pineal regulation of body blanching in amphibian larvae. Pages 489-506 *in* Kappers, J. A., and J. P. Schade (eds.), Structure and function of the epiphysis cerebri. Progress in brain research, Vol. 10. Elsevier, Amsterdam.

Baker, P. C. and K. M. Hoff. 1971. The effect of light and dark background upon melatonin levels in larval *Xemopus*. Comp. Gen. Pharmacol. 2:59-62.

Bashieri, L., F. De Luca, L. Cramarossa, C. De Martino, A. Oliverio, and M. Negri. 1963. Modification of thyroid activity by melatonin. Experientia (Basel) 19: 15-17.

Bern, H. A. 1967. Hormones and endocrine glands of fishes. Sci. 158:455-462.

Breder, C. M., and P. Rasquin. 1947. Comparative studies in the light sensitivity of blind characins from a series of Mexican caves. Bull. Amer. Mus. Nat. Hist. 89:325-351.

Breder, C. M., and P. Rasquin. 1950. A preliminary report on the role of the pineal organ in the control of pigment cells and light reactions in recent teleost fishes. Sci. 111:10-12.

Brick, I. 1962. Relationship of the pineal to the pituitary-melanophore effector system in *Ambystoma opacum*. The Anatomical Record 142:299.

Byrne, J. E. 1970. Locomotor activity responses in juvenile sockeye salmon, *Oncorhynchus nerka*, to melatonin and serotonin. Can. Journ. Zool. 48:1425-1427.

Charlton, H. M. 1969. The pineal gland and color change in *Xenopus lavis* Daudin. Gen. Comp. Endocrinol. 7:384-397.

Dana, C. L., and W. H. Berkeley. 1913. The Medical Record 83:835. (Cited in Addair, J., and F. E. Chidester. 1928. Pineal and metamorphosis. The influence of pineal feeding upon the rate of metamorphosis in frogs. Endocrin., Vol. 12.)

De Prospo, N. D., L. J. Demartino, and E. T. McGuinness. 1968. Melatonin's effect on ^{131}I uptake by the thyroid gland in normal and ovariectomized rats. Life Sci. 7:183-188.

De Vlaming, V. L. 1975. Effects of photoperiod-temperature regimes and pinealectomy on body fat reserves in the Golden Shiner, *Notemigonus crysoleucas*. Fish. Bull. Vol. 73, No. 4.

De Vlaming, V. L., and M. J. Vodicnik. 1977. Effects of pinealectomy on pituitary gonadotrops, pituitary gonadotropin potency and hypothalamic gonadotropin releasing activity in *Notemigonus crysoleucas*. J. Fish. Biol. 10(1):73-86.

Dodt, E. 1963. Photosensitivity of the pineal organ in the teleost *Salmo irideus* (Gobbons). Experientia 19:652-643.

Donofrio, R. J. and R. J. Reiter. 1972. Depressed pituitary prolactin levels in blinded anosmic female rats: role of the pineal gland. J. Repro. Fert. 31:159-162.

Eddy, J. M. P., and R. Strahan. 1968. The role of the pineal complex in the pigmentary effector system of the pampreys, *Mordacia mordax* (Richardson) and *Geotria australis* (Gray). Gen. Comp. Endocrin. 11:528-534.

Fenwick, J. C. 1970*a*. Demonstration and effect of melatonin in fish. Gen. Comp. Endocrin. 14:86-97.

Fenwick, J. C. 1970*b*. The pineal organ. Pages 91-104 *in* Hoar, W. S., and D. J. Randall (eds.), Fish physiology, Vol. 4. Academic Press, New York.

Gern, W. A., D. W. Owens, and C. L. Ralph. 1978. Persistence of the nychthermal rhythm of melatonin secretion in pinealectomized or optic tract-sectioned trout *(Salmo gairdneri)*. J. Exper. Zool. 205:371-376.

Hafeez, M. A., and W. B. Quay. 1970*a*. Pineal acetylserotonin methyltransferase activity in the teleost fishes, *Hesperoleucus symmetricus* and *Salmo gairdneri*, with evidence for lack of effect of constant light and darkness. Comp. Gen. Pharmacol. 1:257-262.

Hafeez, M. A., and W. B. Quay. 1970*b*. The role of the pineal organ in the control of phototaxis and body coloration in rainbow trout (*Salmo gairdneri*, Richardson). Vergl. Physiologie 68:403-416.

Hanya, I., H. Niwa, and T. Tamura. 1969. A slow potential from the epiphysis cerebri of fishes. Vision Research 9: 621-623.

Hill, C. 1894. The epiphysis of teleosts and Amia. J. Morph. 9:237-268.

Hoar, W. S. 1955. Phototactic and pigmentary responses of sockeye salmon smolts following injury to the pineal organ. Fish Res. Bd. Can. 12:178-185

Hoar, W. S. 1959. Endocrine factors in the ecological adaptation of fishes. Pages 1-24 *in* Gorbman, A. (ed.), Comparative endocrinology. John Wiley and Sons, Inc., New York.

Hoar, W. S. 1962. Reproductive behavior in fish. Gen. Comp. Endocrin., suppl. 1:206-216

Hoar, W. S. 1963. The endocrine regulation of migratory behavior in anadromous teleosts. Pages 14-20 *in* Proc. of the Sixteenth International Cong. Zool. Vol. 3. Washington, D. C.

Hoar, W. S., M. H. A. Kennleyside, and R. G. Goodall. 1955. The effects of thyroxine and gonadal steroids on the activity of salmon and goldfish. Can. Journ. Zool. 33:428-439

Hoar, W. S., D. MacKinnon, and A. Redlick. 1952. Effects of some hormones on the behavior of salmon fry. Can. Journ. Zool. 30:273-286

Holmgren, N. 1918. Zum Bau der Epiphyse von *Squalus acanthias*. Arkiv fur zoologi 11:1. (Cited in Wurtman, R. J., and J. Axelrod. 1965. The pineal gland. Sci. Amer. 213:50-64.

Huxley, J. S., and L. T. Hogben. 1922. Experiments on amphibian metamorphosis and pigment responses in relation to internal secretions. Proc. of the Royal Soc., London. 93 (series B):36-54.

Joss, J. M. P. 1973. The pineal complex, melatonin, and color change in the Lampetra. Gen. Comp. Endocrin. 21:188-195.

Joss, J. M. P. 1977. Hydroxyindole-o-methyltransferase (HIOMT) activity and the uptake of ^{3}H-melatonin in the lamprey. *Geotria australis* (Gray). Gen. Comp. Endocrin. 31(3):270-275.

Kamberi, I. A., R. S. Mical, and J. C. Parter. 1971. Effects of melatonin and serotonin on the release of FSH and prolactin. Endocrin. 88:278-284.

Kappers, J. A., and J. P. Schade (eds.). 1965. Structure and function of the epiphysis cerebri. Progress in Brain Research. Elsevier, Amsterdam. Vol. 10. 694 pp.

Kitay, J. I., and M. D. Altschule. 1954. The pineal gland. Harvard Univ. Press, Cambridge. 280 pp.

Kelly, D. E. 1962. Pineal organs: photoreception, secretion, and development. Amer. Scientist 50:597-625.

Klein, D. and J. L. Weller. 1970. Indole metabolism in the pineal gland: a circadian rhythm in N-acetyltransferase. Sci. 169:1093-1095.

Lam, T. J., and W. S. Hoar. 1967. Seasonal effects of prolactin on freshwater osmoregulation of the marine form *(Trachurus)* of the stickleback, *Gasterosteus aculeatus*. Can. Journ. Zool. 45:509-516.

Lam, T. J. 1968. Effect of prolactin on plasma electrolytes of the early-winter marine threespine sticklebacks, *Gasterosteus aculeatus* form *trachurus*, following transfer from sea to freshwater. Can. Journ. Zool. 46:1095-1097.

Lam, T. J., and J. F. Leatherland. 1969. Effects of prolactin on freshwater survival of the marine form *(trachurus)* on the threespine stickleback, *Gaserosteus aculeatus*, in the early winter. Gen. Comp. Endocrin. 12:385-387.

Laurens, H. 1915. The reaction of the melanophores of *Amblystoma* larvae. J. Exper. Zool. 18:577-638.

Lerner, A. B., J. D. Case, Y. Takashaskie, T. H. Lee, and W. Mori. 1958. Isolation of melatonin, the pineal gland factor that lightens melanocytes. J. Amer. Chem. Soc. 80:2587.

Locy, W. A. 1893. The derivation of the pineal eye. Anatomischer Anzeiger. 9: 169-180.

McCord, C. P. 1914. The pineal gland in relation to somatic, sexual and mental development. J. Amer. Med. Assoc. 63: 232-235.

McCord, C. P., and F. P. Allen. 1917. Evidence associating pineal gland function with alteration in pigmentation. J. Exp. Zool. 23:207-224.

Morita, Y. 1966. Entladungsmuster pinealer Neurone der Regenbogenforelle *(Salmo irideus)* bei Belichtung des Zwischenhirns. Pflugers Archiv ges Physiologie 289:155-167.

Morita, Y., and E. Dodt. 1965. Nervous activity of the frog's epiphysis cerebri in relation to illumination. Experientia (Basel) 21:221-222.

Neill, R. M. 1970. On the existence of two types of chromatic behaviour in teleostean fishes. J. Exper. Biol. 17:74-95.

Oishi, T., and J. K. Lauber. 1973. Effect of light and darkness on pineal hydroxyindole-o-methyltransferase (HIOMT) in Japanese quail. Life Sci. 13:1105-1116.

Owens, D. W., W. A. Gern, C. L. Ralph, and T. J. Boardman. 1978. Non-relationship between plasma melatonin and background adaptation in the rainbow trout *(Salmo gairdneri)*. Gen. Comp. Endocrin. 34: 459-467.

Panda, J. N., and C. W. Turner. 1968. The role of melatonin in the regulation of thyrotrophin secretion. Acta Endocrinologica 57:363-373.

Parry, G. 1966. Osmotic adaptation in fishes. Biol. Reviews 41:392-444.

Pickford, G. E., and J. C. Phillips. 1959. Prolactin, a factor in promoting survival of hypothysectomized killfish in freshwater. Sci. 130:454-455.

Quastel, M. R., and R. Rahamimoff. 1965. Effect of melatonin on spontaneous contractions and response to 5-hydroxytryptamine of rat isolated duodenum. Brit. Journ. Pharmacol. and Chemotherapy 24:455-461.

Quay, W. B. 1965. Retinal and pineal hydroxyindole-o-methyltransferase activity in vertebrates. Life Sci. 4:983-991.

Rahamimoff, R., I. Bruderman, and G. Golshani. 1965. Effect of melatonin on 5-hydroxytryptamine induced contraction of isolated cat trachea. Life Sci. 4:2281-2287.

Rahamimoff, R., and I. Bruderman. 1965. Changes in pulmonary mechanics induced by melatonin. Life Sci. 4:1383-1389.

Rasquin, P. 1958. Studies in the control of pigment cells and light reactions in recent teleost fishes. Bull. Amer. Mus. Nat. Hist. 115:1-68.

Reed, B. L. 1968. The control of circadian pigment changes in the pencil fish: a proposed role for melatonin. Life Sci. 7:961-973.

Reiter, R. J. 1973. Comparative physiology: pineal gland. Ann. Rev. Physiol. 35: 305-328.

Relkin, R. 1972. Effects of variations in environmental lighting on pituitary and plasma prolactin levels in the rat. Neuroendocrinology 9:278-284.

Relkin, R. 1973. Effects of sodium deprivation and pinealectomy on pituitary and plasma prolactin levels in the rat. J. Endocrin. 59:383-384.

Relkin, R. 1976. The pineal, Vol. 1. Eden Press, Montreal. 187 pp.

Relkin, R. 1977. The pineal, Vol. 2. Eden Press, Montreal. 187 pp.

Schriebman, M. P., and K. D. Kallman. 1966. Endocrine control of freshwater tolerance in teleosts. Gen. Comp. Endocrin. 6:155.

Shiino, M., A. Arimura, and E. G. Rennels. 1974. Effects of blinding, olfactory bulbectomy, and pinealectomy on prolactin and growth hormone cells of the rat, with special reference to ultrastructure. Amer. Journ. Anat. 139:191-209.

Smith, J. R., and L. F. Weber. 1973. Acetylserotonin methyltransferase (ASMT) activity in the pineal gland and retina of rainbow trout. Proc. of the West. Pharmacol. Soc. 16:191-194.

Smith, J. R., and L. F. Weber. 1974. Diurnal fluctuations in acetylserotonin methyltransferase (ASMT) activity in the pineal gland of the steelhead trout (*Salmo gairdneri*). Proc. for Exper. Biol. and Medicine 147:441-443.

Smythe, G. A., and L. Lazarus. 1974*a*. Growth hormone responses to melatonin in man. Sci. 184:1373-1374.

Smythe, G. A., and L. Lazarus. 1974*b*. Suppression of human growth hormone secretion by melatonin and cyproheptadinel. J. Clinical Invest. 54:106-112.

Swingle, W. W. 1921. The relation of the pars intermedia of the hypophysis and the pineal gland to pigmentation changes in anuran larvae. Proc. XXXVII Meeting Amer. Assoc. Anatomists, Anatomical Record 21:87.

Vodicnik, M. J. 1976. The effect of pinealectomy on pituitary prolactin levels in the goldfish, *Carassius auratus*. Amer. Zool. Vol. 16, p. 254.

von Frisch, K. 1911*a*. Beitrage zur physiologie der pigmentzellen in der fischhaut. Archiv fur der ges ammt physiologie, Bd. 138:319-387. (Cited in Laurens, H., The reactions of the melanophores of Amblystoma. The supposed influence of the pineal organ. J. Exper. Zool. 20:237-261.

von Frisch, K. 1911*b*. Das parietslorgan der fische als functionserendes organ sitzber. Gesannt Morpholigi und Physiologie, Munch 27:16-18. (Cited in Hoar, W. S., and D. J. Randall (eds.). Fish physiology, Vol. 4. Academic Press, New York.)

Wagner, H. H. 1974. Photoperiod and temperature regulation of smolting in steelhead trout *(Salmo gairdneri)*. Can. Journ. Zool. 52:219-234.

Wiener, H. 1968. External chemical messengers IV. Pineal gland. New York State Journ. Med. 68:92-138.

Wolstenholme, G. E. W., and J. Knight (eds.). 1971. The pineal gland. Churchill Livingstone, Edinburgh and London. 401 pp.

Wurtman, R. J., J. Axelrod, and J. E. Fischer. 1964. Melatonin synthesis in the pineal gland: effect of light mediated by the sympathetic nervous system. Sci. 143:1328-1330.

Wurtman, R. J., J. Axelrod, and L. S. Phillips. 1963. Melatonin synthesis in the pineal gland: control by light. Sci. 142:1071-1073.

Wurtman, R. J., J. Axelrod, and D. E. Kelly. 1968. The pineal. Academic Press, New York. 199 pp.

Wurtman, R. J., and M. A. Moskowitz. 1977. The pineal organ. New Eng. Journ. of Med. Part 1, 296(23):1329-1333; Part 2, 296(24):1383-1386.

Wykes, U. 1938. The control of photopigmentary responses in eyeless catfish. J. Exper. Biol. 15:363-370.

Wyman, L. C. 1924. The reactions of the melanophores of embryonic and larva *Fundulus* to certain chemical substances. J. Exper. Zool. 40:161-180.

Young, J. Z. 1935. The photoreceptors of lampreys: II. The functions of the pineal complex. J. Exper. Biol. 12-13: 254-270.

Zweig, et al. 1966. Evidence for a nonretinal pathway of light to the pineal gland of newborn rats. Proc. Nat. Acad. Sci. 56:515-520.

Appendix

Abstracts of Soviet Salmonid Studies

The Growth of Young Sockeye of Different Year Classes and Fluctuations in Primary Production in Lake Dalneye

F. V. Krogius

It has been well established that juvenile sockeye salmon (*Oncorhynchus nerka*) of different year classes grow at different rates during their first year of life in Lake Dalneye. We do not know how to adequately explain the most rapid growth rate we have observed among young of highly abundant and dominant year classes that occur every four years. Genetic effects do not fully explain observed differences in growth rates.

It has been found that a considerable amount of phosphates dissolved in Lake Dalneye, as in other lakes where sockeye spawn, are carried into the lakes with carcasses of spawned fish. It appears that the phosphate regime of the lake can be distinctly improved if there are at least 500-700 carcasses per million cubic meters of lake volume.

A four-year periodicity in fluctuations of the flow has been observed in Lake Dalneye. The minimum annual flow falls on the year of the maximum sockeye run into the lake. Strong year classes of juveniles fatten in the lake in the years of increasing flow. It is believed that increased flow facilitates survival, development and growth of the young.

Analysis shows that the primary productivity of Lake Dalneye increases after a large sockeye run and during a period of increasing water flow. This indicates that water flow and phosphate content regulates primary production in the lake.

Fluctuations in primary production first affect production processes in the second trophic level and are evidently one of the main factors regulating production of juvenile sockeye. Phosphate inflow into the lake during the 1970s was a tenth of the level observed from 1937-47. Primary production during 1960-1976 averaged only 60 percent of that observed from 1937-47. Zooplankton largely regenerate the phosphate when sockeye spawners are scarce.

With increasing numbers of young sockeye grazing in the lake and removing zooplankton from the system, phosphate replenishment in the trophogenic layer through zooplankton will be reduced. This, and other factors, should be taken into account when estimating affects of lake productivity on survival, development and growth of sockeye salmon.

Peculiarities of Cranial Anatomy of Far East Salmon in Postembryonic Ontogenesis

N. S. Romanov

In previous studies of the development of Pacific salmon, we showed that changes in chondrocranium shape continue from the fry stage through the formation of secondary sexual characteristics associated with maturation and breeding. Cartilaginification and growth of bone characterize the cerebral section. Cartilaginification results from the closure of dorsal fontanelles. The cartilaginous bridge between them enlarges four times and thickens. External occipital and antere-, sphenoid- and pterigoid-octic bones exhibit intensive growth. The orbital section decreases in size with maturity. The eyes and orbits decrease in size, accompanied by the shortening of the olfactory cerebral bridge.

The olfactory section changes during marine growth and acquisition of spawning characteristics. Widening and lengthening of the olfactory section in the caudal direction, accompanied with some decrease in the size of the eyes, results from cartilage growth. Considerable enlargement and lengthening of the rostrum and enlargement of the cartilaginous septum between the olfactory capsules are also observed as fish mature.

Integmentary bones of the chondrocranium and visceral skeleton of the head are in close contact with the chondrocranium. A study on sockeye salmon provides information on the scale of these changes in *Oncorhynchus* sp.

Integmentary bones of chondrocranium: During freshwater life of the young, configuration of the forepart of the supra-ethmoid changes. Significant changes in proportions of the supra-ethmoid occur during the sea period of life in both males and females with additional changes in the spawning period occurring mainly in males. The frontal bone undergoes considerable changes in the freshwater period of growth and in the first months of life in the sea. The vomer changes during periods of growth in fresh and marine waters, with minor changes during maturation. Changes in the parasphenoid during postembryonic ontogenesis are restricted mostly to the orbito-ethmoidal part.

Primary upper jaw: Significant changes occur in the proportions of the palete and external pterigoid bones during freshwater growth of sockeye. During life in the sea, the internal and posterior pterigoid bones and square bone undergo considerable change. Changes in the primary upper jaw are modest as fish mature.

Secondary upper jaw: Changes in the freshwater growth period of sockeye are modest. The most noticeable changes in the secondary upper jaw take place during marine growth and spawning.

Lower jaw: Teeth and joining bones change mainly during the first year of freshwater life. Additional changes in size and shape of bones occur during the period of marine growth.

Hypoglossal arch: The size and shape of bones comprising the hypoglossal arch undergo change mainly in the first year of life. While at sea, the most significant changes are observed in the supra-hyoid and corniculatus-hyoid bones. There are no significant changes in the bones of the hypoglossal arch during maturation and spawning.

Gill-cover: Changes during the period of freshwater growth are confined mainly to the cover, precover and undercover bones. Further changes in the gill cover are not very significant during marine life and spawning.

No sexual dimorphism is observed in the integmentary bones of chondrocranium and visceral skeleton of sockeye at the beginning of marine life. Sexual dimorphism becomes quite apparent, however, when fish return to spawning grounds.

Mature dwarf male sockeye do not exhibit changes in the visceral skeleton to the same degree as full-sized males. Mature dwarf males tend to resemble immature fish. Rapid maturation of gonads in dwarf males is not accompanied by accelerated development of head bones. After a year of sea life, the head skeleton of sockeye males assumes an intermediate shape between dwarf and full-sized males.

The duration of morphogenesis of the head skeleton, and the absence of correlation between the rate of development and maturity of gonads, result in considerable variation in the structure of the head skeleton, especially in males. Such variability in Pacific salmon expands their possibilities to adapt to various habitats to a maximum degree.

Materials on the Assortive Crossing and the Role of Sex Ratio during the Spawning Period of the Pacific Salmons

N. A. Chebanov

Prespawning and spawning behavior of chum and pink salmon was studied in specially prepared sections of the Karymaysky Spring within the Bolskoy River basin in Kamchatka. In the experiments with chum, there was a 3:1 predominance of males. The pink salmon sex ratio was equal.

Behavior of male chum could be conventionally divided into four categories:

1. Monogamous leaders
2. Monogamous satellites
3. Polygamous leaders
4. Polygamous satellites

Large males were observed predominantly in the first and third categories, while small males were predominantly in the second and fourth categories. Small males were less aggressive than large males. They tended to wander about the spawning ground more than large males. At time of death, small males tended to retain large quantities of unused reproductive product.

Eggs laid by the female chum were fertilized by one or more large males. Small males usually did not participate in the spawning act until after most of the large males were weak or dead. Females remained on the spawning grounds longer than males. Their longer life is apparently beneficial for protecting redds against other spawners and may encourage the participation of small males in spawning.

Three experiments were conducted with chum and pink salmon at various sex ratios and various densities of spawners. Results of the experiments provide support for the conclusion that chum and pink salmon tend to practice assortative mating. There is a tendency for similar sized fish to mate on spawning grounds. The larger males were observed to spawn both with large and small females.

With the aim to show the importance of assortative crossing for the dynamics of Pacific salmon populations, we carried out a number of crossings with pink and sockeye salmon spawners of various sizes and checked mortality rates of embryos and larvae. Progeny of small adult fish, both males and females, were characterized by highest viability, while progeny from males of medium size mated with large females were considerably less viable. This relationship was observed for both species. Thus, assortative crossing can affect the abundance and viability of progeny. In the absence of other factors, assortative mating can be important.

The sex ratio on the spawning grounds can have a considerable effect on the abundance of generations. It has been shown experimentally that a great surplus of adult pink salmon of either sex can considerably decrease spawning efficiency, mainly because of aggressive behavior by spawners. High density may also be responsible for some increase in males staying on the spawning grounds.

The Main Problems of the Population Biology of Pacific Salmons

S. M. Konovalov

Variability within a species is an important aspect of population biology. This field of study dates from the time when biologists ceased to see differences among individuals comprising a species and aimed their efforts at understanding intraspecific variety.

Biology takes one of the leading places in the development of science over the last 20 to 30 years. Population biology to a great extent accounts for the sharply increasing significance of biology in the life of man. The contribution of population biology to the practical knowledge of man is tremendous indeed. This is largely due to the fact that population biology provides a basis for the rational exploitation of biological resources and the theory of long-term forecasting of the condition of populations. It is only on the basis of the laws of population biology that one can begin to adapt artificial populations to natural conditions.

Pacific salmon are a suitable subject for population research since they play an important part in the economy of the U.S.S.R., U.S.A., Canada and Japan. The importance of Pacific salmon is further enhanced by their suitability for artificial propagation. Successful exploitation of natural and artificially propagated populations is made possible through developing a thorough understanding of Pacific salmon population biology. It becomes necessary therefore, to develop the whole complex of disciplines that contribute to this understanding.

Research on variability in the structure of populations within a species should be regarded as an essential part of population biology. In this regard, Pacific salmon are suitable objects of research on variations in population structure during the reproduction period because populations become separated and confined to certain types of water bodies for spawning. Unfortunately, information on migration and emigration in relation to abundance of spawners and other factors is scarce.

Research on population structure and population homeostasis of a subisolate should be regarded as an integral part of population biology. Disciplines such as population genetics, population physiology, population morphology, population ecology and ethology can provide a detailed description of biological structure.

Studies of various forms of polymorphism, genetic structure of subisolates and the influence of factors such as selection, gene drift, mutation, migration and rate of transformation are included under population genetics. The effective reproductive segment of a subisolate is a most important parameter. While polymorphism of different proteins is being intensively studied by scientists in the U.S.S.R., U.S.A., Japan and Canada, investigations in fluctuations of subisolates and, hence, the calculation of the rate of subisolate transformation under the influence of different factors, are inadequate.

There are few examples of direct physiological investigations in mature salmon which would characterize the structure of a subisolate on the basis of physiological type. Our preliminary investigations have revealed considerable differences in the standard metabolism of fish that are in the sea at different times. We are of the opinion that further investigations will reveal the role of physiological structure and its function in adaptation of subisolates to changing conditions of the environment.

A similar situation pertains to morphological investigations of population structure. Morphological polymorphism within a subisolate is the subject of investigation by TINRO workers. It has been determined that different morphological types of sockeye salmon relate to different periods of time spent at sea, i.e. to ontogenetic variation. Males and females exhibit differences in morphology, for instance the chondrocranium, which characterize different populations. Sockeye of Lake Azabachje for example, fall into six main morphotypes, four in males and two in females. Morphological differences provide qualitative characters for identifying isolates. As a result there are grounds to regulate the share of fish of different morphotypes which are adapted to different conditions of the habitat. These investigations will make it possible to understand the formation of the morphological structure and environmental factors related to this manifestation of morphological homeostasis. These investigations are important for artificial rearing of different morphotypes of Pacific salmon.

It is reasonable to analyze ecological structure on the basis of age and sex structures. Dynamics of the sockeye age structure have been analyzed by different investigators. Results suggest that the environment regulates in part the age and sex composition of a population. However, factors regulating mean length of life in salmon are not fully understood. There is no well established theory on which it would be possible to predict the age and size at return of any given generation and which would describe the relationship between the age structure of parents and their progeny. Consequently, in the near future it is necessary to develop studies aimed at working out a theory of regulation of the age structure in Pacific salmon. Such a theory might make it possible to forecast the age structure in the first and the following generations and to form the optimum age structure in artificial populations.

The ratio of sexes is an important population parameter that is closely connected with the environment. The transition of the sex structure of a subisolate from one state to another involves a transition of each sex into a new state. Our observations show that the formation of a new size-age group in an isolate is always connected with the appearance of males. Therefore adaptation connected with a change in the sex structure as a result of environmental influence is, in our estimation, homeostasis of regulation allowing the subisolate to adapt to new conditions with minimal losses.

The mechanism of sex determination in salmon is not well understood. Today, there are no theoretical grounds for forecasting the sex structure of progeny in relation to that of parents. This is an important problem for both rational exploitation of natural stocks and artificial propagation.

Acquiring food and reproducing are the main components of behavior. Each of these components is realized through the formation of ethological structure from dominant, subdominant and subordinate elements within a population. Obtaining enough food to guarantee the survival of at least part of a population when food is limited can be regarded as a function of population structure. In the case of ethological structure on spawning grounds, the function of the population structure is the reproduction of the best adapted members of the population.

All of the above are elements of a population's biological structure. These elements are characterized by an interrelationship, which is apparent in direct or indirect change of all the substructures if transformation takes place in any of them. Thus, the appearance of an age group is connected with changes in the characteristics of sexual, physiological and other substructures. Population homeostasis is the ability of a population to change several parameters simultaneously with the purpose of adequate response to environmental change.

From the above it follows that natural populations should be exploited without causing disturbances in population homeostasis. Characteristics of population homeostasis have not been determined for many populations. It is necessary therefore, that the main populations of Pacific salmon should be studied by the above enumerated disciplines.

Artificial rearing of Pacific salmon should be aimed at securing the conditions in which all newly formed and artificially maintained populations should be able to adequately react to the environment, i.e. they should have population homeostasis. The peculiarities of the formation of population homeostasis in artificial populations is another major problem to be elaborated.

With consideration of such peculiarities of Pacific salmon biology as stock mixing in the marine period of life and complicated problems facing population biology, it is necessary to come over to interdisciplinary planning of research, creating a material and technical basis for investigations and drawing upon the cooperation of many scientists of various specialities. Consequently, the organization of research in population biology requires long-term planning, investments for many years and creation of large, temporary working teams on a national and, in a number of cases, on an international basis.

Sakhalin Taimen (*Hucho perrui*, Brevoot) of the Kievka River

V. P. Bushuev

The Sakhalin taimen inhabits rivers of the Primorski Territory, Sakhalin; the South Kuril Islands; and North Japan. It is the only anadromous species representative of the genus *Hucho*. However, it does not migrate long distances while at sea and returns to rivers for wintering. This accounts for the considerable genetic isolation of taimen populations found in different rivers. Morphological analysis of taimen from the Kievka River, Primorski, revealed differences from populations found in rivers of Sakhalin and Japan.

Investigations showed that Kievka taimen have the highest growth rate, exceeding the growth of the Sakhalin and Japanese populations approximately two times for length and six to ten times in mass. These comparisons extend through the first spawning which occurs at about seven years of age for all populations.

Some factors limiting the abundance of taimen have been identified. The most important are the limited number of rivers suitable for growing young (up to three years) and the lack of suitable river areas for wintering adult taimens that weigh up to 40 kg.

The high rate of growth and good flesh quality of the Kievka taimen allow it to be regarded as a suitable object for culture.

Peculiarities of Age Structure of Subisolates of Sockeye, *Oncorhynchus nerka* (Walb.), in the First Generation

S. M. Konovalov and V. U. Ostrovsky

Analysis of the age-sex structure of parents and F_1 progeny in subisolates of sockeye in Lake Azabachje led us to make four conclusions. First, the inheritance of the length of freshwater and marine periods of life is linked with the sex. Second, when crossing large fish, more fast-maturing middle-size and small males result. Third, the progeny of middle-size males appear to mature faster than progeny of large males. Fourth, mating of a middle-size early-maturing male with a large late-maturing female produces a large share of early-maturing females.

The population analysis of the age-sex structure of parents and their progeny in different subisolates makes it possible to describe inheritance of characteristics related to length of life. Further study will undoubtedly allow the nature of inheritance of the length of life to be explained more clearly.

Biochemical Genetics of Populations of Pacific Salmon

U. P. Altukhov

The paper deals with the results of population genetics studies with local stocks of pink and sockeye salmon. The studies, which covered a 10-year time span, were made in connection with requirements for information on rational exploitation and artificial reproduction. Populations from Sakhalin, the Kuril Islands, Primorski Territory and Kamchatka were analyzed to determine distribution of gene frequencies in lactate dehydrogenase, malate dehydrogenase, glycerophosphate dehydrogenase, isocytrate dehydrogenase and phosphoglucomutase.

It was determined that stocks could be differentiated and divided into subisolate units. It is important to maintain genetic variety of the population system of stocks to guarantee long-term stock stability for commercial exploitation. The same principle is regarded to be important for artificial reproduction of salmon stocks.

The paper discusses the biological meaning of hereditary polymorphism in populations of bisexual species. Intra-specific divergence is interpreted as a stationary process brought about by mutually balancing pressures of accidental and systematic factors of evolution on the genetic structure of the population system.

Index